Plant Nanotechnology

Chittaranjan Kole · D. Sakthi Kumar
Mariya V. Khodakovskaya
Editors

Plant Nanotechnology

Principles and Practices

Springer

Editors
Chittaranjan Kole
Bidhan Chandra Agricultural University
Mohanpur 741252 West Bengal
India

and

Jacob School of Biotechnology and Bioengineering
Sam Higginbottom Institute of Agriculture, Technology and Sciences
Allahabad 211007
India

D. Sakthi Kumar
Graduate School of Interdisciplinary New Science, Bio Nano Electronics Research Center
Toyo University
Kawagoe
Japan

Mariya V. Khodakovskaya
Department of Applied Science
University of Arkansas at Little Rock
Little Rock, AR
USA

ISBN 978-3-319-42152-0 ISBN 978-3-319-42154-4 (eBook)
DOI 10.1007/978-3-319-42154-4

Library of Congress Control Number: 2016945109

Printed on acid-free paper

This Springer imprint is published by Springer Nature
The registered company is Springer International Publishing AG Switzerland

Preface

Nanotechnology and nanomaterials are increasingly imparting its great influence in our life and environment. During the last two decades, significant amount of research has been conducted in nanotechnology focusing on their application in electronics, energy, mechanics, and life sciences including plant sciences. The impact of nanotechnology and nanomaterials is inevitable in the field of agriculture, and many researches are evidencing their potential in improving the food and agricultural systems through different approaches resulting in the enhancement of agricultural output and development of new food and food products, etc.

The early research investigations in this direction documented absorption, translocation, accumulation, and effects of nanomaterials, mostly metal-based and carbon-based, in several plants including crops. Many of these research studies evidenced for the potential utility of nanomaterials in crop improvement as demonstrated by enhanced germination and seedling parameters in rice, maize, wheat, alfalfa, soybean, rape, tomato, radish, lettuce, spinach, onion, pumpkin, and cucumber; and also enhanced nitrogen metabolism, chlorophyll content, and activities of several enzymes leading to enhanced photosynthesis in maize, soybean, peanut, tomato, and spinach.

There are many investigations reported on nanomaterial-induced improvement in agronomic traits including yield, biomass content, and content of secondary metabolites by direct treatment in soybean, bitter melon, and rice indicating the ability of the nanomaterials in modifying genetic constitution of plants. Nanomaterials have exhibited promise in targeted gene delivery for developing atomically modified plants—a safer and acceptable strategy in contrast to genetic engineering. Interestingly, generational transmission of nanomaterials has been documented in rice and bitter melon.

The usage of these nanomaterials can ultimately land in our food cycle and so a careful study and analysis is pertinent regarding their usage before putting these materials in actual use.

The spurt in the research in this interdisciplinary field that involves primarily the fusion of nanotechnology and plant science may lead to the creation of a new field as "Plantnanomics."

Nanomaterials have also exhibited promise for precise and environmentally safe application of fertilizers and plant protection chemicals using nanoformulations besides plant disease management using nanosensors and nano-based diagnostic kits.

Some concerns have been raised about potential adverse effects of nanomaterials on biological systems and environment although carbon-based nanomaterials, in general, have been found to be safe in many instances.

The book "Plant Nanotechnology" comprises 15 chapters. Chapter 1 clearly lays out the foundation of the book by providing the overview of the concepts, strategies, techniques, and tools of nanobiotechnology and its promises and future prospects. Before using the nanomaterials, we should know its physical and chemical properties. Based on the properties, we can decide the use of the materials in different applications. Chapter 2 deals with the physical and chemical nature of the nanoparticles. After characterizing the nanomaterials, we can employ them in intended applications in plants. While doing that we should know how it could be applied and how we could detect and quantify the uptake of the nanomaterials, translocation, and accumulation. Chapter 3 is devoted to provide the information about the quantification of uptake, translocation, and accumulation of nanomaterials in plants.

For application of any materials anywhere, we should have a clear-cut knowhow, such as how it can be applied and what are the different ways. Chapter 4 describes various methods for using nanomaterials. After the usage of the nanomaterials, naturally we have to look for their impact on plants. The earlier indication of their impact can be assessed by the germination, seedling parameters, and physiological attributes. Chapter 5 deals with the assessment of the impact of nanomaterials on plant growth and development. Chapter 6 provides the information on the effects of nanomaterials on plants with regard to physiological attributes.

After laying a very good foundation toward the characterization and application of nanomaterials and their impact, in general, in plants, we are discussing on the response of plants to nanoparticles at molecular level including changes in gene expression (Chap. 7), and movement and fate of nanoparticles in plants and their generational transmission (Chap. 8).

Recent researches have shown that nanomaterials can be used for the improvement of yield of crops and quality. This finding will lead to the application of nanomaterials in agriculture. For shedding light on the use of nanomaterials in agriculture for different applications, Chap. 9 has been incorporated to elucidate the potential of nanomaterials for the enhancement of yield, plant biomass, and secondary metabolites. A highly promising application potential of nanomaterials for delivery of genetic materials has been deliberated in Chap. 10. Application of agrochemicals including fertilizer and plant protection chemicals using conventional methods leads to less effectivity and even pollution of plant products, soil, water, and air. In contract, use of nanomaterials can lead to precise and targeted delivery of these chemicals. Utilization of nanoparticles for delivery of fertilizers and for plant protection has been deliberated in Chap. 11 and Chap. 12, respectively. We have included another chapter (Chap. 13) to discuss the impact of the nanomaterials in soil-plant systems.

Use of nanomaterials can arouse the concern of safety of their usage with regard to human health and environment. This concern led us to include the Chap. 14 that deals with the concerns of hazards of nanomaterials to human health and environment and also critical views on compliances.

As mentioned earlier, nanotechnology and nanomaterials are increasingly finding their application in the field of agriculture; time has come for the policy makers and researchers to think and depict a road map for the use of nanotechnology in future. Chapter 15 has been specially designed for enumerating on the future road map for plant nanotechnology.

The fifteen chapters of this book have been authored mostly by different teams of scientists dealing with various aspects related to the concepts, strategies, techniques, and tools of plant nanotechnology focusing on the application potential and also on concern for nanotoxicity. Hence, some overlapping contents, particularly on a few fundamental aspects of nanomaterials including their types, natures, and impacts, are obvious. However, the responsibility lies on us as the editors for such redundancy and for addressing them in the future editions of this book.

We believe that our book "Plant Nanotechnology" provides a very precise discussion pertinent to the application of nanotechnology and nanomaterials in plant sciences so that by reading the book, any student, researcher, or policy maker can appreciate the potential and the tremendous application value of this approach and can have a precise and clear idea as to what is going on in this field.

We express our sincere thanks to the 23 scientists beside us for their chapters contributed to this book and their constant cooperation from submission of the first drafts to revision and final fine-tuning of their chapters commensurate with the reviews.

Finally, we wish to extend our thanks to Springer Nature and its entire staff particularly Dr. Christina Eckey and Dr. Jutta Lindenborn involved in publication and promotion of this book that will hopefully be useful to students, scientists, industries, and policy makers.

Mohanpur, India — Chittaranjan Kole
Kawagoe, Japan — D. Sakthi Kumar
Little Rock, USA — Mariya V. Khodakovskaya

Contents

Abbreviations

μ-XANES	Micro-X-ray absorption spectroscopic near-edge structure
μ-XRF	Micro-X-ray fluorescence
2,4-D	2,4-Dichlorophenoxyacetic acid
2-DE	Two-dimensional electrophoresis
3D	Three-dimensional
ADS	Amorpha diene synthase
AES	Atomic absorption spectrometry
AF4	Asymmetrical flow-field flow fractionation
AFM	Atomic force microscopy
AgNP	Silver nanoparticle
Al_2O_3	Aluminum oxide
ALDH	Aldehyde dehydrogenase
APX	Ascorbate peroxidase
ARGOS	Auxin-regulated gene involved in organ size
AuCapped-MSN	Mesoporous silica nanoparticle closed end with gold nanoparticle
AuNP	Gold nanoparticle
AuNR/Ag	Plasmonically active nanorods based on gold cores and silver shells
BET	Brunauer–Emmett–Teller
BP	Bulk particle
BSE	Backscattered electron
BY-2	Tobacco bright yellow-2 cell line
CAT	Catalase
CB	Carbon based
CB NP	Carbon-based nanoparticle
CEC	Cation exchange capacity
CeO_2	Cerium dioxide
CeO_2 NPs	Cerium oxide nanoparticles
Cfu	Colony-forming unit

CLSM	Confocal laser scanning microscopy
CM	Confocal microscopy
CNM	Carbon-based nanomaterial
CNT	Carbon nanotube
CPN	Conjugated polymer nanoparticle
CPS	Counts per second
CSCNT	Cup-stacked carbon nanotube
CS-Se NP	Chitosan-modified selenium nanoparticle
DBR2	Double-bond reductase
DDE	Dichlorodiphenyldichloroethylene
DF-STEM	Dark-field scanning electron microscopy in transmission mode
DLS	Dynamic light scattering
Ebeam	Electron beam
EDAX	Energy dispersive analysis of X-rays
EDX	Energy dispersive X-ray spectrometer
EELS	Energy loss spectroscopy
EM	Electron microscopy
ENM	Engineered nanomaterial
ENP	Engineered nanoparticle
EPA	European Parliament
ER	Endoplasmic reticulum
EXAFS	Extended X-ray absorption fine structure
Fe_3O_4	Magnetite
FEG	Field emission gun
FFF-ICP-MS	Field flow fractionation inductively coupled plasma mass spectrometry
FIFRA	Federal Insecticide, Fungicide and Rodenticide Act
FITC	Fluorescein isothiocyanate
FS	Fullerene soot
FTIR	Fourier-transformed infrared
GA	Gum arabic
GC	Gas chromatography
GLP	Germin-like protein
GMO	Genetically modified organism
GO	Graphene oxide
GPS	Global positioning satellite
GSH	Glutathione
HA	Humic acid
HAP	Hydroxylapatite
HPLC	High-performance liquid chromatography
HRTEM	High-resolution transmission electron microscopy
HS-AFM	High-speed atomic force microscopy
ICDD	International Center for diffraction Data
ICP	Inductively coupled plasma
ICP-MS	Inductively coupled plasma mass spectrometry

ICTA	International Center for Technology Assessment
IDMS	Isotope dilution mass spectrometry
IgG	Immunoglobulin G
In_2O_3	Indium oxide
JCPDS	Joint Committee on Powder Diffraction Standards
LaB6	Lanthanum hexaboride
LA-ICP-MS	Laser ablation inductively coupled plasma mass spectrometry
LC-ESI-MS/MS	Liquid chromatography electrospray ionization tandem mass spectrometry
LM	Light microscopy
MB NP	Metal-based nanoparticle
MeJA	Methyl jasmonate
miRNA	Micro-RNA
MNM	Manufactured nanomaterial
MS	Mass spectrometry
MSN	Mesoporous silica nanoparticle
MSNS	Mesoporous silica nanoparticle system
MWCNT	Multi-walled carbon nanotube
$NaBH_4$	Sodium borohydrate
nAg	Silver nanoparticle
Nano Fe_2O_3	Nano ferric oxide
$nCeO_2$	Cerium dioxide nanoparticle
NDEA	N-nitroso-diethylamine
NGS	Next-generation sequencing
NM	Nanomaterial
NOM	Natural organic matter
NOx	Nitric oxides
NP	Nanoparticle
NR	Nanorod
NS	Nanosphere
NSS	Nanosized Ag–silica hybrid
NT	Nanotechnology
$nTiO_2$	Titanium dioxide nanoparticle
OES	Optical emission spectrometry
OPO	Optical parametric oscillator
PCD	Programmed cell death
PHSN	Porous hollow silica nanoparticle
PIP	Plasma membrane intrinsic protein
PIXE	Particle-induced X-ray emission
PLA	poly(L-lactide)
POX	Peroxidase
PR	Pathogenesis-related
PVP	Polyvinylpyrrolidone

QD	Quantum dot
qRT-PCR	Quantitative real-time polymerase chain reaction
RAPD	Random amplified polymorphic DNA
RER	Rare earth element
ROS	Reactive oxygen species
SA	Salicylic acid
SADS	Selected area (electron) diffraction
SAR	Systemic acquired resistance
SE	Secondary electrons
SEM	Scanning electron microscope
SERS	Surface-enhanced Raman spectroscopy
SiO2	Silicon dioxide
SIP	Small and basic intrinsic protein
SNP	Silver nanoparticle
SOD	Superoxide dismutase
SP-ICP-MS	Single particle inductively coupled plasma mass spectrometry
SPION	Super paramagnetic iron oxide nanoparticle
SPR	Surface plasmon resonance
SQS	Squalene synthase
SR	Synchrotron radiation
Sr31	Wheat stem rust gene
STEM	Scanning electron microscopy in transmission mode
SWCNH	Single-walled carbon nanohorn
SWCNT	Single-walled carbon nanotube
TEM	Transmission electron microscopy
TiO_2	Titanium dioxide
TIP	Tonoplast intrinsic protein
TMA-OH	Tetramethyl ammonium hydroxide
TMAPS/F-MSNs	N-trimethoxysilylpropyl-N,N,Ntrimethylammonium chloride-labeled MSNs
TNB	Temple northeastern Birmingham
TPEM	Two-photon excitation microscopy
TSC	Trisodium citrate
TUNEL	Terminal deoxynucleotidyl transferase-mediated dUTP nick end-labeling
TXM	Transmission X-ray microscopy
Ug99	Uganda99 (race)
USEPA	United States Environmental Protection Agency
USDA	United States Department of Agriculture
XANES	X-ray absorption near-edge structure
XAS	Synchrotron X-ray absorption spectroscopy
XRD	X-ray diffraction

ZnO	Zinc oxide
ZnO NP	Zinc oxide nanoparticle
$ZnTiO_3$	Zinc titanate

Chapter 1
Plant Nanotechnology: An Overview on Concepts, Strategies, and Tools

Joydeep Banerjee and Chittaranjan Kole

Abstract Nanotechnology is the branch of science dealing with manipulation of matter on an atomic, molecular, or supramolecular level. Application of nanoparticles is of great scientific interest due to diverse applications of nanotechnology in the field of life sciences, medicine, electronics, and energy. Since the last couple of decades, several research groups worked on the application of nanoscience in the field of agriculture. Efficient utilization of agrochemicals and manipulation of several physiological parameters of plants are key research areas of agriculture nanotechnology. This introductory chapter presents a brief glimpse on the present global scenario of research on plant nanotechnology and several pros and cons of nanoscience in the fields of plant sciences particularly agriculture.

Keywords Nanoparticles · Agriculture · Nanotechnology · Germination · Translocation · Accumulation · Yield · Agrochemicals · Physiology · Gene expression · Safety issues

J. Banerjee
Survey Selection and Mass Production, Bidhan Chandra Agricultural University, Mohanpur 741252, West Bengal, India

C. Kole (✉)
Bidhan Chandra Agricultural University, Mohanpur 741252, West Bengal, India
e-mail: ckole2012@gmail.com

Present Address:
C. Kole
Jacob School of Biotechnology and Bioengineering, Sam Higginbottom Institute of Agriculture, Technology and Sciences, Allahabad 211007, India

C. Kole et al. (eds.), *Plant Nanotechnology*,
DOI 10.1007/978-3-319-42154-4_1

1.1 Nanoparticles and Nanotechnology

The materials that are lesser than 100 nm, at least one dimension, are referred to as nanomaterials. Hence, the nanoparticles can be zero-dimensional (all dimensions are at nanoscales), one-dimensional (fine rod-shaped), two-dimensional (ultrathin films), or three-dimensional (of any shape) based on their manipulation of matter (Bernhardt et al. 2010; Tiwari et al. 2012). Hence, nanotechnology is the study of different nanoparticles, which are available in 1–100 nm range, at least in one dimension (Love et al. 2005). Nanoparticles categorized on the basis of dimensions, which are not confined to the nanoscale range, are presented in Fig. 1.1. During the last two decades, a significant amount of research has been conducted in nanotechnology focusing on their applications in electronics, energy, medicine, life sciences including plant sciences (Mnyusiwalla et al. 2003; Nair et al. 2010). In the field of agriculture, nanotechnology has been used to improve the food and agricultural systems through different approaches including enhancement of agricultural output, development of new food products, and conservation of foods (Chen 2002). In the course of time, the experiences in the field of nanotechnology facilitated the development of genetically modified crops, chemicals for protecting the plants from biotic stresses, better weed management, and improvement of precision farming techniques. The chapters of this book deliberate on the achievements so far made in plant nanotechnology and the safety issues as well as prospects for fundamental and applied research.

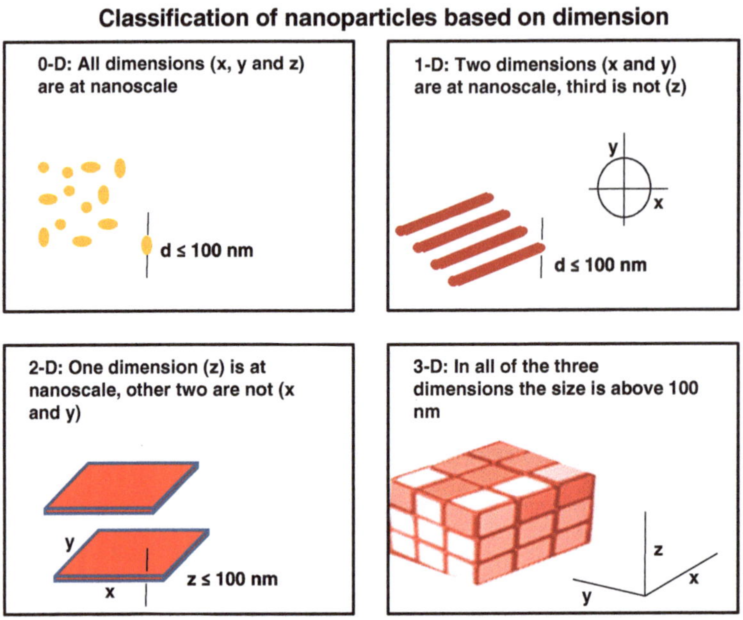

Fig. 1.1 Classification of nanoparticles based on dimension. Four different types of nanoparticles viz., zero dimentional (0-D), one dimentional (1-D), two dimentional (2-D) and three dimentional (3-D) have been mentioned with appropriate diagram

1.2 Use of Nanoparticles in Agriculture, Medicine, and Environment

In the field of agriculture and medicine, the use of nanoparticles (NPs) was found to be effective to combat biotic stresses, to increase the efficacy of agrochemicals including pesticides, and to manage the weeds in a better and eco-friendly manner. To control various bacterial and fungal pathogens, the silver NPs (Ag NPs) were found to be very effective (Nair et al. 2010). To control pathogenic *Candida* species, application of Ag NPs was found to be effective at the concentrations below their cytotoxic limit compared to that of the ionic silver against the tested human fibroblasts (Panáceka et al. 2009). Similarly, silver-based NPs were more effective against gram-negative bacteria compared to the gram-positive bacteria and the larger surface-to-volume ratio was the main reason for the effectiveness of these smaller particles (Singh et al. 2008). Similar to the Ag NPs, silica-based NPs have been widely used in medical as well as agricultural industries. Gold-coated silica has been used for the treatment of benign as well as a malignant tumor. Additionally, lipophilic nano-silica has been used for the treatment of chicken malaria and nuclear polyhedrosis virus infestation of silkworm (*Bombix mori*) (Barik et al. 2008). Other studies documented that the use of surface-modified hydrophobic nano-silica is absorbed into the cuticular layer of the insects and subsequently causes damages to the protective wax layer causing the death of the insects by desiccation (Nair et al. 2010). Such pesticides are not only safer to the plants but also less harmful to the environment compared to the chemical pesticides.

In agricultural field, different agrochemicals are used as fungicides, insecticides, pesticides, or herbicides either by spraying or by broadcasting at various growth stages of plants. A significant amount of the applied chemicals is lost due to various means such as leaching of chemicals, degradation of chemicals by photolysis, hydrolysis, and by microbial degradation. A field study was conducted in cotton plants infested with aphid for estimating the efficacy of nanosphere (NS) formulations compared to a classical suspension used as a reference. The results indicated that compared to the classical suspension, the NS formulations were slower regarding the speed of action and sustained release, but NS formulations were better for enhancing the systemicity of the active ingredient and for improving the penetration through the plant (Boehm et al. 2003). Hence, nano-encapsulated agrochemicals should be designed in such a way that the active ingredients will be released efficiently with improved solubility, stability as well as effectiveness, and finally enhanced targeted activity and reduced ecotoxicity will be achieved. In a similar approach to controlling obnoxious and parasitic weeds, nanocapsule herbicide could be used effectively to reduce the phytotoxicity as mentioned by earlier researchers (Perez-de-Luque and Rubiales 2009). Another class of nanoparticles, namely porous hollow silica nanoparticles (PHSN), was found to provide shielding protection to pesticides from degradation due to UV light exposure (Li et al. 2007). Another study on wheat demonstrated the slow release of fertilizer for regulated, responsive, and timely release of active ingredients using nano- and subnanocomposites (Zhang et al. 2006; Nair et al. 2010). Very recently, a slow-release fertilizer hydrogel nanocomposite has been prepared by free radical polymerization of sodium alginate, acrylic acid, acrylamide, and clinoptilolite using N, N′-methylene bisacrylamide as a crosslinker and ammonium persulfate as an initiator (Rashidzadeh et al. 2014). Additionally, it was found that the swelling of the hydrogels was pH dependent, and the swelling in different salt conditions was significantly lower than the values in distilled water. Moreover, another group showed that plasmonically active nanorods linked with 2,4-D, an auxin growth regulator, can enhance the growth of tobacco (*Nicotiana tabacum*) cells (Nima et al. 2014). In this way, although different NPs are being studied in the agricultural industries, their uptake, accumulation as well as their impact on the yield and different yield attributing characters should be analyzed in detail.

1.3 Types of Nanoparticles and Their Relative Merits

Based on their origin, nanoparticles (NPs) are of three types, namely natural, incidental, and engineered NPs (Monica and Cremonini 2009). Naturally occurring NPs are existing since the beginning of the Earth, and those are available in volcanic dust, lunar dust, terrestrial dust storms, mineral composites, photochemical reactions, forest fires, simple erosion, etc. Incidental NPs are generated mostly by a man-made industrial process like petrol/diesel exhaust, coal combustion, welding fumes, industrial exhausts, etc. (Buzea et al. 2007). Engineered NPs can be categorized into five types including carbon-based NPs (CB NPs), metal-based NPs (MB NPs), magnetic NPs, dendrimers, and composite NPs. Carbon-based NPs include fullerene (C_{70}), fullerol [$C_{60}(OH)_{20}$], single-walled carbon nanotubes (SWCNTs), multiwalled carbon nanotubes (MWCNTs), and single-walled carbon nanohorns (SWCNHs), while MB NPs include gold (Au), silver (Ag), copper (Cu), and iron (Fe)-based nanomaterials. In addition to that, different types of metal oxide-based NPs, such as TiO_2, CeO_2, FeO, Al_2O_3, and ZnO, are extensively studied in agriculture and medical sciences. Magnetic NPs can be manipulated using a magnetic field, and such particles commonly consist of Fe, cobalt (Co) and nickel (Ni) and their compounds. Among different magnetic NPs when ferrite (an iron oxide Fe_2O_3) particles become smaller than 128 nm, they become superparamagnetic (Lu et al. 2007). Dendrimers are nano-sized polymers built from branched units, and they are typically symmetrical around the core part, and mostly, they adopt a spherical three-dimensional structure. Composite NPs are either the combination of different NPs, or the combination of NPs with larger bulk-type materials and those include hybrids. In addition to that, the core–shell nanoparticles are prepared using two or more materials, e.g., silica/inorganic, silica/polymer, or polymer/inorganic combinations. Composite NPs possess improved solubility, easier functionalization, and decreased toxicity compared to the single-component materials (Lin and Xing 2007; Janczak and Aspinwall 2012). NPs are available in different shapes such as spheres, tubes, rods, and prisms.

Among the CB NPs, the significance of fullerene C_{70} and fullerol in agricultural sciences has been extensively studied and reviewed by some researchers (Lin et al. 2009; Kole et al. 2013) and it has been found that these two types of CB NPs get readily accumulated in plants (Rico et al. 2011). An interesting study on rice documented that the individual fullerene C_{70} NPs were possibly entering plant roots through osmotic pressure, capillary forces, pores on cell walls, and intercellular plasmodesmata, or via the highly regulated symplastic route (Lin et al. 2009), whereas another study on onion, *Alium cepa*, reported that the application of hydrophobic fullerenes C_{70}—Natural organic matter in onion cell suspensions—caused negligible NPs uptake by the cells due to blockage of cell wall pores (Chen et al. 2010). In contrast to the fullerenes, C_{70}—another CB NP (SWCNT)—was found to penetrate the cell walls and cell membranes of tobacco cells (Liu et al. 2009). It was demonstrated that CNT can activate water channels in roots as well as seeds and enhance seed germination/plant growth (Khodakovskaya and Biris 2009; Khodakovskaya et al. 2011; Villagarcia et al. 2012; Lahiani et al. 2013). Likewise, another study on tobacco cells demonstrated that the application of MWCNTs in a wide range of concentrations (5–500 μg/mL) could enhance the cell growth significantly compared to the control conditions and a correlation was found between the activation of MWCNT-treated cell growth and the up-regulation of some major genes involved in cell division/cell wall formation and water transport (Khodakovskaya et al. 2013). Lahiani et al. (2013) showed that NPs could successfully activate germination of valuable crops including soybean (*Glycine max*), maize (*Zea mays*), and barley (*Hordeum vulgare)* after deposition of MWCNTs on seed surfaces. Later on, another group confirmed the promising capabilities of carbon nanohorns, another group of CB NPs, in activating the germination of different crop seeds and enhancing growth of plant organs (Lahiani et al. 2015). Furthermore, it was also documented that MWCNTs could improve the water uptake in wheat (*Triticum aestivum*), maize, peanut (*Arachis hypogea*), and garlic (*Allium sativum*) seeds possibly through the creation of new pores (Srivastava and Rao 2014). In contrast to the positive findings on the application of CNTs, another study depicted inhibitory effect on root elongation in tomato (*Solanum lycopersicum*) but enhanced root elongation in onion and cucumber (*Cucumis sativa*) (Cañas et al. 2008). Other studies also evidenced the toxic effect of MWCNTs in plant cells, and application of MWCNTs was found to be deleterious due to the accumulation of reactive oxygen species (ROS) and subsequently decreased cell proliferation and cell death (Tan and Fugetsu 2007; Tan et al. 2009). Based on the positive as well as negative effects of CB NPs, it can be stated that the response of plants or plant cells to NPs varies with the plant species, stages of growth, and the nature of the NPs. Further research on nanosciences is needed to reveal the most efficient and useful combinations of NPs for the betterment of agriculture.

Biogenic nanocrystallines such as Fe, manganese (Mn), zinc (Zn), Cu, Co, selenium (Se) have been extensively used in the agricultural sector due to their participation in different redox processes and their presence in many enzymes as well as complex proteins. Out of these metals, Fe, Cu, and Co with variable valences are highly bioactive in nature and their application in soybean was found to show positive role in germination, growth, and production in a dose-dependent manner (Ngo et al. 2014). Similarly, the application of silver nanoparticles showed their positive impact on germination, biotic stress tolerance, and other physiological parameters of plants (Nair et al. 2010; Savithramma et al. 2012; Sharma et al. 2012). Also, some reviews suggested the importance of typical metals such as gold (Au), platinum (Pt), and palladium (Pa) in agriculture, biosciences, and pharmacology (Abhilash 2010; Agrawal and Rathore 2014). An excellent review has documented the plant uptake, translocation, accumulation as well as toxicity of different NPs including those belonging to metal oxide/hydroxide category, namely TiO_2, ZnO, CeO_2, $Ni(OH)_2$, and Fe_3O_4 (Rico et al. 2011). Although some studies have been carried out on the beneficial role of various metal oxides including CuO, TiO_2, ZnO, $CuZnFe_2O_4$, Fe_3O_4, Fe_2O_3, the adverse effects of some of those metal oxide NPs on soil microbial community and soil structure have also been identified (Frenk et al. 2013). Hence, it is important to research on plant type and soil conditions before applying any specific type of NPs and further experimentation is needed in that regard.

1.4 Impacts of NPs on Germination and Seedling Parameters in Various Crops

Application of NPs was found to have positive as well as negative impact on seed germination and in different stages of growth and development. Khodakovskaya and her group demonstrated the ability of MWCNTs to penetrate tomato seed coat and activate seed germination (Khodakovskaya and Biris 2009; Khodakovskaya et al. 2011). Later, the same group documented that tomato plants grown in soil supplemented with MWCNTs were able to produce two times more flowers and fruits compared to plants grown in control soil (Khodakovskaya et al. 2013). Further studies showed that the positive effect of MWCNTs on germination and growth of corn, soybean, and barley seedlings was reproducible between crop species (Lahiani et al. 2013). An in-depth study was carried out on wheat, maize, peanut, and garlic for knowing the effect of MWCNTs on seed germination and plant growth (Srivastava and Rao 2014). Seeds exposed to nanotubes showed three to four times faster sprouting compared to the controlled condition, and after about 5–10 days of exposure to MWCNTs, a significant enhancement was detected in the plant growth and biomass production of the treated plants compared to the control one. It is to be noted here that the same study also showed evidence on the detrimental effects of MWCNTs at higher doses. Another study on tomato documented the inhibition of root elongation after application of CNTs (Cañas et al. 2008). Application of nanosized TiO_2 (10 ppm concentration) on wheat showed lowest germination time compared to the control condition, while the shoot as well as seedling length was found to be sufficiently higher after application of 2–10 ppm nanosized TiO_2 compared to control and bulk TiO_2-treated plants (Feizi et al. 2012). In addition, it was stated that the higher concentrations of TiO_2-based NPs had inhibitory effect or not any effect on wheat. Similarly, another study reported that the application of nano-TiO_2 in proper concentration accelerated the germination of aged spinach (*Spinacia oleracea*) seeds and enhanced vigor (Zheng et al. 2005). A different study on chickpea (*Cicer arietinum*) demonstrated that the application of hydroxylapatite (HAP) nanorod resulted in better germination and enhanced plant growth. The best performance was observed in presence of 1 mg/ml Hap-nanorod compared to control and other doses (Bala et al. 2014). Soybean seeds treated with superdispersive iron, cobalt, and copper nanocrystalline powders at zerovalent state under laboratory condition showed improved germination frequencies compared to the control condition (Ngo et al. 2014). In addition to that, the application of extra low dose (not more than 300 mg of each metal per hectare) of nanocrystalline powders in field experiment was found to have improvement in different aspects of plant growth and development such as chlorophyll content, number of nodules/root, number of pods/plant, pods weight, 1000-grain weight, and crop yield. Similarly, another study on soybean reported improved germination and growth parameters after application of nano-SiO_2 and nano-TiO_2 mixtures (Lu et al. 2002). Ag NPs are one of the widely used engineered NPs. A comprehensive study was carried out for knowing the effects of Ag NPs on germination and growth on 11 species of wetland plants including *Lolium multiflorum, Panicum virgatum, Carex lurida, C. scoparia, C. vulpinoidea, C. crinita, Eupatorium fistulosum, Phytolacca americana, Scirpus cyperinus, Lobelia cardinalis,* and *Juncus effusus* belonging to six different families, and it was found that different species showed differential response to germination (Yin et al. 2012). Additionally, the root growth was found to be affected more compared to the leaf growth after exposure to Ag. Exposure of tobacco plants to different concentrations of Al_2O_3 (0, 0.1, 0.5, and 1 %) documented that as the exposure to NPs increased, the average root length, average biomass, and leaf count of the NP- exposed plants were significantly decreased compared to the control samples (Burklew et al. 2012). Along with the various reports on the detrimental effect of various NPs on germination and plant growth, some studies reported the genotoxic effect of some NPs. Random amplified polymorphic DNA analysis was carried out for knowing the DNA damage as well as mutations caused by NPs, and it was found that after exposure to CeO_2 NPs on soybean plants, four new bands were detected at 2000 mg L^{-1}, and three new bands were found at 4000 mg L^{-1} treatment (López-Moreno et al. 2010). Another report documented the copper oxide NP-mediated DNA damage in some terrestrial plants. In that study, under controlled condition, strong plant growth inhibitions were recorded for radish (*Raphanus sativus*), perennial ryegrass (*Lolium perenne*), and annual ryegrass (*Lolium rigidum*) and in addition, some oxidatively modified, mutagenic DNA lesions (7,8-dihydro-8-oxoguanine; 2,6-diamino-4-hydroxy-5-formamidopyrimidine; and 4,6-diamino-5-formamidopyrimidine) were found to be accumulated in significant amount under laboratory conditions (Atha et al. 2012). Further experimentation is needed for understanding the probable impacts of NPs in biological systems as well as on their physiological aspects. Some of the chapters of this book are going to address those specific questions in detail.

1.5 Effects of Nanoparticles on Gene Expression

The effect of different NPs on gene expression of animals, human as well as plants has been studied by many workers (Khodakovskaya et al. 2011; Poynton et al. 2011; Lee et al. 2012; Kaveh et al. 2013; Lahiani et al. 2013). Some studies documented that after exposure to nanoparticles, the gene expression of superoxide dismutase (SOD) was altered along with other enzymes in the animal as well as in plant system (Lee et al. 2012; Kaveh et al. 2013; Siddiqi 2014). In addition to that, higher concentration (1 %) of Al_2O_3 nanoparticle stress was found to show significant up-regulation of a number of micro-RNA genes including miR395, miR397, miR398, and miR399 (Burklew et al. 2012). These findings might be analyzed in great detail to understand the global gene expression profiling after the application of NPs. Out of these miRNAs, especially miR398 was found to possess a significant relation to SOD expression (Sunkar et al. 2006; Dugas and Bartel 2008), whereas other miRNAs were involved in other stresses (Sunkar 2010). Microarray-based gene expression analyses were carried out in Arabidopsis (*Arabidopsis thaliana*) for knowing the nanoparticle-specific changes in gene expression after exposure to ZnO, TiO_2, and fullerene soot (Landa et al. 2012). The study reported that after exposure to ZnO and fullerene soot (FS), mostly the biotic (wounding and defense to pathogens) and abiotic stress (oxidative, salt, and water deprivation) responsive genes were up-regulated, whereas ZnO-exposure was responsible for down-regulation of genes involved in cell organization and biogenesis but FS-exposure leads to down-regulation of genes involved in electron transport and energy pathways. Interestingly, after exposure to TiO_2, most of the expressional changes (up-regulation and down-regulation) were detected for genes, which were responsive to abiotic and abiotic stimulus. Another study on Arabidopsis was done by microarray for knowing the changes in gene expression after exposure to AgNPs as well as Ag^+ (Kaveh et al. 2013). Among the up-regulated genes, a major part was associated with the response to metals and oxidative stress (such as cation exchanger, cytochrome P450-dependent oxidase, SOD, and peroxidase), whereas the down-regulated genes were responsive to pathogens and hormonal stimuli such as genes involved in systemic acquired resistance, ethylene signaling, and auxin-regulated gene involved in organ size (ARGOS). On the other hand, among the differentially expressed genes in response to AgNPs only, most remarkable up-regulation (>4.0 fold) was detected in two salt stress-related genes (*AT3G28220* and *AT1G52000*), one gene codes for myrosinase-binding protein (AT1G52040) involved in biotic stress, three genes engaged in the thalianol biosynthetic pathway (*AT5G48010*, *AT5G48000*, and *AT5G47990*), and a gene responsive to wounding (*AT2G01520*). Although it is clear from the above discussions that the exposure of Arabidopsis to ZnO, FS or AgNPs causes similar type of changes in gene expression (Landa et al. 2012; Kaveh et al. 2013), the mechanisms of phytotoxicity are highly specific to the type as well as concentrations of NPs. Interestingly, germins and germin-like proteins belonging to cupin superfamily were found to be involved in various biotic as well as abiotic stresses (Dunwell et al. 2008) and some of the members of this superfamily possessed SOD activity (Dunwell et al. 2008; Banerjee et al. 2010). Very recently, an interesting study on Indian mustard (*Brassica juncea*) showed a correlation between copper oxide nanoparticles induced growth suppression and enhanced lignification as well as modification in root system. It is worthy to mention that a germin-like protein from rice (OsGLP1) was found to have some relation to plant height and SOD-mediated cell wall reinforcement (Banerjee and Maiti 2010; Banerjee et al. 2010). If the proteins belonging to cupin superfamily members are involved in nanoparticle-regulated cascades, there will be a new area of research for understanding such complex plant signaling networks involving various stresses. A variety of NPs was found to have effects on gene expression in plant system as well as in animal systems including humans, and NPs are able to express distinct bioactivity and unique effects with different biological systems. For assessing the potential health risks after exposure to NPs, luciferase reporter system has been used for understanding the gene expression profiles in response to NPs (Ding et al. 2012). Further work is needed in model organisms to specifically identify the signaling cascades or to determine the regulation of a set of genes by specific NPs in a dose-dependent manner.

1.6 Translocation and Accumulation of Nanoparticles in Plant Tissues and Organs

Due to rapid progress in the field of nanosciences and wide applications of nanomaterials (NMs) in medical sciences as well as in agriculture, some researchers started analyzing the potential impacts of NMs along with their translocation and accumulation in tissues. The first study on the uptake, accumulation, and translocation analyses of magnetite (Fe_3O_4) nanoparticles was carried out on pumpkin (*Cucurbita maxima*) (Zhu et al. 2008). The study revealed that the iron oxide NPs (Fe_3O_4) were taken up by pumpkin roots and subsequently translocated through plant tissues. In addition to that, it was also found that almost 45.5 % of fed nanoparticles were accumulated in roots and about 0.6 % of the nanoparticles were detected in leaves. In contrast to that, application of same NPs on another crop, lima bean (*Phaseolus limensis*), did not show any uptake and transport of the NMs as revealed by same researchers.

Among the CB NMs, fullerene C_{70} and fullerols were mostly found to be taken up as well as accumulated in plants (Rico et al. 2011; Kole et al. 2013). An interesting study on uptake and translocation of CB NPs on rice (*Oryza sativa*) established that fullerene C_{70} was easily taken up by roots and transported to shoots compared to MWCNTs (Lin et al. 2009), possibly due to the relatively larger size of MWCNTs than fullerenes. Additionally, in the roots of mature plants, no C_{70} was detected, explaining robust transport of NPs from root to shoot. SWCNTs, another CB NPs, were found to show gradual findings regarding its penetration to plant cells (Liu et al. 2009; Shen et al. 2010). Some study on Bright Yellow (BY-2) cells reported that the water-soluble SWCNTs (<500 nm in length) were able to penetrate the intact cell wall and the cell membrane through fluidic phase endocytosis, whereas another study on cucumber documented no uptake of SWCNTs by the roots upon exposure to CB NPs for 48 h (Cañas et al. 2008). Little is known about the quantity of NPs being delivered inside plant tissues due to less availability of detection methods. Dr. Green and his group showed the ability of the microwave-inducing heating technique to quantify tubular structure CB NPs inside plant tissue (Irin et al. 2012). This method was followed to quantify SWCNHs inside different crop roots system (Lahiani et al. 2015) and MWCNTs inside different plant tissues (Irin et al. 2012).

Application of an aqueous colloidal solution of NaYF4:Yb,Er nanocrystals during watering was found to show uptake and transport of nanocrystals from roots to leaves in moth orchid (*Phalaenopsis* spp.) and Arabidopsis (Hischemoller et al. 2009). Probably that was the first report on uptake kinetics and that illustrated the potential penetration routes of NPs in plant tissues. The route of penetration of the nanocrystals at different period of times in different plant tissues was carried out using confocal laser scanning microscopic analyses. The uptake and accumulation of Cu NPs, Ag NPs, and metal NPs have been described in some recent reviews (Ma et al. 2010), and it was found that the higher application of Cu NPs resulted in higher uptake and accumulation under laboratory condition. Another review has nicely described the uptake and accumulation of metal oxide NPs as well as metal NPs in plant systems (Rico et al. 2011).

Other than the CB NPs, the magnetic NPs (Fe_3O_4) were detected in roots, stems, and leaves of pumpkin plants and the uptake was found to be dependent on the growth medium (Zhu et al. 2008). Among the metal oxide-based NPs, an ultra-small TiO_2 (<5 nm) complexed with Alizarin red S nanoconjugate was found to show uptake and translocation in Arabidopsis plants (Kurepa et al. 2010). The study also documented that the mucilage released by the roots of Arabidopsis formed a pectin hydrogel capsule surrounding the root, which either facilitated or inhibited the entry of TiO_2 complexed with Alizarin red S or sucrose. In contrast to that, another study on maize roots did not show any uptake of TiO_2 NPs (30 nm) probably due to the larger size of the NPs than the pore diameters (Asli and Neumann 2009). Other studies also documented that polysaccharides in mucilage might adsorb and inactivate toxic heavy metals in root rhizosphere and ultimately enhanced the accumulation of aluminum (Watanabe et al. 2008). The uptake and accumulation study of another metal oxide NPs (ZnO) by soybean seedlings demonstrated that at 500 mg L^{-1} concentrations, the uptake of the NPs (8 nm) was significantly higher possibly due to lesser aggregation, whereas at higher concentrations (1000–4000 mg L^{-1}), the passage of NPs through the cell pore walls was difficult probably due to agglomeration, and that caused reduced uptake and accumulation (López-Moreno et al. 2010).

It has been found that after application of CeO_2 at 4000 mg L^{-1}, the concentration of Ce (mg kg^{-1} dry weight biomass) significantly varied between soybean, alfalfa (*Medicago sativa*), maize, and tomato. The concentrations of Ce in corn, soybean, tomato, and alfalfa were found to be approximately 300, 462, 4000, and 6000 mg kg^{-1} dry weight biomass, respectively. The differences in concentrations could be explained by the variations in the root microstructures and the physical as well as chemical interactions between the NPs and root exudates in the rhizosphere of respective plant species (Rico et al. 2011). Due to advancement in the field of nanotechnology, some of the present research papers and review articles are focusing on the shape, size, structure, chemical composition, and surface chemistry of NPs for understanding the nanoparticle aggregation in the environment and subsequently the accumulation and transport of NPs in living systems (Hotze et al. 2010; Albanese et al. 2012). Further research is needed in this context for knowing the uptake capacity and permissible limit of different NPs in agriculture and food industry.

This book is going to describe the physio-chemical properties of different NPs, their merits as well as demerits, the detection, and quantification of NPs along with their involvement in uptake, accumulation, and translocation. Additionally, the chapters of this book will focus on the use of NPs and their impacts on germination, growth, and other physiological aspects as well as yield and quality components. Some of the sections will describe the modern understanding of the gene expressional changes caused by NPs and different modes of transmission of NPs. Later chapters will focus the importance of NPs for gene delivery, fertilizer delivery, and various agrochemicals applications along with their involvement in plant protection. At last but not the least, the possible merits and demerits of various NPs, the effects of NPs on soil, plant and environments and the prospects and policies for nanosciences will be considered.

References

Abhilash M (2010) Potential applications of nanoparticles. Int J Pharma Bio Sci V1(1)

Agrawal S, Rathore P (2014) Nanotechnology pros and cons to agriculture: a review. Int J Curr Microbiol App Sci 3(3):43–55

Albanese A, Tang PS, Chan WCW (2012) The effect of nanoparticle size, shape, and surface chemistry on biological systems. Annu Rev Biomed Eng 14:1–16

Asli S, Neumann M (2009) Colloidal suspensions of clay or titanium dioxide nanoparticles can inhibit leaf growth and transpiration via physical effects on root water transport. Plant, Cell Environ 32:577–584

Atha DH, Wang H, Petersen EJ, Cleveland D, Holbrook RD, Jaruga P, Dizdaroglu M, Xing B, Nelson BC (2012) Copper oxide nanoparticle mediated DNA damage in terrestrial plant models. Environ Sci Technol 46:1819–1827

Bala N, Dey A, Das S, Basu R, Nandy P (2014) Effect of hydroxyapatite nanorod on chickpea (*Cicer arietinum*) plant growth and its possible use as nano-fertilizer. Iran J Plant Physiol 4 (3):1061–1069

Banerjee J, Maiti MK (2010) Functional role of rice germin-like protein1 in regulation of plant height and disease resistance. Biochem Biophys Res Commun 394:178–183

Banerjee J, Das N, Dey P, Maiti MK (2010) Transgenically expressed rice germin-like protein1 in tobacco causes hyper-accumulation of H_2O_2 and reinforcement of the cell wall components. Biochem Biophys Res Commun 402:637–643

Barik TK, Sahu B, Swain V (2008) Nanosilica-from medicine to pest control. Parasitol Res 103:253–258

Bernhardt ES, Colman BP, Hochella JrMF, Cardinale BJ, Nisbet RM, Richardson CJ, Yin L (2010) An ecological perspective on nanomaterial impacts in the environment. J Environ Qual 39:1954–1965

Boehm AL, Martinon I, Zerrouk R, Rump E, Fessi H (2003) Nanoprecipitation technique for the encapsulation of agrochemical active ingredients. J Microencapsul 20:433–441

Burklew CE, Ashlock J, Winfrey WB, Zhang B (2012) Effects of aluminum oxide nanoparticles on the growth, development, and microRNA expression of tobacco (*Nicotiana tabacum*). PLoS ONE 7(5):e34783

Buzea C, Blandino IIP, Robbie K (2007) Nanomaterials and nanoparticles: sources and toxicity. Biointerphases 2(4):MR17–MR172

Cañas JE, Long M, Nations S, Vadan R, Dai L, Luo M, Ambikapathi R, Lee EH, Olszyk D (2008) Effects of functionalized and nonfunctionalized single-walled carbon-nanotubes on root elongation of select crop species. Environ Toxicol Chem 27:1922–1931

Chen H (ed) (2002) Nanoscale science and engineering for agriculture and food systems. In: Proceedings of USDA conference, Washington DC, USA, 18–19 Nov 2002

Chen R, Ratnikova TA, Stone MB, Lin S, Lard M, Huang G, Hudson JS, Ke PC (2010) Differential uptake of carbon nanoparticles by plant and mammalian cells. Small 6:612–617

Ding M, Bowman L, Castranova V (2012) Luciferase reporter system for studying the effect of nanoparticles on gene expression. Meth Mol Biol 906:403–414. doi:10.1007/978-1-61779-953-2_33

Dugas DV, Bartel B (2008) Sucrose induction of Arabidopsis miR398 represses two Cu/Zn superoxide dismutases. Plant Mol Biol 67:403–417

Dunwell JM, Gibbings JG, Mahmood T, Naqvi SMS (2008) Germin and germin-like proteins: evolution, structure, and function. Crit Rev Plant Sci 27:342–375

Feizi H, Moghaddam PR, Shahtahmassebi N, Fotovat A (2012) Impact of bulk and nanosized titanium dioxide (TiO2) on wheat seed germination and seedling growth. Biol Trace Elem Res 146:101–106

Frenk S, Ben-Moshe T, Dror I, Berkowitz B, Minz D (2013) Effect of metal oxide nanoparticles on microbial community structure and function in two different soil types. PLoS ONE 8(12): e84441

Hischemoller A, Nordmann J, Ptacek P, Mummenhoff K, Hasse M (2009) In-vivo imaging of the uptake of unconversion nanoparticles by plant roots. J Biomed Nanotechnol 5:278–284

Hotze EM, Phenrat T, Lowry GV (2010) Nanoparticle aggregation: challenges to understanding transport and reactivity in the environment. J Environ Qual 39:1909–1924

Irin F, Shrestha B, Cañas JE, Saed MA, Green MJ (2012) Detection of carbon nanotubes in biological samples through microwave-induced heating. Carbon 50:4441–4449

Janczak CM, Aspinwall CA (2012) Composite nanoparticles: the best of two worlds. Anal Bioanal Chem 402(1):83–89. doi:10.1007/s00216-011-5482-5

Kaveh R, Li YS, Ranjbar S, Tehrani R, Brueck CL, van Aken B (2013) Changes in *Arabidopsis thaliana* gene expression in response to silver nanoparticles and silver ions. Environ Sci Technol 47:10637–10644

Khodakovskaya MV, Biris AS (2009) Method of using carbon nanotubes to affect seed germination and plant growth. WO 2011059507 A1—patent application

Khodakovskaya MV, de Silva K, Nedosekin DA, Dervishi E, Biris AS, Shashkov EV, Galanzha EI, Zharov VP (2011) Complex genetic, photothermal, and photoacoustic analysis of nanoparticle-plant interactions. Proc Natl Acad Sci USA 108:1028–1033

Khodakovskaya M, Kim B-M, Jong Kim JN, Alimohammadi M, Dervishi E, Mustafa T, Cernigla CE (2013) Carbon nanotubes as fertilizers: effects on tomato growth, reproductive system and soil microbial community. Small 9(1):115–123

Kole C, Kole P, Randunu KM, Choudhary P, Podila R, Ke PC, Rao AM, Marcus RK (2013) Nanobiotechnology can boost crop production and quality: first evidence from increased plant biomass, fruit yield and phytomedicine content in bitter melon (*Momordica charantia*). BMC Biotechnol 13:37

Kurepa J, Paunesku T, Vogt S, Arora H, Rabatic BM, Lu J, Wanzer MB, Woloschak GE, Smalle JA (2010) Uptake and distribution of ultrasmall anatase TiO_2 alizarin red S nanoconjugates in *Arabidopsis thaliana*. Nano Lett. doi:10.1021/nl903518f

Lahiani MH, Dervishi E, Chen J, Nima Z, Gaume A, Biris AS, Khodakovskaya MV (2013) Impact of carbon nanotube exposure to seeds of valuable crops. ACS Appl Mater Interfaces 5: 7965–7973

Lahiani MH, Chen J, Irin F, Puretzky AA, Green MJ, Khodakovskaya MV (2015) Interaction of carbon nanohorns with plants: Uptake and biological effects. Carbon 81:607–619

Landa P, Vankova R, Andrlova J, Hodek J, Marsik P, Storchova H, White JC, Vanek T (2012) Nanoparticle-specific changes in *Arabidopsis thaliana* gene expression after exposure to ZnO, TiO_2, and fullerene soot. J Hazard Mat 241–242:55–62

Lee SH, Pie J-E, Kim Y-R, Lee HR, Son SW, Kim M-K (2012) Effects of zinc oxide nanoparticles on gene expression profile in human keratinocytes. Mol Cell Toxicol 8(2):113–118

Li ZZ, Chen JF, Liu F, Lu AQ, Wang Q, Sun HY, Wen LX (2007) Study of UV shielding properties of novel porous hollow silica nanoparticle carriers for avermectin. Pest Manag Sci 63:241–246

Lin D, Xing B (2007) Phytotoxicity of nanoparticles:inhibition of seed germination and root growth. Environ Pollut 150:243–250

Lin S, Reppert J, Hu Q, Hudson JS, Reid ML, Ratnikova TA, Rao AM, Luo H, Ke PC (2009) Uptake, translocation, and transmission of carbon nanomaterials in rice plants. Small 5: 1128–1132

Liu Q, Chen B, Wang Q, Shi X, Xiao Z, Lin J, Fang X (2009) Carbon nanotubes as molecular transporters for walled plant cells. Nano Lett 9:1007–1010

López-Moreno ML, de la Rosa G, Hernández-Viezcas JÁ, Castillo-Michel H, Botez CE, Peralta-Videa JR, Gardea-Torresdey JL (2010) Evidence of the differential biotransformation and genotoxicity of ZnO and CeO2 nanoparticles on soybean (*Glycine max*) plants. Environ Sci Technol 44(19):7315–7320. doi:10.1021/es903891g

Love JC, Estroff LA, Kriebel JK, Nuzzo RG, Whitesides GM (2005) Self-assembled monolayers of thiolates on metals as a form of nanotechnology. Chem Rev 105:1103–1169

Lu CM, Zhang CY, Wen JQ, Wu GR, Tao MX (2002) Research on the effect of nanometer materials on germination and growth enhancement of *Glycine max* and its mechanism. Soybean Sci 21(3):68–172

Lu A-H, Hui A, Salabas EL, Schüth F (2007) Magnetic nanoparticles: synthesis, protection, functionalization, and application. Angew Chem Int Ed Engl 46(8):1222–1244. doi:10.1002/anie.200602866

Ma X, Geiser-Lee J, Deng Y, Kolmakov A (2010) Interactions between engineered nanoparticles (ENPs) and plants: phytotoxicity, uptake and accumulation. Sci Total Environ 408(16): 3053–3061. doi:10.1016/j.scitotenv.2010.03.031

Mnyusiwalla A, Daar AS, Singer PA (2003) 'Mind the gap": science and ethics in nanotechnology. Nanotechnology 14:R9. doi:10.1088/0957-4484/14/3/201

Monica RC, Cremonini R (2009) Nanoparticles and higher plants. Caryologia Int J Cytol Cytosyst Cytogenet 62(2):161–165. doi:10.1080/00087114.2004.10589681

Nair R, Varghese SH, Nair BG, Maekawa T, Yoshida Y, Kumar DS (2010) Nanoparticulate material delivery to plants. Plant Sci 179:154–163

Ngo QB, Dao TH, Nguyen HC, Tran XT, Nguyen TV, Khuu TD, Huynh TH (2014) Effects of nanocrystalline powders (Fe, Co and Cu) on the germination, growth, crop yield and product quality of soybean (Vietnamese species DT-51). Adv Nat Sci Nanosci Nanotechnol 5:015016

Nima ZA, Lahiani MH, Watanabe F, Xu Y, Khodakovskaya MV, Biris AS (2014) Plasmonically active nanorods for delivery of bio-active agents and high-sensitivity SERS detection in planta. RSC Adv 4:64985–64993

Panáčeka A, Kolářb M, Večeřovåb R, Pruceka R, Soukupováa J, Kryštofc V, Hamalb P, Zbořila R, Kvítek L (2009) Antifungal activity of silver nanoparticles against *Candida* spp. Biomaterials 30(31):6333–6340

Perez-de-Luque A, Rubiales D (2009) Nanotechnology for parasitic plant control. Pest Manag Sci 65:540–545

Poynton HC, Lazorchak JM, Impellitteri CA, Smith ME, Rogers K, Patra M, Hammer KA, Allen HJ, Vulpe CD (2011) Differential gene expression in *Daphnia magna* suggests distinct modes of action and bioavailability for ZnO nanoparticles and Zn ions. Environ Sci Technol 45 (2):762–768

Rashidzadeh A, Ali O, Salari D, Reyhanitabar A (2014) On the preparation and swelling properties of hydrogel nanocomposite based on Sodium alginate-g-Poly (acrylic acid-co- acrylamide)/ Clinoptilolite and its application as slow release fertilizer. J Polymer Res 21:344. doi:10.1007/ s10965-013-0344-9

Rico CM, Majumdar S, Duarte-Gardea M, Peralta-Videa JR, Gardea-Torresdey JL (2011) Interaction of nanoparticles with edible plants and their possible implications in the food chain. J Agric Food Chem 59:3485–3498

Savithramma N, Ankanna S, Bhumi G (2012) Effect of nanoparticles on seed germination and seedling growth of *Boswellia ovalifoliolata*—an endemic and endangered medicinal tree taxon. Nano Vis 2(1–3):61–68

Sharma P, Bhatt D, Zaidi MG, Saradhi PP, Khanna PK, Arora S (2012) Silver nanoparticle-mediated enhancement in growth and antioxidant status of *Brassica juncea*. Appl Biochem Biotechnol 167(8):2225–2233. doi:10.1007/s12010-012-9759-8

Shen CX, Zhang QF, Li J, Bi FC, Yao N (2010) Induction of programmed cell death in *Arabidopsis* and rice by single-wall carbon nanotubes. Am J Bot 97:1–8

Siddiqi NJ (2014) Effect of gold nanoparticles on superoxide dismutase and indoleamine 2, 3-dioxygenase in various rat tissues. Indian J Biochem Biophys 51(2):156–159

Singh M, Singh S, Prasad S, Gambhir IS (2008) Nanotechnology in medicine and antibacterial effect of silver nanoparticles. Digest J Nanomat Biostruct 3(3):115–122

Srivastava A, Rao DP (2014) Enhancement of seed germination and plant growth of wheat, maize, peanut and garlic using multiwalled carbon nanotubes. Eur Chem Bull 3(5):502–504

Sunkar R (2010) MicroRNAs with macro-effects on plant stress responses. Semin Cell Dev Biol 21:805–811

Sunkar R, Kapoor A, Zhu JK (2006) Posttranscriptional induction of two Cu/Zn superoxide dismutase genes in Arabidopsis is mediated by downregulation of miR398 and important for oxidative stress tolerance. Plant Cell 18:2051–2065

Tan XM, Fugetsu B (2007) Multi-walled carbon nanotubes interact with cultured rice cells: evidence of a self-defense response. J Biomed Nanotechnol 3:285–288

Tan XM, Lin C, Fugetsu B (2009) Studies on toxicity of multi-walled carbon nanotubes on suspension rice cells. Carbon 47:3479–3487

Tiwari JN, Tiwari RN, Kim KS (2012) Zero-dimensional, one-dimensional, two-dimensional and three-dimensional nanostructured materials for advanced electrochemical energy devices. Prog Mater Sci 57:724–803

Villagarcia H, Dervishi E, de Silva K, Biris AS, Khodakovskaya M (2012) Specific surface chemistry of carbon nanotubes can determine their biological effects in planta. Small 8 (15):2328–2334

Watanabe T, Misawa S, Hiradate S, Osaki M (2008) Root mucilage enhances aluminum accumulation in *Melastoma malabathricum*, an aluminum accumulator. Plant Sig Behav 3:603–605

Yin L, Colman BP, McGill BM, Wright JP, Bernhardt ES (2012) Effects of silver nanoparticle exposure on germination and early growth of eleven wetland plants. PLoS ONE 7(10):e47674

Zhang F, Wang R, Xiao Q, Wang Y, Zhang J (2006) Effects of slow/controlled-release fertilizer cemented and coated by nano-materials on biology. II. Effects of slow/controlled-release fertilizer cemented and coated by nano-materials on plants. Nanoscience 11:18–26

Zheng L, Hong F, Lu S, Liu C (2005) Effect of nano-TiO_2 on strength of naturally aged seeds and growth of spinach. Biol Trace Elem Res 104:83–91

Zhu H, Han J, Xiao JQ, Jin Y (2008) Uptake, translocation, and accumulation of manufactured iron oxide nanoparticles by pumpkin plants. J Environ Monit 10:713–717

Chapter 2
Physical and Chemical Nature of Nanoparticles

Sanmathi Chavalmane Subbenaik

Abstract Nanoparticles have some specific features, including physical properties, chemical properties, merits, and demerits, which have drawn much attention for their application in nanobiotechnology. This chapter explains the state of the art of different properties of nanoparticles and their potential beneficial roles. In addition, this chapter discusses on the research on nanoparticles essentiality for plants and describes the current knowledge concerning the key nanoparticles with important studies for their future applications.

Keywords Nanoparticles · Physiochemical nature · Merits and demerits

2.1 Introduction

Nanoparticles in general refer to particles having internal structural measurement or external dimensions within the size range of a few nanometers, preferable up to 100 nm size. According to the European Committee for Standardization, nanomaterials are defined as the materials with any external dimension at the nanoscale, or that possess nanoscale internal or surface structures. Nanoscale describes the size range from approximately 1–100 nm (ISO/TS 27687: 2008) (Lövestam et al. 2010). It is most frequently used as a specific size description (usually <100 nm, though sometimes <50 nm), and this book chapter will use the term nanoparticle to refer to particles of <100 nm.

Nanoparticles have been developed for use in the area of agriculture (Nair et al. 2010; Campos et al. 2014), where they can increase the efficiency and productivity of crops. To properly assign the mechanisms for the application of nanoscale materials in plants, their synthesis and characterization must be well understood. Scientists have many methods to synthesize NPs of different size, shape, and

S.C. Subbenaik (✉)
Nano Research Facility (NNIN-NSF), School of Engineering and Applied Science, Washington University in St. Louis, St. Louis, MO 63130, USA
e-mail: sanmathi.nrf.wustl@gmail.com

C. Kole et al. (eds.), *Plant Nanotechnology*,
DOI 10.1007/978-3-319-42154-4_2

surface properties. The major synthesis routes are liquid phase, gas-phase, and biological methods (Klaus et al. 1999; Konishi et al. 2007; Raliya and Tarafdar 2012; Mittal et al. 2013). The main liquid phase syntheses of inorganic NPs are coprecipitation, solgel processing, micro-emulsions, hydrothermal or solvothermal methods, template synthesis, and biometric synthesis (Cushing et al. 2004). The biological method can be approached for synthesis of NPs, which is rapid and cost-effective. (Gilaki 2010; Raliya and Tarafdar 2012). Besides these synthesis methods, the gas-phase synthesis methods are of interest because they allow elegant way to control process parameter in order to be able to produce size-, shape-, and chemical composition-controlled nanostructures, and also can be used to prepare the large quantity of NPs (Jiang et al. 2007; Thimsen et al. 2008).

Nanoparticles are of two types: non-engineered and engineered NPs. Non-engineered NPs present in the environment are derived from natural events such as terrestrial dust storms, erosion, volcanic eruption, and forest fires (Nowack and Bucheli 2007). Engineered NPs (ENPs) are intentionally produced by man using many different materials, such as metals (including Au, Ag, Zn, Ni, Fe, and Cu) (Fedlheim and Foss 2001), metal oxides (TiO_2, Fe_2O_4, SiO_2, CeO_2, and Al_2O_3) (Fernández-García and Rodriguez 2011), nonmetals (silica and quantum dots) (Ehrman et al. 1999), carbon (graphene and fullerene) (Endo et al. 2013), polymers (alginate, chitosan, hydroxyethylcellulose, polyhydroxyalkanoates and polyhydroxyalkanoates, and poly-E-caprolactone) (Paques et al. 2014) (Rao and Geckeler 2011), and lipids (soybean lecithin and stearic acid) (Ekambaram et al. 2012).

Engineered NPs are able to enter into plants cells and leaves and also can transport DNA and chemicals into plant cells (Galbraith 2007; Tripathi et al. 2011; Raliya et al. 2015). The unique physical and chemical properties of nanoparticles could boost plant metabolism (Nair et al. 2011; Brew and Strano 2014). Here, we describe the physical and chemical nature of the NPs and compare their merits and demerits during application. Figure 2.1 shows the different physical and chemical nature of NPs.

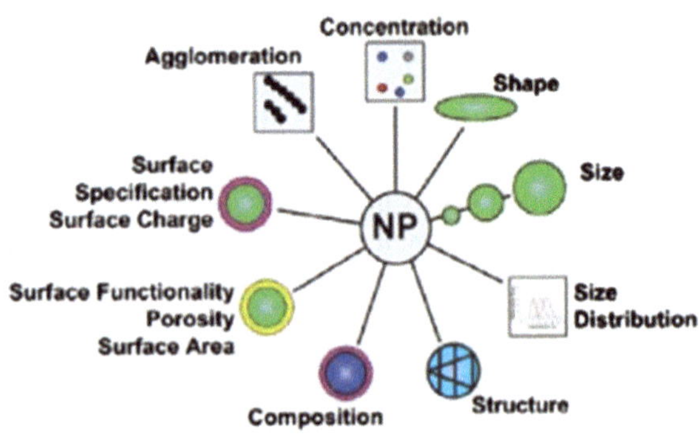

Fig. 2.1 Physical and chemical nature of nanoparticles

2.2 Physical Properties of Nanoparticles

Physical properties of NPs include shape, size, specific surface area, agglomeration/aggregation, state of size distribution, surface morphology/topography, and structure including crystallinity, defect structure, and solubility (Cadden 1987; Rao and Biswas 2009). The size, shape, surface area, and size distribution of NPs are important deciding factors controlling their uptake by plants as it is highly dependent on cell wall pores and size of stomata (Eichert et al. 2008; Schreck et al. 2012; Wang et al. 2013). The following section will describe the basics of each physical property of NPs.

2.2.1 Size and Shape

The size and shape can be identified as the most important parameter to define the nanomaterial in general. Jolivet et al. (2004) and Jolivet et al. (2004) postulated that NPs below 20–30 nm in size are characterized by an excess of energy at the surface and are thermodynamically unstable because of the interfacial tension, acting as a driving force, which leads to a spontaneous reduction of the surface area. However, most types of particles have a critical size of about 30 nm below which NPs exhibit their typical "nano" properties from their bulk material. When the size of a nanoparticle decreases, the amount of molecules present at the particle's surface increases in an exponential trend. Slomberg and Schoenfisch (2012) studied the size-dependent effects of silica particles on Arabidopsis (*Arabidopsis thaliana*) plants. The studies showed reduced development of plants for treatment with 50 and 200 nm silica NPs. Figure 2.2 shows the effect of different size of silica NPs on growth of Arabidopsis plant.

The design of NPs has gained a lot of attention, resulting in particles with various shapes such as spheres, rods, tubed, fibers, and disks, and more extraordinary geometries such as worms, squares, urchins, and ellipsoids. The optical properties of NPs also depend on its size and shape. Figure 2.3 exemplifies the difference in the optical properties of gold NPs for different shapes (synthesis and characterization done in Nano Research Facility, Washington University, in Saint Louis using the article (Wu et al. 2010).

2.2.2 Surface and Size Distribution of Nanoparticles

The surface morphology and surface area of NPs can be analyzed using scanning electron microscope (SEM) and Brunauer–Emmett–Teller (BET), respectively (Hayat 1974). To get a higher resolution of approximately 0.2 nm, atomic force microscopy (AFM) can be used. It provides real topographical images of sample

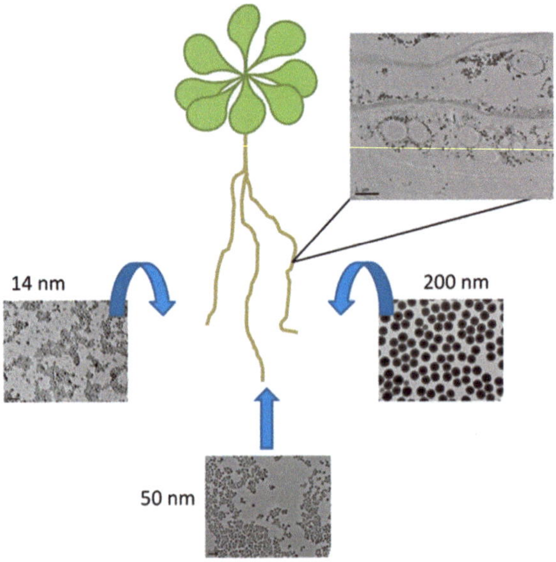

Fig. 2.2 Effect of different sized (14, 50, and 200 nm) silica nanoparticles on growth of *Arabidopsis thaliana* plant (reprinted with permission from Slomberg and Schoenfisch 2012 copyright of American Chemical Society)

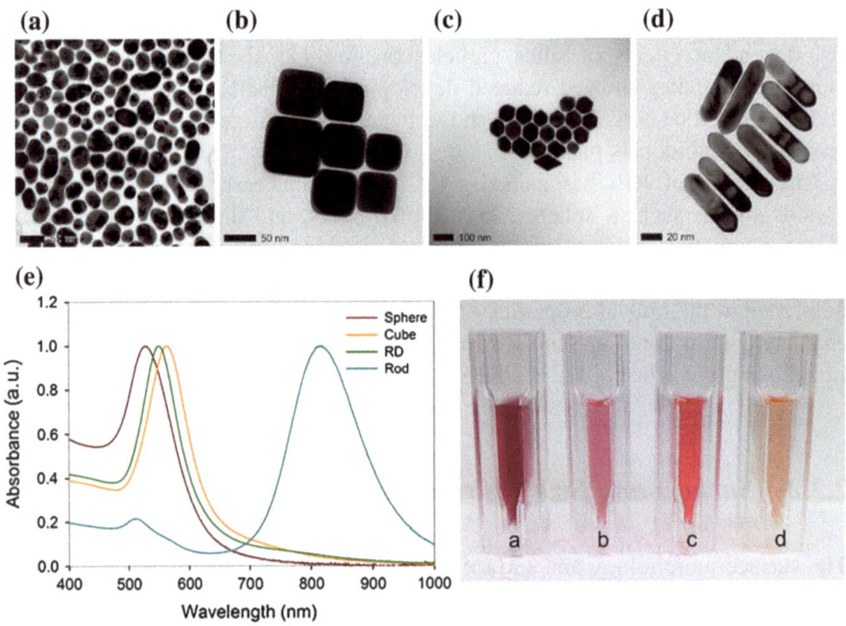

Fig. 2.3 Characterization of gold NPs. TEM images of gold nanostructure **a** spheres, **b** truncated cubes, **c** rhombic dodecahedra, and **d** rods **e** UV–visible absorption spectra showing characteristic absorption peak(s) for each nanostructre **f** Pictured from *left* **a** to *right* **d**: nanospheres, truncated cubes, rhombic dodecahedra, and rods as synthesized in aqueous suspension

surfaces. Dynamic light scattering (DLS) enables evaluation of the size distribution of NPs, and zetasizer can be used to determine the surface charge of NPs. The attachment of NPs to cell membrane seems to be most affected by the surface charge of the NPs (Honary and Zahir 2013). The plant cellular uptake is usually a prerequisite and is also governed by surface hydrophobicity and charge in addition to size. Compared to NPs having a neutral or negative charge, positively charged NPs are taken up at faster rate. The dispersion status of NPs in aqueous media principally depends on the surface charge of the given NPs. A number of in vitro studies with different NPs have been published in which the effect of the different parameters such as dispersion, surface properties, and agglomeration and de-agglomeration can be controlled using ultrasonication, ionic strength and pH of aqueous solutions, physiological buffers, and cell culture media (Orts-Gil et al. 2011; Barreto et al. 2015).

2.2.3 *Structural and Species Specifics of Nanoparticles*

Determinations of purity of NPs are important in biological application. X-ray diffraction (XRD) is the most essential tool used to characterize crystal structures (Warren 1969). The most commonly used database for the identification of crystal structures is the Joint Committee on Powder Diffraction Standards—International Center for Diffraction Data (JCPDS-ICDD) system. Detailed profile analysis of experimental XRD patterns provides information about a given material's space group and structural parameters. (Jain et al. 2009; Kumar and Yadav 2009). Syu et al. (2014) studied the effects of size and shape of silver NPs on growth and gene expression in Arabidopsis plants and found that the application could result in a complex physiological response in the treated tissues. The literature reported the species specificity of NPs in which their effects vary with the type of NPs and type and nature of biological systems that got treated with NPs (Zhang et al. 2013; Song et al. 2014).

2.3 Chemical Properties of Nanoparticles

Chemical properties include the elemental composition of nanomaterials and its surface chemistry such as zeta potential and photocatalytic properties (Cadden 1987; Rao and Biswas 2009). The chemical properties of a material are determined by the type of motion of its electrons. There is a wide range of NPs contributing to many different chemical properties (Schmid 2011). Here, we describe the chemical characteristics separately with different kinds of NPs.

2.3.1 Metallic Nanoparticles

Compared with other nanostructures, metallic NPs have been proven to be the most flexible nanostructures owing to the synthetic control of their size, shape, composition, structure, assembly, and encapsulation, as well as the resulting tunability of their optical properties. Compared with other metallic nanostructures, colloidal gold and silver NPs are especially promising in nanobiotechnology because of their simple and fast preparation and bioconjugation. The attraction of surface plasmon excitations for the applications typically arises from the large electromagnetic field enhancement near the metal surface and the dependence of the resonance wavelength on the size, shape, and local dielectric properties of NPs. Such nanoparticles work as platform materials for biomolecular ultrasensitive detection, hyperthermal treatment for cancer, cell and protein labeling, and targeted delivery of therapeutic agents within the cells. Whereas silver NPs have a comparatively high cytotoxicity (Greulich et al. 2009), gold NPs are biologically almost inert (Mahl et al. 2011) and have a remarkable role on seed germination and antioxidant systems in Arabidopsis and altered levels of micro-RNAs expression that regulates various morphological, physiological, and metabolic processes in plants (Kumar et al. 2013).

2.3.2 Metal Oxide Nanoparticles

Metal oxide NPs can exhibit unique chemical properties due to their limited size and a high density of corner or edge surface sites. Particle size is expected to influence important groups of basic properties in any material. The properties such as structural characteristics, namely the lattice symmetry, cell parameters, and effect of size, are related to the electronic properties of the oxide, and structural and electronic properties obviously drive the chemical properties of the solid and also by size in a simple classification (Ayyub et al. 1995). Metal oxide particles serve many functions in the various field of plant technology (Picó and Blasco 2012; Raliya and Tarafdar 2013; Tarafdar et al. 2013). For example, nanosized silicon dioxide (SiO_2) treatments in proper concentration increased the percentage germination (Siddiqui and Al-Whaibi 2014). It was also reported that alumina NPs increased the root growth of plants (Lin and Xing 2007). Magnetic NPs exhibit a wide variety of attributions, which make them highly promising connection with biological system and bioapplications usually exists or can be prepared in the form of either single domain or superparamagnetic magnetite (Fe_2O_3) or greigite (Fe_3S_4). Due to their favorable beneficial effects, magnetic NPs approved for clinical use by Food and Drug Administration.

2.3.3 Quantum Dots

The size effects in metal oxide chemistry have frequently two interrelated faces, structural/electronic quantum-size and size-defect or non-stoichiometry effects. Structurally quantum dots (QDs) consist of a variety of metal complexes such as semiconductors, metals, and magnetic transition metals. The bioactivity of QDs can be improved by suitable surface coating with biocompatible material and/or modification with desired functional groups.

Depending on their size, it fluoresces with different colors and QD's composed of cadmium selenide core wrapped in zinc sulfide shell is such of a kind (Chan and Nie 1998; Kloepfer et al. 2003). To make them biologically compatible/active, newly synthesized QDs are functionalized or given secondary coatings, which improves water solubility. Studies also reported the effects of QDs on plant system showing both positive and negative effects (Nair et al. 2011).

2.3.4 Carbon Nanoparticles

The fullerene provided an exciting insight into carbon nanostructure and how architectures built from sp^2 carbon units based on simple geometrical principles can change the physical and chemical properties. Carbon nanotubes (CNTs) represent the more evident example. About decade after discovery, the knowledge available increased the interest in biological and biomedical applications of carbon nanotubes (Liu et al. 2007; Prato et al. 2007). There is a certain duality in the nanotubes. On the one hand, single-walled nanotubes consist of a single graphite sheet seamlessly wrapped into a cylindrical tube. Multiwalled nanotubes comprise an array of such nanotubes that are concentrically nested like rings of a tree trunk. Several of the enabling technology required for nanotube application are under development, for example, the ability to disperse individual multiwalled nanotubes uniformly into a polymer matrix and controlling its alignment within a composite material (Qian et al. 2000; Andrews et al. 2002). Recently, several works have been reported with the use of CNTs as smart delivery system for the delivery of desired molecules/chemicals to animal and plant cells. Another carbon modification from the NPs group are graphene oxide (GO) and their best known representative precursor for chemical preparation of graphene (Zhang et al. 2011). Recently, graphene/GOs have been extensively explored as imaging agents, drugs carries, and tissue engineering materials. The main advantage of these material is biodistribution and pharmacokinetics properties that can be turned by controlling the size, the surface chemistry, and the targeting ligand and highest Young's modulus among any known materials and the ability to increase the tensile strength of other materials. The combination of these advantages makes graphene an ideal platform for multimodal application in the biotechnological fields. Apart from these advantages, the important challenge and current limitations in this area are still the potential long-term toxicity. It was reported that fullerene and carbon nanotubes increased the water-retaining capacity, biomass, and fruit yield in plants up to 118 % which is highly remarkable (Husen and Siddiqi 2014).

2.3.5 *Polymeric Nanoparticles*

Polymer NPs have attracted the interest of many plant research groups. The term polymer nanoparticle is given for any type of polymer NPs but specifically for nanospheres and nanocapsules. These are obtained from synthetic such as from synthetic polymers, such as polycaprolactone (Bilensoy et al. 2009), polyacrylamide (Bilensoy et al. 2009), and polyacrylate (Turos et al. 2007), or natural polymers, albumin, DNA and chitosan (Martínez et al. 2011), gelatin, and poly (L-lactide) (PLA) (Mainardes et al. 2010; Saraogi et al. 2010). The various polymer NPs had been used to improve the pharmacokinetics and pharmacodynamics properties of various drugs, for example, chitosan polymer used as a carrier of plant extract (Bhatia et al. 2011).

Figure 2.4 shows the selective uptake, translocation, and biotransformation pathway of different NPs in plant organs. According to the scientist, data about NPs uptake by plants are still not conclusive (Rico et al. 2011).

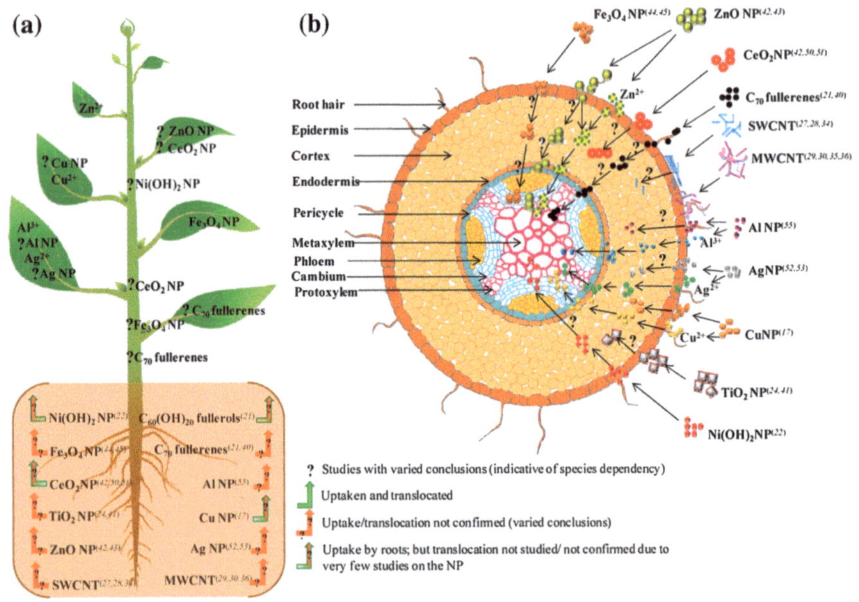

Fig. 2.4 Selective uptake, translocation, and biotransformation pathway of different NPs in plant organs (reprinted with permission from Cyrén and Hermansson 2012, copyright of American Chemical Society)

2.4 Merits and Demerits of Nanoparticles

Due to instability of the NPs, retaining the size and shape of NPs is highly challenging. As the kinetics associated with NPs is rapid and is highly reactive, they inherently interact with impurities. In addition, encapsulation of NPs becomes necessary when they are synthesized in a solution. Synthesis of pure NPs becomes highly difficult. Hence, retaining high purity in NPs can become a challenge hard to overcome. It is noticeable that most experimental studies with NPs have been carried out with aggregates/agglomerates of NPs. This has significantly repercussions on the biokinetics of the material. Several questions can be raised: What is the size distribution of the aggregates/agglomerates and what is the portion of the particles present as a monodispersed material?

Current research work revealed that the uptake, translocation, and accumulation of NPs depend on the species of plant and the size, chemical composition, functionalization, and stability of the NPs (Kole et al. 2013; Raliya et al. 2015). Among the carbon-based NPs, only the fullerene C_{70} and fullerols were shown to get readily accumulated in plants (Rico et al. 2011; Nair et al. 2012). Most of the data corresponding to the germination stage and cell culture, because the protocols for quantification of NPs within tissues, are not well defined yet.

The discussion of the current research is more oriented to the effect of the NPs on plants. A very few of the NPs to the next generation of plants exposed to NPs is unknown.

2.5 Conclusion

The major physical and chemical properties and comparative merits and demerits of NPs are discussed. NPs are capable of penetrating living plant tissues and migrating to different organs of the plant, although detailed study of their nature is very important. These studies allow us to constitute an important step forward in elucidating the mechanisms of interaction between plant cells and NPs and thus in designing strategies for using NPs for targeted delivery of substances. Although there are many exciting potential applications of NPs, considerable challenges and issues remain to be resolved. For example, nanomaterial remains a major problem, and it is hard to precisely control the number of functional molecules on the surface of NPs. Researchers need to develop better strategies for producing NPs that have precise composition, uniform surface modification, and reproducible functionalization. For applications, the purity, dispersity, and stability of the NPs in a physiological environment are highly important. Therefore, it is necessary to further study and explore physical and chemical properties for creating successful nanobiotechnology. Also, more studies are needed to explore the mode of action of NPs and their interaction and status in plant biomass.

References

Andrews R, Jacques D, Minot M, Rantell T (2002) Fabrication of carbon multiwall nanotube/polymer composites by shear mixing. Macromol Mater Eng 287:395–403

Ayyub P, Palkar V, Chattopadhyay S, Multani M (1995) Effect of crystal size reduction on lattice symmetry and cooperative properties. Phys Rev B 51:6135

Barreto Â, Luis LG, Girão AV, Trindade T, Soares AM, Oliveira M (2015) Behavior of colloidal gold nanoparticles in different ionic strength media. J Nanopart Res 17:1–13

Bhatia A, Shard P, Chopra D, Mishra T (2011) Chitosan nanoparticles as carrier of immunorestoratory plant extract: synthesis, characterization and immunorestoratory efficacy. Int J Drug Deliv 3:381–385

Bilensoy E, Sarisozen C, Esendağlı G, Doğan AL, Aktaş Y et al (2009) Intravesical cationic nanoparticles of chitosan and polycaprolactone for the delivery of Mitomycin C to bladder tumors. Int J Pharmaceut 371:170–176

Brew JA, Strano MS (2014) Plant nanobionics approach to augment photosynthesis and biochemical sensing. Nat Mater 13:400–408

Cadden A (1987) Comparative effects of particle size reduction on physical structure and water binding properties of several plant fibers. J Food Sci 52:1595–1599

Campos EVR, de Oliveira JL, Fraceto LF (2014) Applications of controlled release systems for fungicides, herbicides, acaricides, nutrients, and plant growth hormones: a review. Adv Sci Eng Med 6:373–387

Chan WC, Nie S (1998) Quantum dot bioconjugates for ultrasensitive nonisotopic detection. Science 281:2016–2018

Cushing BL, Kolesnichenko VL, O'Connor CJ (2004) Recent advances in the liquid-phase syntheses of inorganic nanoparticles. Chem Rev 104:3893–3946

Cyrén B, Hermansson B (2012) Linear actuator assembly. U.S. Patent Application No. 14/359,669.

Ehrman SH, Friedlander SK, Zachariah MR (1999) Phase segregation in binary SiO2/TiO2 and SiO2/Fe2O3 nanoparticle aerosols formed in a premixed flame. J Mater Res 14:4551–4561

Eichert T, Kurtz A, Steiner U, Goldbach HE (2008) Size exclusion limits and lateral heterogeneity of the stomatal foliar uptake pathway for aqueous solutes and water-suspended nanoparticles. Physiol Planta 134:151–160

Ekambaram P, Sathali AAH, Priyanka K (2012) Solid lipid nanoparticles: a review. Sci Rev Chem Commun 2:80–102

Endo M, Iijima S, Dresselhaus MS (eds) (2013) Carbon Nanotubes. Elsevier, Shinshu University, Japan

Fedlheim DL, Foss CA (2001) Metal Nanoparticles: Synthesis, Characterization, and Applications. CRC Press, Boca Raton, FL, USA

Fernández-García M, Rodriguez JA (2011) Metal oxide nanoparticles. Encycl Inorg Bioinorg Chem. Instituto de Catálisis y Petroleoquímica, CSIC, Madrid, Spain and Brookhaven National Laboratory, Upton, NY, USA

Galbraith DW (2007) Nanobiotechnology: silica breaks through in plants. Nat Nanotechnol 2:272–273

Gilaki M (2010) Biosynthesis of silver nanoparticles using plant extracts. J Biol Sci 10:465–467

Greulich C, Kittler S, Epple M, Muhr G, Köller M (2009) Studies on the biocompatibility and the interaction of silver nanoparticles with human mesenchymal stem cells (hMSCs). Langenbeck's Arch Surg 394:495–502

Hayat MA (1974) Principles and Techniques of Scanning Electron Microscopy. Biological Applications,vol 1. Van Nostrand Reinhold Company, New Jersey, USA

Honary S, Zahir F (2013) Effect of zeta potential on the properties of nano-drug delivery systems-a review (Part 1). Trop J Pharmaceut Res 12:255–264

Husen A, Siddiqi KS (2014) Carbon and fullerene nanomaterials in plant system. J Nanobiotechnol 12:16

Jain D, Daima HK, Kachhwaha S, Kothari S (2009) Synthesis of plant-mediated silver nanoparticles using papaya fruit extract and evaluation of their anti microbial activities. Digest J Nanomater Biostruct 4:557–563

Jiang J, Chen D-R, Biswas P (2007) Synthesis of nanoparticles in a flame aerosol reactor with independent and strict control of their size, crystal phase and morphology. Nanotechnology 18:285603

Jolivet J-P, Froidefond C, Pottier A, Chanéac C, Cassaignon S et al (2004) Size tailoring of oxide nanoparticles by precipitation in aqueous medium. A Semi-Quant Model J Mater Chem 14:3281–3288

Klaus T, Joerger R, Olsson E, Granqvist C-G (1999) Silver-based crystalline nanoparticles, microbially fabricated. Proc Natl Acad Sci USA 96:13611–13614

Kloepfer J, Mielke R, Wong M, Nealson K, Stucky G, Nadeau J (2003) Quantum dots as strain-and metabolism-specific microbiological labels. Appl Environ Microb 69:4205–4213

Kole C, Kole P, Randunu KM, Choudhary P, Podila R et al (2013) Nanobiotechnology can boost crop production and quality: first evidence from increased plant biomass, fruit yield and phytomedicine content in bitter melon (*Momordica charantia*). BMC Biotechnol 13:37

Konishi Y, Ohno K, Saitoh N, Nomura T, Nagamine S et al (2007) Bioreductive deposition of platinum nanoparticles on the bacterium (*Shewanella algae*). J Biotechnol 128:648–653

Kumar V, Guleria P, Kumar V, Yadav SK (2013) Gold nanoparticle exposure induces growth and yield enhancement in *Arabidopsis thaliana*. Sci Total Environ 461:462–468

Kumar V, Yadav SK (2009) Plant-mediated synthesis of silver and gold nanoparticles and their applications. J Chem Technol Biotechnol 84:151–157

Lin D, Xing B (2007) Phytotoxicity of nanoparticles: inhibition of seed germination and root growth. Environ Pollut 150:243–250

Liu Z, Sun X, Nakayama-Ratchford N, Dai H (2007) Supramolecular chemistry on water-soluble carbon nanotubes for drug loading and delivery. ACS Nano 1:50–56

Lövestam G, Rauscher H, Roebben G, Klüttgen BS, Gibson N et al (2010) Considerations on a definition of Nanomaterial for regulatory purposes. Publications Office of the European Union

Mahl D, Diendorf J, Meyer-Zaika W, Epple M (2011) Possibilities and limitations of different analytical methods for the size determination of a bimodal dispersion of metallic nanoparticles. Colloids Surf A Physicochem Eng Asp 377:386–392

Mainardes RM, Khalil NM, Gremião MPD (2010) Intranasal delivery of zidovudine by PLA and PLA–PEG blend nanoparticles. Int J Pharmaceut 395:266–271

Martínez A, Iglesias I, Lozano R, Teijón J, Blanco M (2011) Synthesis and characterization of thiolated alginate-albumin nanoparticles stabilized by disulfide bonds. Evaluation as drug delivery systems. Carbohyd Polym 83:1311–1321

Mittal AK, Chisti Y, Banerjee UC (2013) Synthesis of metallic nanoparticles using plant extracts. Biotechnol Adv 31:346–356

Nair R, Mohamed MS, Gao W, Maekawa T, Yoshida Y et al (2012) Effect of carbon nanomaterials on the germination and growth of rice plants. J Nanosci Nanotechnol 12:2212–2220

Nair R, Poulose AC, Nagaoka Y, Yoshida Y, Maekawa T, Kumar DS (2011) Uptake of FITC labeled silica nanoparticles and quantum dots by rice seedlings: effects on seed germination and their potential as biolabels for plants. J Fluores 21:2057–2068

Nair R, Varghese SH, Nair BG, Maekawa T, Yoshida Y, Kumar DS (2010) Nanoparticulate material delivery to plants. Plant Sci 179:154–163

Nowack B, Bucheli TD (2007) Occurrence, behavior and effects of nanoparticles in the environment. Environ Pollut 150:5–22

Orts-Gil G, Natte K, Drescher D, Bresch H, Mantion A et al (2011) Characterisation of silica nanoparticles prior to in vitro studies: from primary particles to agglomerates. J Nanopart Res 13:1593–1604

Paques JP, van der Linden E, van Rijn CJ, Sagis LM (2014) Preparation methods of alginate nanoparticles. Adv Colloid Interf Sci 209:163–171

Picó Y, Blasco C (2012) Nanomaterials in food, which way forward? Analys Risk Nanomater Environ Food Samp 59:305

Prato M, Kostarelos K, Bianco A (2007) Functionalized carbon nanotubes in drug design and discovery. Accounts Chem Res 41:60–68

Qian D, Dickey EC, Andrews R, Rantell T (2000) Load transfer and deformation mechanisms in carbon nanotube-polystyrene composites. Appl Phys Lett 76:2868–2870

Raliya R, Nair R, Chavalmane S, Wang W-N, Biswas P (2015) Mechanistic evaluation of translocation and physiological impact of titanium dioxide and zinc oxide nanoparticles on the tomato (*Solanum lycopersicum* L.) plant. Metallomics 7:1584–1594

Raliya R, Tarafdar J (2012) Novel approach for silver nanoparticle synthesis using *Aspergillus terreus* CZR-1: mechanism perspective. J Bionanosci 6:12–16

Raliya R, Tarafdar J (2013) ZnO nanoparticle biosynthesis and its effect on phosphorous-mobilizing enzyme secretion and gum contents in Clusterbean (*Cyamopsis tetragonoloba* L.). Agric Res 2:48–57

Rao C, Biswas K (2009) Characterization of nanomaterials by physical methods. Annu Rev Analyt Chem 2:435–462

Rao JP, Geckeler KE (2011) Polymer nanoparticles: preparation techniques and size-control parameters. Prog Polymer Sci 36:887–913

Rico CM et al (2011) Interaction of nanoparticles with edible plants and their possible implications in the food chain. J Agric Food Chem 59:3485–3498

Saraogi GK, Gupta P, Gupta U, Jain N, Agrawal G (2010) Gelatin nanocarriers as potential vectors for effective management of tuberculosis. Int J Pharmaceut 385:143–149

Schmid G (ed) (2011) Nanoparticles: From Theory to Application. Wiley-VCH, Weinheim, Germany

Schreck E, Foucault Y, Sarret G, Sobanska S, Cécillon L et al (2012) Metal and metalloid foliar uptake by various plant species exposed to atmospheric industrial fallout: Mechanisms involved for lead. Sci Total Environ 427:253–262

Siddiqui MH, Al-Whaibi MH (2014) Role of nano-SiO_2 in germination of tomato (*Lycopersicum esculentum* seeds Mill.). Saudi J Biol Sci 21:13–17

Song L, Connolly M, Fernández-Cruz ML, Vijver MG, Fernández M et al (2014) Species-specific toxicity of copper nanoparticles among mammalian and piscine cell lines. Nanotoxicology 8:383–393

Slomberg DL, Schoenfisch MH (2012) Silica nanoparticle phytotoxicity to Arabidopsis thaliana. Environ Sci Technol 46(18):10247–10254

Syu Y-y, Hung J-H, Chen J-C, Chuang H-W (2014) Impacts of size and shape of silver nanoparticles on Arabidopsis plant growth and gene expression. Plant Physiol Biochem 83:57–64

Tarafdar A, Raliya R, Wang W-N, Biswas P, Tarafdar J (2013) Green synthesis of TiO2 nanoparticle using Aspergillus tubingensis. Adv Sci Eng Med 5:943–949

Thimsen E, Rastgar N, Biswas P (2008) Nanostructured TiO_2 films with controlled morphology synthesized in a single step process: Performance of dye-sensitized solar cells and photo watersplitting. J Phys Chem C 112:4134–4140

Tripathi S, Sonkar SK, Sarkar S (2011) Growth stimulation of gram (*Cicer arietinum*) plant by water soluble carbon nanotubes. Nanoscale 3:1176–1181

Turos E, Shim J-Y, Wang Y, Greenhalgh K, Reddy GSK et al (2007) Antibiotic-conjugated polyacrylate nanoparticles: New opportunities for development of anti-MRSA agents. Bioorg Med Chem Lett 17:53–56

Wang W-N, Tarafdar JC, Biswas P (2013) Nanoparticle synthesis and delivery by an aerosol route for watermelon plant foliar uptake. J Nanopart Res 15:1–13
Warren BE (1969) X-ray Diffraction. Courier Dover Publications. Reprint of the Addison- Wesley Publishing Compnay, Inc,. Reading Massachusettes, USA, 1969 edn
Wu H-L, Kuo C-H, Huang MH (2010) Seed-mediated synthesis of gold nanocrystals with systematic shape evolution from cubic to trisoctahedral and rhombic dodecahedral structures. Langmuir 26:12307–12313
Zhang P, Ma Y, Zhang Z, He X, Li Y et al (2013) Species-specific toxicity of ceria nanoparticles to *Lactuca* plants. Nanotoxicology 9:1–8
Zhang Y, Ren L, Wang S, Marathe A, Chaudhuri J, Li G (2011) Functionalization of graphene sheets through fullerene attachment. J Mater Chem 21:5386–5391

Chapter 3
Biophysical Methods of Detection and Quantification of Uptake, Translocation, and Accumulation of Nanoparticles

Illya A. Medina-Velo, Nubia Zuverza-Mena, Wenjuan Tan, Jose A. Hernandez-Viezcas, Jose R. Peralta-Videa and Jorge L. Gardea-Torresdey

Abstract Manufactured nanomaterials (MNMs) are more frequently found in consumer products as well as in industrial and agricultural applications. The high volume of production, use, and disposal of MNM-containing wastes increase the probability of release of these products to the environment. An ever-increasing number of articles have shown that MNMs impact plants and other organisms in different ways. In this chapter, we discuss the biophysical methods currently used to measure the uptake, translocation, accumulation, and speciation of MNMs within plants. We included methods used to analyze plants exposed to carbon-based and metal-based MNMs. Advantages and disadvantages of each analytical technique are discussed.

Keywords Nanoparticles · Plants · Absorption · Detection · Quantification · Microscopy · Spectroscopy

---

I.A. Medina-Velo · W. Tan · J.A. Hernandez-Viezcas · J.R. Peralta-Videa · J.L. Gardea-Torresdey (✉)
Department of Chemistry, The University of Texas at El Paso, 500 West University Ave., El Paso, TX 79968, USA
e-mail: jgardea@utep.edu

I.A. Medina-Velo
e-mail: iamedinavelo@utep.edu

W. Tan
e-mail: wtan@miners.utep.edu

J.A. Hernandez-Viezcas
e-mail: jahernandez19@utep.edu

J.R. Peralta-Videa
e-mail: jperalta@utep.edu

C. Kole et al. (eds.), *Plant Nanotechnology*,
DOI 10.1007/978-3-319-42154-4_3

3.1 Introduction

From the beginning of the twenty-first century, technologies in manufacturing, electronics, communications, medicine, water and wastewater treatment, agriculture and food packaging, among others are considered nanoenabled technologies (Bandyopadhyay et al. 2013; Roco and Bainbridge 2013). This is due to the unique properties of manufactured nanomaterials (MNMs), unavailable in bulk counterparts, which allow for a multiplicity of applications (Salamon et al. 2010). Nanoenabled technologies currently include carbon-based [fullerenes, graphene, and carbon nanotubes (CNTs)] and metal-based (quantum dots, metal, and metal oxide) MNMs (Klaine et al. 2008; Peralta-Videa et al. 2011).

The profuse use of MNMs in personal care, industrial, food and agricultural products, as well as in soil and water remediation technologies, has raised concerns about contamination of ecosystems and food supply (Gardea-Torresdey et al. 2014). Concerns are higher in agricultural sectors where food production takes place in areas either exposed to the direct use of MNMs for the delivery of agricultural products or amended with biosolids. Approximately four million dry tons of biosolids are applied to agricultural soils in the United States (Lu et al. 2012), and some biosolids are unintentionally loaded with MNMs (Colman et al. 2013; Gardea-Torresdey et al. 2014). These MNM-loaded biosolids originate at wastewater remediation facilities fed with water containing MNMs from end user products or that use supported and unsupported MNMs (Trujillo-Reyes et al. 2014).

Concerns about possible contamination of food with MNMs have encouraged the search for accurate methods to determine and quantify the presence of MNMs in agricultural plants. One of the first studies showing the uptake of MNMs by plants was reported by Zhu et al. (2008). These researchers exposed for 20 days pumpkin (*Cucurbita maxima*) plants to magnetic Fe_3O_4 nanoparticles (NPs) of 20 nm diameter. At harvest, they measured the concentrations of Fe_3O_4 particles with a vibrating sample magnetometer (VSM, LakeShore 7400). The sample was vibrated in an external magnetic field, and magnetization was quantified through the voltage measured in a pickup coil. According to the authors, "the magnetization of pure Fe_3O_4 particles is 53.19 emu g^{-1}, and one memu correlates to 8.48×10^{11} particles

J.R. Peralta-Videa · J.L. Gardea-Torresdey
Environmental Science and Engineering Ph.D. Program, The University of Texas at El Paso, 500 West University Ave., El Paso, TX 79968, USA

N. Zuverza-Mena
Metallurgical and Materials Engineering Department, The University of Texas at El Paso, 500 West University Ave., El Paso, TX 79968, USA
e-mail: nzuverza@miners.utep.edu

I.A. Medina-Velo · N. Zuverza-Mena · W. Tan · J.A. Hernandez-Viezcas ·
J.R. Peralta-Videa · J.L. Gardea-Torresdey
University of California Center for Environmental Implications of Nanotechnology (UC CEIN), The University of Texas at El Paso, 500 West University Ave., El Paso, TX 79968, USA

(assuming Fe_3O_4 particles have a density of 5.17 g cm^{-3})." With this method, Zhu et al. (2008) were able to determine that the Fe_3O_4 NPs were taken up/adsorbed by roots (45.4 %) and translocated to the leaves (0.6 %) of pumpkin plants. Another pioneering study aimed to measure the uptake of MNMs by plants was performed by Cañas et al. (2008) who exposed functionalized and non-functionalized single-walled CNTs (SWCNTs) to six crop plants. A year later, Lin et al. (2009), by using scanning electron microscopy (SEM) and transmission electron microscopy (TEM), demonstrated that C_{70} fullerene and multiwalled carbon nanotubes (MWCNTs) were taken up through the roots and translocated to the leaves in rice (*Oryza sativa*) plants. However, quantification of the uptake was still elusive. This question was solved by Larue et al. (2012b) who quantified the uptake and translocation of MWCNTs in wheat and rapeseed plants using TEM.

Efforts have also been made to detect metal-based MNMs in plants. By using X-ray absorption spectroscopic near edge structure (XANES), Lopez-Moreno et al. (2010a) demonstrated, for the first time, that cerium dioxide NPs ($nCeO_2$) are taken up and stored in roots of soybean (*Glycine max* L.). A year later, Larue et al. (2011) used micro-X-ray fluorescence (μ-XRF) to map titanium (Ti) within the roots of wheat plants and, by using XANES, they determined that the Ti was in the form of titanium dioxide nanoparticles ($nTiO_2$). Subsequently, Servin et al. (2013) used μ-XRF combined with micro-X-ray absorption spectroscopic near edge structure (μ-XANES) to show that cucumber (*Cucumis sativus* L.) plants can absorb $nTiO_2$ through the roots and translocate them into the fruit. Also, Hernandez-Viezcas et al. (2013) used μ-XRF, μ-XANES, and transmission X-ray microscopy (TXM) to demonstrate that $nCeO_2$ accumulate in all soybean plant organs, including pods and

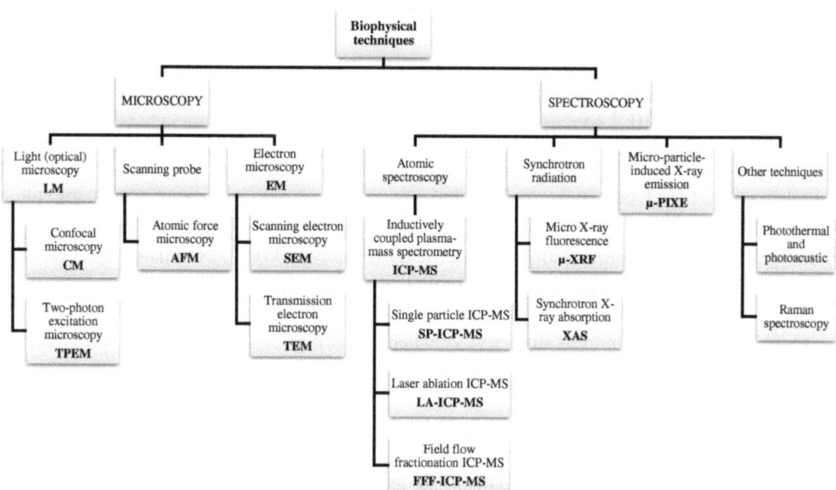

Fig. 3.1 Biophysical methods for detection and quantification of uptake, translocation, and accumulation of nanomaterials in plants

seeds. The above-mentioned methods and other spectroscopy and microscopy techniques utilized to determine the uptake and translocation of MNMs in plants are highlighted in this chapter (Fig. 3.1).

3.2 Microscopy Methods

Microscopy is a biophysical method that provides submicron resolution to detect MNMs in plant tissues. There are three basic techniques for monitoring MNMs in plants using microscopy methods that are discussed in this chapter: light beam source, scanning probe, and electron beam microscopy (Fig. 3.1). Each one of these techniques provides different information levels on the interaction and physical transformation of both the plant cells and the nanomaterials (NMs). Several researchers have used these techniques and have depicted challenges to detect and analyze NPs in soft and organic matter, such as in food and agriculture (Tiede et al. 2008; Bandyopadhyay et al. 2012; Mbundi et al. 2014; Grillo et al. 2015). This section provides an overview of microscopy techniques and examples for the detection, translocation, and accumulation of NMs in organic matrices.

3.2.1 Light (Optical) Microscopy

Conventional optical or light microscopy (LM) allows the observation of objects in the range of submicrometer to micrometer size. However, two variants of light

Table 3.1 Comparison of advantages and disadvantages of electron microscopy and light microscopy (Inoue 2010; Larue et al. 2014a)

	Electron microscopy	Light microscopy
Sample requirement and preparation	Completely dry sample;	Living specimens can be observed
	Coating with thin metal layer (usually gold) before detection if sample is non-conductive or poor-conductive;	No coating needed
	Staining is unnecessary when preparing nanomaterial suspension	Non-autofluorescence materials have to be stained
Operation environment	Vacuum environment (except environmental electron microscopes)	Ambient environment
Sample damage	Greater sample damage by electron beams	Lower sample damage by photon beams
Image quality	Black/white image	Colorful image

microscopy, confocal microscopy (CM) and two-photon excitation microscopy (TPEM), have shown to be effective for the detection of MNMs in plants. For detailed information about the imaging processes using these techniques, the reader is referred to Inoué (2010).

Light microscopy imaging has several advantages, compared with electron microscopy (EM) imaging (Table 3.1). The most striking difference between these two techniques is that LM allows the observation of MNMs in living specimens, which is not possible with EM. However, non-autofluorescence MNMs have to be stained before the exposure to plants in order to be detected with LM. The study of the uptake of MNMs by plant cells/roots by LM goes back to 2009. Since then, a few studies have shown the use of this technique to corroborate the uptake of CNTs and metallic NMs by plant roots. Examples and their respective references are shown in Table 3.2. A brief description of each procedure is shown in the next sections.

3.2.1.1 Confocal Microscopy (CM)

Confocal microscopy images are obtained by scanning the specimen with a point of light in a raster pattern (Inoué 2010). Since it was patented in 1957, CM has been widely used in the biological and biomedical fields due to its 3D resolution capability, in situ/in vivo imaging, less phototoxicity, and higher optical resolution, compared to EM (Inoué 2010). CM has shown great capabilities to study the impact of nanomaterials in human organs (Lee et al. 2014), and some researchers have shown its potential for monitoring the uptake of MNMs by plants.

One of the first reports about the use of CM to verify the uptake of NPs was published by Hischemöller et al. (2009). These researchers exposed moth orchid (*Phalaenopsis* spp.) and Arabidopsis (*Arabidopsis thaliana*) plants to a colloidal solution of $NaYF_4$:Yb,Er for a few days. Subsequently, they observed root samples with confocal laser scanning microscopy and detected fluorescent nanocrystals in tissues of the velamen radicum (root epidermis) and in the stele of roots, demonstrating that the NPs were taken up and translocated within the plant. The same year, Liu et al. (2009) used CM to demonstrate that fluorescein isothiocyanate (FITC)-stained SWCNTs penetrated both cell walls and cell membranes of tobacco Bright Yellow (BY-2) cells' line in a time- and temperature-dependent manner.

The CM has also been used to detect the uptake of widely used metallic NPs by plants. Ma et al. (2010) exposed four weeks Arabidopsis seedlings to silver nanoparticles (*n*Ag) colloid (40 nm). After exposure, they observed, with a confocal/multiphoton microscope, that the majority of the *n*Ag accumulated in the columella (cells of the root cap arranged longitudinally), but some of them were able to reach the vasculature of the seedlings and consequently, potentially translocated to the upper plant parts. Zhao et al. (2012a) exposed the roots of one-month-old corn (*Zea mays*) plants to FITC-stained ZnO ENPs and observed the samples with a CM after 48 h of exposure. Confocal images showed that the stained NPs were accumulated in the cell walls in root cortex and most of them retained at

Table 3.2 Examples of the use of confocal and two-photon excitation microscopy for the detection of ENMS in plants

Nanomaterials	Particle size (nm)	Plant	Concentration (ppm)	Mode of exposure	Growth media	Accumulation	Detection methods	Reference
$NaYF_4$:Yb,Er		Moth orchid (*Phalaenopsis* spp.)		Root	NPs suspension	NaYF4:Yb,Er nanoparticles was translocated from velamen radicum to passage cells, and eventually to vascular tissues	CM	Hischemoller et al. (2009)
MWCNT, CeO_2, TiO_2	MWCNT (diameter is between 110 and 170, length is up to 9 μm), CeO_2 (<25), TiO_2 (100)	Wheat (*Triticum* spp.)	100	Root	NPs suspension	Only MWCNTs have capabilities to pierce the root epidermal cell	TPEM	Wild and Jones (2009)
ZnO	380	Maize (*Zea mays*)	100, 200, 400, 800	Root	Sandy loam soil	NP aggregates pierced corn roots epidermis and cortex through apoplastic and symplastic pathways	CM	Zhao et al. (2012a)
CeO_2	8 ± 1	Maize (*Zea mays*)	100, 200, 400, 800	Root	Sandy loam soil/organic soil	Uncoated $nCeO_2$ have preferential translocation by corn roots than coated ones and higher concentration in organic soil than in unenriched soil Converse results have shown in corn shoots	CM	Zhao et al. (2012b)

(continued)

Table 3.2 (continued)

Nanomaterials	Particle size (nm)	Plant	Concentration (ppm)	Mode of exposure	Growth media	Accumulation	Detection methods	Reference
Mn	100	Mung bean (*Vigna radiata var.* Sonali)	0.05, 0.1, 0.5, 1	Root	NPs suspension	Mn NPs were observed in the root (cortical and stellar) and leaves (stomata and mesophyll)	CM	Pradhan et al. (2013)
Mesoporous silica nanoparticles (MSNs)	20	Maize (*Zea mays*), wheat (*Triticum* spp.), lupin (*Lupinus* spp.) Arabidopsis (*Arabidopsis thaliana*)	200, 500, 1000, 2000, 10000, 20000	Root	NPs suspension	MSNs were observed in the roots, stems and leaves of lupin, wheat, maize, and *Arabidopsis thaliana* The accumulation percentage of MSN was between 25 and 37.5 % in the root of maize	CM, μ-PIXE	Sun et al. (2014)

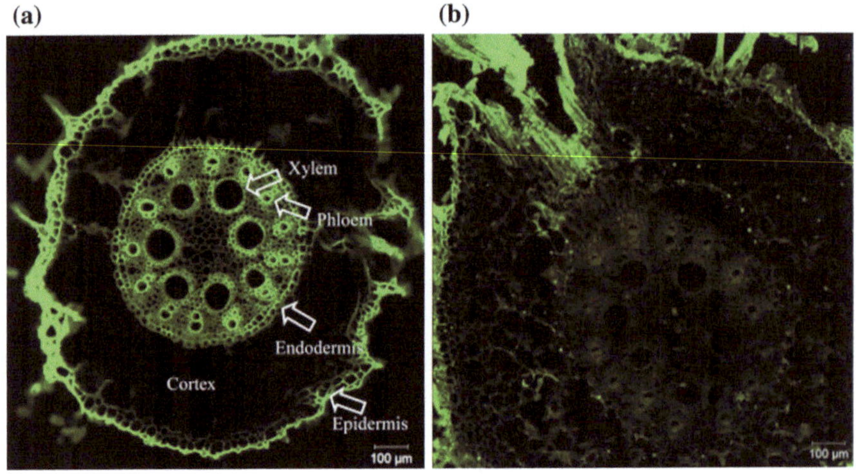

Fig. 3.2 Confocal images of cross sections of root treated for 24 h with 200 mg/L FITC (**a**) and FITC-stained *n*ZnO (**b**) suspensions. Stained NP aggregates were observed in root epidermis, cortex, endodermis, and xylem. The transport was restrained by the Casparian strip. (Adopted from Zhao et al. (2012a). Copyright @ 2012 American Chemical Society)

the endodermis, but some of them reached the transport system. Subsequently, Zhao et al. (2012b) exposed the corn roots to FITC-stained CeO_2 NPs and corroborated the previous results found with ZnO NPs (Figs. 3.2 and 3.3). Authors hypothesized that the ENPs entered through the Casparian band at the emission points of the lateral roots, where it was not fully formed. Pradhan et al. (2013) observed, after 15 days of exposure, that FITC-labeled Mn NPs were taken up by cortical and stellar root tissues and translocated to leaves (stomata and mesophyll) of mung beans (*Vigna radiata* var. Sonali). Sun et al. (2014) investigated the uptake of mesoporous silica nanoparticles (MSNs) in four plant species, maize (*Zea mays*), wheat (*Triticum* spp.), lupin (*Lupinus* spp.), and Arabidopsis. After five days of treatment, MSNs were observed in the roots of maize and in roots, stems, and leaves of lupin and wheat. Afterward, they incubated Arabidopsis seedlings in FITC-stained MSNs for 12 h and obtained confocal images of the NPs in chloroplasts, corroborating the capability of CM for the observation of ENM uptake and translocation in plants.

3.2.1.2 Two-Photon Excitation Microscopy (TPEM)

TPEM is another variation of LM that offers less sample phototoxicity, greater depth penetration (down to millimeter scale), and 3D resolution, which allows in vivo and situ observation of living plant cells. Different from traditional CM, which is one-photon excitation, TPEM only requires half energy and does not need a pinhole to block the background signal in the detection pathway (Zipfel et al.

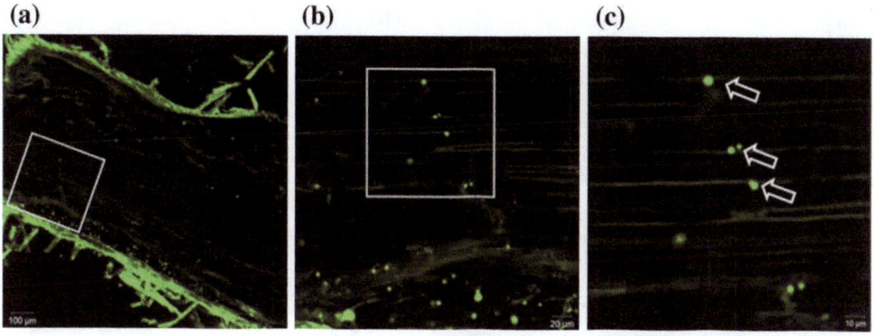

Fig. 3.3 Confocal images of a longitudinal section of corn roots treated with 200 mg/L of *n*ZnO/FITC-stained for 24 h. Images (**a**) and (**b**) showed NP aggregates in cortex and within xylem cells. (**c**) *Arrows* showed NP aggregates in the vascular cylinder adhered with the xylem vessel walls; (**b**) is a magnification of the *gray* square area indicated in (**a**); (**c**) is a magnification of the *gray* square area in (**b**). (Adopted from Zhao et al. (2012a). Copyright @ 2012 American Chemical Society)

2003; Rubart 2004; Stutzmann and Parker 2005; Fahrni 2009; Wild and Jones 2009). To the best of authors' knowledge, only one report has shown the capability of TPEM to image the uptake of MNMs by plants. Wild and Jones (2009) employed TPEM to determine the uptake of MWCNTs, TiO_2, and CeO_2 ENPs by wheat roots. These researchers exposed the plants for 28 days and observed single and aggregate MWCNTs in epidermal cells (Fig. 3.4). They also found that some MWCNTs penetrated cell walls and entered up to 4 μm into the cytoplasm, but they were not found to enter fully into the cells, perhaps due to the size of MWCNTs (diameter between 110 and 170 nm). TiO_2 NP and CeO_2 NP aggregates were just found adhered onto the root surface.

Compared with EM, light imaging microscopy has outstanding advantages, such as unsophisticated sample preparation and less sample damage. It provides images with high resolution and has a high potential for tracking the fate of non-fluorescence emission MNMs in plants. However, few reports include the use of this technique to investigate the uptake and translocation of nanomaterials by the whole plant. A possible reason is that non-fluorescence MNMs have to be stained before exposure to plants, which might modify their surface properties; additionally, unpredictable artificial contaminants might be introduced. As a result, microscopes and observation platforms without any staining are urgently needed. We believe this will be achievable shortly (Min et al. 2011).

3.2.2 *Atomic Force Microscopy (AFM)*

Atomic force microscopy (AFM) is a non-destructive technique that utilizes a fine tip (usually of a 5–20 nm radius) mounted at the end of a spring-like lever that

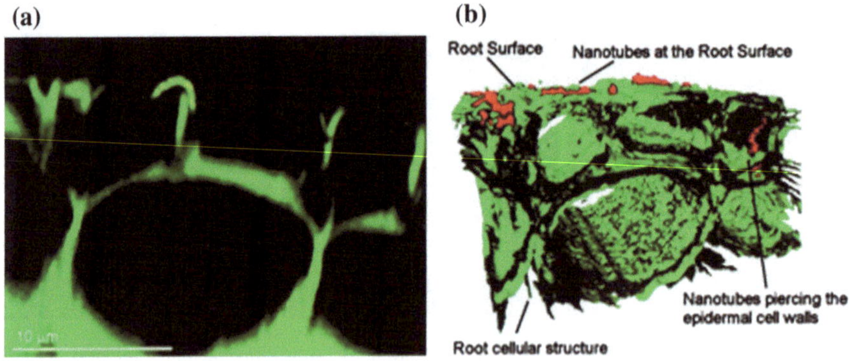

Fig. 3.4 **a** TPEM image of both ends of individual MWCNT penetrating the root epidermal cell (*green*). **b** A TPEM image of MWCNT aggregates (*red*) adhered on the root surface and penetrating the epidermal cell (*green*). (Adopted from Wild and Jones (2009). Copyright @ 2009 American Chemical Society)

raster scans of a specimen's surface. A description of the operation of the AFM is beyond the scope of this chapter, and the reader is referred to McPherson et al. (2000), Krautbauer et al. (2003), and Yang et al. (2007). Figure 3.5 provides a representation of a basic AFM setup. Originally, AFM was developed to characterize surfaces' roughness. Topographical maps with information regarding irregularities on the surface (flat or bumpy areas) can be obtained. Benitez et al. (2004) characterized tomato cutin (a component of the cuticule) of young versus mature fruits by contact mode AFM topographical maps, finding the surface of young tomatoes smoother than those of ripe fruit. Besides the high magnification provided, AFM became a promising tool to study biological systems (Gerber and Lang 2006; Cohen and Bitler 2008; Gaczynska and Osmulski 2008; Müller and Dufrene 2008) under different environmental conditions. Molecules and molecular structures can be visualized, and their interactions can be captured (Alessandrini and Facci 2005; Ando et al. 2008; Whited and Park 2014). However, one of the limitations of AFM is the scanning time, and biological processes may take place faster than the AFM can scan and capture the reaction. Ando et al. (2013) described in a review how the visualization of biomolecular processes has been improved by the development of high-speed AFM (HS-AFM) and the limitations yet to overcome.

Different approaches have been developed for the analysis of soft specimens. Lenaghan and Zhang (2012) used AFM to image a nanocomposite adhesive secretion from English ivy (*Hedera helix*). Abraham et al. (2013) studied the sorption of NPs onto environmental surfaces and corroborated the presence of Ag NPs on the surface of *Ficus benjamina* leaf disks by AFM imaging; they were also able to measure the NPs and observe their morphology. Although MNMs can be captured under the AFM probe, measurements often come accompanied with a degree of uncertainty, depending on the matrix (Klapetek et al. 2011); therefore, the quantification of NPs' parameters such as size and size distribution by AFM is often combined with other techniques such as electron microscopy and light scattering techniques.

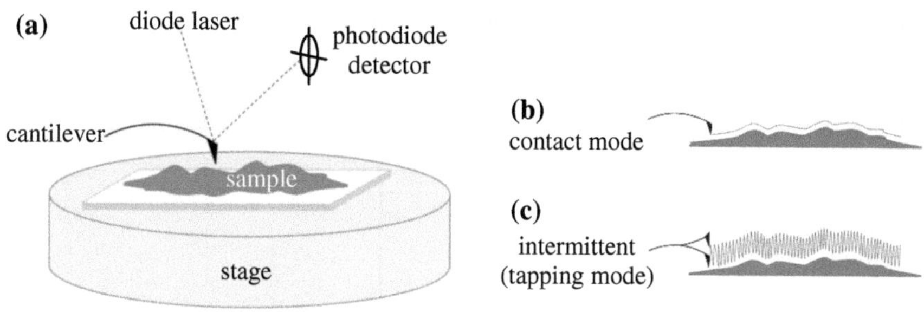

Fig. 3.5 Overall illustration of basic AFM setup. **a** A laser points toward the tip of the probe, and its motion is sensed by a photodiode detector for the creation of an image. **b** Contact mode: the cantilever passes dragging the tip over the surface of the sample. **c** Intermittent mode: the cantilever vibrates up and down at the time of scanning

3.2.3 *Electron Microscopy (EM)*

There are two primary types of electron microscopy instruments: scanning electron microscope (SEM) and transmission electron microscope (TEM). Both instruments have an energy electron source, known as the electron gun, which creates an electron beam (ebeam). The electron gun is typically composed of a tungsten (W) filament, although modern electron microscopes are fitted with a lanthanum hexaboride (LaB_6) or field emission gun (FEG) that provide a brighter beam for better resolution. Electron microscopes utilize a vacuum column through which the ebeam passes vertically, promoting the straight travel of electrons (Roming 1986). The electrons will then either interact with the sample in the case of SEM or pass through the ultrathin sample (for biological specimens up to 1 μm thick) in the case of TEM. While SEM can give rise to apparent three-dimensional (3D) images, the flat portraits' resolution of TEM can be higher than SEM by one order of magnitude (Luykx et al. 2008).

3.2.3.1 Scanning Electron Microscopy (SEM)

With SEM, it is possible to obtain images of the sample's surface as well as material composition information. While scanning, the data gathered is based on the resulting energy emitted from the sample hit by the electron beam. Most commonly, the electrons that bounce as backscattered or secondary electrons (BSE or SE) are detected for imaging. EMs equipped with energy dispersive X-ray spectrometer (named EDXS, EDAX, EDS, or EDX) can provide compositional information of the sample by the X-rays produced to identify the chemical nature of the elements present. EDX can also provide an elemental mapping of the atomic composition. Mapping is usually represented by colored dots, and their distribution in the image elucidates the element location within the sample (Fig. 3.6a). Abd-Alla et al. (2016) imaged different tissues of faba bean plant previously exposed to Ag NPs by using

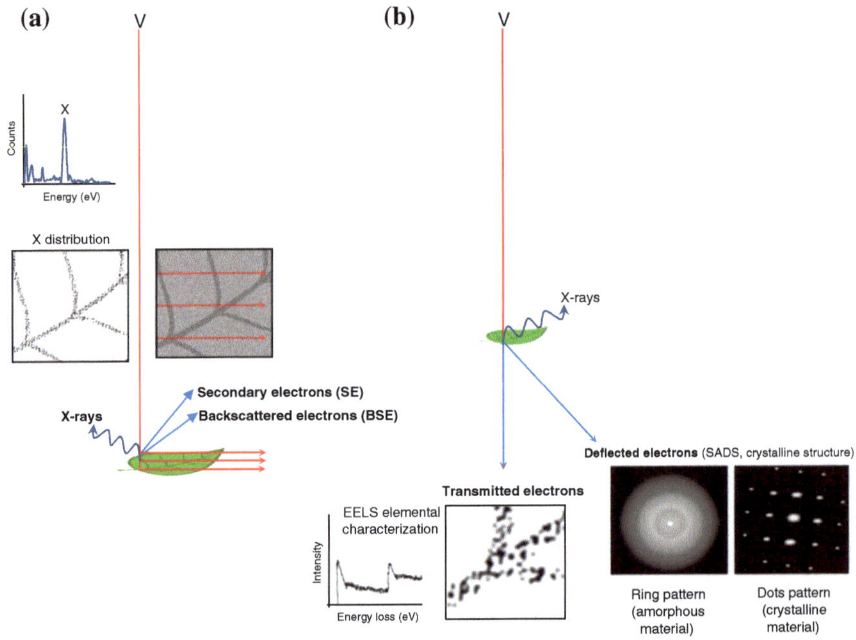

Fig. 3.6 Common ebeam–sample interaction effects used for (**a**) SEM and (**b**) TEM imaging. The sketch simulates nanoparticles "X" encountered in the vascular system of the plant. **a** SEM—electron beam (primary electrons, *red line* in sketch) at low acceleration voltage (~5–15 kV) scans over the sample's surface. The ebeam hitting the sample generates signals that, if detected, create images. Backscattered electrons, secondary electrons, and X-rays are usually detected for the acquisition of a micrograph. Backscattered and secondary electrons produce surface images by differences in atomic number (z) represented in shades of *gray*. In Figure (**a**), the X-ray spectrum confirms the presence of an "X" NP, while the mapping shows the distribution of "X" as dots within the vascular system of a leaf. **b** TEM—electron beam at high acceleration voltage (~80–200 kV) passes through an ultrathin sample resulting in transmitted or deflected electrons. Transmitted electrons create an image of the trespassed sample; if the electron energy lost in the path is quantified and related to the elements in the sample, the material composition can be identified. The deflected electrons provide information about the structure of the sample; ring pattern for amorphous materials or dots pattern for crystalline materials

EDX coupled to SEM. The authors reported clear accumulation of Ag in roots, shoots, and nodules. However, even when the quantification was reported, the published data correspond to Ag concentration, with no information about particle size. In a similar way, Yan et al. (2013) exposed the leaves of soybeans to droplets containing Cs NPs and were able to identify the presence of NPs in the outer surface of the leaves as well as in the pods, roots, and stems, confirming the uptake and translocation of the NPs. Moreover, they performed EDX analysis on the particles to evidence the content of Cs. Cañas et al. (2008) studied CNTs in plants exposing different food crops to the nanomaterial for 0, 24 and 48 h. Their SEM analysis revealed CNTs in root surfaces but not in inner tissues of the roots. Perhaps the exposure time was too short for the uptake.

3.2.3.2 Transmission Electron Microscopy (TEM)

In TEM, the ebeam hits and trespasses an ultrathin sample. The electrons transmitted through the sample create an image, and these electrons lose energy while passing through the sample. The difference in energy before and after crossing the specimen is correlated with the material's composition,. known as electron energy loss spectroscopy (EELS). Therefore, TEMs with EELS detectors can provide elemental analysis of the sample. Transmitted electrons that are deflected exhibit structural information about the specimen's components. For example, NPs can be characterized by TEM with selected-area electron diffraction (SADS), which provides information about the nanocrystalline structure. Figure 3.6b shows an overview of TEM.

Transmission electron microscopy allows the detection and quantification of carbonaceous materials in plant cells. Lin et al. (2009) confirm the uptake of carbon-based MNMs evidenced by TEM micrographs showing C_{70} in vacuoles and leaves' cell walls. Lahiani et al. (2015) proposed the use of single-walled carbon nanohorns (SWCNHs) as plant growth regulators after studying the effects of SWCNHs in different edible plant species, confirming the presence of SWCNH in the roots and seeds of tomato and tobacco cells by TEM. Larue et al. (2012b) used $MW^{14}CNT$ and TEM to quantify the uptake and translocation of gum Arabic (GA), and humic acid (HA) stabilized MWCNTs in wheat (*Triticum aestivum*) and rapeseed (*Brassica napus*) plants. They reported the presence of CNTs in the leaves of both plant species, though at low concentration. In rapeseed, total accumulation was 140 ± 32 and 108 ± 47 μg CNTs/kg, and in wheat, the uptake was 200 ± 83 and 43 ± 15 μg CNTs/kg dry biomass for GA and HA stabilized CNTs, respectively. In a very recent study, Le Van et al. (2016) utilized TEM to localize CuO NPs in cotton leaves. These researchers found that CuO NPs aggregated on the epidermis of leaves in non-modified plants, while they penetrated the cells after endocytosis on transgenic cotton plants. After correlating CuO uptake and toxicity to gene expression, the authors were able to conclude that CuO NPs enhanced the expression of a gene related to cotton insect resistance.

Several studies have shown the versatility of TEM for the detection of the uptake and translocation of NMs in plants. High-resolution TEM (HRTEM) is a variant of TEM that is suitable for visualizing materials where contrast is not an issue to overcome. The HRTEM can provide substantial information on the crystalline structure of materials utilizing a higher acceleration voltage than conventional TEM. By using this technique Gardea-Torresdey et al. (2002, 2003) showed, for the first time, the formation of Au and Ag NPs in alfalfa (*Medicago sativa*) plants grown in agar medium enriched with either potassium tetrachloroaurate or silver nitrate. With HRTEM, Gardea-Torresdey et al. (2002) measured Au particles of 4, 20, and 40 nm along the alfalfa stem, which endorsed them to hypothesize the continuous growth of the Au NPs within alfalfa plant. Low-magnification TEM and HRTEM also allowed Gardea-Torresdey et al. (2003) to show images of alfalfa shoot with icosahedral silver nanoparticles ranging from 2 to 3 nm in size. These studies opened the door for the detection of metal NPs in plants using TEM. Other studies have also shown the capability of TEM for localizing NPs within the ultrastructure of plant cells. Taylor et al. (2014) studied the effects of $K(AuCl_4)$ and $AuCl_3$ in *A. thaliana*. They grew alfalfa in agar medium enriched with ionic gold and Au NPs of 5 and 100 nm. The researchers found Au NPs in the roots of plants treated with ionic Au, leading them to conclude that the plants did not take up the NPs.

The SEM and TEM have been used together for complementing the analysis of NPs in plants. For example, Du et al. (2011) detected Ti in periderm cells of wheat root by SEM-X act analysis and with TEM they observed TiO_2 NPs (20 $\pm$ 5 nm) in cortex cells of the root. Li et al. (2013), using SEM found TiO_2 NPs accumulated on the *Lemna minor* leaves, and TEM micrographs showed no cellular uptake of TiO_2 NPs.

3.2.3.3 Scanning Transmission Electron Microscopy (STEM)

The STEM combines capabilities of SEM and TEM. Modified SEM and TEM equipped with additional detectors are enabled to run in STEM mode. The SEM in transmission mode (SEM/STEM) provides images with better spatial resolution (http://www.fei.com/introduction-to-electron-microscopy/stem/) still using the relatively low accelerating voltages. Bandyopadhyay et al. (2015) used dark-field STEM (DF-STEM) to study tissues of alfalfa plants exposed to ZnO NPs. DF-STEM, which is sensitive to atomic number due to Z-contrast, showed that striations accumulated along the cell walls of stem cells were formed by small particles of 9–12 nm. Punctual EDX spectroscopy confirmed that these high-contrast structures corresponded to Zn and O combined, showing the aggregation of ZnO NPs.

3.2.3.4 Sample Preparation for EM Analysis

The detection and analysis of NPs inside organic matrices are difficult. One of the biggest challenges of EM analysis is sample preparation of biological specimens because it is time-consuming, and the process itself can damage or contaminate the sample, introducing artifacts. Moreover, studying NMs' morphology, size, and other properties may vary depending on the instrument being used, the image acquisition, image analysis, and the selected sample for examination (Dudkiewicz et al. 2015). Analytes of living organisms have to withstand the vacuum environment needed for the analysis, which is not possible in their native state for conventional instruments. However, variants in electron microscopes such as cryo-EMs and environmental-EMs allow for lesser sample preparation and a more suitable ambient for living specimens. Moreover, the relatively recent development of SEM capsules enables the observation of biological samples with ongoing metabolic activity by protecting the sample from the harsh conditions of the SEM (Kokina et al. 2013). When analyzing NPs in plants, minimum specimen disturbance is desired in order to analyze the exact location of the nanomaterial [that may change according to the matrix environment (Kumari et al. 2011)] and to elucidate the effects produced in the plant.

Protocols vary, but the main objective is to preserve the integrity of the sample without disturbing the morphology of cells and structure of their components while being analyzed (Pathan et al. 2008; Wu et al. 2012). Ensikat et al. (2010) presented and evaluated sample preparation methods that are not routinely used but are feasible for the visualization of even fresh plant surfaces in conventional SEM. Dudkiewicz et al. (2011) reviewed the EM technologies that have been applied to characterize NPs in food matrices. Their work is a useful reference for the analysis of NPs in the agricultural field, as it involves the study of MNMs inside soft and moistened organic matrix.

The above literature shows that EM is a state of the art technique for the detection of ENMs in plants. However, this technique does not allow quantification of the MNMs within plant tissues.

3.3 Spectroscopy Methods

Currently, analytical methods based on the interaction of matter with various types of radiation, broadly called spectroscopy, have become state of the art methods for the detection, quantification of the uptake, translocation, and accumulation of MNMs in plants (Fig. 3.1). Some of these techniques are qualitative and others quantitative. In addition, it is usual to use more than one technique for a complete assessment of the uptake, distribution, and speciation of MNMs exposed to plants.

3.3.1 Atomic Spectroscopy

Atomic spectroscopy is the most used method for the determination and quantification of trace elements in environmental samples. In this method, high heat is used to decompose the sample in atoms and ions (a process called atomization) that are measured by specific detectors. Atoms are detected and quantified based on the emission (optical emission spectrometry, OES) or absorption (atomic absorption spectrometry, AES) of light, while ions are separated based on mass-to-charge ratios (mass spectrometry, MS) (Skoog et al. 1998).

Currently, atomization for OES is produced by a type of discharge called plasma, which is supported by argon and usually called "inductively coupled plasma (ICP)" (Boss and Fredeen 2004). According to Skoog et al. (1998) "plasma is an electrical conducting gaseous mixture containing a significant concentration of cations and electrons, at similar concentrations." Argon ions in the mixture can absorb a big amount of energy from an external source to maintain highly elevated temperatures (as high as 10,000 K). The basic principle of ICP consists in the introduction of a liquid sample into the argon plasma in the form of an aerosol produced by a nebulizer. The droplets forming the aerosol are carried along with some vapor that undergoes atomization to produce free atoms and ions that can be measured and quantified. Two main techniques, optical emission spectrometry (OES) and mass spectrometry (MS), coupled to the ICP (ICP-OES/MS), are among the most used for uptake determination and quantification of different elements. ICP-OES provides lower cost analysis when compared to ICP-MS; thus, it is widely used as quantification tool. However, this technique can only be used to determine the elemental composition or dissolution of the NMs (Elzey 2010) and metallic elements of the NMs taken up by plants (Larue et al. 2012a). Moreover, with this technique, it is not possible to discriminate between the amounts of NMs adsorbed/absorbed by plant roots (Larue et al. 2012a).

On the other hand, ICP-MS is widely used to measure the number of single charged ions in a sample and separate them according to the mass-to-charge ratio (Boss and Fredeen 2004). According to Thomas (2013), ICP-MS is the "fastest growing trace element analysis technique" currently available. ICP-MS offers high sensitivity for metals/metalloids with detection limits ranging from the parts per million (ppm) to the parts per trillion (ppt) levels (Schaumann et al. 2015; Arruda et al. 2015). The ICP-MS alone can only be used to determine concentration and composition. However, in the last decade, ICP-MS has been coupled with other analytical techniques, allowing the measurement of other variables, including particle size and distribution, thus increasing its capabilities for NMs determination in environmental samples. Examples of ICP-MS coupled techniques used to measure the uptake of ENMs by plants include single particle analysis (SP-ICP-MS), field flow fractionation (FFF-ICP-MS), and laser ablation (LA-ICP-MS) (Fig. 3.1).

3.3.1.1 Single Particle (SP-ICP-MS)

Single particle analysis is a technique that "quantifies the number of particles in a volume of fluid" (Degueldre and Favarger 2003). It was developed to detect individual particles in aqueous suspensions. In this technique, the analyte is spatially concentrated, allowing the introduction of only one particle into the ICP, where atoms or ions are detected as a single pulse. The number of counts is related to the number of atoms, and the frequency of the pulses is proportional to the concentration of particles (Laborda et al. 2011). ICP-MS operated in the single particle mode allows the possibility of analyzing individual NPs, thanks to the reading of thousands of signals within a very short dwell time ($\sim$10 ms) (Mitrano et al. 2012). The dwell time is a key element in this analytical technique; normally, as the dwell time decreases, the resolution is higher.

Few references describe the use of this technique to determine the size and size distribution of NPs in a sample. Advantages of this technique include the following: (1) It allows working with dilute solutions, reduces or eliminates sample preparation, avoids NP agglomeration, and it is faster than microscopy techniques; (2) it has relatively high sensitivity because it is able to discriminate particles of 10 nm diameter. However, SP-ICP-MS requires a lot of statistical work and multiple running (Arruda et al. 2015), and many diluted solutions can represent a challenge when working with complex matrices. After several improvements, this technique has shown capabilities for the detection and quantification of MNMs in biological tissues. Dan et al. (2015) quantified the uptake and translocation of Au NPs in tomato (*Solanum lycopersicum* L.) plants by SP-ICP-MS. These researchers exposed polyvinylpyrrolidone (PVP)-coated Au NPs (40 nm) to hydroponically grown tomato plants for four days. Then, they digested the samples with Macerozyme R-10, a multicomponent enzyme mixture that contains cellulose (0.1 unit/mg), hemicellulose (0.25 unit/mg), and pectinase (0.5 unit/mg). Dan et al. (2015) analyzed the digests by using SP-ICP-MS and were able to determine "the size, size distribution, particle concentration, and dissolved Au concentration." The authors reported that 20 nm was the required size for quantification of the Au NPs with the SP-ICP-MS, while the concentration detection limit was 1000 NPs/mL.

3.3.1.2 Laser Ablation (LA-ICP-MS)

This hyphenated technique allows the analysis of solids without the need of chemical dissolution. It provides less contamination risk in small samples, reduced time for sample preparation, and increased sample throughput with less spectral interferences (Mokgalaka and Gardea-Torresdey 2006; Koelmel et al. 2013). One of the first reports about the use of LA-ICP-MS for the determination of the uptake of NPs by plants was performed in tobacco (*Nicotiana tabacum*) exposed to Au NPs of different sizes (Judy et al. 2011). Using this technique, the researchers determined the presence of Au in the mesophyll of tobacco leaves harvested from plants treated with Au NPs of 5, 10, and 15 nm. They used a "LSX-213 laser ablation system that removed 400 × 400 μm^2 craters, the depth of which ranged from 8 to 10 μm as measured using a Nikon Eclipse 90i light microscope." According to the authors, "gold concentration reported as log counts per second (CPS) of m/z 197 (Au) normalized by CPS for m/z 66 (Zn) to account for the mass of tissue removed from each laser burst." However, this study only proved the presence of Au within tissues, with no mention of the Au form. The technique was improved by Koelmel et al. (2013) that used a culture system with no presence of ionic Au. They fed rice plants with surface modified Au NPs proven to be stable. The use of LA-ICP-MS allowed these researches to show the uptake and spatial distribution of Au NPs in shoots and roots of rice plants. Koelmel et al. (2013) were able to quantify the concentration of Au NPs in different tissues, separated by particle surface change. This study reported that negatively charged Au NPs were more abundant in rice shoots, compared to neutral or positive charged Au particles. However, the authors concluded that the technique can only be used to determine the uptake of insoluble nanoparticles.

3.3.1.3 Field Flow Fractionation (FFF-ICP-MS)

Field flow fractionation is a group of techniques that allows separation and sizing of molecules through the application of different fields and modes of operation (Mitrano et al. 2012). The basis of FFF is, thus, physical separation of the particles. The analytes are passed through a channel that does not involve the use of a stationary phase. The channel with laminar flow is subjected to a certain field (sedimentation, flow, electrical, or thermal) that allows the separation of particles based on size or mass. The advantages of this technique include a high resolution for size fractionation, which varies from 1 μm up to 1 nm (Dubascoux et al. 2010), and its capability for analyzing nanoparticles in complex matrices when coupled to a high-resolution detector, such as ICP-MS (Artiaga et al. 2015). A variant of FFF is asymmetrical flow field flow fractionation (AF4), a widely used technique for environmental analysis of both natural and manufactured NPs that allows the possibility of performing "multi-element analysis when coupled to MS" (Mitrano et al. 2012). Palomo-Siguero et al. (2015) used AF4-ICP-MS and TEM to detect chitosan-modified selenium NPs (CS-Se NPs) in radish (*Raphanus sativus*) plants. The CS-Se NPs were extracted from the root by using 0.1 % chitosan, 0.034 M ascorbic acid, and 0.24 M acetic acid as an extracting solution. The extracts were centrifuged at 10,000 rpm for 10 min, and the supernatant was injected into the AF4-UV-ICP-MS that showed CS-Se NPs extracted from lateral roots. TEM images show the presence of spherical Se NPs with an estimated particle diameter of 25 ± 8 nm, mainly interconnected, assembled or aggregated.

3.3.2 *Synchrotron Radiation Techniques*

Synchrotron techniques have emerged as a powerful tool to study the speciation and distribution of metal and metalloids in plants exposed to nanomaterials. These techniques are based on the electromagnetic radiation produced when a magnetic field alters the direction of particles moving at nearby the speed of light. Synchrotron facilities produce high-intensity photons with brilliance that is several orders of magnitude higher than that produced by conventional X-ray sources. High brilliance provides unique capabilities on experiments requiring a high photon flux and a small beam size. The tunability of synchrotron radiation (SR) allows the study of samples with different techniques (e.g., X-ray fluorescence and X-ray absorption spectroscopy). Some specific advantages offered by SR for the study of metals/metalloids in plants are as follows:

(1) Samples can be analyzed with little or no pretreatment.
(2) Sensitivity limit at femtogram level and spatial resolution at micro- and nanoscale levels (Sarret et al. 2013).
(3) Potential to identify the chemical forms of the element of interest. (i.e., speciation and coordination environments).

Two SR techniques seem to be more suitable to study the effects of nanomaterials in plants: X-ray fluorescence and X-ray absorption spectroscopy. Both techniques can be used in tandem to obtain the distribution and speciation of elements of interest. Several reviews have focused on the use of SR techniques in plants (Lombi and Susini 2009; Lombi et al. 2011; Donner et al. 2012; Majumdar et al. 2012; Sarret et al. 2013). This section provides a brief description of the SR techniques used to study plants exposed to MNMs and emphasizes results.

3.3.2.1 Micro-X-Ray Fluorescence (μ-XRF)

This technique is based on the emission of characteristic X-rays from atoms excited by the SR. When incident X-rays eject core level electrons from the atom into the continuum, a core-hole vacancy is created. Electrons from higher energy states fill the core-hole creating photons (fluorescence) of specific energies.

The multi-elemental μ-XRF technique has been predominantly used to create bi-dimensional maps in plants. The high spatial resolution in μ-XRF maps is provided by special optics (e.g., Kirkpatrick-Baez mirrors, Fresnel zone plates) that generate focused beams with sizes that can reach less than 1 μm. Due to the high penetration of hard X-rays, thin sections are recommended when analyzing plant tissue, in order to avoid signal originating from different depths (Scheckel et al. 2007; Lombi et al. 2011; Hernandez-Viezcas et al. 2013; Majumdar et al. 2014). Multiple elements can be analyzed and mapped simultaneously with the present technique. Metal oxide MNMs have been at the forefront of plant nanotoxicity research due to their unique properties and high production (Kahru and Dubourguier 2010; Hendren et al. 2011). Nanoparticles of CeO_2, ZnO, TiO_2, and FeO_x are some of the most studied metal oxides MNMs (Piccinno et al. 2012). Larue et al. (2011) exposed wheat plants to anatase TiO_2 NPs (12 nm) in hydroponics for seven days. The μ-XRF analysis showed Ti in the parenchyma and vascular cylinder of the roots, suggesting root absorption and translocation of TiO_2 NPs. Servin et al. (2013) cultivated cucumber in soil amended with 750 mg/kg of TiO_2 NPs and analyzed the fruit with an X-ray beam of 0.3×0.7 μm^2 generating a fluorescence map. The synchrotron μ-XRF map showed Ti in the fruit (Fig. 3.7). The transfer of TiO_2 NPs to lettuce (*Lactuca sativa*) leaves was also evaluated with μ-XRF. This spectroscopic technique provided spatial localization of Ti in all lettuce tissues; however, no phytotoxicity was observed (Larue et al. 2014a). In recent studies with plants exposed to $nCeO_2$ (8 nm), SR μ-XRF showed the presence of Ce in the vascular tissue of kidney bean (*Phaseolus vulgaris*) roots, cucumber leaves, rice roots, and soybean pods (Zhao et al. 2013; Rico et al. 2013; Hernandez-Viezcas et al. 2013; Majumdar et al. 2014). Zhao et al. (2014, 2015) used μ-XRF to determine the effects $nCeO_2$ on micro- and macronutrients in cucumber and corn (*Zea mays*). The authors reported that $nCeO_2$ did not change the nutrient element distribution in cucumber plant and found a reduced Ca translocation and elements' redistribution in kernels of $nCeO_2$ treated corn plants. In plants exposed to ZnO NPs, μ-XRF maps have shown increased Zn concentration

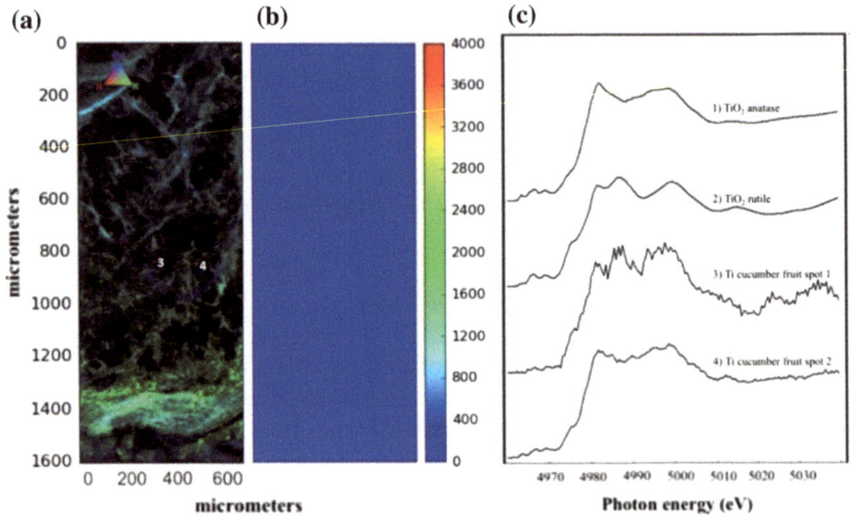

Fig. 3.7 **a** Tricolor μ-XRF images of the cross sections of cucumber fruit treated with $n\mathrm{TiO_2}$, **b** Ti temperature map, **c** μ-XANES spectra, spots of interest (3–4) were chosen from image (**a**). (Adopted from Servin et al. (2013). Copyright @ 2013 American Chemical Society)

in roots of cowpea (*Vigna unguiculata*), the root and leaves of mesquite (*Prosopis juliflora velutina*), and the stem and pods of soybean. Nevertheless, no major signs of toxicity were found in these plant species (Hernandez-Viezcas et al. 2011, 2013; Wang et al. 2013).

Elemental NPs' production is smaller when compared to metal oxide NPs; however, there is a need to study their potential plant nanotoxicity (Piccinno et al. 2012). μ-XRF, in combination with other analytical techniques, was used to demonstrate that Au NPs (3.5–18 nm) were taken up by roots of tobacco plants and subsequently translocated to the aerial plan parts (Sabo-Attwood et al. 2012). Larue et al. (2014a) exposed, through the leaves, lettuce plants to several concentrations of Ag NPs and analyzed the leaves with several techniques. μ-XRF showed that the Ag NPs were entrapped in the cuticle, and some of them penetrated the leaves.

3.3.2.2 Synchrotron X-Ray Absorption Spectroscopy (XAS)

XAS is a spectroscopic technique that provides information about the chemistry of the element of interest in a sample. The obtained XAS spectra can provide the oxidation state, interatomic distances, coordination number, and species of the atoms surrounding the analyte. This technique requires a high photon flux and tunability, which makes it almost exclusive to synchrotron facilities (Lombi and Susini 2009). As with XRF, the XAS phenomenon is dominated by the photoelectric effect.

In XAS measurements, incident photons progressively increase their energy while impacting the sample, starting at 50 eV below the binding energy of the analyte and finishing a few hundreds to over a thousand eV above it. The XAS spectrum indicates the energy absorption of the element due to the ejection of the photoelectron to the continuum. Also, oscillations caused by the interferences of the photoelectron with neighboring atoms are indicated in the spectrum. The XAS spectrum is divided into two parts, X-ray absorption near edge structure (XANES) and extended X-ray absorption fine structure (EXAFS). XANES encompasses the region approximately 50 eV below and above the absorption edge. This portion of the spectrum provides information about the oxidation state, local symmetry, and molecular species, when compared to model (pure) compounds. The EXAFS part of the spectrum extends from 50 to 1000 eV above the absorption edge and contains information about the coordination environment of the analyte. Detailed explanations about the XAS principles can be found in excellent reference reviews (Fendorf et al. 1994; Bertsch and Hunter 2001; Lombi and Susini 2009; Sarret et al. 2013). The incident beam that excites the elements of interest in the sample can have different sizes depending on the optics used: focused beam on the order of micrometer or submicrometer, and "bulk" beams on the order of millimeters.

Lopez-Moreno et al. (2010a, b) investigated the speciation of Ce in hydroponically grown soybean, corn, cucumber, alfalfa, and tomato plants exposed to CeO_2 NPs. Root samples were freeze-dried and homogenized with mortar and pestle, loaded into aluminum sample holders, covered with Kapton film, and analyzed by XAS. By comparing the obtained XAS spectra with the spectra of model compounds, authors reported that CeO_2 NPs were absorbed and stored within the roots with no modification. In another study, ZnO NPs were exposed to tumbleweed (*Salsola tragus*), mesquite, and palo verde (*Parkinsonia florida*). XAS studies showed no presence of ZnO NPs within the roots tissues and Zn was present as Zn (II) (Lopez-Moreno et al. 2010a; De La Rosa et al. 2011). Wang et al. (2013) performed XAS on cowpea grown in soil amended with ZnO NPs and corroborated that ZnO NPs are not stored within tissues. Further experiments concluded that ZnO NPs is not stable in the soil.

Several studies have shown that μ-XRF and μ-XAS can be used together to provide complementary information about the uptake, distribution, and speciation of NPs within plant tissues. After a μ-XRF map is created, specific areas in the image can be analyzed by μ-XAS to determine the oxidation state of the analyte. The distribution and speciation of rare earth nano-oxides ($n$$La_2O_3$, $n$$Yb_2O_3$, and $n$$CeO_2$) were studied with μ-XRF and μ-XAS in cucumber tissues. A small portion of $n$$La_2O_3$ and $n$$Yb_2O_3$ was present in the root cells as $LaPO_4$ and $YbPO_4$, respectively, whereas a portion of the $n$$CeO_2$ was biotransformed into $CePO_4$ in the root and $Ce(CH_3COO)_3$ in the shoot (Ma et al. 2011; Zhang et al. 2012a, b). Similarly, Cui et al. (2014) found that 6 % of $n$$CeO_2$ biotransformed into Ce(III) carboxylates in the root of lettuce exposed to $n$$CeO_2$. Hernandez-Viezcas et al. (2013) used μ-XRF to localize Ce in the soybean pod and by using μ-XAS, they found that most of the Ce remained as Ce(IV) in the form of CeO_2 NPs. Servin et al. (2012) also used μ-XRF and μ-XAS to study cucumber plants exposed to $n TiO_2$ (anatase 82 %, rutile 18 %) (Fig. 3.7). The results showed that TiO_2 NPs can be absorbed and translocated to the aerial parts in cucumber. Interestingly, the anatase phase remained in the root, while the rutile phase was found in the aerial parts of the plant.

3.3.2.3 Synchrotron-Based X-Ray Microscopy and Tomography

Synchrotron-based X-ray microscopy measures the absorbance above and below the edge energy of the analyte of interest, and the differences are used to generate 2D images. The samples can be rotated, reanalyzed, and the resulting images used to reconstruct 3D images. This technique can achieve resolutions as high as 20–40 nm. A disadvantage of plant analysis is the need for high concentrations in the samples. Fluorescence tomography, on the other hand, uses a fluorescence detector to produce 2D images; the sample is then rotated and reanalyzed to render a 3D image. This type of tomography has a higher sensitivity than contrast absorption measurements (Lombi et al. 2011). Patty et al. (2009) used transmission X-ray microscopy (TXM) to create absorption contrast transmission 2D and 3D images of cordgrass (*Spartina foliosa*) root exposed to mercury (Fig. 3.8). The results of the study show Hg NPs in the root tissue. Ma et al. (2011) and Zhang et al. (2012a) also used TXM to study rare earth nano-oxides, $n La_2O_3$, $n Yb_2O_3$, and $n CeO_2$ in cucumber tissues (Zhang et al. 2012c). Their studies suggest that the NPs distribute inside the cucumber tissues at varying degrees. Later on, Hernandez-Viezcas et al. (2013) employed TXM to investigate the behavior of ZnO NPs in the soybean; the technique showed clusters of Zn in the pod tissue but no presence of ZnO NPs.

Synchrotron radiation techniques have proven to be a powerful tool to study the interactions of MNM and plants. With the challenges of nanotechnology, life sciences, and engineering, synchrotron facilities tend to provide beamlines capable of micro- and nanoscale analysis. Nevertheless, XAS can only provide a certain amount of information, and complementary techniques should be used to improve the data collection. Table 3.3 summarizes recent studies on the use of SR to assess of the effects of NMs in plants.

3.3.3 *Micro-Particle-Induced X-Ray Emission (μ-PIXE)*

The μ-PIXE technique is a non-destructive elemental localization and quantification analysis. This technique uses a highly focused ion beam (usually protons) to excite the desired analyte, afterward the fluorescence emission from the relaxed atoms is detected. μ-PIXE offers higher microscopic analysis capability by using a focused

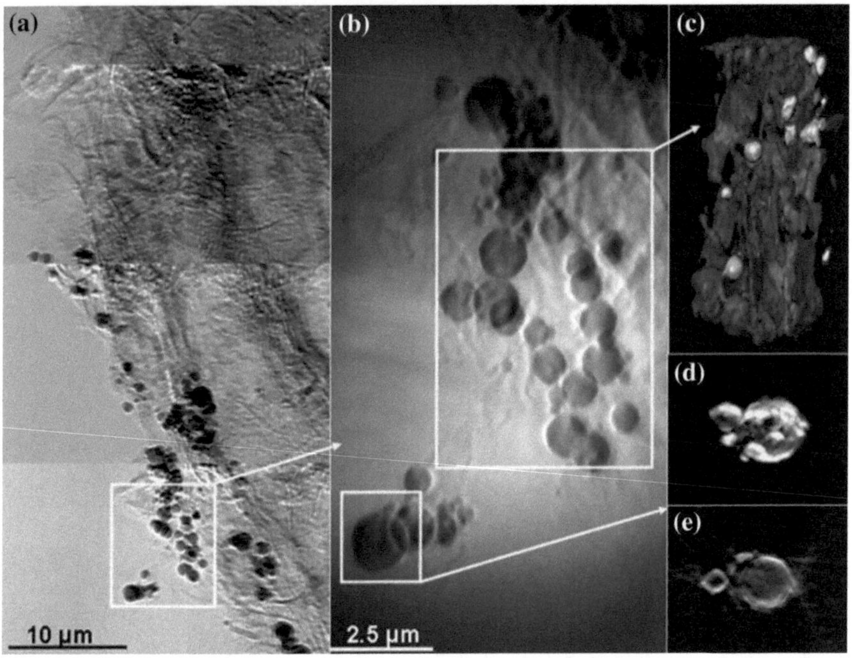

Fig. 3.8 TXM mosaic transmission image of *S. foliosa* roots taken at 9 keV in absorption contrast shows dark particles and dark channels due to absorption by Hg (**a**). Blowup (**b**) shows greater detail. 2D stills from tomography of particles from (**b**) show particles with greatest absorption (lightest), possibly surrounded by biofilms (**c**). 2D tomographic still (**d**) and reconstructed slice (**e**) of large particle indicate that highest Hg concentrations (lightest intensity) are on the outside of the fairly hollow particles. (Adopted from Patty et al. (2009). Copyright @ 2009 American Chemical Society)

ion beam (normally, two μ) (Lombi et al. 2011). μ-PIXE also has 3D imaging capability, which shows great potential when various elements are present in different layers (micron scale) of samples (Karydas et al. 2007). A disadvantage of μ-PIXE is that samples could be charged during analysis; however, this problem can be avoided by applying a conductive coating to the sample. Larue et al. (2012a) prepared a sample for μ-PIXE analysis by cryofixing in LN_2, cooling in isopentane, and then embedding in Tissue Tek resin. By using μ-PIXE imaging, Larue et al. (2011, 2012b) quantified and localized Ti in leaves and roots of wheat and rapeseed after foliar exposure to $n\text{TiO}_2$. Results revealed that rapeseed accumulated a higher content of Ti (22 mg Ti/kg dry matter) than wheat (2 mg Ti/kg dry matter). In addition, Sun et al. (2014) used μ-PIXE to show heavy accumulation of Si in the roots of maize plants exposed to MSNs. As per the literature, μ-PIXE has shown great potential for the detection and quantification of trace elements in various living systems.

Table 3.3 Summary of the study of NP–plant interactions by using X-ray spectroscopy techniques

NPs	Particle size (nm)	Plant	Concentration	Mode of exposure	Growth media	Accumulation	Speciation	Detective methods	Reference
TiO_2	4	Lettuce (*Lactuca sativa*)	0.125, 1.25, 12.5	Leaf	NP suspension	$nTiO_2$ agglomerates were detected in all types of tissues	No biotransformation of $nTiO_2$ was detected	μ-XRF, μ-XANES, μ-PIXE	Larue et al. (2014b)
	27 ± 4	Cucumber (*Cucumis sativus*)	0, 50, 250, 500, 1000, 2000, 4000	Root	Nutrient solution	Ti was detected from the roots to the leaf trichomes, and rutile $nTiO_2$ had preferential translocation in cucumber tissues	No biotransformation of $nTiO_2$ was detected	μ-XRF, μ-XANES	Servin et al. (2012)
			0, 250, 500, 750	Root	Sandy loam soil	$nTiO_2$ accumulated by the root and translocated to the fruit	No biotransformation of $nTiO_2$ was detected.	μ-XRF, μ-XANES	Servin et al. (2013)
	14–655	Wheat (*Triticum* spp.)	100	Root	NP suspension	14 nm $nTiO_2$ translocated from root to all the plant tissues	No dissolution and speciation of $nTiO_2$ was observed	μ-XRF, μ-XANES, μ-PIXE	Larue et al. (2012a)
	12 ± 3, 25 ± 7	Wheat (*Triticum* spp.)	100	Root	NP suspension	12 nm anatase $nTiO_2$ were taken up by roots rather than 25 nm	No dissolution and speciation of $nTiO_2$ was observed	μ-XRF, μ-XANES, μ-PIXE	Larue et al. (2011)
	14, 25	Wheat (*Triticum* spp.), rapeseed (*Brassica napus*)	100	Root and leaf	NP suspension	$nTiO_2$ translocation from roots to leaves and smaller NPs had preferential translocation; Rapeseeds uptake more NPs than wheat in the foliar treatments		μ-XRF, μ-PIXE	Larue et al. (2012b)

(continued)

Table 3.3 (continued)

NPs	Particle size (nm)	Plant	Concentration	Mode of exposure	Growth media	Accumulation	Speciation	Detective methods	Reference
CuO	<50	Wheat (*Triticum* spp.)	500	Root	Sand	Cu^{2+} were translocated into plant tissues and formed new complexes with organic ligands	Both CuO and Cu_2S complexes were detected	μ-XANES, EXAFS	Dimkpa et al. (2013)
ZnO	<100	Wheat (*Triticum* spp.)	500	Root	Sand	*n*ZnO was dissolved and accumulate in the shoots	Zn-phosphate species was detected in the shoots	μ-XANES, EXAFS	Dimkpa et al. (2013)
	10	*Soybean* (*Glycine max*)	500	Root	Soil	Zn was translocated to the nodules, stems, and pods.	Zn was mainly detected in plant tissues as Zn-citrate rather than *n*ZnO	μ-XRF, μ-XANES, TXM	Hernandez-Viezcas et al. (2013)
	30 ± 5	Maize (*Zea mays L.*)	2, 5, 10, 15, 20, 40, 60, 80, 100	Root	Nutrient solution	*n*ZnO were localized in epidermis, cortex, root tip cells, and vascular tissues	Zn was mainly detected in plant tissues as Zn-phosphate.	μ-XRF, μ-XANES, μ-EXAFS	Lv et al. (2015)
	67 ± 2	Cowpea (*Vigna unguiculata*)	25	Root	Nutrient solution	*n*ZnO mainly accumulated on the root surface.	Zn was in the forms of Zn-citrate, phytate, and histidine	μ-XRF, XANES, EXAFS	Wang et al. (2013)
			200		Natural soil	No upward translocation of NPs from roots to shoots was observed			

(continued)

Table 3.3 (continued)

NPs	Particle size (nm)	Plant	Concentration	Mode of exposure	Growth media	Accumulation	Speciation	Detective methods	Reference
CeO_2	8	Soybean (*Glycine max*)	1000	Root	Soil	Ce was localized in the epidermis, nodule, and pods	A small portion of Ce(IV) was biotransformed into Ce(III) as Ce acetate	μ-XRF, μ-XANES	Hernandez-Viezcas et al. (2013)
		Rice (*Oryza sativa*)	62.5, 125, 250, 500	Root	Potting soil	Ce was detected in the vascular tissues of the roots at 500 ppm		μ-XRF	Rico et al. (2013)
		Cucumber (*Cucumis sativus*)	400	Root	Soil	Ce moves in all plant tissues by transpiration and a small fraction of Ce accumulated in the fruits.		μ-XRF	Zhao et al. (2013)
						Cu, Mn and Zn were mainly accumulated in the seeds.		μ-XRF	Zhao et al. (2014)
	6.9 ± 0.4	Cucumber (*Cucumis sativus*)	2000	Root	Nutrient solution	Ce was mainly detected in the roots with similar concentration in leaves and stems	Ce was found in the roots as CeO_2 and $CePO_4$ while as CeO_2 and cerium carboxylate in the shoots	μ-XANES, STXM	Zhang et al. (2012b)

(continued)

Table 3.3 (continued)

NPs	Particle size (nm)	Plant	Concentration	Mode of exposure	Growth media	Accumulation	Speciation	Detective methods	Reference
	8 ± 1	Kidney bean (*Phaseolus vulgaris*)	62.5, 125, 250, 500	Root	Nutrient solution	Ce was found in vascular tissues and moved to aerial parts with time.	A small percentage of Ce (IV) was biotransformed into Ce(III) compounds	μ-XRF, μ-XANES	Majumdar et al. (2014)
	7, 25	Lettuce (*Lactuca sativa*)	2000	Seed	NP suspension	$nCeO_2$ showed various degree of toxicity to three kinds of lettuce plants.	A small portion of Ce(IV) was in the form of Ce(III) compounds in roots	μ-XANES	Zhang et al. (2015)
Ag	38.6	Lettuce (*Lactuca sativa*)	1, 10, 100	Leaf	Soil	nAg were entrapped by the cuticle and penetrated in the leaf tissue through stomata.	16 % of Ag^0 was in form of Ag^+, Ag-GSH and AgCl might be the secondary species.	μ-XRF, μ-XANES	Larue et al. (2014a)
Yb_2O_3	62 ± 8	Cucumber (*Cucumis sativus*)	0.32, 0.8, 2, 5, 20, 200, 2000	Seed	NP suspension	After treated with nYb_2O_3, bulk Yb_2O_3 and $YbCl_3{\cdot}6H_2O$, only nYb_2O_3 translocated into the cytoplasm of root cells.	nYb_2O_3, bulk Yb_2O_3 and $YbCl_3{\cdot}6H_2O$ were all biotransformed into $YbPO_4$	STXM, NEXAFS	Zhang et al. (2012a)

3.3.4 Other Spectroscopic Techniques

Other spectroscopic techniques have shown to be useful to determine the uptake of MNMs by plants. For example, Khodakovskaya et al. (2011) developed a photothermal and photoacoustic scanning cytometry platform to observe MWCNTs in tomato plants. The device works "on the basis of an invert microscope, a spectrally tunable optical parametric oscillator (OPO) with increased pulse rate of up to 100 Hz, and automated." With the photothermal and photoacustic amplifier, authors detected the CNTs through nanobubbles produced by laser overheating. With this technique, Khodakovskaya et al. (2011) detected CNTs in leaves and tomato fruits. In addition, Khodakovskaya et al. (2013) used Raman spectroscopy to detect CNTs in flowers of tomato plants grown in soil amended with MWCNTs at 50 and 200 μg/L. They found a peak at 1587 cm^{-1} in the surface of flowers from the CNT-exposed plants, which is characteristic of the MWCNTs.

In summary, several microscopy and spectroscopy methods have proven to be useful for the detection, and quantification of the uptake, translocation, and accumulation of MNMs in plants. These include microscopy (light, scanning probe, and electron microscopes) and spectroscopic techniques (atomic spectroscopy, synchrotron radiation, μ-particle-induced X-ray emission, Raman, and photothermal/photoacustic techniques). However, the literature has shown that a combination of techniques provides a more complete panorama of the interaction of MNMs with plants.

Acknowledgments This material is based upon work supported by the National Science Foundation and the Environmental Protection Agency under Cooperative Agreement Number DBI-0830117. Any opinions, findings, and conclusions or recommendations expressed in this material are those of the author(s) and do not necessarily reflect the views of the National Science Foundation or the Environmental Protection Agency. This work has not been subjected to EPA review, and no official endorsement should be inferred. This work was also supported by Grant 2G12MD007592 from the National Institutes on Minority Health and Health Disparities (NIMHD), a component of the National Institutes of Health (NIH). Authors also acknowledge the USDA grant number 2011-38422-30835 and the NSF Grants # CHE-0840525 and DBI 1429708. Partial funding was provided by the NSF ERC on Nanotechnology-Enable Water Treatment (EEC-1449500). J. L. Gardea-Torresdey acknowledges the Dudley family for the Endowed Research Professorship, the Academy of Applied Science/US Army Research Office, Research and Engineering Apprenticeship program (REAP) at UTEP, grant # W11NF-10-2-0076, sub-grant 13-7, and STARs programs of the University of Texas System. N. Zuverza-Mena and I.A. Medina-Velo thank the support of Consejo Nacional de Ciencia y Tecnologia of Mexico (CONACyT).

References

Abd-Alla MH, Nafady NA, Khalaf DM (2016) Assessment of silver nanoparticles contamination on faba bean-*Rhizobium leguminosarum* bv. *viciae-Glomus aggregatum* symbiosis: implications for induction of autophagy process in root nodule. Agr Ecosyst Environ 218:163–177. doi:10.1016/j.agee.2015.11.022

Abraham PM, Barnikol S, Baumann T, Kuehn M, Ivleva NP, Schaumann GE (2013) Sorption of silver nanoparticles to environmental and model surfaces. Environ Sci Technol 47(10):5083–5091. doi:10.1021/es303941e

Alessandrini A, Facci P (2005) AFM: a versatile tool in biophysics. Meas Sci Technol 16(6):R65–R92. doi:10.1088/0957-0233/16/6/R01

Ando T, Uchihashi T, Kodera N, Yamamoto D, Miyagi A, Taniguchi M, Yamashita H (2008) High-speed AFM and nano-visualization of biomolecular processes. Pflugers Arch Eur J Physiol 456(1):211–225. doi:10.1007/s00424-007-0406-0

Ando T, Uchihashi T, Kodera N (2013) High-speed AFM and applications to biomolecular systems. Annu Rev Biophys 42:393–414. doi:10.1146/annurev-biophys-083012-130324

Arruda SCC, Silva ALD, Galazzi RM, Azevedo RA, Arruda MAZ (2015) Nanoparticles applied to plant science: a review. Talanta 131:693–705. doi:10.1016/j.talanta.2014.08.050

Artiaga G, Ramos K, Ramos L, Cámara C, Gómez-Gómez M (2015) Migration and characterisation of nanosilver from food containers by AF4-ICP-MS. Food Chem 166:76–85. doi:10.1016/j.foodchem.2014.05.139

Bandyopadhyay S, Peralta-Videa JR, Plascencia-Villa G, José-Yacamán M, Gardea-Torresdey JL (2012) Comparative toxicity assessment of CeO_2 and ZnO nanoparticles towards *Sinorhizobium meliloti*, a symbiotic alfalfa associated bacterium: Use of advanced microscopic and spectroscopic techniques. J Hazard Mater 241–242:379–386. doi:10.1016/j.jhazmat.2012.09.056

Bandyopadhyay S, Peralta-Videa JR, Gardea-Torresdey JL (2013) Advanced analytical techniques for the measurement of nanomaterials in food and agricultural samples: a review. Environ Eng Sci 3030(3):118–125. doi:10.1089/ees.2012.0325

Bandyopadhyay S, Plascencia-Villa G, Mukherjee A, Rico CM, José-Yacamán M, Peralta-Videa JR, Gardea-Torresdey JL (2015) Comparative phytotoxicity of ZnO NPs, bulk ZnO, and ionic zinc onto the alfalfa plants symbiotically associated with Sinorhizobium meliloti in soil. Sci Total Environ 515–516:60–69. doi:10.1016/j.scitotenv.2015.02.014

Benıtez JJ, Matas AJ, Heredia A (2004) Molecular characterization of the plant biopolyester cutin by AFM and spectroscopic techniques. J Struct Biol 147(2):179–184. doi:10.1016/j.jsb.2004.03.006

Bertsch PM, Hunter DB (2001) Applications of synchrotron-based X-ray microprobes. Chem Rev 101(6):1809–1842. doi:10.1021/cr990070s

Boss CB, Fredeen KJ (2004) Concepts, instrumentation and techniques in inductively coupled plasma optical emission spectrometry. Perkin Elmer

Cañas JE, Long M, Nations S, Vadan R, Dai L, Luo M, Ambikapathi R, Lee EH, Olszyk D (2008) Effects of functionalized and nonfunctionalized single-walled carbon nanotubes on root elongation of select crop species. Environ Toxicol Chem 27(9):1922–1931. doi:10.1897/08-117.1

Cohen SR, Bitler A (2008) Use of AFM in bio-related systems. Curr Opin Colloid Interface Sci 13 (5):316–325. doi:10.1016/j.cocis.2008.02.002

Colman BP, Arnaout CL, Anciaux S, Gunsch CK, Hochella MF Jr, Kim B, Lowry GV, McGill BM, Reinsch BC, Richardson CJ (2013) Low concentrations of silver nanoparticles in biosolids cause adverse ecosystem responses under realistic field scenario. PLoS ONE 8(2): e57189. doi:10.1371/journal.pone.0057189

Cui D, Zhang P, Ma Y, He X, Li Y, Zhang J, Zhao Y, Zhang Z (2014) Effect of cerium oxide nanoparticles on asparagus lettuce cultured in an agar medium. Environ Sci Nano 1(5):459–465. doi:10.1039/c4en00025k

Dan Y, Zhang W, Xue R, Ma X, Stephan C, Shi H (2015) Characterization of gold nanoparticle uptake by tomato plants using enzymatic extraction followed by single-particle inductively coupled plasma–mass spectrometry analysis. Environ Sci Technol 49(5):3007–3014. doi:10.1021/es506179e

De La Rosa G, López-Moreno ML, Hernandez-Viezcas JA, Montes MO, Peralta-Videa J, Gardea-Torresdey J (2011) Toxicity and biotransformation of ZnO nanoparticles in the desert plants *Prosopis juliflora-velutina*, *Salsola tragus* and *Parkinsonia florida*. Int J Nanotechnol 8 (6):492–506. doi:10.1504/IJNT.2011.04019

Degueldre C, Favarger PY (2003) Colloid analysis by single particle inductively coupled plasma-mass spectroscopy: a feasibility study. Colloid Surf A 217(1–3):137–142. doi:10.1016/S0927-7757(02)00568-X

Dimkpa CO, Latta DE, McLean JE, Britt DW, Boyanov MI, Anderson AJ (2013) Fate of CuO and ZnO nano-and microparticles in the plant environment. Environ Sci Technol 47(9):4734–4742. doi:10.1021/es304736y

Donner E, Punshon T, Guerinot ML, Lombi E (2012) Functional characterisation of metal (loid) processes in plants through the integration of synchrotron techniques and plant molecular biology. Anal Bioanal Chem 402(10):3287–3298. doi:10.1007/s00216-011-5624-9

Du W, Sun Y, Ji R, Zhu J, Wu J, Guo H (2011) TiO_2 and ZnO nanoparticles negatively affect wheat growth and soil enzyme activities in agricultural soil. J Environ Monit 13(4):822–828. doi:10.1039/C0EM00611D

Dubascoux S, Le Hecho I, Hassellov M, Von Der Kammer F, Potin Gautier M, Lespes G (2010) Field-flow fractionation and inductively coupled plasma mass spectrometer coupling: history, development and applications. J Anal Atom Spectrom 25(5):613–623. doi:10.1039/B927500B

Dudkiewicz A, Tiede K, Loeschner K, Jensen LHS, Jensen E, Wierzbicki R, Boxall ABA, Molhave K (2011) Characterization of nanomaterials in food by electron microscopy. Trends Anal Chem 30(1):28–43. doi:10.1016/j.trac.2010.10.007

Dudkiewicz A, Boxall ABA, Chaudhry Q, Mølhave K, Tiede K, Hofmann P, Linsinger TPJ (2015) Uncertainties of size measurements in electron microscopy characterization of nanomaterials in foods. Food Chem 176:472–479. doi:10.1016/j.foodchem.2014.12.071

Elzey SR (2010) Applications and physicochemical characterization of nanomaterials in environmental, health, and safety studies. Ph.D. (Doctor of Philosophy) Thesis, University of Iowa, USA. http://ir.uiowa.edu/etd/494

Ensikat HJ, Ditsche-Kuru P, Barthlott W (2010) Scanning electron microscopy of plant surfaces: simple but sophisticated methods for preparation and examination. In: Méndez-Vilas A, Díaz J (eds) Microscopy: science, technology, applications and education, FORMATEX, pp 248–255

Fahrni CJ (2009) Fluorescent probes for two-photon excitation microscopy. In: Reviews in fluorescence 2007, Springer, pp 249–269. doi:10.1007/978-0-387-88722-7_11

Fendorf SE, Sparks DL, Lamble GM, Kelley MJ (1994) Applications of X-ray absorption fine structure spectroscopy to soils. Soil Sci Am J 58(6):1583–1595

Gaczynska M, Osmulski PA (2008) AFM of biological complexes: what can we learn? Curr Opin Colloid Interface Sci 13(5):351–367. doi:10.1016/j.cocis.2008.01.004

Gardea-Torresdey JL, Parsons JG, Gomez E, Peralta-Videa J, Troiani HE, Santiago P, Yacaman MJ (2002) Formation and growth of Au nanoparticles inside live alfalfa plants. Nano Lett 2(4):397–401. doi:10.1021/nl015673+

Gardea-Torresdey JL, Gomez E, Peralta-Videa JR, Parsons JG, Troiani H, Jose-Yacaman M (2003) Alfalfa sprouts: a natural source for the synthesis of silver nanoparticles. Langmuir 19 (4):1357–1361. doi:10.1021/la020835i

Gardea-Torresdey JL, Rico CM, White JC (2014) Trophic transfer, transformation, and impact of engineered nanomaterials in terrestrial environments. Environ Sci Technol 48(5):2526–2540. doi:10.1021/es4050665

Gerber C, Lang HP (2006) How the doors to the nanoworld were opened. Nat Nanotechnol 1(1):3–5. doi:10.1038/nnano.2006.70

Grillo R, Rosa AH, Fraceto LF (2015) Engineered nanoparticles and organic matter: a review of the state-of-the-art. Chemosphere 119:608–619. doi:10.1016/j.chemosphere.2014.07.049

Hendren CO, Mesnard X, Dröge J, Wiesner MR (2011) Estimating production data for five engineered nanomaterials as a basis for exposure assessment. Environ Sci Technol 45(7):2562–2569. doi:10.1021/es103300g

Hernandez-Viezcas JA, Castillo-Michel H, Servin AD, Peralta-Videa JR, Gardea-Torresdey JL (2011) Spectroscopic verification of zinc absorption and distribution in the desert plant

Prosopis juliflora-velutina (velvet mesquite) treated with ZnO nanoparticles. Chem Eng J 170 (1–3):346–352. doi:10.1016/j.cej.2010.12.021

Hernandez-Viezcas JA, Castillo-Michel H, Andrews JC, Cotte M, Rico C, Peralta-Videa JR, Ge Y, Priester JH, Holden PA, Gardea-Torresdey JL (2013) In situ synchrotron X-ray fluorescence mapping and speciation of CeO_2 and ZnO nanoparticles in soil cultivated soybean (Glycine max). ACS Nano 7(2):1415–1423. doi:10.1021/nn305196q

Hischemöller A, Nordmann J, Ptacek P, Mummenhoff K, Haase M (2009) In-vivo imaging of the uptake of upconversion nanoparticles by plant roots. J Biomed Nanotechnol 5(3):278–284. doi:10.1166/jbn.2009.1032

Inoué S (2010) Foundations of confocal scanned imaging in light microscopy. In: Pawley JB (ed) Handbook of biological confocal microscopy, 3rd edn. Springer, New York, NY, USA, pp 1–19

Judy JD, Unrine JM, Bertsch PM (2011) Evidence for biomagnification of gold nanoparticles within a terrestrial food chain. Environ Sci Technol 45(2):776–781. doi:10.1021/es103031a

Kahru A, Dubourguier H-C (2010) From ecotoxicology to nanoecotoxicology. Toxicology 269(2–3):105–119. doi:10.1016/j.tox.2009.08.016

Karydas AG, Sokaras D, Zarkadas C, Grlj N, Pelicon P, Zitnik M, Schutz R, Malzer W, Kanngie (2007) 3D Micro PIXE—a new technique for depth-resolved elemental analysis. J Anal At Spectrom 22(10):1260–1265. doi:10.1039/b700851c

Khodakovskaya MV, de Silva K, Nedosekin DA, Dervishi E, Biris AS, Shashkov EV, Galanzha EI, Zharov VP (2011) Complex genetic, photothermal, and photoacoustic analysis of nanoparticle-plant interactions. Proc Natl Acad Sci USA 108(3):1028–1033. doi:10.1002/smll.201201225

Khodakovskaya MV, Kim BS, Kim JN, Alimohammadi M, Dervishi E, Mustafa T, Cernigla CE (2013) Carbon nanotubes as plant growth regulators: effects on tomato growth, reproductive system, and soil microbial community. Small 9(1):115–123. doi:10.1002/smll.201201225

Klaine SJ, Alvarez PJ, Batley GE, Fernandes TF, Handy RD, Lyon DY, Mahendra S, McLaughlin MJ, Lead JR (2008) Nanomaterials in the environment: behavior, fate, bioavailability, and effects. Environ Toxicol Chem 27(9):1825–1851. doi:10.1897/08-090.1

Klapetek P, Valtr M, Nečas D, Salyk O, Dzik P (2011) Atomic force microscopy analysis of nanoparticles in non-ideal conditions. Nanoscale Res Lett 6(1):1–9. doi:10.1186/1556-276X-6-514

Koelmel J, Leland T, Wang H, Amarasiriwardena D, Xing B (2013) Investigation of gold nanoparticles uptake and their tissue level distribution in rice plants by laser ablation-inductively coupled-mass spectrometry. Environ Pollut 174:222–228. doi:10.1016/j.envpol.2012.11.026

Kokina I, Gerbreders V, Sledevskis E, Bulanovs A (2013) Penetration of nanoparticles in flax (*Linum usitatissimum* L.) calli and regenerants. J Biotechnol 165:127–132

Krautbauer R, Rief M, Gaub HE (2003) Unzipping DNA oligomers. Nano Lett 3(4):493–496. doi:10.1021/nl034049p

Kumari M, Khan SS, Pakrashi S et al (2011) Cytogenetic and genotoxic effects of zinc oxide nanoparticles on root cells of *Allium cepa*. J Hazard Mater 190(1–3):613–621. doi:10.1016/j.jhazmat.2011.03.095

Laborda F, Jimenez-Lamana J, Bolea E, Castillo JR (2011) Selective identification, characterization and determination of dissolved silver(i) and silver nanoparticles based on single particle detection by inductively coupled plasma mass spectrometry. J Anal Atom Spectrom 26 (7):1362–1371. doi:10.1039/C0JA00098A

Lahiani MH, Chen J, Irin F, Puretzky AA, Green MJ, Khodakovskaya MV (2015) Interaction of carbon nanohorns with plants: uptake and biological effects. Carbon 81:607–619. doi:10.1016/j.carbon.2014.09.095

Larue C, Khodja H, Herlin-Boime N, Brisset F, Flank A, Fayard B, Chaillou S, Carriere M (2011) Investigation of titanium dioxide nanoparticles toxicity and uptake by plants. J Phys: Conf Ser 304:012057. doi:10.1088/1742-6596/304/1/012057

Larue C, Laurette J, Herlin-Boime N, Khodja H, Fayard B, Flank A-M, Brisset F, Carriere M (2012a) Accumulation, translocation and impact of TiO_2 nanoparticles in wheat (*Triticum aestivum* spp): influence of diameter and crystal phase. Sci Total Environ 431:197–208. doi:10.1016/j.scitotenv.2012.04.073

Larue C, Pinault M, Czarny B, Georgin D, Jaillard D, Bendiab N, Mayne-L'Hermite M, Taran F, Dive V, Carrière M (2012b) Quantitative evaluation of multi-walled carbon nanotube uptake in wheat and rapeseed. J Hazard Mater 227:155–163. doi:10.1016/j.jhazmat.2012.05.033

Larue C, Castillo-Michel H, Sobanska S, Cécillon L, Bureau S, Barthès V, Ouerdane L, Carrière M, Sarret G (2014a) Foliar exposure of the crop *Lactuca sativa* to silver nanoparticles: evidence for internalization and changes in Ag speciation. J Hazard Mater 264(98–106):106. doi:10.1016/j.jhazmat.2013.10.053

Larue C, Castillo-Michel H, Sobanska S, Trcera N, Sorieul S, Cécillon L, Ouerdane L, Legros S, Sarret G (2014b) Fate of pristine TiO_2 nanoparticles and aged paint-containing TiO_2 nanoparticles in lettuce crop after foliar exposure. J Hazard Mater 273:17–26. doi:10.1016/j.jhazmat.2014.03.014

Le Van N, Ma C, Shang J, Rui Y, Liu S, Xing B (2016) Effects of CuO nanoparticles on insecticidal activity and phytotoxicity in conventional and transgenic cotton. Chemosphere 144:661–670. doi:10.1016/j.chemosphere.2015.09.028

Lee P-L, Chen B-C, Gollavelli G, Shen S-Y, Yin Y-S, Lei S-L, Jhang C-L, Lee W-R, Ling Y-C (2014) Development and validation of TOF-SIMS and CLSM imaging method for cytotoxicity study of ZnO nanoparticles in HaCaT cells. J Hazard Mater 277:3–12. doi:10.1016/j.jhazmat.2014.03.046

Lenaghan SC, Zhang M (2012) Real-time observation of the secretion of a nanocomposite adhesive from English ivy (*Hedera helix*). Plant Sci 183:206–211. doi:10.1016/j.plantsci.2011.08.013

Li L, Sillanpää M, Tuominen M, Lounatmaa K, Schultz E (2013) Behavior of titanium dioxide nanoparticles in *Lemna minor* growth test conditions. Ecotoxicol Environ Saf 88:89–94. doi:10.1016/j.ecoenv.2012.10.024

Lin S, Reppert J, Hu Q, Hudson JS, Reid ML, Ratnikova TA, Rao AM, Luo H, Ke PC (2009) Uptake, translocation, and transmission of carbon nanomaterials in rice plants. Small 5 (10):1128–1132. doi:10.1002/smll.200801556

Liu Q, Chen B, Wang Q, Shi X, Xiao Z, Lin J, Fang X (2009) Carbon nanotubes as molecular transporters for walled plant cells. Nano Lett 9(3):1007–1010. doi:10.1021/nl803083u

Lombi E, Susini J (2009) Synchrotron-based techniques for plant and soil science: opportunities, challenges and future perspectives. Plant Soil 320(1–2):1–35. doi:10.1007/s11104-008-9876-x

Lombi E, Scheckel KG, Kempson IM (2011) In situ analysis of metal(loid)s in plants: state of the art and artefacts. Environ Exp Bot 72(1):3–17. doi:10.1016/j.envexpbot.2010.04.005

López-Moreno ML, de la Rosa G, Hernández-Viezcas JÁ, Castillo-Michel H, Botez CE, Peralta-Videa JR, Gardea-Torresdey JL (2010a) Evidence of the differential biotransformation and genotoxicity of ZnO and CeO_2 nanoparticles on soybean (*Glycine max*) plants. Environ Sci Technol 44(19):7315–7320. doi:10.1021/es903891g

López-Moreno ML, de la Rosa G, Hernández-Viezcas JA, Peralta-Videa JR, Gardea-Torresdey JL (2010b) X-ray absorption spectroscopy (XAS) corroboration of the uptake and storage of CeO_2 nanoparticles and assessment of their differential toxicity in four edible plant species. J Agric Food Chem 58(6):3689–3693. doi:10.1021/jf904472e

Lu Q, He ZL, Stoffella PJ (2012) Land application of biosolids in the USA: a review. App Environ Soil Sci 2012. doi:10.1155/2012/201462

Luykx DM, Peters RJ, van Ruth SM, Bouwmeester H (2008) A review of analytical methods for the identification and characterization of nano delivery systems in food. J Agri Food Chem 56 (18):8231–8247. doi:10.1021/jf8013926

Lv J, Zhang S, Luo L, Zhang J, Yang K, Christie P (2015) Accumulation, speciation and uptake pathway of ZnO nanoparticles in maize. Environ Sci Nano 2:68–77. doi:10.1039/c4en00064a

Ma X, Geiser-Lee J, Deng Y, Kolmakov A (2010) Interactions between engineered nanoparticles (ENPs) and plants: phytotoxicity, uptake and accumulation. Sci Total Environ 408(16):3053–3061. doi:10.1016/j.scitotenv.2010.03.031

Ma Y, He X, Zhang P, Zhang Z, Guo Z, Tai R, Xu Z, Zhang L, Ding Y, Zhao Y, Chai Z (2011) Phytotoxicity and biotransformation of La_2O_3 nanoparticles in a terrestrial plant cucumber (*Cucumis sativus*). Nanotoxicology 5(4):743–753. doi:10.3109/17435390.2010.545487

Majumdar S, Peralta-Videa JR, Castillo-Michel H, Hong J, Rico CM, Gardea-Torresdey JL (2012) Applications of synchrotron μ-XRF to study the distribution of biologically important elements in different environmental matrices: A review. Anal Chim Acta 755:1–16. doi:10.1016/j.aca.2012.09.050

Majumdar S, Peralta-Videa JR, Bandyopadhyay S, Castillo-Michel H, Hernandez-Viezcas J-A, Sahi S, Gardea-Torresdey JL (2014) Exposure of cerium oxide nanoparticles to kidney bean shows disturbance in the plant defense mechanisms. J Hazard Mater 278:279–287. doi:10.1016/j.jhazmat.2014.06.009

Mbundi L, Gallar-Ayala H, Rizwan Khan MR, Barber JL, Losada S, Busquets R (2014) Chapter two—advances in the analysis of challenging food contaminants: nanoparticles, bisphenols, mycotoxins, and brominated flame retardants. Adv Mol Toxicol 8:42–45

McPherson A, Malkin A, Kuznetsov YG (2000) Atomic force microscopy in the study of macromolecular crystal growth. Annu Rev Biophys Biomol Struct 29(1):361–410

Min W, Freudiger CW, Lu S, Xie XS (2011) Coherent nonlinear optical imaging: beyond fluorescence microscopy. Annu Rev Phys Chem 62:507. doi:10.1146/annurev.physchem.012809.103512

Mitrano DM, Barber A, Bednar A, Westerhoff P, Higgins CP, Ranville JF (2012) Silver nanoparticle characterization using single particle ICP-MS (SP-ICP-MS) and asymmetrical flow field flow fractionation ICP-MS (AF4-ICP-MS). J Anal Atom Spectrom 27(7):1131–1142. doi:10.1039/C2JA30021D

Mokgalaka NS, Gardea-Torresdey JL (2006) Laser ablation inductively coupled plasma mass spectrometry: Principles and applications. Appl Spectrosc Rev 41(2):131–150. doi:10.1080/05704920500510703

Müller DJ, Dufrene YF (2008) Atomic force microscopy as a multifunctional molecular toolbox in nanobiotechnology. Nat Nanotechnol 3(5):261–269. doi:10.1038/nnano.2008.100

Palomo-Siguero M, Lopez-Heras MI, Camara C, Madrid Y (2015) Accumulation and biotransformation of chitosan-modified selenium nanoparticles in exposed radish (*Raphanus sativus*). J Anal Atom Spectrom 30:1237–1244. doi:10.1039/C4JA00407H

Pathan AK, Bond J, Gaskin RE (2008) Sample preparation for scanning electron microscopy of plant surfaces—horses for courses. Micron 39(8):1049–1061. doi:10.1016/j.micron.2008.05.006

Patty C, Barnett B, Mooney B, Kahn A, Levy S, Liu Y, Pianetta P, Andrews JC (2009) Using X-ray microscopy and Hg L3 XANES to study Hg binding in the rhizosphere of *Spartina* cordgrass. Environ Sci Technol 43(19):7397–7402. doi:10.1021/es901076q

Peralta-Videa JR, Zhao L, Lopez-Moreno ML, de la Rosa G, Hong J, Gardea-Torresdey JL (2011) Nanomaterials and the environment: a review for the biennium 2008–2010. J Hazard Mater 186(1):1–15. doi:10.1016/j.jhazmat.2010.11.020

Piccinno F, Gottschalk F, Seeger S, Nowack B (2012) Industrial production quantities and uses of ten engineered nanomaterials in Europe and the world. J Nanopart Res 14(9):1–11. doi:10.1007/s11051-012-1109-9

Pradhan S, Patra P, Das S, Chandra S, Mitra S, Dey KK, Akbar S, Palit P, Goswami A (2013) Photochemical modulation of biosafe manganese nanoparticles on *Vigna radiata*: a detailed molecular, biochemical, and biophysical study. Environ Sci Technol 47(22):13122–13131. doi:10.1021/es402659t

Rico CM, Morales MI, McCreary R, Castillo-Michel H, Barrios AC, Hong J, Tafoya A, Lee W-Y, Varela-Ramirez A, Peralta-Videa JR, Gardea-Torresdey JL (2013) Cerium oxide nanoparticles modify the antioxidative stress enzyme activities and macromolecule composition in rice seedlings. Environ Sci Technol 47(24):14110–14118. doi:10.1021/es4033887

Roco MC, Bainbridge WS (2013) The new world of discovery, invention, and innovation: convergence of knowledge, technology, and society. J Nanopart Res 15(9):1–17. doi:10.1007/s11051-013-1946-1

Roming AD (1986) Electron optical methods. In: Materials characterization asm handbook, 9th edn. ASM International, The Materials Information Society, vol. 10. pp 429–536

Rubart M (2004) Two-photon microscopy of cells and tissue. Circ Res 95:1154–1166. doi:10.1161/01.RES.0000150593.30324.42

Sabo-Attwood T, Unrine JM, Stone JW, Murphy CJ, Ghoshroy S, Blom D, Bertsch PM, Newman LA (2012) Uptake, distribution and toxicity of gold nanoparticles in tobacco (*Nicotiana xanthi*) seedlings. Nanotoxicology 6(4):353–360. doi:10.3109/17435390.2011.579631

Salamon AW, Courtney P, Shuttler I (2010) A primer. In: Frequently asked questions, nanotechnology and engineered material, Available via Perkin Elmer http://www.perkinelmer.com/Content/Manuals/GDE_NanotechnologyPrimer.pdf. Accessed 5 Feb 2015

Sarret G, Smits EAHP, Michel HC, Isaure MP, Zhao FJ, Tappero R (2013) Chapter 1—Use of synchrotron-based techniques to elucidate metal uptake and metabolism in plants. In: Donald LS (ed) Adv Agron 119:1–82. doi:10.1016/B978-0-12-407247-3.00001-9

Schaumann GE, Philippe A, Bundschuh M, Metreveli G, Klitzke S, Rakcheev D, Grün A, Kumahor SK, Kühn M, Baumann T, Lang F, Manz W, Schulz R, Vogel H-J (2015) Understanding the fate and biological effects of Ag − and TiO_2-nanoparticles in the environment: the quest for advanced analytics and interdisciplinary concepts. Sci Total Environ 535(1):3–19. doi:10.1016/j.scitotenv.2014.10.035

Scheckel K, Hamon R, Jassogne L, Rivers M, Lombi E (2007) Synchrotron X-ray absorption-edge computed microtomography imaging of thallium compartmentalization in Iberis intermedia. Plant Soil 290(1–2):51–60. doi:10.1007/s11104-006-9102-7

Servin AD, Castillo-Michel H, Hernandez-Viezcas JA, Diaz BC, Peralta-Videa JR, Gardea-Torresdey JL (2012) Synchrotron micro-XRF and micro-XANES confirmation of the uptake and translocation of TiO_2 nanoparticles in cucumber (*Cucumis sativus*) plants. Environ Sci Technol 46(14):7637–7643. doi:10.1021/es300955b

Servin AD, Morales MI, Castillo-Michel H, Hernandez-Viezcas JA, Munoz B, Zhao L, Nunez JE, Peralta-Videa JR, Gardea-Torresdey JL (2013) Synchrotron verification of TiO_2 accumulation in cucumber fruit: a possible pathway of TiO_2 nanoparticle transfer from soil into the food chain. Environ Sci Technol 47(20):11592–11598. doi:10.1021/es403368j

Skoog D, Holler F, Nieman T (1998) Principles of instrumental analysis. 5th edn. Harcourt Brace, Orlando, FL, USA, pp 231–235, 262–263

Stutzmann GE, Parker I (2005) Dynamic multiphoton imaging: a live view from cells to systems. Physiology 20(1):15–21. doi:10.1152/physiol.00028.2004

Sun D, Hussain HI, Yi Z, Siegele R, Cresswell T, Kong L, Cahill DM (2014) Uptake and cellular distribution, in four plant species, of fluorescently labeled mesoporous silica nanoparticles. Plant Cell Rep 33(8):1389–1402. doi:10.1007/s00299-014-1624-5

Taylor AF, Rylott EL, Anderson CW, Bruce NC (2014) Investigating the toxicity, uptake, nanoparticle formation and genetic response of plants to gold. PLoS ONE 9(4):e93793. doi:10.1371/journal.pone.0093793

Thomas R (2013) Practical guide to ICP-MS: a tutorial for beginners. CRC Press, Boca Raton, FL, USA, pp 1–4

Tiede K, Boxall AB, Tear SP, Lewis J, David H, Hassellöv M (2008) Detection and characterization of engineered nanoparticles in food and the environment. Food Addit Contam 25(7):795–821. doi:10.1080/02652030802007553

Trujillo-Reyes J, Peralta-Videa J, Gardea-Torresdey J (2014) Supported and unsupported nanomaterials for water and soil remediation: are they a useful solution for worldwide pollution? J Hazard Mater 280:487–503. doi:10.1016/j.jhazmat.2013.11.067

Wang P, Menzies NW, Lombi E, McKenna BA, Johannessen B, Glover CJ, Kappen P, Kopittke PM (2013) Fate of ZnO nanoparticles in soil and cowpea (*Vigna unguiculata*). Environ Sci Technol 47(23):13822–13830. doi:10.1021/es403466p

Whited AM, Park PSH (2014) Atomic force microscopy: a multifaceted tool to study membrane proteins and their interactions with ligands. Biochim Biophys Acta 1838(1):56–68. doi:10.1016/j.bbamem.2013.04.011

Wild E, Jones KC (2009) Novel method for the direct visualization of in vivo nanomaterials and chemical interactions in plants. Environ Sci Technol 43(14):5290–5294. doi:10.1021/es900065h

Wu S, Baskin TI, Gallagher KL (2012) Mechanical fixation techniques for processing and orienting delicate samples, such as the root of *Arabidopsis thaliana*, for light or electron microscopy. Nat Protocols 7(6):1113–1124. doi:http://www.nature.com/nprot/journal/v7/n6/abs/nprot.2012.056.html#supplementary-information. Accessed 5 June 2015

Yan D, Zhao Y, Lu A, Wang S, Xu D, Zhang P (2013) Effects of accompanying anions on cesium retention and translocation via droplets on soybean leaves. J Environ Radioact 126:232–238

Yang H, Wang Y, Lai S, An H, Li Y, Chen F (2007) Application of atomic force microscopy as a nanotechnology tool in food science. J Food Sci 72(4):R65–R75. doi:10.1111/j.1750-3841.2007.00346.x

Zhang M, Wu S-C, Zhou W, Xu B (2012a) Imaging and measuring single-molecule interaction between a carbohydrate-binding module and natural plant cell wall cellulose. J Phys Chem B 116(33):9949–9956. doi:10.1021/jp304686q

Zhang P, Ma Y, Zhang Z, He X, Guo Z, Tai R, Ding Y, Zhao Y, Chai Z (2012b) Comparative toxicity of nanoparticulate/bulk Yb_2O_3 and $YbCl_3$ to cucumber (*Cucumis sativus*). Environ Sci Technol 46(3):1834–1841. doi:10.1021/es2027295

Zhang P, Ma Y, Zhang Z, He X, Zhang J, Guo Z, Tai R, Zhao Y, Chai Z (2012c) Biotransformation of ceria nanoparticles in cucumber Plants. ACS Nano 6(11):9943–9950. doi:10.1021/nn303543n

Zhang P, Ma Y, Zhang Z, He X, Li Y, Zhang J, Zheng L, Zhao Y (2015) Species-specific toxicity of ceria nanoparticles to Lactuca plants. Nanotoxicology 9(1):1–8. doi:10.3109/17435390.2013.855829

Zhao L, Peralta-Videa JR, Ren M, Varela-Ramirez A, Li C, Hernandez-Viezcas JA, Aguilera RJ, Gardea-Torresdey JL (2012a) Transport of Zn in a sandy loam soil treated with ZnO NPs and uptake by corn plants: electron microprobe and confocal microscopy studies. Chem Eng J 184:1–8. doi:10.1016/j.cej.2012.01.041

Zhao L, Peralta-Videa JR, Varela-Ramirez A, Castillo-Michel H, Li C, Zhang J, Aguilera RJ, Keller AA, Gardea-Torresdey JL (2012b) Effect of surface coating and organic matter on the uptake of CeO_2 NPs by corn plants grown in soil: insight into the uptake mechanism. J Hazard Mater 225:131–138. doi:10.1016/j.jhazmat.2012.05.008

Zhao L, Sun Y, Hernandez-Viezcas JA, Servin AD, Hong J, Niu G, Peralta-Videa JR, Duarte-Gardea M, Gardea-Torresdey JL (2013) Influence of CeO_2 and ZnO nanoparticles on cucumber physiological markers and bioaccumulation of Ce and Zn: a life cycle study. J Agri Food Chem 61(49):11945–11951. doi:10.1021/jf404328e

Zhao L, Peralta-Videa JR, Rico CM, Hernandez-Viezcas JA, Sun Y, Niu G, Servin A, Nunez JE, Duarte-Gardea M, Gardea-Torresdey JL (2014) CeO_2 and ZnO nanoparticles change the nutritional qualities of cucumber (*Cucumis sativus*). J Agri Food Chem 62(13):2752–2759. doi:10.1021/jf405476u

Zhao L, Sun Y, Hernandez-Viezcas JA, Hong J, Majumdar S, Niu G, Duarte-Gardea M, Peralta-Videa JR, Gardea-Torresdey JL (2015) Monitoring the environmental effects of CeO_2 and ZnO nanoparticles through the life cycle of corn (*Zea mays*) plants and *in Situ* μ-XRF mapping of nutrients in kernels. Environ Sci Technol 49(5):2921–2928. doi:10.1021/es5060226

Zhu H, Han J, Xiao JQ, Jin Y (2008) Uptake, translocation, and accumulation of manufactured iron oxide nanoparticles by pumpkin plants. J Environ Monit 10(6):713–717. doi:10.1039/B805998E

Zipfel WR, Williams RM, Webb WW (2003) Nonlinear magic: multiphoton microscopy in the biosciences. Nat Biotechnol 21(11):1369–1377. doi:10.1038/nbt899

Chapter 4
Methods of Using Nanoparticles

M. Sheikh Mohamed and D. Sakthi Kumar

Abstract Though moderate, the advances of nanotechnology in the field of plant sciences have been steadily making its mark as a technology to reckon with. Unlike in electronics, energy harvesting, or medical sciences where nanotechnology has initiated a revolution of events, the effects on plants and related disciplines have been limited, to say the least. Though reasons can be stacked up in this regard, the major concern remains as to how this technology should be employed. The ambassadors of this technology, the various nanomaterials currently available, pose a peculiar problem of the modes in which they should be allowed to interact with the plant species and their microenvironment. Problems associated with the toxicity, bioavailability, and consequential effects depend primarily on the methods employed for the administration of these nanomaterials. The mode of nanomaterial administration decides to a large extent how and where they will interact with the plants and their subsequent fate. This chapter deals with the diverse methods adopted by researchers over the years in their pursuit to develop efficient and reliable ways in which the nanomaterials can be delivered to the plant system to assess their beneficial or detrimental effects thereof.

Keywords Nanotechnology · Nanomaterial · Plants · Toxicity · Bioavailability

4.1 Introduction

Nanomaterial (NM) exposure to plants has gained much attention for some time now. With the willing and inadvertent release of NMs to the environment, numerous studies on the toxicity as well as beneficial aspects of these foreign compounds on the plant system have been conducted. Though much data have been accumulated on their toxic aspects, increasing evidence of their beneficial role has

M. Sheikh Mohamed · D. Sakthi Kumar (✉)
Bio-Nano Electronics Research Centre, Toyo University,
2100, Kujirai, Kawagoe, Saitama 350-8585, Japan
e-mail: sakthi@toyo.jp

C. Kole et al. (eds.), *Plant Nanotechnology*,
DOI 10.1007/978-3-319-42154-4_4

created a curious possibility of their prospective applications in the field of plant bio-nanotechnology (Pavel et al. 1999; Liu et al. 2002a, b, c; Cotae and Creanga 2005; Joseph and Morrison 2006). This profitable angle of the NM–plant interaction has diverted focus on enhancing the compatibility and bioavailability of NMs to the host system.

Though the NM–plant interactions seem lucrative, the ultimate availability, translocation, accumulation, and subsequent effects of NMs depend primarily on the mode of their administration in addition to the element's availability, uptake and storage capacity of plants. It is also of extreme importance to consider that uptake and accumulation of nanoparticles (NPs) in plants represent an important pathway for potential, consequential human exposure to NPs. Therefore, the strategies employed toward attaining efficient NP–plant relationship need also to consider a wider perspective of consequential events.

The current chapter aims to provide an overview of the varied NM exposure routes exercised on plants with additional notes on the culture methods employed for the purpose and the ultimate objectives of the research. The major modes of NM application have been found to encompass traditional techniques as direct seed and seedling exposure through in vitro culture media or soil, spraying, hydroculture, etc. as well as more modern approaches as isolated cell, protoplast incubation, and biolistics (Fig. 4.1).

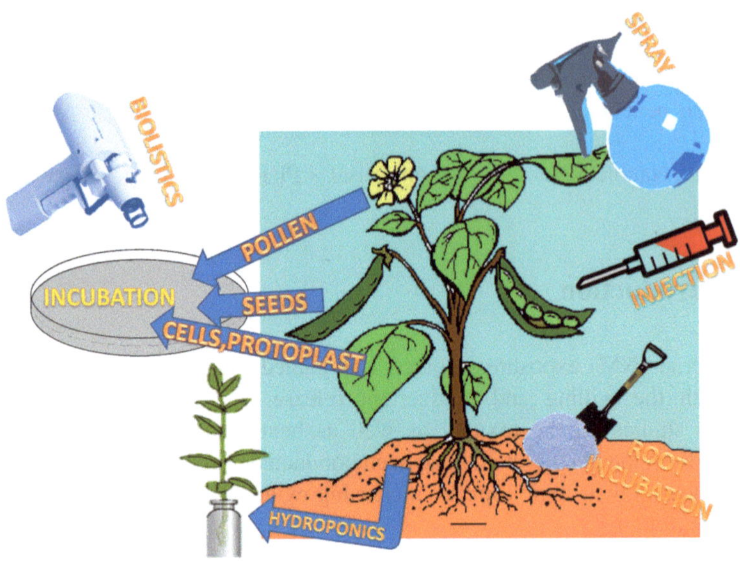

Fig. 4.1 Various modes of nanoparticle exposure to plant system

4.2 Incubation

4.2.1 Via Seeds

Germination is generally defined as the physiological process of water and nutrient imbibition by seeds leading to the emergence of radicle and plumule by puncturing the seed coat (Kordan 1992). Protection of the most important part of seed, the embryo, from diverse biotic and abiotic factors lies with the seed coat due to its robust nature and selective permeability (Wierzbicka and Obidzinska 1998). This selective permeability of seed coat may pertain to the size, charge, shape, etc. of foreign materials trying to gain access to the most sensitive parts of the seed. The process of seed germination, though primarily depends on the species, origin, and certain attributes as recalcitrance, dormancy, etc. remains highly reliant on the water and nutrient composition of the germination medium. This particular aspect is an essential indicator for the studies relating to foreign elements as NMs on the early stages of plant life. Numerous works have been conducted to analyze the effects of various NMs on the germination and uptake parameters of seeds. The primary method of NM administration to seeds remains either by soaking the seeds in NM suspension or by germinating them directly in nanoparticle-spiked media or soil.

Lin and Xing (2007) duo studied the effects of five types of NPs (multi-walled carbon nanotube (MWCNT), aluminum, alumina, zinc, and zinc oxide) on a variety of seeds viz., radish (*Raphanus sativus*), rape (*Brassica napus*), ryegrass (*Lolium perenne*), lettuce (*Lactuca sativa*), maize (*Zea mays*), and cucumber (*Cucumis sativus*) by soaking and incubating seeds in NP suspensions, soaking seeds in NP suspensions prior to transferring them to Petri dishes with deionized water, and germinating the seeds in Petri dishes with NP suspensions after being soaked in deionized water. Stampoulis et al. (2009) compared the effects of five types of commonly used NPs (MWCNTs, Ag, Cu, Si, and Zn oxide) with their bulk material counterparts on parameters as germination, root elongation, and biomass of the agricultural plant zucchini (*Cucurbita pepo*). Pre-sterilized seeds of zucchini were germinated in 3 mL of respective NP or bulk material solution (1000 mg/L) in a Petri dish on an orbital shaker. The NPs in this case were suspended in 0.2 % sodium dodecyl sulfate (SDS), which acts as a surfactant and stabilizing agent preventing NP aggregation. Surfactant-mediated emulsification of cell membranes and related lipid-containing cellular constituents have been previously established (Spurrier and Jackobs 1955; Temple and Hilton 1963; Ernst et al. 1971). Though the NP dissolution was achieved, surfactant-induced phytotoxicity was clearly evident in the test material, warranting their use with caution and control.

The biotransformation of ZnO and CeO_2 NPs on soybean (*Glycine max*) plants, their impact on DNA stability, and the effects on germination and seedling growth were studied by Lopez-Moreno et al. (2010a). The high biomass production and ease of cultivation have made soybean a perfect model for metal accumulation studies. Post-sterilization, seeds were incubated on filter papers soaked with 5 mL of hexagonal ZnO NPs (8 nm) or cubic CeO_2 NPs (7 nm) and subsequent observations were made. NMs have been shown to be much beneficial compared to their bulk counterparts, but very few studies have been focused on addressing the underlying mechanism of such physiological outcomes. In their study, Moon et al. (2014) performed seed germination and root elongation tests on cucumber seeds treated with bulk copper oxide (CuO) and CuO NPs. Post-surface sterilization with 5 % sodium hypochlorite, the seeds were saturated with distilled water, CuO, and CuO NP solutions for 6 h. The soaked seeds were placed on NP solution (5 mL) added filter papers in Petri plates, parafilm-sealed, and placed in an incubator. Seed germination and root elongation were measured each day after incubation for seven days. Kim et al. (2015) found that exposure of *Arabidopsis thaliana* to nano zerovalent iron (nZVI) triggered high plasma membrane H^+-ATPase activity along with increased leaf area and wider stomatal aperture. The nZVI particles were washed with 99 % ethanol and degassed deionized water to prepare an NP slurry. This slurry was mixed with autoclaved soil at a final concentration of 0.5 g/kg, with the initial water content of the nZVI mixed soil being 60–70 %. Arabidopsis was cultivated on this nZVI-spiked soil in the greenhouse with 16-h/8-h light/dark at 20–22 °C. This was the first study implicating the role of nZVI in the enhancement of stomatal opening by inducing the activation of plasma membrane H^+-ATPase, leading to the possibility of increased CO_2 uptake. The same group (Kim et al. 2014) showed the effect of nZVI on root elongation in Arabidopsis by inducing OH radical-induced cell wall loosening. The surface-sterilized Arabidopsis were grown on half-strength Murashige and Skoog (MS) medium spiked with 0.5 g/L of nZVI.

One of the major NMs currently investigated are the graphene-based NMs, especially graphene oxide (GO). With the use of GOs in various research- and application-oriented fields, it is imperative to understand their relationship with the ecosystem, especially plants. (Zhao et al. 2015b) investigated the effects of GO at environmentally relevant concentrations on Arabidopsis plants (Fig. 4.2). Post-sonication, GO was added to the growth medium before autoclaving. The seeds were surface-sterilized and soaked in distilled water at 4 °C in the dark for 24 h and subsequently transferred to Petri plates containing GO-spiked MS media. To facilitate the root growth along the surface of the media, the plates were positioned vertically. GO-MS media-cultured 2-week-old seedlings were transferred to Hoagland nutritive fluid containing GO. Freshly harvested seedlings were blotted on an absorbent paper and used for assays. It was found that even though GO could be easily absorbed by Arabidopsis root hairs, their translocation pattern from root to stem or leaves was highly influenced by the resistance mechanisms of the plant.

Lin et al. (2009) provided the first evidence on the uptake, accumulation, and generational transmission of NOM (natural organic matter)-suspended carbon NPs in rice (*Oryza sativa*) plants. Freshly harvested rice seeds were incubated in Petri dishes containing 15 mL of C_{70}-NOM and MWCNT-NOM at varied concentrations, in rice germination buffer. Post-germination, 2-week-old seedlings were transferred to a greenhouse and grown till maturity in soil devoid of nanoparticles, which was subsequently referred to as the first generation. To investigate generational transmission of nanomaterials, mature seeds from the control plants and C_{70}-treated plants were harvested 6 months after germination and planted in Petri dishes with

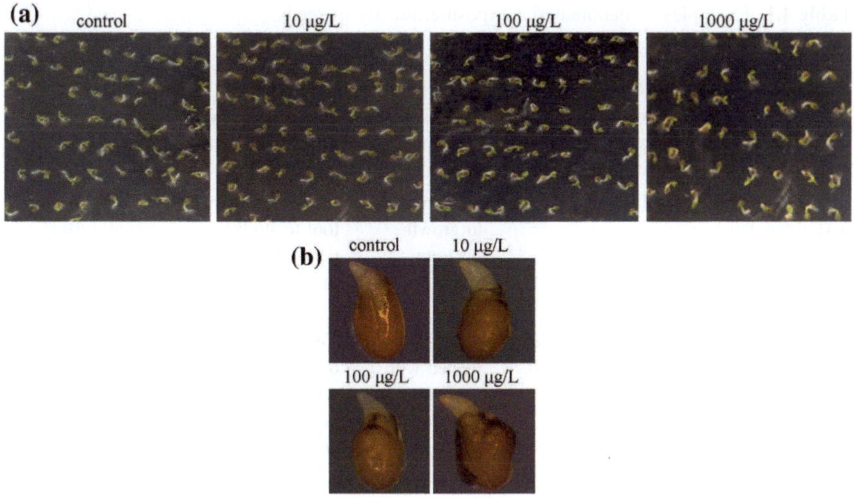

Fig. 4.2 Effects of GO on seed germination and development in *Arabidopsis*. **a** Effects of GO on seed germination at 48 h. **b** Effects of GO on seed sprouting at 36 h. Reprinted from Zhao et al. (2015b), with permission from Elsevier

rice germination buffer for 2 weeks, subsequently known as the second generation. Anderson et al. (2014) prepared nanoparticles of poly(epsilon-caprolactone) containing the herbicide atrazine and evaluated their herbicidal activity. Seeds of *Brassica sp.* and maize were sown in pots filled with 600 g of Orgam Biomix. After four days, the nanoformulations were applied at a concentration of 2.5 kg/ha. The same amount of NPs was also applied in the absence of the herbicide atrazine to assess possible effects on the plants. Some more examples of NM exposure to seeds are provided in Table 4.1.

4.2.2 Via Roots

The root, especially root hair, is the most essential plant organ dealing with the uptake of nutrients. Nutrient ions are transported to the core of the root (stele) for them to reach the conducting xylem and phloem tissues. Xylem is responsible for the movement of water and inorganic molecules, whereas phloem accounts for organic molecule transportation across the plant body. Nanoparticulate exposure through the roots is thought to provide direct access to the conductive tissues, which would facilitate the efficient translocation of NPs to the desired tissues or throughout the plant body.

2,4-dichlorophenoxyacetic acid (2,4-D)-induced leaf senescence and its inhibition by silver (Ag^+) ions in the form of silver nitrate ($AgNO_3$) or silver nanoparticles (Ag NPs) were analyzed in 8-day-old mung bean (*Vigna radiata*) seedlings.

Table 4.1 Examples of nanomaterial exposure directly to seeds

Nanoparticles	Plant	Method	Objective	Reference
NanoCeO_2, nano-La_2O_3, nano-Gd_2O_3 and nano- Yb_2O_3	Radish, rape, tomato, lettuce, wheat, cabbage and cucumber	Seeds soaked in NP solutions	Evaluation of phytotoxicity-root growth	Ma et al. (2010)
Activated carbon (AC), few-layer graphene structures, multi-wall and single-wall CNTs	Tomato	NPs are added to growth medium	Integrated analysis tool to study nano-plant interactions	Khodakovskaya et al. (2011)
Nanoceria, CeO_2	Alfalfa, maize, cucumber, and tomato	Seeds were treated with nanoceria	Cerium uptake and oxidation state within tissues, germination rate, and root elongation	Lopez-Morino et al. (2010b)
Gold NPs (Au NPs)	Soybean	Seeds were exposed to NP solution	Au NP translocation and accumulation	Falco et al. (2011)
Aluminum Oxide (Al_2O_3) NPs	Tobacco	Seeds exposed to NPs supplemented growth media	Growth and development of tobacco seedlings	Burklew et al. (2012)
Cerium Oxide (CeO_2) NPs	Tomato	Seeds germinated in different concentration of NPs	Plant growth and fruit yield. NPs uptake and translocation to shoots and edible tissues	Wang et al. (2012)
CeO_2 NPs	Maize	Maize plants grown in soil spiked with CeO_2 NPs and alginate	Physiological effects of NPs on maize plants	Zhao et al. (2014a)
Au NPs	Arabidopsis	Plants exposed to Au NPs via germination media	Physiological and free radical scavenging activity analysis	Kumar et al. (2013)
TiO_2 NPs and nanotubes (TiO_2-NTs)	Paddy microcosm	NPs were added to the soil sediments	Ti levels in the organisms, bioaccumulation	Yeo and Nam (2013)
ZnO NPs	Green Peas	Seeds exposed to NPs mixed soil	Accumulation of Zn and physiological effects	Mukherjee et al. (2014)
Au nanoclusters	Bean sprouts	Seeds were germinated in medium supplemented with NPs	Sensing ferric ions	(Su et al. 2015)

(continued)

Table 4.1 (continued)

Nanoparticles	Plant	Method	Objective	Reference
CeO_2 and ZnO NPs	Cucumber	NPs mixed with soil and seeded	Analysis of carbohydrate, protein, antioxidant contents, and calorific value of food	Zhao et al. (2014b)
CeO_2 NPs	Radish	Plants were cultivated in potting soil treated with $nCeO_2$	Physiological and nutritional parameters	Corral-Diaz et al. (2014)
Ag NPs	Rice	Seeds soaked in NPs before potting	Toxicity, germination, and growth	Thuesombat et al. (2014)
Carbon nanohorns	Barley, maize, rice, soybean, switchgrass, tomato, and tobacco cell culture	Seeds were cultured in MS medium supplemented with NPs	Seed germination, cell growth, and stress response	Lahiani et al. (2015)
Mn NPs	Mung bean	Seeds imbibed in NP solution	Nitrogen metabolism in plants, biochemical and molecular analysis	Pradhan et al. (2014)
Ceria NPs	Cucumber	Seeds were soaked in NP solution	Quantification of NP uptake by fluorescence	Gui et al. (2015)
Iron pyrite (FeS_2) NPs	Spinach	Seeds exposed to NP before sowing	Biomass production	Srivastava et al. (2014)
Apatite NPs	Soybean	Seeds exposed to NPs mixed with soil	Biomass production—growth rate and seed yield	Liu and Lal (2014)
CdS: Mn/ZnS quantum dots	Green peas	Seeds were exposed to QDs before sowing	Germination and growth of seeds	Das et al. (2015)
CeO_2 and ZnO NPs	Maize	Plants were cultivated in soil amended with $nCeO_2$ or nZnO	Toxicity—photosynthesis and respiration rate, bioaccumulation, cob yield	Zhao et al. (2015a)

Seeds were germinated on double-distilled water-moistened cotton pads for 48 h. Eight-day-old seedlings received a single application of 18 mL (twice a day) solutions of 2,4-D, indole acetic acid (IAA), $AgNO_3$, and Ag NPs, either alone or in combination. After eight days, seedlings were harvested and used for analysis (Karuppanapandian et al. 2011).

Peralta-Videa et al. (2014) determined the nutrient composition in soybean plants cultivated in farm soil amended with $nCeO_2$ at 0–1000 mg/kg and nZnO at 0–500 mg/kg. 10 nm nZnO and 8 nm $nCeO_2$ NPs in powder form were mixed with soil approximately 24 h before planting. 2.4 kg of individual soil samples were placed in polyethylene bags with drainage holes and placed within 4-L polyethylene/polypropylene blend garden pots previously bottom-filled with 400 g of washed gravel (1.25–2.5 cm). Dwarf soybean seeds (Early Hakucho, variety product no. 5555) were prior-germinated in peat pellets and transferred to the NP soil pots, 18 days after planting. The plants were harvested after 48 days of growth in the NP-treated soil and respective samples used for analysis.

In their research, Torre-Roche et al. (2013) analyzed the effect of nanoparticle, bulk, or ionic Ag exposure on dichlorodiphenyldichloroethylene (p, p′-DDE; DDT metabolite) accumulation by soybean and zucchini. Before transplanting for exposure assay, the seeds were pre-germinated in vermiculite for 5–7 days. 125-mL jars with 12 g of dry vermiculite (approximately 80 mL) were amended with 20 or 40 mg of bulk or NP Ag to achieve concentrations of 500 and 2000 mg/L (based on 40 mL of added solution) and mixed thoroughly. Zucchini and soybean seedlings were gently planted in the vermiculite (one plant per jar). The soil mixture was then supplemented with 40 mL of 25 % Hoagland solution containing 100 μg/L DDE, yielding Ag bulk and NP concentrations of 500 or 2000 mg/L. Post-19 days of growth period, the plants were harvested for further experiments. The same group (Torre-Roche et al. 2012) performed similar experiments to assess the effect of C_{60} fullerene exposure on the accumulation of p, p′-DDE by zucchini, soybean, and tomato. The plants were grown in 125-mL jars of vermiculite amended with 0 or 40 mg of C_{60} fullerenes.

The effects of Ag NPs on two important crop plants, mung bean and sorghum (*Sorghum bicolor*) were analyzed in both agar and soil media (Lee et al. 2012). Ag NPs were selected for this study due to their designation as an OECD priority nanomaterial. Also to note is that, sorghum and mung bean are recommended by the OECD (Paris, France) and the American Society for Testing Materials (ASTM). Pre-sterilized seeds were germinated on moist cotton, and after 24 h, the sprouted seeds were used in the tests. Petri plates with 30 mL of agar culture media and specified concentrations of nanoparticles were prepared. Ten seedlings were inoculated just above the surface of the agar plates and incubated at culture conditions for two days, after which the plants were separated from the agar, washed thoroughly, and used for further analysis.

Nair and Chung (2015) investigated the impact of CuO NPs in mustard (*Brassica juncea*) plants. Pre-sterilized seeds were germinated on wet Whatman No. 1 paper. Different concentrations of CuO NPs (0, 20, 50, 100, 200, 400, and 500 mg/L) were added to the pre-autoclaved, half-strength, semi-solid MS medium and mixed thoroughly by vortexing. Ten germinated seeds were placed in culture vessels with the spiked media and cultured for 14 days. Enhanced lignification of both shoot and root cells was observed and was speculated to have been the result of an increase in cell wall rigidity and hormonal imbalances as well as inhibition of

Table 4.2 Examples of nanomaterial-exposed to roots of seedlings

Nanoparticles	Plant	Method	Objective	Reference
Ag NPs	Onion	*A. cepa* root tip cells were treated with Ag NPs	Cytotoxic and genotoxic impacts	Kumar et al. (2009)
ZnO NPs	Onion	2- to 3-cm-long rooted bulbs were exposed to NP solution	Cytogenetic and genotoxic effects	Kumar et al. (2011)
CdSe/ZnS quantum dots	Ryegrass, chrysanthemum, and onion	Plants uprooted and the roots submerged in 1 mL of NP solution	Uptake of QDs	Al-Salim et al. (2011)
Ag NPs	Fava bean	Lateral roots of seedlings exposed to Ag NPs	Genotoxicity analysis	Patlolla et al. (2012)

mineral uptake to the plants due to the NP exposure. It has to be acknowledged that despite the ever-increasing publications addressing NPs–plant interactions, the implications of NPs on the nutritional parameters of food crops have been limited. Rico et al. (2013) analyzed the quality of rice grains harvested from plants cultivated in cerium oxide NPs ($nCeO_2$) supplemented soil. Three rice varieties (high, medium, and low amylose) were grown in soil supplemented with $nCeO_2$ at 0 and 500 mg kg^{-1} soil. Plastic pots (24 cm diameter × 25 cm high) containing 5 kg of soil (Earthgro potting soil) and $nCeO_2$ suspension at a final concentration of 500 mg kg^{-1} were readied. The mixture was equilibrated for three days before rice seedling transplant. Regular irrigation of the soil–NP mixture was done with distilled water to maintain saturation. Thirty-day-old seedlings were transplanted into the pots, fertilized with 200 mL of Yoshida nutrient solution per week, and placed in a greenhouse. After 135 days, the grains were harvested and dried at 80 °C and brown rice, obtained by removing the rice hull, was powdered, sieved to pass mesh number 4, and utilized for experimental analysis. Rico et al. (2014) conducted similar experiments with $nCeO_2$ on growth and yield attributes and nutritional composition in wheat (*Triticum aestivum*). Wheat was cultivated till maturity and grain production in soil spiked with 0, 125, 250, and 500 mg of $nCeO_2$/kg. At harvest, grains and tissues were analyzed for mineral, fatty acid, and amino acid content. Additional examples of NM exposure to roots are tabulated in Table 4.2.

4.2.3 Via Pollen

Kiwifruit (*Actinidia deliciosa*) pollen was obtained from plants of the male genotype (cv. Tomuri) of kiwi and rehydrated for 30 min at 30 °C under 100 % relative humidity. Germination was performed by suspending pollen in liquid basal medium (1 mg/mL) containing 0.29 M sucrose and 0.4 mM H_3BO_3, in Petri dishes. The pollen was exposed to either nanoparticulate or soluble Pd(II). Post-sonication of Pd NP suspension for 45 min, the cultures were treated with the same, either at the beginning of incubation or to pollen incubated for 45 min. 0.05 % Triton X-100 was added to cultures to facilitate disaggregation of grains and tubes before observation. Germination rate and tube length were evaluated by scoring at least 1500 pollen grains and 500 tubes per sample (Speranza et al. 2010).

4.2.4 Via Cells

Plant cells are protected by rigid cell walls that are primarily composed of carbohydrate polymers, and are semipermeable (Campbell 1990). The average pore size of typical plant cell walls, through which various molecules pass, has been determined to be less than several nanometers (Carpita et al. 1979). Thus, foreign materials with morphological parameters greater than the pore capacity would have restricted ability to penetrate into an active plant cell. However, with further understanding of NM–plant interactions, it is becoming evident that nanoparticles may also penetrate into plant cells by going through or bypassing the cell wall and in some cases with nanotubes, creating new channels.

Liu et al. (2009) investigated the capability of single-walled carbon nanotubes (SWNTs) to penetrate the cell wall and cell membrane of intact plant cells. Tobacco (*Nicotina tabacum*) cells [cv Bright Yellow 2 (BY-2)] were cultured in MS media on an orbital shaker at 130 rpm and 26 °C in dark. Prior to incubation with SWNTs, 3- to 4-day-cultured BY-2 cells were filtered through a cell strainer and rinsed with sterile culture medium for a well-dispersed cell suspension. 0.5 mL of this cell suspension was supplemented with 0.1 mL SWNT/FITC or 0.15 mL SWNT/DNA, and the cells were settled on a shaker at 130 rpm and incubated in dark at either 26 or 4 °C (incubation on ice). Post-incubation, cells were washed thoroughly to remove excess SWNT conjugates or reagents, and re-suspended in the culture medium for immediate confocal imaging. 33 μM wortmannin (from 1 mM stock solution in DMSO) was added to 0.5 mL of the cell suspension and incubated with the cells for 30 min followed by addition of 0.1 mL SWNT/FITC. After incubation at 26 °C for 60 min, the cells were washed and imaged to obtain the quantitative cellular fluorescence intensity of NP uptake (Fig. 4.3).

Torney et al. (2007) developed a honeycomb mesoporous silica nanoparticle (MSN) system with 3-nm pores that has the potential ability to transport DNA and chemicals into isolated plant cells and intact leaves. The MSNs were loaded with a gene and its chemical inducer and capped at the ends with gold NPs to prevent cargo escape. Protoplasts isolated from 6- to 8-week-old tobacco plants (cv Petite Havana) were aseptically grown on MS media without any plant growth hormones. The protoplasts were mixed with MSNs in W5 media using 106 cells for 10 mg of MSN and incubated overnight in 3 mL of W5 in 6-well flat-bottom plates. Non-internalized MSNs were removed by centrifugation at 100 g for 10 min at room temperature. The protoplasts at the interphase were collected and evaluated.

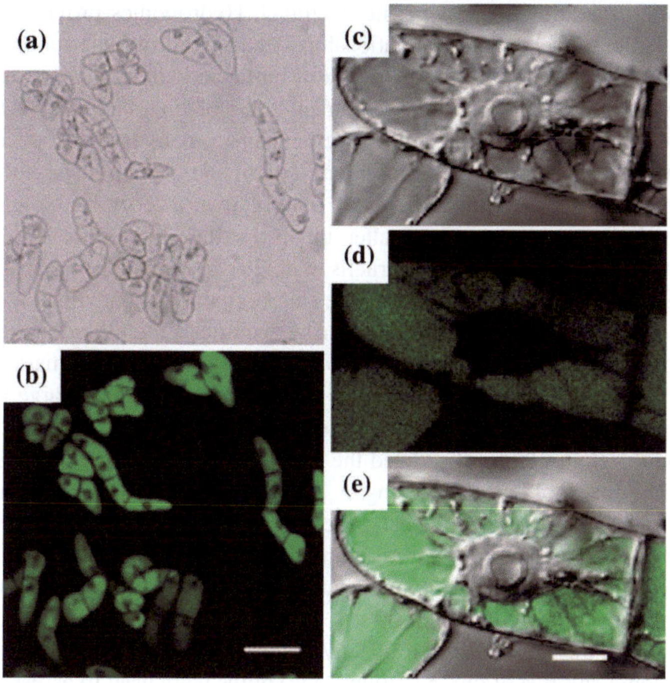

Fig. 4.3 Confocal microscopy images of BY-2 incubated with SWNT/FITC: **a** bright field image; **b** fluorescence image; **c** DIC image under high magnification; **d** fluorescence image under high magnification; **e** overlay of C and D. Scale bars are 100 μm for (**a**) and (**b**) and 10 μm for (**c**–**e**). Reprinted with permission from Liu et al. (2009). Copyright (2009) American Chemical Society

As not much is known about the effects of aluminum oxide NPs on plants at the cellular level, Poborilova et al. (2013) initiated a study on the effects of these NPs on the plant cell model tobacco BY-2 cell suspension culture. *Nicotina tabacum* L. cv. Bright Yellow-2 (BY-2) suspension-cultured cells were cultured in modified liquid MS medium under constant shaking (130 rpm) at 27 °C in the dark in 250 ml Erlenmeyer flasks. Exponential growth phase cells were subsequently transferred to fresh cultivation media with Al_2O_3 NPs (<50 nm) at concentrations of 0, 10, 20, 50, and 100 μg mL^{-1}. Additionally, to assess the possible impact of the size of the NPs, Al_2O_3 microparticles (5 μm) were applied at the same concentrations. Cells were subsequently cultivated for 96 h and counted at defined time intervals.

4.3 Hydroponic Treatment

Hydroponics is a type of hydroculture involving mineral nutrient solutions to grow plants, in water, without soil. Though more suitable for semiaquatic plants, terrestrial plants may also be grown with their roots immersed in the nutrient solution or in an inert medium, such as perlite or gravel. Hydroponics provides an opportunity to better understand the relationship between nutritional status and plant growth, in addition to assessing the impact of biotic and abiotic factors on the plant development. Hydroponic systems are ideal for isolating factors affecting plant growth and to establish relationship of elements from hydro/pedosphere to biosphere. In soil–plant systems, the activity of a particular element inevitably depends on numerous other environmental factors than its own activity in the interfaces of roots, which in many cases may alter the properties and ultimately the actual signature effects of the elements (Cornelis et al. 2012). Many a times, in the case of NMs, the materials may interact with the microorganisms, enzymatic factors, pH changes, etc. in the soil, which can neither be accurately monitored nor controlled. In the case of a closed system like hydroponics, an accurate maintenance of essential parameters could be achieved, which is evident from the large number of works being carried out with this system to analyze various NM–plant interactions.

Stampoulis et al. (2009) elucidated the effects of MWCNTs, Ag, Cu, Si, and Zn oxide on biomass of zucchini using a batch hydroponic experimental system. Germination was induced on moist paper, and 4-day-old seedlings were subsequently transferred to amber vials with 7.5 mL of 25 % Hoagland solution. Cultures were maintained at optimum conditions for 14 days and further transferred to 40-mL amber vials containing 39 mL of solution with either nanoparticles or corresponding bulk materials at 1000 mg/L.

Hydroponic cultures of onion (*Allium cepa*) bulbs were used to assess the effects of cobalt and zinc oxide NPs on the root elongation, root morphology, and cell morphology, as well as their adsorption potential (Ghodake et al. 2011). Healthy and fresh onions were washed under running tap water, and scales were removed. Onions were grown for three days without the NPs and when the length of the roots reached between 1.3 and 1.5 cm, the plants were transferred to fresh solutions of CoO and ZnO NPs. The physiological parameters were measured at different intervals over a span of three days, and average values were recorded. Significant adsorption of CoO NPs into the root system was observed.

Feichtmeier et al. (2015) analyzed the reversibility and effects of citrate-coated Au NPs on barley by cultivating the seeds in Au NP-containing nutrient solution for two weeks with subsequent transfer of the seedlings to Au NP-free media. The stability of Au NPs in the cultivation media was also investigated over a period of two weeks. Barley (*Hordeum vulgare*) was cultivated for two weeks in a hydroculture media on a floating layer consisting of low-density polyethylene granulate (LD-PE). The granulate is used to provide necessary anchorage to the plant roots and access to sufficient nutrient supply from the medium. Au NP hydrosol (1, 3, 5, 8, and 10 $\mu g\ mL^{-1}$) was supplemented with 0.022 g of an MS basal medium. For initial germination of barley seeds, a room temperature, dark chamber incubation was arranged. After two days, the cultures were set in a chamber with a 16-h light and 8-h dark period cycle at 21.5 °C. Barley plants were harvested after 14-day exposure and further used for determination of fresh biomass and other parameters.

It is claimed that Ag NPs are safe and efficient agents against disease causatives in agriculture. To understand the protein populations and sub-populations along with the environmental Ag NPs stresses, Mirzajani et al. (2014) employed a proteomic approach on rice. The seeds were germinated in sterile water filled Petri dishes, for seven days. Post-germination, seedlings were transferred to rice-specific growth cultivation media and cultured in a phytotron under suitable growth conditions, with the renewal of hydroponic media every five days. Ten-day-old plants were treated with Ag NPs (18.34 nm) colloidal solution at concentrations of 0, 30, and 60 mg/mL for 20 days. Post-incubation plants were removed from the NP-augmented solution and thoroughly washed with water prior to analysis.

Koo et al. (2015) explored the impact of surface-modified quantum dots on stability, uptake, and translocation in Arabidopsis, and subsequent transfer to primary consumers, cabbage looper (*Trichoplusia ni*). Arabidopsis samples were exposed to CdSe/CdZnS QDs with three different surface coatings: poly(acrylic acid-ethylene glycol) (PAA-EG), polyethylenimine (PEI), and poly(maleic anhydride-alt-1-octadecene)- poly(ethylene glycol) (PMAO–PEG), which are anionic, cationic, and relatively neutral, respectively. Arabidopsis seeds were sown in 1/16 strength Hoagland solution fortified with 0.8 % agar in a seed holder hydroponic system (Kit 140 HD, Aquaponics, Belgium). Post-germination, Arabidopsis plants were transferred to 15-mL conical tubes for four weeks followed by a subsequent transfer to 1/16 strength Hoagland solution amended with either PAA-EG or PEI QDs at 10 μg/mL for seven days before analysis.

Schwabe et al. (2015) investigated the uptake of cerium dioxide NPs by hydroponically grown wheat, pumpkin (*Cucurbita pepo*), and sunflower (*Helianthus annuus* var. Iregi) plants. Pre-sterilized seeds were germinated on water-irrigated paper and, after 6–8 days of soaking, were subsequently transferred to a hydroponic system with 1 L 20 % Hoagland solution and cultured for 26 days. Plants were then exposed for 6 days to CeO_2-NPs suspension at 100 mg L^{-1}, dispersed by ultrasonication in 1 L of 20 % Hoagland medium (Fig. 4.4). For stabilization of the nanoparticles in suspension, gum arabicum (GA) was supplemented at a concentration of 60 mg L^{-1}.

The determination of unique characteristics of NPs after their entry into the plant system is highly restricted with currently available technologies. Therefore, Dan et al. (2015) developed an enzymatic digestion and single-particle inductively coupled plasma-mass spectrometry (SP-ICP-MS) analysis, for simultaneous determination of Au NP size, size distribution, particle concentration, and dissolved Au concentration in tomato plant tissues. For this purpose, pre-sterilized tomato (*Solanum lycopersicum*) seeds were initially germinated on deionized water-moistened filter paper for seven days. Seedlings of uniform features were transferred to 50 mL of quarter-strength Hoagland solution in polypropylene centrifuge tubes. Post-20 days of culture in the nutrient solution, plants were transferred to fresh 50-mL centrifuge tubes containing only deionized water for a further two days. The water was subsequently replaced with solutions of different concentrations of PVP-coated 40 nm Au NPs. The plants were exposed to the NPs solution for a period of 4 days before harvesting for enzyme extraction and SP-ICP-MS analysis.

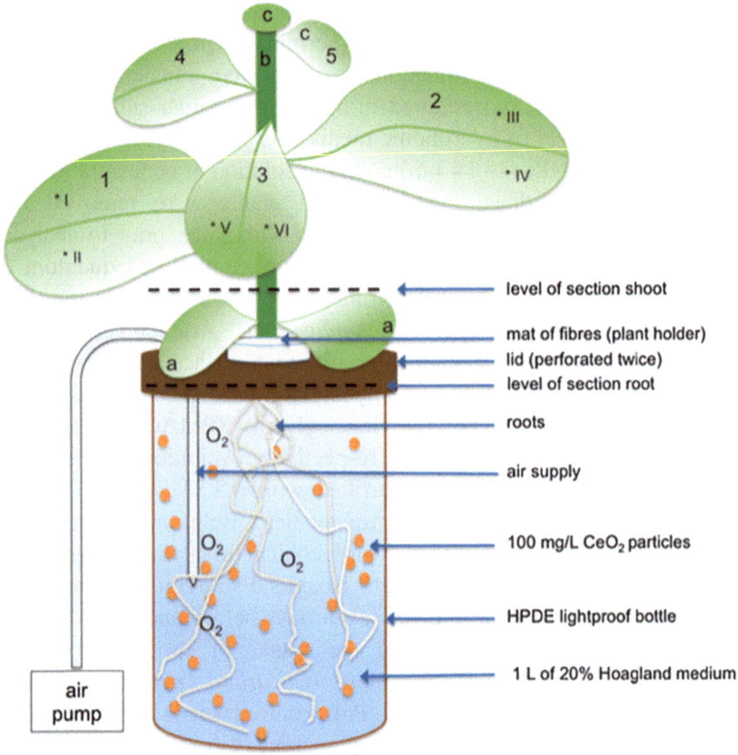

Fig. 4.4 Scheme of the experimental setup for hydroponic system. **a** Cotyledons were removed before harvest; **b** stem; **c** meristem incl. newest leaf not older than 7 d; *1–5* leaf count; **I–*VI* marked points for chlorophyll measurements. Reproduced from Schwabe et al. (2015), with permission from Royal Society of Chemistry

Zhang et al. (2014) reported on the accumulation and elimination of CuO NPs and CdS/ZnS QDs in aquatic mesocosms with *Schoenoplectus tabernaemontani* cultivated in hydroponic mesocosms. *S. tabernaemontani* rhizomes were thoroughly washed and acclimatized in 25 % strength Hoagland nutrient solution for four weeks. The plants were eventually transferred to 2-L vessels containing Hoagland nutrient solution spiked with CuO NPs and CdS QDs and cultured for 21 days prior to analysis.

Six-inch cuttings of poplar plants (*Populus deltoides* × *nigra*, DN-34) were used to evaluate the vegetative uptake and subsequent translocation and transport of commercially available Au NPs into plant cells in a hydroponic culture system. Plants were grown for 25 days in 1/2 strength Hoagland solution prior to NP exposure. Different concentrations of Au NPs (3 mL of 498 ± 50.5, 247 ± 94.5, and 263 ± 157 ng/mL) and Au(III) ions (5.0, 10.0, and 20.0 mg/L) were added to 200 mL of deionized water in 250-mL glass conical flask test reactors with a PTFE-faced septum sampling port. The reactors were covered with aluminum foil and kept at 23 ± 1 °C under optimum culture conditions. Deionized water saturated with oxygen was injected into the reactors twice per day to compensate for the evapotranspiration loss (Zhai et al. 2014). Further cases of hydroponic application of NMs to plants are presented in Table 4.3.

4.4 Direct Injection

As discussed earlier, nanomaterials face numerous challenges and hurdles during the course of gaining entry into the plant systems. Some may arise due to intrinsic limitations of the NMs with respect to size, charge, shape, etc., but majority of obstacles are posed during the application and contact of NMs with the microenvironment of the target exposure tissues. Even when the NMs are successful in gaining entry into the system, either by force or by active/passive diffusion, still, gaining access to the conductive tissues of the plants viz., xylem and phloem remains a challenge. One solution, taken from the field of medicine, is the age-old tradition of direct injection of substances into the tissues. Though, when applied to the plant systems, this technique provides direct access to the desired tissues bypassing the discussed barriers, it remains a matter of debate when considering the limitations of large-scale application.

Pumpkin plants were selected by Gonzalez-Melendi et al. (2008) due to their large-sized vessels, which are expected to facilitate efficient transport of NPs through the vascular system. Seeds were germinated in Petri dishes on moistened filter paper, and 15-day-old seedlings with a radicle length of about 4–5 cm were transferred to a bag, with Hoagland nutrient solution. The bags were suspended vertically in a controlled environment chamber. Carbon-coated iron NPs were dispersed by sonication in gelafundin, a commercial succinated gel to obtain a biocompatible magnetic fluid. The aim of this work was to analyze the capacity of a magnetic field to retain the particles in specific parts of the plant in addition to studying their penetration and movement in plant cells. The bioferrofluid was directly injected inside the internal hollow of the leaf petiole to deliver instant access to the nanoparticles to the vascular system for faster and efficient translocation and distribution in the plant body (Fig. 4.5). The experimental setup consisted of small magnets placed on the petiole of the leaf opposite to the injection point and on some of the roots. It was observed how the bioferrofluid can be concentrated in the desired areas by using magnets. No particles greater than 50 nm were detected inside the tissues, implying a possible size-based selection mechanism, probably involving a barrier of cell walls and waxes.

Corredor et al. (2009) performed similar experiments to observe the subcellular localization of carbon-coated nanoparticles under the influence of external magnets. The employability of magnetic nanoparticles in a large-scale scenario is somewhat a debatable prospect as placing magnets in annual extensive crops (e.g., cereals) is impractical. However, the possibility of using such a system under greenhouse/controlled conditions would be possible for specific treatments in fruit trees (e.g.,

Table 4.3 Examples of hydroponically cultured plant exposure to NPs

Nanoparticle	Plant	Method	Objective	Reference
TiO_2 NPs	Onion and tobacco	Germinated onion bulbs and fourth-leaf stage tobacco plantlets were exposed to TiO_2 NPs	Genotoxicity	Ghosh et al. (2010)
ZnO NPs	Velvet mesquite	Seedlings were exposed to NP solution	Zn accumulation in tissues and physiological evaluation	Hernandez-Viezcas et al. (2011)
CdSe/ZnS QDs	Arabidopsis	Seedlings exposed to media supplemented QD solution	Uptake mechanism	Navarro et al. (2012)
Ag NPs	Brahmi plant	Week old seedlings were exposed to hydroponic medium supplemented with Ag NPs	Plant growth metabolism and biochemical parameters	Krishnaraj et al. (2012)
Au NPs	Tobacco and wheat	30 days germinated seedlings of tobacco and 7 days germinated seedlings of wheat were exposed to Au NPs	Bioaccumulation/Uptake of different size and surface coated Au NPs	Judy et al. (2012)
Au NPs	Rice, radish, pumpkin, and perennial ryegrass	Exposure of seedlings to Au NPs	Effect of surface charge on the uptake and distribution of Au NPs	Zhu et al. (2012)
Au NPs	Rice	9-day-old germinated seedlings were exposed to NPs dispersed in MilliQ water	Tissue level uptake and spatial distribution of Au NPs in rice roots and shoots	Koelmel et al. (2013)
Magnetite NPs	Soybean	Seeds were exposed to nutrient media containing different conc. of NPs	Uptake, translocation, and effect on chlorophyll content in soybean, under hydroponic conditions	Ghafariyan et al. (2013)
CeO_2 NPs	Wheat and pumpkin	NPs were added to the hydroponic culture medium	Influence of organic matter on NP–plant interaction	Schwabe et al. (2013)
ZnO NPs	Mustard	Seeds were germinated under a hydroponic condition with varying concentrations of ZnO NPs	Estimation of plant biomass, biochemical parameters, and bioaccumulation of ZnO	Rao and Shekhawat (2014)

(continued)

Table 4.3 (continued)

Nanoparticle	Plant	Method	Objective	Reference
ZnO NPs	Maize	7-day-old seedlings were exposed to NPs in the hydroponic culture medium	Root metabolic activities, accumulation in roots, biotransformation, phytotoxicity	Lv et al. (2015)
CeO NPs	Kidney bean	Plants were exposed to suspensions of $\sim 8 \pm 1$ nm $nCeO_2$ for 15 days in hydroponic conditions	Primary indicators of stress, chlorophyll contents, bioaccumulation, translocation, and cellular homeostasis	Majumdar et al. (2014)
$NaYF_4$:Yb,Er upconversion nanocrystals	Pumpkin	6-day-old seedlings in hydroponic culture were exposed to NPs	Kinetics of the uptake and the translocation of NPs	Nordmann et al. (2015)
Fe (nZVI) NPs	Arabidopsis	Plants were exposed to ½ strength MS media fortified with NPs	Plasma membrane H^+-ATPase activity—stomatal opening and gene expression	Kim et al. (2015)
ZnO NPs	Aquatic plant mesocosm—*Schoenoplectus tabernaemontani*	Plant exposure to NPs augmented hydroponic nutrient medium	Phytotoxicity and bioaccumulation	Zhang et al. (2015)
Cu NPs	Lettuce and alfalfa	10- to 15-day-old hydroponically grown plants exposed to nCu, bulk Cu, nCuO, bulk CuO, $Cu(OH)_2$ (CuPRO 2005, Kocide 3000), and $CuCl_2$	Phytotoxicity-root length, catalase and ascorbate peroxidase measurement, nutrient content	Hong et al. (2015)

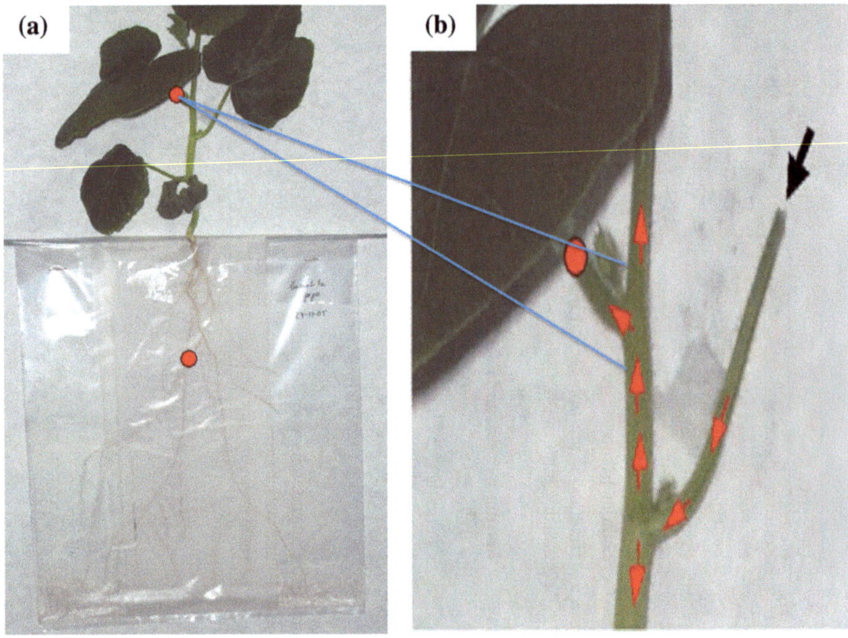

Fig. 4.5 **a** Pumpkin plant growing in the polyethylene bag system. *Red circles* indicate positions of magnets. **b** Detail showing the point of application of the bioferrofluid (*black arrow*) and further expected movement of NPs through the vascular system (*red arrows*). Reprinted from Gonzalez-Melendi et al. (2008), with permission from Oxford University Press

olive (*Olea europaea*) trees) or high-input crops. Still, the major scope of using this model, as of now, would be for laboratory-scale research applications since it allows a very precise localization of particles.

4.5 Spraying

Currently, most of the studies on plants involve exposure of NPs through roots, either directly to established roots or during the germination stage. However, it has to be noted that plants also interact with atmospheric NPs through the leaves, foliar pathway, and knowledge on their response to this contact is limited. Significant effects of NPs on plant foliage are inevitable due to deposition of atmospheric particles or application of purpose-made NPs.

Corredor et al. (2009) apart from studying the localization of carbon-coated NPs by direct injection also employed spray technique to analyze the progressive penetration of NPs in the plant tissues. This methodology closely resembles that employed by agronomist and breeders in field applications. Droplets of ferrofluid were placed on the leaf surface, close to the insertion point of the petiole, in an attempt to emulate spraying of a nanoparticle solution onto a cultivated plant. The method used in this work is similar to the large-scale and hands-on spraying procedures, which are used by breeders and coordinators of phyto-sanitary control. This spray technique seemed to be a more practical approach from an agronomic perspective, as large-scale and hands-on modules currently exist for the spraying of pesticides and chemical fertilizers.

To control the pathogenic fungi in green zucchini plants infected with powdery mildew, Park et al. (2006) tested nanosized silica-silver (nano-silver combined with silica molecules and water-soluble polymer) by uniformly spraying the nanocomposite onto green squash plants infected with powdery mildew at 0.3 ppm concentration. After the nanosized silica-silver had been applied, the progress of powdery mildew was observed for three weeks. In addition to the anti-fungal assay, the possible chemical injuries to plants due to the application of nanosized silica-silver were assessed with the spray of undiluted and 10, 100 and 1000 times diluted solutions of the nanocomposite on the surface of squash leaves including new leaves of cucumber and pansy (*Viola tricolor*). After three days, chemical injuries on plants were observed.

Birbaum et al. (2010) focused on the quantitative investigation of uptake and translocation of ceria NPs into maize plants using ICP-MS. Various scenarios were simulated, wherein ceria NPs were introduced to leaves as airborne aerosols and aqueous suspensions. The NP exposure to plants was conducted under artificial daylight, to facilitate stomatal opening, and under dark conditions, where stomata are closed. A 10 μg/mL of CeO_2 NP suspensions were diluted from freshly prepared stock suspension. Three- to five-week-old, greenhouse-grown maize plants (Birko) were used for NP exposure studies in an in-house-built glove box of 2.25 m^3, wherein the NPs were produced (Fig. 4.6). The exposition unit consisted of NP production unit and an exposure chamber, hosting the plants. Once the NP production was initiated, a total exposure of 0.4 g of NPs was achieved in 1 min. A fan was used to disperse the NPs homogeneously over the plants for an exposure time of 20 min. During harvesting, the leaves were separately rinsed with deionized water and abraded with a glove, simulating a possible naturally occurring washing procedure by rain and the wind. Post-exposure, a batch of NP exposed plants was returned to the greenhouse for a further 12 weeks to determine translocation of NPs into newly developed leaves.

To find an effective solution against the soilborne Oomycete *Pythium aphanidermatum*, the causal agent of one of the most serious threats of rhizome rot disease to turmeric (*Curcuma longa*) crops, Anusuya and Sathiyabama (2015) developed β-D-Glucan nanoparticles (GNPs) with β-Glucan isolated from *P. aphanidermatum* mycelium. Rhizomes of turmeric were thoroughly cleaned under running tap water and surface-sterilized. Two–three rhizomes each with three nodes were planted in earthen pots containing soil and manure and maintained under glass house condition. A foliar spray of GNPs (0.1 %, w/v) was applied to 30-day-old plants (5 mL/plant) at intervals of 30 days till 210 days after which the leaves were excised and rhizomes left for another 30 days for harvest.

Fig. 4.6 Cerium dioxide nanoparticles were exposed as in situ-prepared aerosol. Aerosol exposure was performed in a setup developed for application at the air–liquid interface. **a** The nanoparticles production unit. **b** The ventilator for the well-characterized and homogeneous particles deposition and **c** the maize plants within the totally enclosed setup. Reprinted with permission from Birbaum et al. (2010). Copyright (2010) American Chemical Society

In their study, Hong et al. (2014) aerially treated hydroponically grown cucumber plants with nanoceria powder ($nCeO_2$). Fifteen days after treatment, the test plants were assayed for Ce uptake by using inductively coupled plasma optical emission spectrometry (ICP-OES) and transmission electron microscopy (TEM). Post-surface sterilization and soaking in deionized water for 24 h, the seeds were placed on the edge of wet germination paper towels, rolled, and supplemented with 10 drops of antimycotic/antibiotic solution, and kept in Mason jars with distilled water at the bottom, set in the dark for four days, and, after that, exposed to light for one day. Seven-day-old plants were transferred to hydroponic jars with 300 mL of a modified Hoagland solution. After two weeks of growth, the young plants were transferred to separate chambers and treated with 0, 0.98 and 2.94 g/m^3 NP concentrations. CeO_2 NPs were kept on a tray in front of a fan (120 V, 0.25 A, 60 Hz) inside the chamber, and blowing times were fixed at 15 and 45 min. In another set of experiments, leaves of 2-week-old plants were sprayed with CeO_2 suspended in distilled water at concentrations 0, 40, 80, 160, and 320 mg/L, with a handheld sprayer bottle (Fig. 4.7). A total of 100 mL of respective concentrations were sprayed every 4 h. At defined time points after treatment, samples of leaves, stems, and roots were collected and prepared for analysis.

Fig. 4.7 **a** Hydroponic setup for cucumber plants treated with CeO_2 NPs. **b** Nanoparticle application as powder and **c** suspension. The plants were cultivated for 15 days before NP treatment and harvested 15 days after treatment. Reprinted with permission from Hong et al. (2014). Copyright (2014) American Chemical Society

The uptake of Ag NPs in lettuce after foliar exposure and possible NP biotransformation and phytotoxic effects were studied by Larue et al. (2014). Lettuce was chosen as model species because of its large foliar surface and cosmopolitan occurrence in gardens or farmlands making it an ideal model to assess the foliar transfer of NPs. Young lettuce plantlets were exposed to 1, 10, or 100 μg Ag NPs. Plants were harvested after a 7-day exposure.

4.6 Biolistics

Torney et al. (2007) coated DNA onto Type-II MSNs for endocytosis experiments by incubating 1 mg of purified plasmid DNA with 10 mg MSN in 50 ml water for 2 h. The MSNs were thoroughly washed with W5 media prior to incubation with isolated protoplasts. Bio-Rad Biolistic PDS-1000/He particle delivery system was used for MSN delivery into plant cells. To coat DNA onto gold-capped MSNs, a standard protocol was followed similar for biolistic gun with minor modifications. The particle bombardment parameters for tobacco plantlets were 650 p.s.i. rupture disk, 10-cm gap distance, and 10-cm target distance. A sterile 150-mm mesh was used between the macro-carrier and the target tissue. Bombardment of maize immature embryos was performed using the standard laboratory procedure. After bombardment, the embryos were placed on N6-30 medium for 10 days to avoid callus autofluorescence interfering with the GFP expression evaluation. Plants were germinated and grown in Y-segmented Petri dishes. Plants were grown for 48 h after bombardment before being subjected to evaluation.

4.7 Discussion

Any mode of nanoparticle application to plants, either under laboratory conditions or in field scenario, must take into account certain essential aspects as the accessibility of the nanoformulations to target tissues, minimum concentrations with maximum effect, scalability of the approach, and, of course, the economics of the whole process. Transportation of molecules across plant cells is more complicated due to the cell wall, which is mostly made up of cross-linked polysaccharides, and cell membrane, which pose a great challenge for nanoparticulate movement. This particular aspect is one of the main distinguishing factors between plant and animal cells, the main reason why nanotechnology has moved at a rapid pace in medicine and still at infancy in the agro-related field. The complexity to overcome these phyto-cellular barriers due to their complex architecture has limited the use of many mammalian cells-applicable nanomaterials in plants. Therefore, preprocessing of plant cells to pave the way for smoother nanomaterial access has been attempted. Protoplasts, plant cells whose cell wall is removed enzymatically together with certain cell surface proteins, have been employed to show the internalization of certain nanomaterials such as silicon NPs and polystyrene nanospheres. However, uncertainties regarding the similarities between isolated protoplasts and intact cells raise questions on this mode of NP administration. The technique along with the more modern biolistic approach, though seem lucrative under in vitro settings, may not be feasible in the long run due to major limitations as high cost and restrictions in large-scale applications.

Foliar application of nanoparticles as powders dispersed manually or mechanically and aqueous suspension by spray technique are being widely practiced as plant leaves provide a wider canvas for the NPs to exercise their effects. Nanoparticles applied on leaf surfaces are found to gain entry through stomatal openings, cuts, and wounds, or through the bases of trichomes and subsequently get translocated to various tissues (Eichert et al. 2008; Navarro et al. 2008; Fernandez and Eichert 2009; Uzu et al. 2010). Such a mode of aerial application of NPs on leaf and stem surface has not been found to alter essential physiological processes as photosynthesis or respiration in several groups of horticultural and crop plants. Also, they have not been responsible for alteration of gene expressions in insect trachea and have been qualified for use as nanobiopesticides. One successful example is of amorphous silica, which is designated as safe for humans by World Health Organization and US Department of Agriculture (Barik et al. 2008). Still, this cannot be taken as a general statement for all classes of nanoparticles as it is well known that the characteristics of elements vastly differ at the nanoscale compared to the bulk and are dependent on the constitutional make of individual or composite NPs. There have been reports of foliar heating due to the accumulation of NPs on photosynthetic surface leading to alterations in gas exchange due to stomatal obstruction resulting in various physiological and cellular functional impairments in plants (Da Silva et al. 2006).

Small size and huge surface energy of nanoparticles renders them susceptible to aggregation in aqueous media, which in turn may modulate their bioavailability. Root tips and hairs are found to secrete large amounts of mucilage (Campbell 1990) composed of highly hydrated polysaccharides, which might contribute to the aggregation of NPs causing clogging of pore channels and arresting NP entry. Changes in zeta potentials of tested NPs also are indicators of the ability of root exudates to change the property and behavior of NPs.

One major emerging field of plant cultivation is hydroponics. The hydroponic strategy has many advantages as plants can be grown anywhere and it uses only 1/20th of water compared to traditional (soil-based) gardening. This technique does not require the use of pesticides, fertilizers, and other chemicals, as the risk of soilborne pathogens is negated. However, care has to be taken when considering infections posed by air and waterborne diseases. As hydroponic system is a closed and controlled system, crops grow faster. The NP exposure to the plant system can be constantly monitored, and numerous extraneous interactions, which would otherwise hinder in the efficiency of the experiments, can be avoided. Apart from the usual hydroponic cultures, there have also been studies on the extension of shelf life of post-harvest products as in horticulture using hydro-treatment. One example is from the experiments conducted by Solgi et al. (2009) who utilized silver nanoparticles and essential oils as novel antimicrobial agents for extending the vase life of gerbera (*Gerbera jamesonii* cv. 'Dune') flowers. Flowers were harvested from the plants and incubated in solutions of silver nanoparticles with or without essential oils. Similarly, the vase life of Rose (*Rosa hybrida* L.), one of the major cut flower crops, was significantly improved by treatment with biologically synthesized Ag NPs (Hassana et al. 2014).

4.8 Conclusion

The secure application of nanomaterials to agriculture and related crop cultivation has been extensively delayed when compared to the rapid strides in fields as electronics, energy harvesting, and medicine. Though one main reason for this scenario is the limited understanding of plant–NP interactions, as unlike the mammalian cell system, the growth and related phenotypic/genotypic expressions in plants are a lengthier process, one other limiting aspect is the choice of methods employed for the application of NPs. It does depend on factors as the running cost, the area of application, availability of related resources, disease, and pest management, etc., still, a major emphasis has to be given to the select methods that allow for ready access to plant tissues in consideration for optimal desired effects. Though a lot of research has focused on this aspect, conclusions have varied. A solid framework for the NP–plant interaction has to be designed for the development of fool proof and highly efficient nanoparticle administration to crops.

References

Al-Salim N, Barraclough E, Burgess E, Clothier B, Deurer M, Green S, Malone L, Weir G (2011) Quantum dot transport in soil, plants, and insects. Sci Total Environ 409:3237

Anderson ES, Grilloa PR, Nathalie FS, Melloa Rosab AH, Fraceto LF (2014) Application of poly (epsilon-caprolactone) nanoparticles containing atrazine herbicide as an alternative technique to control weeds and reduce damage to the environment. J Hazard Mater 268:207

Anusuya S, Sathiyabama M (2015) Foliar application of β-D-Glucan nanoparticles to control rhizome rot disease of turmeric. Int J Biol Macromol 72:1205

Barik TK, Sahu B, Swain V (2008) Nanosilica-from medicine to pest control. Parasitol Res 103:253

Birbaum K, Brogioli R, Schellenberg M, Martinoia E, Stark WJ, Gunther L, Limbach LK (2010) No evidence for cerium dioxide nanoparticle translocation in maize plants. Environ Sci Technol 44:8718

Burklew CE, Ashlock J, Winfrey WB, Zhang B (2012) Effects of aluminum oxide nanoparticles on the growth, development and microRNA expression of tobacco (*Nicotiana tabacum*). PLoS ONE 7:e34783

Campbell NA (1990) Biology, 2nd edn. The Benjamin/Cummings Publishing Company, Redwood City

Carpita N, Sabularse D, Montezinos D, Delmer DP (1979) Determination of the pore size of cell walls of living plant cells. Science 205:1144

Cornelis J-T, Kruyts N, Dufey JE, Delvaux B, Opfergelt S (2012) Understanding root uptake of nutrients, toxic and polluting elements in hydroponic culture. In: Asao T (ed) Hydroponics—a standard methodology for plant biological researches. ISBN 978-953-51-0386-8, InTech, Rijeka, Croatia, p 153

Corral-Diaz B, Peralta-Videa JR, Alvarez-Parrilla E, Rodrigo-García J, Morales MI, Osuna-Avila P, Niu G, Hernandez-Viezcas JA, Gardea-Torresdey JL (2014) Cerium oxide nanoparticles alter the antioxidant capacity but do not impact tuber ionome in *Raphanus sativus* (L). Plant Physiol Biochem 84:277

Corredor E, Testillano PS, Coronado MJ, González-Melendi P, Fernández-Pacheco R, Marquina C, Ibarra MR, de la Fuente JM, Rubiales D, Pérez- de-Luque A, Risueño MC (2009) Nanoparticle penetration and transport in living pumpkin plants: in situ subcellular identification. BMC Plant Biol 9: 45

Cotae V, Creanga I (2005) LHC II system sensitivity to magnetic fluids. J Magn Magn Mater 289:459

Da Silva LC, Oliva MA, Azevedo AA, De Araujo JM (2006) Responses of restinga plant species to pollution from an iron pelletization factory. Water Air Soil Pollut 175:241

Dan Y, Zhang W, Xue R, Ma X, Stephan C, Shi H (2015) Characterization of gold nanoparticle uptake by tomato plants using enzymatic extraction followed by single-particle inductively coupled plasma-mass spectrometry analysis. Environ Sci Technol 49:3007

Das S, Wolfson BP, Tetard L, Tharjur J, Bazata J, Santra S (2015) Effect of N-acetyl cysteine coated CdS: Mn/ZnS quantum dots on seed germination and seedling growth of snow pea (*Pisum sativum* L.): imaging and spectroscopic studies. Environ Sci NANO 2:203

Eichert T, Kurtz A, Steiner U, Goldbach HE (2008) Size exclusion limits and lateral heterogeneity of the stomatal foliar uptake pathway for aqueous solutes and water-suspended nanoparticles. Physiol Plant 134:151

Ernst R, Arditti J, Healey PL (1971) Biological effects of surfactants. I. Influence on the growth of orchid seedlings. New Phytol 70:457

Falco WF, Botero ER, Falcao EA, Santiago EF, Bagnato VS, Caires ARL (2011) In vivo observation of chlorophyll fluorescence quenching induced by gold nanoparticles. J of Photochem Photobiol A 225:65

Feichtmeier NS, Walther P, Leopold K (2015) Uptake, effects, and regeneration of barley plants exposed to gold nanoparticles. Environ Sci Pollut Res Int 22:11

Fernandez V, Eichert T (2009) Uptake of hydrophilic solutes through plant leaves: current state of knowledge and perspectives of foliar fertilization. Crit Rev Plant Sci 28:36

Ghafariyan MH, Malakouti MJ, Dadpour MR, Stroeve P, Mahnoudi M (2013) Effects of magnetite nanoparticles on soybean chlorophyll. Environ Sci Technol 47:10645

Ghodake G, Seo YD, Lee DS (2011) Hazardous phytotoxic nature of cobalt and zinc oxide nanoparticles assessed using *Allium cepa*. J Hazard Mater 186:952

Ghosh M, Banyopadhyay M, Mukherjee A (2010) Genotoxicity of titanium dioxide (TiO_2) nanoparticles at two trophic levels: Plant and human lymphocytes. Chemosphere 81:1253

Gonzalez-Melendi P, Fernandez-Pacheco R, Coronado, E. Corredor MJ, Testillano PS, Risueno MC, Marquina C, Ibarra MR, Rubiales D and Perez-De-Luque A (2008) Nanoparticles as smart treatment-delivery systems in plants: assessment of different techniques of microscopy for their visualization in plant tissues. Ann Bot 101: 187

Gui X, He X, Ma Y, Zhang P, Li Y, Ding Y, Yang K, Li H, Rui Y, Chai Z, Zhao Y, Zhang Z (2015) Quantifying the distribution of ceria nanoparticles in cucumber roots: the influence of labeling. RSC Adv 5:4554

Hassana FAS, Ali EF, El-Deeb B (2014) Improvement of postharvest quality of cut rose cv. 'First Red' by biologically synthesized silver nanoparticles. Sci Hort 179:340

Hernandez-Viezcas JA, Castillo-Michel M, Servin AD, Peralta-Videa JR, Gardea-Torresday JL (2011) Spectroscopic verification of zinc absorption and distribution in the desert plant *Prosopis juliflora-velutina* (velvet mesquite) treated with ZnO nanoparticles. Chem Eng J 170:346

Hong J, Rico CM, Zhao L, Adeleye AS, Keller AA, Peralta-Videa JR, Gardea-Torresday JL (2015) Toxic effects of copper-based nanoparticles or compounds to lettuce (*Lactuca sativa*) and alfalfa (*Medicago sativa*). Environ Sci: Process Impacts 17:177

Hong J, Peralta-Videa RJ, Rico C, Sahi S, Viveros MN, Bartonjo J, Zhao L, Gardea-Torresdey JL (2014) Evidence of translocation and physiological impacts of foliar applied CeO_2 nanoparticles on cucumber (*Cucumis sativus*) plants. Environ Sci Technol 48:4376

Joseph T, Morrison M (2006) Nanotechnology in agriculture and food www.nanoforum.org

Judy JD, Unrine JM, Rao W, Wirick S, Bertsch PM (2012) Bioavailability of gold nanomaterials to plants: importance of particle size and surface coating. Environ Sci Technol 46:8467

Karuppanapandian T, Wang HW, Prabakaran N, Jeyalakshmi K, Won MK, Manoharan K, Kim W (2011) 2,4-dichlorophenoxyacetic acid-induced leaf senescence in mung bean (*Vigna radiata* L. Wilczek) and senescence inhibition by co-treatment with silver nanoparticles. Plant Physiol Biochem 49:168

Khodakovskaya MV, de Silva K, Nedosekin DA, Dervishi E, Biris AS, Shashkov EV, Galanzha EI, Zharov VP (2011) Complex genetic, photothermal, and photoacoustic analysis of NP-plant interactions. Proc Natl Acad Sci USA 108:1028

Kim JH, Lee Y, Kim EJ, Gu S, Sohn EJ, Seo YS, An HJ, Chang YS (2014) Exposure of iron nanoparticles to *Arabidopsis thaliana* enhances root elongation by triggering cell wall loosening. Environ Sci Technol 48:3477

Kim JH, Oh Y, Yoon H, Hwang I, Chang YS (2015) Iron nanoparticle-induced activation of plasma membrane H + -ATPase promotes stomatal opening in *Arabidopsis thaliana*. Environ Sci Technol 49:1113

Koelmel J, Leland T, Wang H, Amarasiriwardena D, Xing B (2013) Investigation of gold nanoparticles uptake and their tissue level distribution in rice plants by laser ablation-inductively coupled-mass spectrometry. Environ Pollut 174:222

Koo Y, Wang J, Zhang Q, Zhu H, Chehab EW, Colvin VL, Alvarez PJJ, Braam J (2015) Fluorescence reports intact quantum dot uptake into roots and translocation to leaves of arabidopsis thaliana and subsequent ingestion by insect herbivores. Environ Sci Technol 49:626

Kordan HA (1992) Seed viability and germination: a multi-purpose experimental system. J Biol Educ 26:247

Krishnaraj C, Jagan EG, Ramachandran R, Abirami SM, Mohan N, Kalaichelvan PT (2012) Effect of biologically synthesized silver nanoparticles on *Bacopa monnieri* (Linn.) Wettst plant growth metabolism. Process Biochem 47:651

Kumar M, Khan SS, Parakshi S, Mukherjee A, Chandrasekaran N (2011) Cytogenetic and genotoxic effects of zinc oxide nanoparticles on root cells of *Allium cepa*. J Hazard Mater 190:613

Kumar M, Mukherjee A, Chandrasekaran N (2009) Genotoxicity of silver nanoparticles in *Allium cepa*. Sci Total Environ 407:5243

Kumar V, Guleria P, Kumar V, Yadav SK (2013) Gold nanoparticle exposure induces growth and yield enhancement in *Arabidopsis thaliana*. Sci Total Environ 461–462:462

Lahiani MH, Chen J, Irin F, Puretzky AA, Green MJ, Khodakovskaya MV (2015) Interaction of carbon nanohorns with plants: uptake and biological effects. Carbon 81:607

Larue C, Castillo-Michel H, Sobanska S, Cécillon L, Bureau S, Barthès V, Ouerdane L, Carrière M, Sarret G (2014) Foliar exposure of the crop *Lactuca sativa* to silver nanoparticles: evidence for internalization and changes in Ag speciation. J Hazard Mater 264:98

Lee W-M, Kwak JI, An YJ (2012) Effect of silver nanoparticles in crop plants *Phaseolus radiatus* and *Sorghum bicolor*: media effect on phytotoxicity. Chemosphere 86:491

Lopez-Moreno ML, Rosa GD, Ahernandez-Viezcas J, Michel HC, Botez CE, Videa RP, Gardea-Torresdey JL (2010a) Evidence of the differential biotransformation and genotoxicity of ZnO and CeO_2 nanoparticles on soybean (*Glycine max*) plants. Environ Sci Technol 44:7315

Lopez-Moreno ML, Rosa Gdl, Hernandez-Viezcas JA, Perlta-Videa JR, Gardea-Torresdey JL (2010b) X-ray absorption spectroscopy (XAS) corroboration of the uptake and storage of CeO_2 nanoparticles and assessment of their differential toxicity in four edible plant species. J Agric Food Chem 58:3689

Lin D, Xing B (2007) Phytotoxicity of nanoparticles: Inhibition of seed germination and root growth. Environ Pollut 150:243

Lin S, Reppert J, Hu Q, Hudson JS, Reid ML, Ratnikova TA, Rao AM, Luo H, Ke PC (2009) Uptake, translocation and transmission of carbon nanomaterials in rice plants. Small 5:1128

Liu R, Lal R (2014) Synthetic apatite nanoparticles as a phosphorus fertilizer for soybean (*Glycine max*). Sci Rep 4:5686

Liu Y, Laks P, Heiden P (2002a) Controlled release of biocides in solid wood. I. Efficacy against brown rot wood decay fungus (*Gloeophyllum trabeum*). J Appl Polymer Sci 86:596

Liu Y, Laks P, Heiden P (2002b) Controlled release of biocides in solid wood. II. Efficacy against *Trametes versicolor* and *Gloeophyllum trabeum* wood decay fungi. J Appl Polymer Sci 86: 608

Liu Y, Laks P, Heiden P (2002c) Controlled release of biocides in solid wood. III. Preparation and characterization of surfactant-free nanoparticles. J Appl Polymer Sci 86: 615

Liu Q, Chen B, Wang Q, Shi X, Xiao Z, Lin J, Fang X (2009) Carbon nanotubes as molecular transporters for walled plant cells. Nano Lett 9:1007

Lv J, Zhnag S, Luo L, Zhang J, Yang K, Christie P (2015) Accumulation, speciation and uptake pathway of ZnO nanoparticles in maize. Environ Sci: Nano 2:68

Ma Y, Kuang L, He X, Bai W, Ding Y, Zhang Z, Zhao Y, Chai Z (2010) Effects of rare earth oxide nanoparticles on root elongation of plants. Chemosphere 78:273

Majumdar S, Peralta-Videa J, Bandyopadhay S, Castillo-Michel H, Hernandez-Viezcas JA, Sahi SV (2014) Exposure of cerium oxide nanoparticles to kidney bean shows disturbance in the plant defense mechanisms. J Hazard Mater 278:279

Mirzajani F, Askari H, Hamzelou S, Schober Y, Römpp A, Ghassempour A, Spengler B (2014) Proteomics study of silver nanoparticles toxicity on *Oryza sativa* L. Ecotox Environ Safe 108:335

Moon YS, Park ES, Kim TO, Lee HS, Lee SE (2014) SELDI-TOF MS-based discovery of a biomarker in *Cucumis sativus* seeds exposed to CuO nanoparticles. Environ Toxicol Pharmacol 38:922

Mukherjee A, Peralta-Videa JR, Bandyopadhyay S, Rico CM, Zhao L, Gardea-Torresdey JL (2014) Physiological effects of nanoparticulate ZnO in green peas (*Pisum sativum* L.) cultivated in soil. Metallomics 6: 132

Nair PMG, Chung IM (2015) Study on the correlation between copper oxide nanoparticles induced growth suppression and enhanced lignification in Indian mustard (*Brassica juncea L.*). Ecotox Environ Safe 113: 302

Navarro DA, Bisson MA, Aga DS (2012) Investigating uptake of water-dispersible CdSe/ZnS quantum dot nanoparticles by *Arabidopsis thaliana* plants. J Hazard Mater 211–212:427

Navarro E, Baun A, Behra R, Hartmann NB, Filser J, Miao AJ, Quigg A, Santschi PH, Sigg L (2008) Environmental behaviour and ecotoxicity of engineered nanoparticles to algae, plants and fungi. Ecotoxicology 17:372

Nordmann J, Buczka S, Voss B, Hasse M, Mummenhoff K (2015) In vivo analysis of the size- and time-dependent uptake of NaYF4:Yb, Er upconversion nanocrystals by pumpkin seedlings. J Mater Chem B 3:144

Park H-J, Kim SH, Kim HJ, Choi SH (2006) A new composition of nanosized silica-silver for control of various plant diseases. Plant Pathol J 22:295

Patlolla AK, Berry A, May L, Tchounwou PB (2012) Genotoxicity of silver nanoparticles in Vicia faba: a pilot study on the environmental monitoring of nanoparticles. Int J Environ Res Public Health 9:1649

Pavel A, Trifan M, Bara II, Creanga DE, Cotae C (1999) Accumulation dynamics and some cytogenetical tests at *Chelidonium majus* and *Papaver somniferum* callus under the magnetic liquid effect. J Magn Magn Mater 201:443

Peralta-Videa JR, Hernandez-Viezcas JA, Zhao L, Diaz BC, Ge Y, Priester JH, Holden PA, Gardea-Torresdey JL (2014) Cerium dioxide and zinc oxide nanoparticles alter the nutritional value of soil cultivated soybean plants. Plant Physiol Biochem 80:128

Poborilova Z, Opatrilova R, Babula P (2013) Toxicity of aluminium oxide nanoparticles demonstrated using a BY-2 plant cell suspension culture model. Environ Exp Bot 91:1

Pradhan S, Patra P, Mitra S, Dey KK, Jain S, Sarkar S, Roy S, Palit P, Goswami A (2014) Manganese nanoparticles: impact on non-nodulated plant as a potent enhancer in nitrogen metabolism and toxicity study both in vivo and in vitro. J Agric Food Chem 62:8777

Rico CM, Lee SC, Rubenecia R, Mukherjee A, Hong J, Peralta-Videa JR, Gardea-Torresdey JL (2014) Cerium oxide nanoparticles impact yield and modify nutritional parameters in wheat (*Triticum aestivum L.*). J Agric Food Chem 62:9669

Rico CM, Morales MI, Barrios AC, McCreary R, Hong J, Lee W-J, Nunez J, Peralta-Videa JR, Gardea-Torresdey JL (2013) Effect of cerium oxide nanoparticles on the quality of rice (*Oryza sativa L.*) grains. J Agric Food Chem 61:11278

Schwabe F, Schulin R, Limbach LK, Stark W, Burge D, Nowack B (2013) Influence of two types of organic matter on interaction of CeO_2 nanoparticles with plants in hydroponic culture. Chemosphere 91:512

Schwabe F, Tanner S, Schulin R, Rotzetter A, Stark W, Quadtc AV, Nowack B (2015) Dissolved cerium contributes to uptake of Ce in the presence of differently sized CeO_2-nanoparticles by three crop plants. Metallomics 7:466

Solgi M, Kafi M, Taghavi TS, Naderi R (2009) Essential oils and silver nanoparticles (SNP) as novel agents to extend vase-life of gerbera (*Gerbera jamesonii* cv. 'Dune') flowers. Postharvest Biol Technol 53:155

Speranza A, Leopold K, Maier M, Taddei AR, Scoccianti V (2010) Pd-nanoparticles cause increased toxicity to kiwifruit pollen compared to soluble Pd(II). Environ Pollut 158:873

Spurrier EC, Jackobs JA (1955) Effects of an anionic sodium sulfonate type surfactant upon plant growth. Agron J 47:462

Srivastava G, Das CK, Das A, Singh SK, Roy M, Kim H, Sethy N, Kumar A, Sharma RK, Singh SK, Philip D, Das M (2014) Seed treatment with iron pyrite (FeS_2) nanoparticles increases the production of spinach. RSC Adv 4:58495

Stampoulis D, Sinha SK, White JC (2009) Assay dependent phytotoxicity of nanoparticles to plants. Environ Sci Technol 43:9473

Su Y, Qi L, Mu X, Wang M (2015) A fluorescent probe for sensing ferric ions in bean sprouts based on L-histidine-stabilized gold nanoclusters. Anal Methods 7:684

Rao S, Shekhawat GS (2014) Toxicity of ZnO engineered nanoparticles and evaluation of their effect on growth, metabolism and tissue specific accumulation in *Brassica juncea*. J Environ Chem Eng 2:105

Temple RE, Hilton HW (1963) The effect of surfactants on the water solubility of herbicides and the foliar phytotoxicity of surfactants. Weeds 11:297

Thuesombat P, Hannongbua S, Akasit S, Chadchawan S (2014) Effect of silver nanoparticles on rice (*Oryza sativa L.* cv. KDML 105) seed germination and seedling growth. Ecotox Environ Safe 104: 302

Torney F, Trewyn BG, Lin VSY, Wang K (2007) Mesoporous silica nanoparticles deliver DNA and chemicals into plants. Nat Nanotechnol 2:295

Torre-Roche RDL, Hawthorne J, Deng Y, Xing B, Cai W, Newman LA, Wang C, Ma X, White JC (2012) Fullerene-enhanced accumulation of p, p′-DDE in agricultural crop species. Environ Sci Technol 46:9315

Torre-Roche RDL, Hawthorne J, Musante C, Xing B, Newman LA, Ma X, White JC (2013) Impact of Ag nanoparticle exposure on p, p′-DDE bioaccumulation by Cucurbita pepo (Zucchini) and Glycine max (Soybean). Environ Sci Technol 47:718

Uzu G, Sobanska S, Sarret G, Munoz M, Dumat C (2010) Foliar lead uptake by lettuce exposed to atmospheric pollution. Environ Sci Technol 44:1036

Wang Q, Ma X, Zhang W, Pei H, Chen Y (2012) The impact of cerium oxide nanoparticles on tomato (*Solanum lycopersicum L.)* and its implications for food safety. Metallomics 4: 1105

Wierzbicka M, Obidzinska J (1998) The effect of lead on seed imbibition and germination in different plant species. Plant Sci 137:155

Yeo M-K, Nam D-H (2013) Influence of different types of nanomaterials on their bioaccumulation in a paddy microcosm: a comparison of TiO_2 nanoparticles and nanotubes. Environ Pollut 178:166

Zhai G, Walters KS, Peate DW, Alvarez PJJ, Schnoor JL (2014) Transport of gold nanoparticles through plasmodesmata and precipitation of gold ions in woody poplar. Environ Sci Technol Lett 1:146

Zhao L, Peratla-Videa JR, Peng B, Bandyopadhyay S, Corral-Diaz B, Osuna P, Montes MO, Keller AA, Gardea-Toreesday JL (2014a) Alginate modifies the physiological impact of CeO_2 nanoparticles in corn seedlings cultivated in soil. J Environ Sci 26:382

Zhao L, Peralta-Videa JR, Rico CM, Hernandez-Viezcas JA, Sun Y, Niu G, Servin A, Nunez JE, Duarte-Guardea M, Gardea-Torresdey JL (2014b) CeO_2 and ZnO nanoparticles change the nutritional qualities of cucumber (*Cucumis sativus*). J Agric Food Chem 62:2752

Zhao L, Sun Y, Hernandez-Viezcas JA, Hong J, Majumdar S, Niu G, Duarte-Guardea M, Peralta-Videa JR, Gardea-Torresday JL (2015a) Monitoring the environmental effects of CeO_2 and ZnO nanoparticles through the life cycle of corn (Zea mays) plants and in situ μ-XRF mapping of nutrients in kernels. Environ Sci Technol 49:2921

Zhao S, Wang Q, Zhao Y, Ruia Q, Wang D (2015b) Toxicity and translocation of graphene oxide in *Arabidopsis thaliana*. Environ Toxicol Pharmacol 39:145

Zhang D, Hua T, Xiao F, Chen C, Gersberg RM, Liu Y, Ng WJ, Tan SK (2014) Uptake and accumulation of CuO nanoparticles and CdS/ZnS quantum dot nanoparticles by *Schoenoplectus tabernaemontani* in hydroponic mesocosms. Ecol Eng 70:114

Zhang D, Hua T, Xiao F, Chen C, Gersberg RM, Liu Y, Stuckey D, Ng WJ, Tan SK (2015) Phytotoxicity and bioaccumulation of ZnO nanoparticles in *Schoenoplectus tabernaemontani*. Chemosphere 120:211

Zhu Z-J, Wang H, Yan B, Zheng H, Jiang Y, Miranda OR, Rotello VM, Xing B, Vachet RW (2012) Effect of surface charge on the uptake and distribution of gold nanoparticles in four plant species. Environ Sci Technol 46:12391

Chapter 5
Effects of Nanoparticles on Plant Growth and Development

Remya Nair

Abstract Nanomaterials provide great opportunities in the field of agriculture because of their unique physicochemical properties. The interaction of nanoparticles with plants results in several physiological, morphological, and genotoxic changes, and their understanding is important for the effective use of nanotechnology in agriculture. Researchers suggested both positive and negative responses of nanoparticles on plant growth and development depending upon the properties of nanomaterials, mode of application as well as plant species. Studies on the uptake, translocation and biotransformation, and risks of application of nanomaterials on agriculturally important crops are recent research focus for understanding the physiological, biochemical, and molecular mechanisms of plants in relation to nanoparticles.

Keywords Nanoparticles • Plants • Uptake • Translocation • Growth • Phytotoxicity

5.1 Introduction

The increased use of engineered nanoparticles (ENPs) in different fields results in their accidental release to terrestrial, aquatic, and atmospheric environments. Plants are important component of all ecosystems, and interaction of ENPs with plants results in uptake and accumulation into plant biomass and decides the fate and transport of nanoparticles (NPs) in the environment. Recently, an increased number of studies have been highlighting the potential effects of ENPs to plants (Ruffini and Roberto 2009; Nair et al. 2010; Xingmao et al. 2010; Karl-Josef and Simone 2011). NPs get adsorbed on different plant surfaces, and their subsequent uptake occurs

R. Nair (✉)
Nano Research Facility, Washington University in St. Louis, One Brookings Drive, St. Louis, MO 63130, USA
e-mail: rnair@wustl.edu; remya.d.nair@gmail.com

C. Kole et al. (eds.), *Plant Nanotechnology*,
DOI 10.1007/978-3-319-42154-4_5

through micrometer- and nanometer-scaled plant openings. NPs can get into plant body using different paths, and the uptake rate depends on the size, shape, concentration, and surface charge of NPs (Tarafdar et al. 2012). The aerial part facilitates the interaction of airborne NPs on shoot surfaces. Also, there is a chance for the portion of engineered NPs released into environment to get dispersed by wind and reaching the leaves of plants (Espinosa and Oliva 2006). Plants interact with atmospheric NPs through the leaves, and hence, leaf stomata and hydathodes play an important role in the foliar uptake of NPs. The stomatal pathway is highly capacitive due to its large size exclusion limit above 10 nm and its high transport velocity; however, the chances of variability in permeability make this pathway highly unpredictable (Eichert et al. 2008). Sometimes, only very less percentage of stomata contributes to the uptake process which can be increased by the repeated wetting and drying of a foliar-applied solution. The NPs that penetrate leaf surface through the stomata or hydathodes traverse cell walls of palisade parenchyma and reach the leaf phloem and translocate to other plant parts. NPs can also get deposited on the cell walls of substomatal cavity or nearby cells. Other pathways for association of nanomaterials with aerial parts include cuticle, bark surfaces, and stigma. The epicuticular structures increase the deposition of NPs in epicuticular cavities. Also, the trichomes on shoot surfaces provide greater chance of NP deposition. The size of particles and its concentration play an important role in the uptake and distribution of NPs in plant system. Small lipophilic NPs can be taken up into apolar fluid areas of the cuticle that contains both apolar and polar uptake pathways. The uptake of larger particles occurs through cuticle-free areas such as stomata, hydathodes, and stigma of flowers.

The soil NPs interact with root system of plants, and the adsorption of NPs on plant roots facilitates their incorporation into cell wall and further uptake into cell. Most of the primary roots have a suberized exodermis and endodermis, and the apoplastic bypass flow of solutes and water from the soil to the central cylinder is prevented by suberized exodermis. However, apoplastic bypass is possible with the newly formed lateral roots that break through the cortex. Hence, the NPs could enter at the region of lateral roots formation into the xylem through the cortex and the central cylinder. The plant cells with negative surface charge permit the transport of negative charge surface compounds into apoplast (Nowack and Bucheli 2007). It was hypothesized that the negatively charged NPs could enter the apoplasm of the root cortex and eventually into xylem, but they are not taken up by the cells. Compounds can also enter into xylem through wounded cells or holes and can be further transported to shoots. This has been reported as a dominant process for the uptake of metal complexes and their subsequent translocation to shoots (Nowack et al. 2006; Tandy et al. 2006). This chapter reviews the studies on the interaction of engineered NPs with different plant species, uptake, translocation, and its phytotoxic effects with special mention to germination effects and seedling parameters.

5.2 Effects of Different Nanoparticles in Plant System

5.2.1 Effects of Metallic Nanoparticles

Metallic NPs have great potential in nanotechnology and have wide range of applications in engineering and biomedical sciences, and hence, it is important to study the fate and long-term effects of these nanomaterials on the environment. There are different ways in which the metal NPs can be taken up by plants. They can either be imported as NPs itself or the metal NPs can be oxidized to metal ions in soil solution and imported as ions followed by their reduction in plant system (Rico et al. 2011).

5.2.1.1 Effects of Gold Nanoparticles

The emergence of improved technologies with gold NPs provides great promise for future applications in various fields. However, the interaction of these NPs with biological systems creates new concerns on toxicological effects due to their unique physiochemical properties. Even though gold is not an essential nutrient for plant growth, there is greater chance for the contamination of soils with gold due to their increased application and hence greater exposure of plants to gold at significantly higher levels.

Understanding the interaction of NPs with plants is important for accessing their toxicity. Zhu et al. (2012) studied the effects of surface charge on the uptake and distribution of gold NPs in four different plant species, rice (*Oryza sativa*), radish (*Raphanus sativus*), pumpkin (*Cucurbita mixta* cv. white), and ryegrass (*Lolium perenne*). Plant seedlings were exposed to NP solution hydroponically for 5 days, and it was observed that the uptake and distribution of NPs were dependent on the surface charge and plant species in which the positively charged NPs were taken up by the plant roots and negatively charged NPs were translocated efficiently to plant shoots. Higher accumulation of NPs was observed in ryegrass and radish roots than in rice and pumpkin roots. The effects of NP properties on the uptake and translocation showed a strong interaction between NPs and cellular biomolecules. The uptake of NPs is also size selective as it was reported that the gold NP aggregates were observed in root cytoplasm of tobacco (*Nicotiana xanthi*) when exposed to 3.5-nm gold nanospheres and not when exposed to 18-nm gold NPs (Sabo-Attwood et al. 2012). Seedling toxicity experiments were also carried out using Arabidopsis (*Arabidopsis thaliana* L.) plants with different concentrations of $KAuCl_4$ (Taylor et al. 2014). It was observed that the germination was not affected by any of the concentrations of gold, whereas root growth was inhibited with increased $KAuCl_4$ concentration. The physiological and genetic responses of Arabidopsis to gold NPs (AuNPs) were also investigated. It was reported that the root lengths of the seedlings grown on nutrient agar media with 100 ppm AuNPs were reduced by 75 %. Roots and shoots showed the presence of oxidized gold, whereas reduced gold as NPs were observed only in root tissues.

The growth profile and yield of Indian mustard (*Brassica juncea*) on treatment with five different concentrations of AuNPs by foliar spray were studied under field conditions (Arora et al. 2012). It was reported that the germination percentage of Indian mustard was improved up to treatment with 25 ppm Au NPs than the control seeds and it could be due to the increased seed capsule permeability, which in turn allowed more water and di-oxygen into the cells. An increase in the number of leaves per plant was also observed with the treatment of AuNPs, and maximum increase was observed on treatment with 10 ppm AuNPs. It was also observed that the average leaf area was not increased with treatment of NPs. An improved growth profile was occurred for treated plants with increased plant height, average stem diameter, and number of branches per plant, and maximum increase was reported for treatment with 10 ppm AuNPs. An increased average yield with increased number of pods per plant was also reported on NP treatment.

A few works have been reported with tobacco (*Nicotiana tabacum* cv. xanthi) and wheat (*Triticum aestivum*) plants describing their interaction with gold nanoparticles. Tobacco and wheat seedlings were treated hydroponically with 10-, 30-, and 50-nm-diameter gold nanomaterials coated with tannate or citrate for 3 or 7 days for wheat and tobacco, respectively, for studying the bioavailability of NPs with different size and surface chemistries to plants (Judy et al. 2012). This study provided information on the role of plant cell wall pores in the uptake of NPs and also investigated the influence of soil components on the intrinsic properties of nanomaterials and their further transport to plants.

5.2.1.2 Effects of Silver Nanoparticles

Silver has wide range of industrial applications as well as for medical and antibacterial purposes and hence more likely to get exposed to humans and environment which highlights the needs to study the effects of silver nanoparticles (AgNPs) in plants. The phytotoxicity of AgNPs (29 nm) on cucumber (*Cucumis sativus*) and lettuce (*Lactuca sativa*) was studied with seed germination tests, and a reduced effect on germination index was reported for cucumber seeds (Barrena et al. 2009). For lettuce seeds, the germination index seemed to be comparable with the controls. In the same study, a positive effect was observed with AuNPs of size 10 nm. The impact of AgNPs on the root elongation of greenhouse-grown radish and lettuce with barley (*Hordeum vulgare*) as the reference plant was investigated under hydroponics and soil conditions (Gruyer et al. 2014). Under hydroponics condition, a positive response on root elongation was observed in barley at low concentration of AgNPs, whereas a significant reduction in root length was observed on treatment with higher concentration of AgNPs. For lettuce, a reduction in root length was observed, and for radish, no significant variation had been reported on treatment. No negative effects were seen on treatment with AgNPs for the root length of all the three plants exposed to soil. The growth parameters of common bean (*Phaseolus vulgaris*) and maize (*Zea mays*) were studied with different concentrations of AgNPs in which enhanced growth was observed at low concentrations and inhibitory effects with higher concentrations of NPs (Salama 2012). An increased root length, shoot length, and chlorophyll content were reported up to treatment with 60 ppm AgNPs after which declined growth parameters and chlorophyll content for higher concentrations. Similar observations were also reported for mung bean (*Vigna radiata*) and sorghum (*Sorghum bicolor*) plants on treatment with AgNPs (Namasivayam and Chitrakala 2011). Studies on the phytotoxic effects of AgNPs on rice plants reported uptake of NPs through roots, thus resulted in intracellular damage (Mazumdar and Ahmed 2011). The effects of AgNPs of three different sizes ranging from 1 to 20 nm and concentration ranging from 1 to 100 ppm on the germination of ryegrass, barley, and flax (*Linum usitatissimum*) were studied, and differently sized NPs affected differently on plant species. The smallest sized particle had an inhibitory effect on ryegrass even at very low concentration, whereas the intermediately sized particles had less inhibitory effects at low concentration but greater at higher concentration particularly with barley. Flax seeds were not at all affected by any type and/or any concentrations of AgNPs (El-Temsah and Joner 2010). Since different types of plant species behaved differently to same type of NPs of different size and concentration, seed germination tests solely cannot be relied for the analysis of environmental impact of AgNPs. Size-dependent toxicity studies of AgNPs were also carried out with Italian ryegrass (*Lolium multiflorum*), and it was reported that smaller AgNPs significantly reduced the growth with shorter roots and shoots and less biomass as compared to plants treated with larger NPs of similar concentration (Yin et al. 2011). This showed that the toxicity of AgNPs is highly influenced by total NP surface area. This study also reported that on exposing the seedlings to 40 ppm of gum arabic (GA)-coated AgNPs, the seedlings failed to develop root hairs with vacuolated and collapsed cortical cells and broken root cap which might be due to the loss of gravitropism in roots due to reduced auxin transport. Studies on the phytotoxicity of AgNPs with mung bean and sorghum reported adverse effects on seedling growth (Lee et al. 2012). Experiments were carried out both in agar and in soil media for highlighting the importance of media effect in nanotoxicity. Nanoparticle concentration-dependent growth inhibition was observed for both the plants in agar media; however, mung bean plants were not affected much in soil media, whereas a slight reduction in growth rate was observed for sorghum plants, which clearly showed the influence of exposure media on dissolution, bioavailability, and phytotoxicity of NPs. Wang et al. (2013a, b) reported stimulatory effects on root elongation, fresh weight, and evapotranspiration of poplars (*Populus tremula*) and Arabidopsis on using a narrow range of sublethal concentration of AgNPs, whereas above a certain level of concentration all forms of silver were phytotoxic and accumulation of silver in different plant parts varied with the plant species.

Developmental responses of maize and cabbage (*Brassica oleracea*) exposed to citrate-coated AgNPs and zinc oxide (ZnO) NPs were evaluated, and comparative toxicity profiles were developed with their corresponding ionic salts (Pokhrel and Dubey 2013). The toxicity due to NPs on the germination and root elongation of both maize and cabbage was observed to be lesser than that occurred due to free ions. Anomalies were observed in maize root anatomy on treatment with citrate nano-silver and nano-ZnO, and this confirmed that the NPs might cause different biological interactions and their toxicity outcome is different from those occurred due to their specified ion effect. Uptake of silver by maize seedlings was found to be higher for treatment with $AgNO_3$ than with citrate nano-silver. Changes in the gene expression of plants on treatment with AgNPs and silver ions were also studied with Arabidopsis for analyzing the molecular mechanism of plant response to different contaminants. Both upregulation and downregulation of significant genes had been observed in response to AgNPs and silver ions which are associated with several stress responses in plants in which the upregulated genes are connected with metal and oxidative stresses and downregulated genes to pathogen and hormonal stimuli (Kaveh et al. 2013). These studies highlight the need for more studies to get enough information for better explaining the toxicity differences between metal-based NPs and their free ions. Effects on plant physiological responses and gene expression with NPs of different morphology were also investigated with AgNPs of triangle, spherical, and decahedral shape on Arabidopsis plants (Syu et al. 2014). This study reported the highest degree of root growth promotion and lowest accumulation of Cu/Zn super oxide dismutase with decahedral AgNPs, whereas the spherical AgNPs induced null effect on root growth promotion with highest level of Cu/Zn superoxide dismutase and anthocyanin accumulation. Protein accumulations were induced by all three morphologies of AgNPs and also activated gene expression in Arabidopsis. The phytotoxic and genotoxic effects of AgNPs on germinating wheat seedlings were also investigated, and negative effects were observed with higher concentrations of AgNPs which resulted in reduced seedling growth and morphological variations in root tip cells (Vannini et al. 2014). This study also suggested that the toxicity of AgNPs had come from the release of silver ions from AgNPs, which supported many of the above-mentioned reports. No DNA polymorphism was observed in wheat seedlings with NP treatment in the studied range of concentrations, whereas variations were observed in the expression of several proteins controlling primary metabolism and defense mechanisms. The fate and transport of AgNPs in the aquatic environment was studied by investigating their effects on aquatic plants. An important study was conducted with two different types of AgNPs, PVP-AgNPs (polyvinylpyrrolidone-coated silver nanoparticles), and GA-AgNPs (gum Arabic-coated AgNPs), on a mixed wetland plant community (Yin et al. 2012). Direct exposure and soil exposure experiments were carried out, and magnitude of variation on germination rate was more pronounced under direct exposure with higher concentration of GA-AgNPs treatment. The response of plant growth with NP treatment also varied with plant species. Higher rate of growth inhibition with increased concentrations of AgNPs was also reported for the aquatic plant swollen duckweed (*Lemna gibba*) which demonstrated toxicity due to the accumulation of AgNPs in the aquatic environment (Oukarroum et al. 2013).

5.2.2 *Effects of Metal Oxide Nanoparticles in Plants*

5.2.2.1 Effects of TiO_2, ZnO, SiO_2, Al_2O_3, CeO_2, and CuO Nanoparticles

The extensive production of TiO_2 worldwide resulted in their greater release to the environment with consequent contamination of plants, soils, and water systems. The end of life cycle of NPs was simulated, and it was reported that the greatest concerns were associated with the use of TiO_2 NPs and their fate and effects on ecosystem. Several studies reported the impact of TiO_2 NPs on plant development and physiology with contradictory results. The size-dependent distribution of TiO_2 NPs in wheat plants was reported by Larue et al. (2012a, b). It was suggested that NPs had not accumulated in plant roots above a threshold diameter of 140 nm and the size was limited to 36 nm for easy translocation to shoots from roots. It was also reported that the accumulation of NPs was not affecting the germination of seeds and or total biomass of plants. The same group also studied the effects of TiO_2 NPs on wheat and rapeseed (*Brassica napus*) plantlets grown under hydroponic conditions (Larue et al. 2012a, b). The plants were exposed to NPs both through root and through leaf exposure. Nanoparticle agglomerates were observed in plantlets under both root and leaf exposure methods, and higher Ti amount was found in rapeseed than in wheat. An increased root elongation was observed at early development stages on treatment with TiO_2 NPs; however, null effect on germination, evapotranspiration, and total plant biomass was observed. An increase in the catalase amount and decrease in the ascorbate peroxidase were reported on growing cucumber plants in sandy loam soils treated with 750 and 500 mg/kg of TiO_2, respectively (Servin et al. 2013). Translocation of TiO_2 from roots to fruits without biotransformation was also reported in this study. Enhanced seed germination and vigor of wheat were reported by employing proper concentrations of nano-TiO_2, whereas inhibitory effects were reported with bulk TiO_2 and neutral effects for higher concentration of nano-TiO_2 (Feizi et al. 2012). Larue et al. (2011) also studied the effects of TiO_2 NPs on wheat, rapeseed, and Arabidopsis, and results showed the uptake of NPs by plants. They also reported that the germination rate and root elongation were not affected with the studied type of NPs. Hence, the effects of titanium on plants show significant size, concentration, and species dependence. Studies also investigated the effects of TiO_2 NPs on agronomic traits such as plant height, ear weight, ear number, seed number, final yield, biomass, gluten, and starch content under water-deficit conditions, and improved agronomic traits were reported for TiO_2 NPs at 0.02 % (Jaberzadeh et al. 2013). In maize plants, the changes of photosynthetic pigments upon nano-TiO_2 spraying at different stages of plant growth and development were studied, and it was observed that nano-TiO_2 has significant effects of total content of chlorophyll, carotenoids, and anthocyanin and maximum effect was recorded on spraying nano-TiO_2 at reproductive stage of the plants (Morteza et al. 2013).

Du et al. (2011) studied the effects of TiO_2 and ZnO NPs on wheat growth and soil enzyme activities. The wheat plants were harmed by both the NPs with reduced plant biomass and affected soil environment by decreasing the activity of some of the soil enzymes such as protease, catalase, and peroxidase. The mechanism of toxicity induced by both the NPs is different. The TiO_2 NPs were retained in soil for longer time and hence greater probability to get attached to root cell wall, thus caused changes in the microenvironment and generated reactive oxygen species (ROS), which might damage cells. For ZnO NP treatment, greater uptake of Zn by plants had occurred due to high solubility of ZnO NPs in soil. The effects of nano-TiO_2 and nano-ZnO on rice seed germination showed no reduction in the percent seed germination from both NPs; however, nano-ZnO produced some detrimental effects on rice roots at early seedling stage with stunted and reduced number of roots (Boonyanitipong et al. 2011). No negative effects were observed with nano-TiO_2 on root growth. On exposing the tomato (*Solanum lycopersicum*) plants to TiO_2 NPs and Fe_3O_4 NPs in the same experiment, an abnormal proliferation of root hairs was observed for seedlings exposed to higher concentration of TiO_2 NPs when compared to seedlings exposed to Fe_3O_4 NPs and control seedlings. No changes in shoot morphology and no seedling toxicity were observed on NP treatment in this experiment (Giordani et al. 2012).

Arabidopsis is a model plant to study the developmental phytotoxicity of metal oxide NPs. Four different metal oxides such as aluminum oxide (nAl_2O_3), silicon dioxide ($nSiO_2$), magnetite (nFe_3O_4), and zinc oxide (nZnO) have been selected for studying their effects on three important toxicity indicators such as seed germination, root elongation, and number of leaves (Lee et al. 2010). nZnO was found to be the most phytotoxic metal oxide NP followed by nFe_3O_4, $nSiO_2$, and nAl_2O_3. A significant reduction in germination was observed for nZnO NPs which could be related to their small size, monodispersity, and greater solubility facilitating their transport to intracellular spaces through seed coat pores. The greater particle size of other metal oxide NPs limited their inhibitory effect on seed germination. This study highlighted the significance of elemental composition and particle diameter on developmental phytotoxicity. Increased root elongation was observed with nAl_2O_3 in all the tested concentrations, whereas inhibitory effects were observed with other metal oxide NPs except for low concentration of $nSiO_2$. Hence, the same type of NP could cause different effects on same plant species with varied concentrations as reported in case of $nSiO_2$. This is one of the first reports for positive effects of nAl_2O_3 on plant growth, whereas all previous reports highlighted their negative or neutral effects on plant growth (Yang and Watts 2005; Lin and Xing 2007). The degree to which NPs affect seed germination was also studied with cucumber seeds using Fe_3O_4, TiO_2, and carbon nanoparticles up to 5000 μg/ml, and inhibitory effects were observed with reduced root elongation especially at higher concentrations (Mushtaq 2011). Studies on the effects of nanoscale ZnO on the germination, seedling vigor, plant growth, flowering, chlorophyll content, pod yield, and root growth of the peanut (*Arachis hypogea*) plants showed positive effects on all the studied parameters at a concentration of 1000 ppm ZnO and inhibitory effects at higher concentrations of 2000 ppm which revealed the judicious usage of these particles on plants (Prasad et al. 2012). The effects of ZnO NPs and Zn^{2+} ions were also studied in alfalfa (*Medicago sativa*), tomato, and cucumber (*Cucumis sativus*) (De La Rosa et al. 2013). It was noticed that at higher concentrations of 1600 ppm of ZnO NPs, there was an increase in cucumber germination by 10 %, whereas tomato and alfalfa germination were reduced by 20 and 40 %, respectively. There was wide variation in the toxicity level of each plant species to different concentrations of NPs and difference in plant species might be the reason for differences in NPs uptake, tolerance, and toxicity. This study highlighted the phytotoxicity and uptake of ZnO NPs and Zn^{2+} ions by plants and also investigated the possible ZnO and Zn^{2+} biotransformation in plant tissues (Figs. 5.1 and 5.2; Table 5.1).

Mukherjee et al. (2014) also studied the physiological effects of ZnO NPs in soil cultivated green peas. Toxicological effects were measured by analyzing various parameters such as plant growth, Zn accumulation, chlorophyll production, activity of stress enzymes. An increased root elongation was observed for all studied concentrations of ZnO NPs, whereas the shoot growth was unaffected. The chlorophyll content in the leaves got reduced on treatment with ZnO NPs compared to the

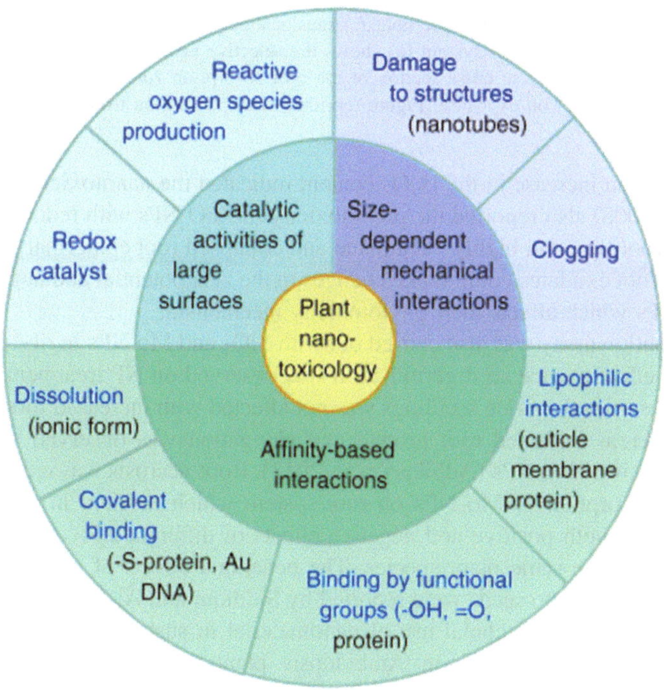

Fig. 5.1 Categorization of nanoparticle-dependent toxicity mechanisms. Different shades of the same color indicate related mechanisms subsumed to (*i*) size-dependent mechanical interactions; (*ii*) catalytic activities of large surfaces; and (*iii*) affinity-based interactions. Abbreviation: Au, gold [Reprinted with permission from Karl-Josef and Simone (2011)]

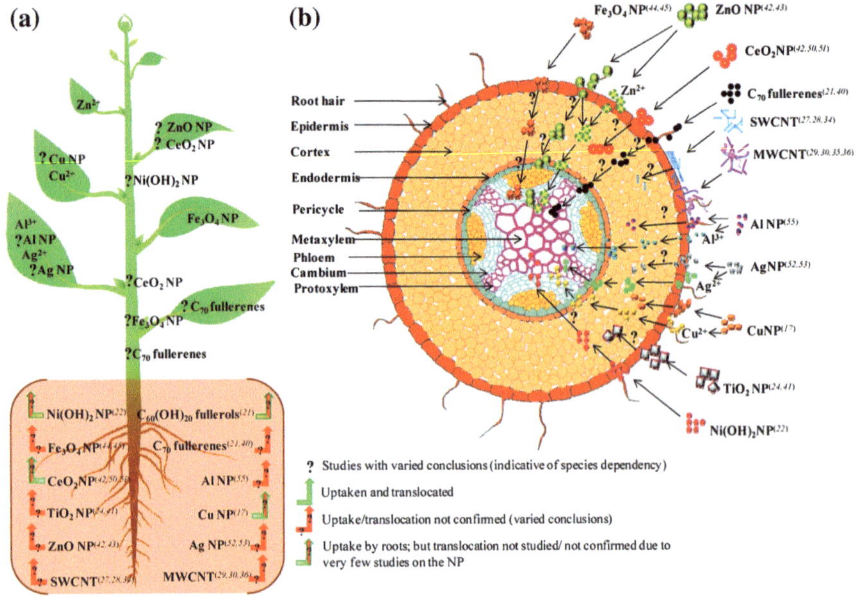

Fig. 5.2 Schematic representation of uptake, translocation, and biotransformation pathway of different nanoparticles in plant system (**a**) shows the selective uptake of nanoparticles by plants and (**b**) shows the transverse cross section of the root absorption zone showing the differential nanoparticle interaction on exposure [Figure reprinted with permission from Rico et al. (2011)]

controls and an increase in the H_2O_2 content indicated the nanotoxicity effects. Lin and Xing (2008) also reported the phytotoxicity of ZnO NPs with reduced biomass, shrunken root tips, and highly vacuolated and collapsed root epidermal and cortical cells. The root exudates could make changes in the zeta potential and aggregate size of ZnO NPs which affects their phytotoxicity mechanism.

Germination assay was also carried out with SiO_2 and Mo NPs in rice plants, and a positive effect on rice seed germination was observed on NP treatment. For SiO_2 NPs, further growth of rice seedlings was not affected with increased root and shoot length, whereas inhibited root growth and elongation was observed with higher concentrations of Mo NPs (Adhikari et al. 2013). Root necrosis was reported due to increased adsorption of Mo NPs on root system which resulted in toxicity. This study showed both positive and negative effects of different NPs at different concentrations on the same plant species. The beneficial effects of nano-SiO_2 on the germination of tomato seeds were reported by Siddiqui and Al-Whaibi (2014). This study supported the beneficial use of nanomaterial in sustainable agriculture. The phytotoxicity of SiO_2 NPs on Arabidopsis plants grown hydroponically was studied, and SiO_2 NPs did not exhibit severe phytotoxicity except for pH-dependent phytotoxicity with reduced development and chlorosis for plants exposed to NPs with high negative zeta potential. Accumulation of NPs was observed in Arabidopsis root cells in a size-dependent manner (Slomberg and Schoenfisch

Table 5.1 Phytotoxicity of important NPs on some major food and agricultural crops

Crops	Type of nanoparticles	Effects on plants	Reference
Mustard greens	AuNPs	Improved germination rate (with 25 ppm AuNPs) and improved total growth profile (at 10 ppm AuNPs)	Arora et al. (2012)
Arabidopsis	AuNPs	Reduced root length	Taylor et al. (2014)
Barley, Ryegrass	Ag NPs	Reduction in germination and shoot length with small-sized particles	El-Temsah and Joner (2010)
Cucumber	AgNPs	Reduced germination	Barrena et al. (2009)
Lettuce, Barley	AgNPs	Reduced root length for lettuce Increased root elongation for barley at low concentration AgNPs and reduced root length at higher concentration AgNPs	Gruyer et al. (2014)
Common bean, Corn, Mung bean, and Sorghum	AgNPs	Improved growth parameters at low concentration and inhibitory effects at high concentration	Salama (2012), Namasivayam and Chitrakala (2011)
Mung bean, Sorghum	AgNPs	Reduced seedling growth	Lee et al. (2012)
Onion, Tomato, and Radish	N-TiO_2	Positive effect on germination at 100 and 200 ppm N-TiO_2	Stampoulis et al. (2009)
Wheat	TiO_2	Not affecting germination and total biomass	Larue et al. (2012a, b)
Wheat	TiO_2 and ZnO	Reduced plant biomass	Du et al. (2011)
Corn	Al_2O_3	Reduced root length	Lin and Xing (2007)
Carrot, Cabbage, Cucumber, and Maize	Al_2O_3	Reduced root growth	Yang and Watts (2005)
Arabidopsis	Al_2O_3	Increased root elongation	Lee et al. (2010)
Soybean	ZnO	Reduced growth	Yoon et al. (2014)
Peanut	ZnO	Improved germination and seedling vigor at 1000 ppm and inhibitory effects at 2000 ppm	Prasad et al. (2012)
Tomato, Alfalfa	ZnO	Reduced germination	de la Rosa et al. (2013)
Green peas	ZnO	Increased root elongation	Mukherjee et al. (2014)
Rice	SiO_2	Improved seed germination and seedling growth	Adhikari et al. (2013)

(continued)

Table 5.1 (continued)

Crops	Type of nanoparticles	Effects on plants	Reference
Tomato	CeO_2	Promoted plant growth and fruit maturity at low concentrations	Wang et al. (2012a, b)
Arabidopsis	CeO_2	Reduced chlorophyll at higher concentration	Ma et al. (2013)
Cucumber	CeO_2, ZnO	No negative impact on whole-plant life cycle	Zhao et al. (2013)
Maize	CuO	No inhibition on seed germination	Wang et al. (2012a, b)
Sunflower	Fe_3O_4	Reduced photosynthetic pigments	Ursache-Oprisan et al. (2011)
Soybean	SPIONs	Increased chlorophyll content	Ghafariyan et al. (2013)
Tomato	CNTs	Increased number of flowers and fruits	Khodakovskaya et al. (2013)
Rice	SWCNTs, MWCNTs, C_{60}	Improved seed germination, water uptake, healthier seedlings	Nair et al. (2012)
Barley, Soybean	MWCNTs	Improved seed germination	Lahiani et al. (2013)
Onion, Cucumber	Poly-3-amino benzenesulfonic acid Functionalized SWCNTs	Enhanced root elongation	Canas et al. (2008)
Lettuce	Poly-3-amino benzenesulfonic acid Functionalized SWCNTs	Inhibited root elongation	Canas et al. (2008)
Maize	MWCNTs	Improved growth	Tiwari et al. (2014)
Arabidopsis, Rice protoplasts	SWCNTs	Programmed cell death	Shen et al. (2010)
Maize	C_{60} Fullerenes	Reduced biomass	Torre-Roche et al. (2013)
Bitter melon	Fullerol	Increased biomass, fruit yield, and improved phytomedicines content	Kole et al. (2013)
Zucchini	MWCNTs	Reduced biomass	Stampoulis et al. (2009)
Lettuce	MWCNTs	Reduced root length	Lin and Xing (2007)

2012). Studies also reported increased germination percentage, dry weight, silica accumulation, and nutrient alleviation in seeds exposed to nano-SiO_2 under hydroponic conditions (Suriyaprabha et al. 2012). This highlighted the use of nano-SiO_2 as a highly utilizable source for plants.

The biotransformation of ZnO and CeO_2 nanoparticles on soybean was investigated, and its effects on germination and seedling growth along with the impact on DNA stability were studied (Lopez-Moreno et al. 2010a, b). Soybean germination was not affected by either of the NPs except at 2000 mg/L of CeO_2. However, both the NPs differentially affected the elongation of roots. Increased root elongation was observed for treatment with CeO_2 NPs, whereas treatment with ZnO NPs showed maximum elongation at 500 mg/L and minimum at 4000 mg/L. This study also reported the uptake of both NPs with the highest Zn accumulation occurred at 500 mg/L ZnO NPs and Ce accumulation increased with increase in the external concentration of CeO_2 NPs. The phenotypic response of tomato plants from seed germination to fruit maturity to low concentrations of CeO_2 NPs were documented, and the results indicated that the CeO_2 NPs at the studied concentrations promoted plant growth and fruit production (Wang et al. 2012a, b). However, translocation of Ce from roots to shoots and fruits presented a high level of risk to human health through dietary exposure. Uptake and toxicity studies of nano-ceria on alfalfa, maize, cucumber, and tomato plants reported reduced germination rate of maize, tomato, and cucumber at higher concentrations of nano-ceria (Lopez-Moreno et al. 2010a, b). It was also observed that the root growth was promoted by nano-ceria in cucumber and maize, whereas reduced in alfalfa and tomato and nano-ceria promoted shoot elongation in all studied plant species with all studied concentrations.

It is important to analyze the phytotoxic effects of NPs through foliar application also as most of the nanotoxicity studies are focusing on root exposure to NPs. Hong et al. (2014) studied the translocation and physiological impacts of foliar-applied CeO_2 NPs on cucumber plants. Hydroponically grown cucumber plants were treated with nano-ceria powder through aerial application. Cerium was detected in different tissues of treated plants suggesting the translocation of Ce from leaves to other plant parts. This is an important study which showed that atmospheric NPs could be taken up by plants which poses a threat to environment and health. However, another study reported that there was no evidence of translocation of CeO_2 NPs in maize plants as no nanoparticles were detected in the newly grown leaves of already treated plants. This suggests that the biological barriers of plants are more resistant toward easy entry and translocation of NPs than the mammalian barriers (Birbaum et al. 2010). Ma et al. (2013) investigated the physiological and molecular responses of CeO_2 and indium oxide (In_2O_3) NPs on Arabidopsis. This study is the first report investigating differential regulatory response through changes in the expression of glutathione and sulfated metabolic pathways in response to exposure to rare earth oxide NPs. In this study, it was also reported that the chlorophyll content was reduced at higher concentrations of CeO_2 NPs; however, it was unaffected on exposure to In_2O_3 NPs. Zhao et al. (2012) reported that the treatment of CeO_2 NPs on maize plants increased the accumulation of H_2O_2 in phloem, xylem, bundle sheath cells, and shoot epidermal cells. The integrity of membranes was not compromised on NP treatment as no ion leakage of reported in either roots or shoots. The net photosynthetic rate of the leaves, transpiration, and conductance of stomata were also not affected on CeO_2 NP treatment. Increased production of stress-related parameters in maize plants on NP treatment helped them to survive against nanotoxicity. Changes in the nutritional property of cilantro (*Coriandrum sativum*) with significant uptake and translocation on treatment with CeO_2 NPs were also reported which showcased their entry and impacts in the food chain (Morales et al. 2013). The transgenerational studies with CeO_2 NPs showed that the second-generation seedlings grown from the seeds obtained from CeO_2-treated tomato plants were smaller and weaker with lesser biomass, lower water transpiration, and higher reactive oxygen species content and also accumulated higher amount of ceria (Wang et al. 2013a, b). This study demonstrated the multigenerational effects of engineered NPs on plants.

The toxicity of CuO NPs to maize was studied by germination tests, and no inhibition was observed on the germination of seeds. Their transport and redistribution were also investigated, and it was found that the NPs were transported to shoots via xylem and back-translocated from shoots to roots through phloem (Wang et al. 2012a, b). The fate of metal oxide NP as a function of size by comparing the behavior of CuO and ZnO NPs with corresponding microparticles in sand matrix with and without wheat plants was studied and greater root toxicity was observed on interaction with smaller particles. It was noticed that several factors from sand and plant modified the aggregation and dissolution of both NPs and microparticles, which decides their route of accumulation and fate in the environment (Dimkpa et al. 2013).

5.2.2.2 Effects of Magnetite Nanoparticles

Increased biological applications of magnetic NPs as multifunctional agents for targeted cell delivery and medicinal imaging have opened avenues for their applications in plant biology. However, studies on their toxicological effects on plants and bioaccumulation in food chain still need to be more addressed for their improved use as smart treatment delivery vehicles in plants. The effects of magnetic NPs coated with perchloric acid on the early ontogenic stages of maize plants were studied, and a slight inhibitory effect on plantlet growth was observed. The toxicity symptoms led to the development of brown spots on leaf surface and excess iron treatment generated oxidative stress on leaf cells which in turn affected photosynthesis with decreased rate of metabolism (Racuciu and Creanga 2009). In sunflower (*Helianthus annuus*) seedlings, it was observed that the chlorophyll content was reduced up to 50 % on the application of low concentrations of magnetic NPs. Magnetic NPs negatively affected the biosynthesis of photosynthetic pigments thus affecting the chlorophyll content (Ursache-Oprisan et al. 2011). The effect of SPIONs (super paramagnetic iron oxide nanoparticles) on soybean has been studied, and it was reported that the SPIONs, which were translocated in soybean, increased the chlorophyll levels with no trace of toxicity (Ghafariyan et al. 2013). The total and nitric nitrogen content of lettuce due to treatment with magnetic nanofluids showed that the treatment influenced both the total and nitric nitrogen content. The total nitrogen content was found to be higher in the treated plants, and the nitric nitrogen in treated plants is lower when compared to the control plants (Pirvulescu and Sala 2012). Feizi et al. (2013) carried out experiments to study the biological responses of muskmelon (*Cucumis melo*) to magnetic field and AgNPs in comparison with commercial fertilizers under field conditions, and the results indicated that the plants treated with AgNPs in magnetic field had the highest fruit yield with improved early ripening.

Zhu et al. (2008) studied the uptake of magnetic NPs (20 nm in size) by pumpkin seedlings in hydroponic solutions and the signals for magnetic NPs were detected in roots, stems, and leaves of pumpkin plants using vibrating sample magnetometer. However, there was difference in the uptake of NPs in different growth medium, and no uptake was observed in soil medium, whereas reduced uptake of NPs by plants was found when grown in sand. This might be due to the difference in the adherence of magnetic NPs to soil and sand. The physiological effects of magnetite NPs on perennial grass and pumpkin plants grown under hydroponic conditions were also investigated, and it was found that the tested NPs were not translocated to the plants as no magnetization was detected in the shoots of treated plants (Wang et al. 2011). The size of NPs used in this test was larger than the cell wall pores which limited their entry. The uptake of magnetic carbon-coated bioferrofluid through the roots of four crop plants, pea, sunflower, tomato, and wheat was studied, and it was reported that the ferrofluid reached the vascular cylinder, moved through the xylem vessels, and reached the entire aerial portions of the plants in less than 24 h (Cifuentes et al. 2010). The same group also reported the penetration and transportation of magnetic carbon-coated NPs through the aerial parts of cucumber (González-Melendi et al. 2008; Corredor et al. 2009). Krystofova et al. (2013) studied the effects of magnetic NPs and modified magnetic NPs on tobacco BY-2 cell suspension cultures. They studied the effects of NPs on growth, proteosynthesis, and antioxidant activity of cells, and it was observed that the effects of magnetic NPs on growth of cell suspension culture were moderate, whereas noticeable changes were detected in all biochemical parameters. Hao et al. (2013) reported the use of magnetic gold NPs as a carrier for the delivery of fluorescein isothiocyanate (FITC) and plasmids into canola cells with and without cell wall which would benefit the development of transgenic plants. All such reports on the uptake and distribution of magnetic-based NPs in plant system opened up great opportunities to explore them for site targeted delivery of chemicals and other substances with an external control using strong magnets.

5.2.3 *Effects of Carbon-based Nanomaterials*

Carbon-based nanomaterials, such as single-walled carbon nanotubes (SWCNTs), multiwalled carbon nanotubes (MWCNTs), buckyballs (C_{60}), have several unique mechanical and structural properties and hence having potential applications in biomedical engineering and medicinal chemistry rather than its large-scale applications in electronics. However, concerns on the toxicity of these nanomaterials are the major limiting factor for its large-scale applications in medicine and agriculture.

Several works were reported with conflicting results for the interaction between carbon nanomaterials and biological systems, especially with animals (Cui et al. 2005; Fabbro et al. 2012; Das et al. 2013) but very limited works on plant system.

The effects of carbon nanotubes on plant phenotype and soil microbial community were studied, and it was observed that the tomato plants grown in soil supplemented with carbon nanotubes produced twice the amount of flowers and fruits when compared to control plants. The soil microbial community was also checked, and phylogenetic analyses indicated that the relative abundances of *Bacteroidetes* and *Firmicutes* got increased, and a decrease was observed for *Proteobacteria* and *Verrucomicrobia* with increasing concentration of CNTs (Khodakovskaya et al. 2013). The effects of engineered carbon nanomaterials of various dimensionalities on rice seed germination were studied, and an increase in germination rate with increased water uptake was observed for treated seeds than the control seeds (Nair et al. 2012). The treated seedlings also appeared to be healthier than the control plants in the studied range of concentration of carbon nanomaterials. In barley and soybean, it was observed that MWCNTs accelerated the seed germination and no negative effects were observed on further development of plants grown from exposed seeds (Lahiani et al. 2013), and it was observed that the expression of genes encoding water channel proteins increased in treated seeds than the control seeds. Canas et al. (2008) functionalized SWCNTs with poly-3-amino benzenesulfonic acid and studied the effects of both functionalized and non-functionalized SWCNTs on root growth of six crop plants, cabbage, carrot (*Daucus carota*), cucumber, lettuce, onion (*Allium cepa*), and tomato. Root elongation was enhanced in onion and cucumber and inhibited in tomato with non-functionalized nanotubes and functionalized nanotubes inhibited root elongation in lettuce. Cabbage and carrots were not affected by both types of nanotubes. Nanotubes were found to be adsorbed on the surface of roots with little uptake in this study. Studies on the effects of MWCNTs on red spinach (*Amaranthus dubius*), lettuce, rice, chili (*Capsicum* spp.) cucumber, okra (*Abelmoschus esculentus*), and soybean showed varied effects on root and shoot growth of different plant species and toxicity of nanotubes on seed germination, and growth was observed at higher concentrations and little effect was observed on chili, soybean, and okra (Begum et al. 2012). The beneficial effects of MWCNTs at low concentrations to maize plants were studied, and growth enhancement was correlated with improved water delivery by MWCNTs (Tiwari et al. 2014). Similar effects were reported for mustard and gram plants too (Mondal et al. 2011; Tripathi et al. 2011). Lin et al. (2009) studied the uptake and translocation of natural organic matter (NOM)-modified C_{70} and MWCNTs, and aggregates of NOM-C_{70} were found near the vascular system of stem which suggested their uptake along with water and nutrients though xylem; however, only minimal uptake of MWCNTs which is limited to roots. The uptake of water, nutrients, and overall plant development could hinder at higher concentrations of MWCNTs due to the blockage of plant roots and roots hairs by the surface adsorbed nanotubes. Khodakovskaya et al. (2012) demonstrated that the growth of tobacco cell culture had been enhanced with MWCNTs in a wide range of concentrations. The expression of tobacco aquaporin gene and the production of corresponding protein increased in cells exposed to MWCNTs when compared to control. Also the expression of marker gene for cell division and cell wall extension was upregulated on treatment with MWCNTs. These results suggested the role of CNTs in regulating cell division and plant growth with applications in enhanced production of plant cell cultures in plant biotechnology and pharmaceutical industry. The adverse cellular responses of SWCNTs to Arabidopsis and rice protoplasts were investigated, and it was found that the oxidative stress generated had led to programmed cell death and the survival of cells was highly dose dependent (Shen et al. 2010). The effects of fullerene exposure on the uptake and accumulation of dichlorodiphenyldichloroethylene (p, p′-DDE, a common agricultural contaminant) by three different plants, zucchini (*Cucurbita pepo*), soybean, and tomato were investigated (Torre-Roche et al. 2012). An increased contaminant level in shoots was observed for zucchini, whereas decreased p, p′-DDE level in soybean shoots and not much change has been observed for tomato. However, the total plant p, p′-DDE level got increased for all plant species on exposure to fullerene which calls for more studies on nanoparticle-contaminant interactions.

The ability of SWCNTs to traverse across the plant cell wall and cell membrane was first reported by Liu et al. (2009). This has opened novel methods to deliver DNA and other molecules to intact plant cells. Liu and his group also studied changes in the cell wall of tobacco cells under the repression of water soluble carboxy-fullerenes. Disruption in cell wall and cell membrane was observed on the adsorption of fullerenes which led to complete inhibition of cell growth (Liu et al. 2013). An increased glycosyl residue was observed in the cell wall of fullerene-treated plants cells with elevated levels of reactive oxygen species. Serag et al. (2011a, b) investigated the ability of FITC-labeled MWCNTs to penetrate the cell membrane of periwinkle (*Catharanthus roseus*) protoplasts, and their internalization mechanism was studied with the help of confocal imaging and TEM techniques. The direct penetration mode helped MWCNTs to bypass endosomes and hence opens new avenues in designing endosomes escaping nanotransporters for plant cells. A size-dependent translocation of MWCNTs to different cellular structures such as nucleus and plastids was also observed which can be utilized for delivering molecular cargoes specifically into target compartments. They also explained a functional approach for the controlled subcellular distribution of FITC-labeled SWCNTs and studied the nature of vacuolar uptake, cytoplasmic accumulation in different subcellular structures, and finally the cellular elimination. Such studies on trafficking of SWCNTs through subcellular membranes are important in site specific delivery of biomolecules for plants that are currently recalcitrant to genetic transformation (Serag et al. 2011a, b). The same group also investigated the ability of cup stacked carbon nanotubes (CSCNTs) with cellulase immobilized on its side walls and tips to penetrate plant cell walls by producing local lesions with the help of cellulase. CNTs can hence be successfully utilized as nanotransporters to plant cells without completely removing the cell wall of plants (Serag et al. 2012a, b). The role of carbon nanotubes in oxidative cross-linking of monolignols during lignin biosynthesis in plant cells was also studied, and this provided information on the post-uptake behavior of CNTs inside the cell which can be more helpful in plant defense research and possible detoxification mechanisms in cells (Serag et al. 2012a, b). Torre-Roche et al. (2013) studied the effects of MWCNTs or C_{60} fullerenes on the uptake of weathered pesticides by maize, zucchini, tomato, and soybean, and the results showed that the pesticide accumulation varied with the type of plant species, type of nanomaterial, and its concentration. Studies on the effects of fullerol on the biomass, fruit yield, and phytomedicine content of bitter melon (*Momordica charantia*) reported increased biomass with large and bigger fruits with improved content of anticancerous phytomedicines (Kole et al. 2013). Recently, the researchers from MIT reported the engineering of plant chloroplasts with SWCNTs in which the nanotubes were passively transported and interacted with the lipid bilayers of plant chloroplasts. A triple fold increase in photosynthetic activity was reported with enhanced electron transport rates. They also demonstrated the use of plants as biochemical detectors with the help of interaction of plants with modified nanotubes (Giraldo et al. 2014). This novel research area called nanobionics could bring more applications of nanotechnology in plant biology.

5.3 Conclusion

For the sustainable development of nanotechnology, it is important to understand the ecotoxicological effects of engineered nanomaterials on environment. The current chapter reviewed the uptake, translocation, accumulation, and phytotoxic effects of different nanoparticles depending on the plant species and size, type, chemical composition, functionalization, concentration, and stability of nanoparticles. Nanoagriculture could utilize nanotechnology in the best possible ways for the improved growth and development of plants. However, still there is a big gap in the knowledge about effects of different nanomaterials in plants as it is depending upon several interrelated factors such as different properties of nanoparticles and also the type of plant species. Some plants are capable of uptaking nanoparticles and accumulating them in different plant tissues. Their effects in plants vary with plant growth stage, time of exposure, method of uptake, and also various physical and chemical properties of plants. Researchers reported both positive and negative effects of nanomaterials on plant system. Some nanoparticles improved the seed germination and stimulated growth parameters in some plants, however, produced contradictory effects on others. Several studies have reported significant phytotoxicity due to the direct exposure to specific type of nanoparticles, and this emphasizes the need for ecologically responsible disposal of nanoparticle containing wastes. This highlights the necessity for more experimental studies extending over several generations of plants that are required for understanding the long-term effects of nanoparticles on ecosystem and for the safe and effective use of nanomaterials at judicious concentrations.

References

Adhikari T, Kundu S, Rao SA (2013) Impact of SiO_2 and Mo nanoparticles on seed germination of rice (*Oryza Sativa* L.). Intl J Agri Food Sci Technol 4:809–816

Arora S, Sharma P, Kumar S, Nayan R, Khanna PK, Zaidi MGH (2012) Gold nanoparticle induced enhancement in growth and seed yield of *Brassica juncea*. Plant Growth Regul 66:303–310

Barrena R, Casals E, Colón J, Font X, Sánchez A, Puntes V (2009) Evaluation of the ecotoxicity of model nanoparticles. Chemosphere 75:850–857

Begum P, Ikhtiari R, Fugetsu B, Matsuoka M, Akasaka T, Watari F (2012) Phytotoxicity of multi-walled carbon nanotubes assessed by selected plant species in the seedling stage. Appl Surf Sci 262:120–124

Birbaum K, Brogioli R, Schellenberg M, Martinoia E, Stark WJ, Günther D, Limbach LK (2010) No evidence for cerium dioxide nanoparticle translocation in maize plants. Environ Sci Technol 44:8718–8723

Boonyanitipong P, Kositsup B, Kumar P, Baruah S, Dutta J (2011) Toxicity of ZnO and TiO_2 nanoparticles on germinating rice seed. Intl J Biosci Biochem Bioinform 1:282–285

Canas JE, Long M, Nations S, Vadan R, Dai L, Luo M, Ambikapathi R, Lee H, Olszyk D (2008) Effects of functionalized and nonfunctionalized single-walled carbon nanotubes on root elongation of select crop species. Environ Toxicol Chem 27:1922–1931

Cifuentes Z, Custardoy L, de la Fuente JM, Marquina C, Ibarra MR, Rubiales D, Pérez-de-Luque A (2010) Absorption and translocation to the aerial part of magnetic carbon-coated nanoparticles through the root of different crop plants. J Nanobiotechnol 8:26–33

Corredor E, Testillano PS, Coronado MJ, González-Melendi P, Fernández R, Marquina C, Ibarra MR, de la Fuente JM, Rubiales D, Pérez-de LA, Risueño MC (2009) Nanoparticle penetration and transport in living pumpkin plants: in situ subcellular identification. BMC Plant Biol 9:45–56

Cui D, Tian F, Ozkan CS, Wang M, Gao H (2005) Effect of single wall carbon nanotubes on human HEK293 cells. Toxicol Lett 155:73–85

Das M, Singh RP, Datir SR, Jain S (2013) Intranuclear drug delivery and effective in vivo cancer therapy via estradiol–PEG-appended multiwalled carbon nanotubes. Mol Pharm 10:3404–3416

de la Rosa G, López-Moreno ML, de Haro D, Botez CE, Peralta-Videa JR, Gardea-Torresdey JL (2013) Effects of ZnO nanoparticles in alfalfa, tomato, and cucumber at the germination stage: root development and X-ray absorption spectroscopy studies. Pure Appl Chem 85:2161–2174

Dimkpa CO, Latta DE, McLean JE, Britt DW, Boyanov MI, Anderson AJ (2013) Fate of CuO and ZnO nano and microparticles in the plant environment. Environ Sci Technol 47:4734–4742

Du W, Sun Y, Ji R, Zhu J, Wu J, Guo H (2011) TiO_2 and ZnO nanoparticles negatively affect wheat growth and soil enzyme activities in agricultural soil. J Environ Monit 13:822–828

Eichert T, Kurtz A, Steiner U, Goldbach HE (2008) Size exclusion limits and lateral heterogeneity of the stomatal foliar uptake pathway for aqueous solutes and water-suspended nanoparticles. Physiol Plant 134:151–160

El-Temsah YS, Joner EJ (2010) Impact of Fe and Ag nanoparticles on seed germination and differences in bioavailability during exposure in aqueous suspension and soil. Environ Toxicol 27:42–49

Espinosa AJF, Oliva SR (2006) The composition and relationships between trace element levels in inhalable atmospheric particles (PM_{10}) and in leaves of *Nerium oleander* L. and *Lantana camara* L. Chemosphere 62:1665–1672

Fabbro C, Ali-Boucetta H, Ros TD, Kostarelos K, Bianco A, Prato M (2012) Targeting carbon nanotubes against cancer. Chem Commun 48:3911–3926

Feizi H, Rezvani MP, Shahtahmassebi N, Fotovat A (2012) Impact of bulk and nanosized titanium dioxide (TiO_2) on wheat seed germination and seedling growth. Biol Trace Elem Res 146:101–106

Feizi H, Pour SJ, Rad KH (2013) Biological response of muskmelon (*Cucumis melo* L.) to magnetic field and silver nanoparticles. Annu Rev Res Biol 3:794–804

Ghafariyan MH, Malakouti MJ, Dadpour MR, Stroeve P, Mahmoudi M (2013) Effects of magnetite nanoparticles on soybean chlorophyll. Environ Sci Technol 47:10645–10652

Giordani T, Fabrizi A, Guidi L, Natali L, Giunti G, Ravasi F, Cavallini A, Pardossi A (2012) Response of tomato plants exposed to treatment with nanoparticles. Environ Qual 8:27–38

Giraldo JP, Landry MP, Faltermeier SM, McNicholas TP, Iverson NM, Boghossian AA, Reuel NF, Hilmer AJ, Sen F, Brew JA, Strano MS (2014) Plant nanobionics approach to augment photosynthesis and biochemical sensing. Nat Mater 13:400–408

González-Melendi P, Fernández-Pacheco R, Coronado MJ, Corredor E, Testillano PS, Risueño MC, Marquina C, Ibarra MR, Rubiales D, Pérez-de- Luque A (2008) Nanoparticles as smart treatment-delivery systems in plants: assessment of different techniques of microscopy for their visualization in plant tissues. Ann Bot 101:187–195

Gruyer N, Dorais M, Bastien C, Dassylva N, Triffault-Bouchet G (2014) Interaction between silver nanoparticles and plant growth. In: International symposium on new technologies for environment control, energy-saving and crop production in greenhouse and plant factory—Greensys, Jeju, Korea, 6–11 Oct 2013

Haghighi M, da Silva TJA (2014) Effect of N-TiO_2 on tomato, onion and radish seed germination. J Crop Sci Biotechnol 17(2014):221–227

Hao Y, Yang X, Shi Y, Song S, Xing J, Marowitch J, Chen J (2013) Magnetic gold nanoparticles as a vehicle for fluorescein isothiocyanate and DNA delivery into plant cells. Botany 91:457–466

Hong J, Peralta-Videa JR, Rico C, Sahi S, Viveros MN, Bartonjo J, Zhao L, Gardea-Torresdey JL (2014) Evidence of translocation and physiological impacts of foliar applied CeO_2 nanoparticles on cucumber (*Cucumis sativus*) plants. Environ Sci Technol 48:4376–4385

Jaberzadeh A, Moaveni P, Moghadam HRT, Zahedi H (2013) Influence of bulk and nanoparticles titanium foliar application on some agronomic traits, seed gluten and starch contents of wheat subjected to water deficit stress. Notulae Bot Hort Agrobo 41:201–207

Judy JD, Unrine JM, Rao W, Wirick S, Bertsch PM (2012) Bioavailability of gold nanoparticles to plants: importance of particle size and surface coating. Environ Sci Technol 46:8467–8474

Karl-Josef D, Simone H (2011) Plant nanotoxicology. Trends Plant Sci 16:582–589

Kaveh R, Li YS, Ranjbar S, Tehrani R, Brueck CL, Aken BV (2013) Changes in *Arabidopsis thaliana* gene expression in response to silver nanoparticles and silver ions. Environ Sci Technol 47:10637–10644

Khodakovskaya MV, de Silva K, Biris AS, Dervishi E, Villagarcia H (2012) Carbon nanotubes induce growth enhancement of tobacco cells. ACS Nano 6:2128–2135

Khodakovskaya MV, Kim BS, Kim JN, Alimohammadi M, Dervishi E, Mustafa T, Cernigla CE (2013) Carbon nanotubes as plant growth regulators: effects on tomato growth, reproductive system, and soil microbial community. Small 9:115–123

Kole C, Kole P, Randunu KM, Choudhary P, Podila R, Ke PC, Rao AM, Marcus RK (2013) Nanobiotechnology can boost crop production and quality: first evidence from increased plant biomass, fruit yield and phytomedicine content in bitter melon (*Momordica charantia*). BMC Biotechnol 13(13):37

Krystofova O, Sochor J, Zitka O, Babula P, Kudrle V, Adam V, Kizek R (2013) Effect of magnetic nanoparticles on tobacco BY-2 cell suspension culture. Int J Environ Res Public Health 10:47–71

Lahiani MH, Dervishi E, Chen J, Nima Z, Gaume A, Biris AS, Khodakovskaya MV (2013) Impact of carbon nanotube exposure to seeds of valuable crops. ACS Appl Mater Interfaces 5:7965–7973

Larue C, Khodja H, Herlin-Boime N, Brisset F, Flank AM, Fayard B, Chaillou S, Carrière M (2011) Investigation of titanium dioxide nanoparticles toxicity and uptake by plants. J Phys Conf Ser 304:012057

Larue C, Laurette J, Herlin-Boime N, Khodja H, Fayard B, Flank AM, Brisset F, Carriere M (2012a) Accumulation, translocation and impact of TiO_2 nanoparticles in wheat (*Triticum aestivum* spp.): influence of diameter and crystal phase. Sci Total Environ 431:197–208

Larue C, Veronesi G, Flank A-M, Surble S, Herlin-Boime N, Carrière M (2012b) Comparative uptake and impact of TiO_2 nanoparticles in wheat and rapeseed. J Toxicol Environ Health Part A 75:722–734

Lee CW, Mahendra S, Zodrow K, Li D, Tsai YC, Braam J, Alvarez PJJ (2010) Developmental phytotoxicity of metal oxide nanoparticles to *Arabidopsis thaliana*. Environ Toxicol Chem 29:669–675

Lee W-M, Kwak J II, An Y-J (2012) Effect of silver nanoparticles in crop plants *Phaseolus radiates* and *Sorghum bicolor*: media effect on phytotoxicity. Chemosphere 86:491–499

Lin D, Xing B (2007) Phytotoxicity of nanoparticles: inhibition of seed germination and root growth. Environ Pollut 150(2):243–250

Lin D, Xing B (2008) Root uptake and phytotoxicity of ZnO nanoparticles. Environ Sci Technol 42:5580–5585

Lin S, Reppert J, Hu Q, Hudson JS, Reid ML, Ratnikova TA, Rao AM, Luo H, Ke PC (2009) Uptake, translocation and transmission of carbon nanomaterials in rice plants. Small 5:1128–1132

Liu Q, Chen B, Wang Q, Shi X, Xiao Z, Lin J, Fang X (2009) Carbon nanotubes as molecular transporters for walled plant cells. Nano Lett 9:1007–1010

Liu Q, Zhang X, Zhao Y, Lin J, Shu C, Wang C, Fang X (2013) Fullerene-induced increase of glycosyl residue on living plant cell wall. Environ Sci Technol 47:7490–7498

Lopez-Moreno ML, de la Rosa G, Hernandez-Viezcas JA, Peralta-Videa JR, Gardea-Torresdey JL (2010a) X-ray absorption spectroscopy (XAS) corroboration of the uptake and storage of CeO_2 nanoparticles and assessment of their differential toxicity in four edible plant species. J Agric Food Chem 58:3689–3693

López-Moreno ML, de la Rosa G, Hernández-Viezcas JA, Castillo-Michel H, Botez CE, Peralta-Videa JR, Gardea-Torresdey JL (2010b) Evidence of the differential biotransformation and genotoxicity of ZnO and CeO_2 nanoparticles on soybean (*Glycine max*) plants. Environ Sci Technol 44:7315–7320

Ma C, Chhikara S, Xing B, Musante C, White JC, Dhankher OP (2013) Physiological and molecular response of *Arabidopsis thaliana* (L.) to nanoparticle cerium and indium oxide exposure. ACS Sustain Chem Eng 1:768–778

Mazumdar H, Ahmed GU (2011) Phytotoxicity effects of silver nanoparticles on *Oryza sativa*. Int J ChemTech Res 3:1494–1500

Mondal A, Basu R, Das S, Nandy P (2011) Beneficial role of carbon nanotubes on mustard plant growth: an agricultural prospect. J Nanopart Res 13:4519–4528

Morales MI, Rico CM, Hernandez-Viezcas JA, Nunez JE, Barrios AC, Tafoya A, Flores-Marges JP, Peralta-Videa JR, Gardea-Torresdey JL (2013) Toxicity assessment of cerium oxide nanoparticles in cilantro (*Coriandrum sativum* L.) plants grown in organic soil. J Agric Food Chem 61:6224–6230

Morteza E, Moaveni P, Farahani HA, Kiyani M (2013) Study of photosynthetic pigments changes of maize (*Zea mays* L.) under nano TiO_2 spraying at various growth stages. Springer Plus 2:247–251

Mukherjee A, Peralta-Videa JR, Bandyopadhyay S, Rico CM, Zhao L, Gardea-Torresdey JL (2014) Physiological effects of nanoparticulate ZnO in green peas (*Pisum sativum* L.) cultivated in soil. Metallomics 6:132–138

Mushtaq YK (2011) Effect of nanoscale Fe_3O_4, TiO_2 and carbon particles on cucumber seed germination. J Environ Sci Health A Tox Hazard Subst Environ Eng 46:1732–1735

Nair R, Varghese SH, Nair BG, Maekawa T, Yoshida Y, Kumar DS (2010) Nanoparticulate material delivery to plants. Plant Sci 179:154–163

Nair R, Mohamed SM, Gao W, Maekawa T, Yoshida Y, Ajayan PM, Kumar DS (2012) Effect of carbon nanomaterials on the germination and growth of rice plants. J Nanosci Nanotechnol 12:2212–2220

Namasivayam SKR, Chitrakala K (2011) Ecotoxicological effect of *Lecanicillium lecanii* (Ascomycota: Hypocreales) based silver nanoparticles on growth parameters of economically important plants. J Biopesticides 4:97–101
Nowack B, Bucheli TD (2007) Occurrence, behavior and effects of nanoparticles in the environment. Environ Pollut 150:5–22
Nowack B, Schulin R, Robinson BH (2006) Critical assessment of chelant-enhanced metal phytoextraction. Environ Sci Technol 40:5225–5232
Oukarroum A, Barhoumi L, Pirastru L, Dewez D (2013) Silver nanoparticles toxicity effect on growth and cellular viability of the aquatic plant *Lemna gibba*. Environ Toxicol Chem 32:902–907
Pirvulescu A, Sala F (2012) Nitrogen content in lettuce under the influence of magnetic nanofluids. J Hort Biotechnol 16:63–66
Pokhrel LR, Dubey B (2013) Evaluation of developmental responses of two crop plants exposed to silver and zinc oxide nanoparticles. Sci Total Environ 452–453:321–332
Prasad TNVK, Sudhakar P, Sreenivasulu Y, Latha P, Munaswamy V, Reddy KR, Sreeprasad TS, Sajanlal PR, Pradeep T (2012) Effect of nanoscale zinc oxide particles on the germination, growth and yield of peanut. J Plant Nutr 35:905–927
Racuciu M, Creanga DE (2009) Biocompatible magnetic fluid nanoparticles internalized in vegetal tissue. Rom J Phys 54:115–124
Rico CM, Majumdar S, Duarte-Gardea M, Peralta-Videa JR, Gardea-Torresdey JL (2011) Interaction of nanoparticles with edible plants and their possible implications in the food chain. J Agric Food Chem 59:3485–3498
Ruffini CM, Roberto C (2009) Nanoparticles and higher plants. Caryologia 62:161–165
Sabo-Attwood T, Unrine JM, Stone JW, Murphy CJ, Ghoshroy S, Blom D, Bertsch PM, Newman LA (2012) Uptake, distribution and toxicity of gold nanoparticles in tobacco (*Nicotiana xanthi*) seedlings. Nanotoxicology 6:353–360
Salama HMH (2012) Effects of silver nanoparticles in some crop plants, common bean (*Phaseolus vulgaris* L.) and corn (*Zea mays* L.). Int Res J Biotechnol 3:190–197
Serag MF, Kaji N, Gaillard C, Okamoto Y, Terasaka K, Jabasini M, Tokeshi M, Mizukami H, Bianco A, Baba Y (2011a) Trafficking and subcellular localization of multiwalled carbon nanotubes in plant cells. ACS Nano 5:493–499
Serag MF, Kaji N, Venturelli E, Okamoto Y, Terasaka K, Tokeshi M, Mizukami H, Braeckmans K, Bianco A, Baba Y (2011b) Functional platform for controlled subcellular distribution of carbon nanotubes. ACS Nano 5:9264–9270
Serag MF, Kaji N, Tokeshi M, Baba (2012a) Introducing carbon nanotubes into living walled plant cells through cellulase-induced nanoholes. RSC Adv 2:398–400
Serag MF, Kaji N, Tokeshi M, Biancoe A, Baba Y (2012b) The plant cell uses carbon nanotubes to build tracheary elements. Integr Biol 4:127–131
Servin AD, Morales MI, Castillo-Michel H, Hernandez-Viezcas JA, Munoz B, Zhao L, Nunez JE, Peralta-Videa JR, Gardea-Torresdey JL (2013) Synchrotron verification of TiO_2 accumulation in cucumber fruit: a possible pathway of TiO_2 nanoparticle transfer from soil into the food chain. Environ Sci Technol 47:11592–11598
Shen CX, Zhang QF, Li J, Bi FC, Yao N (2010) Induction of programmed cell death in *Arabidopsis* and rice by single-wall carbon nanotubes. Am J Bot 97:1602–1609
Siddiqui MH, Al-Whaibi MH (2014) Role of nano-SiO_2 in germination of tomato (*Lycopersicum esculentum* seeds Mill.). Saud J Biol Sci 21:13–17
Slomberg DL, Schoenfisch MH (2012) Silica nanoparticle phytotoxicity to *Arabidopsis thaliana*. Environ Sci Technol 46:10247–10254
Stampoulis D, Sinha SK, White JC (2009) Assay dependent phytotoxicity of nanoparticles to plants. Environ Sci Technol 43:9473–9479
Suriyaprabha R, Karunakaran G, Yuvakkumar R, Rajendran V, Kannan N (2012) Silica nanoparticles for increased silica availability in maize (*Zea mays* L.) seeds under hydroponic conditions. Curr Nanosci 8:902–908

Syu YY, Hung JH, Chen JC, Chuang HW (2014) Impacts of size and shape of silver nanoparticles on Arabidopsis plant growth and gene expression. Plant Physiol Biochem 83:57–64

Tandy S, Schulin R, Nowack B (2006) Uptake of metals during chelant-assisted phytoextraction with EDDS related to the solubilized metal concentration. Environ Sci Technol 40:2753–2758

Tarafdar JC, Xiang Y, Wang W-N, Dong Q, Biswas P (2012) Standardization of size, shape and concentration of nanoparticle for plant application. Appl Biol Res 14:138–144

Taylor AF, Rylott EL, Anderson CWN, Bruce NC (2014) Investigating the toxicity, uptake, nanoparticle formation and genetic response of plants to gold. PLoS ONE 9:e93793

Tiwari DK, Dasgupta-Schubert N, Cendejas LMV, Villegas J, Montoya LC, Garcia SEB (2014) Interfacing carbon nanotubes (CNT) with plants: enhancement of growth, water and ionic nutrient uptake in maize (*Zea mays*) and implications for nanoagriculture. Appl Nanosci 4:577–591

Torre-Roche RDL, Hawthorne J, Deng Y, Xing B, Cai W, Newman LA, Wang C, Ma X, White JC (2012) Fullerene-enhanced accumulation of p, p′-DDE in agricultural crop species. Environ Sci Technol 46:9315–9323

Torre-Roche RDL, Hawthorne J, Deng Y, Xing B, Cai W, Newman LA, Wang Q, Ma X, Hamdi H, White JC (2013) Multiwalled carbon nanotubes and C_{60} fullerenes differentially impact the accumulation of weathered pesticides in four agricultural plants. Environ Sci Technol 47:12539–12547

Tripathi S, Sonkar SK, Sarkar S (2011) Growth stimulation of gram (*Cicer arietinum*) plant by water soluble carbon nanotubes. Nanoscale 3:1176–1181

Ursache-Oprisan M, Focanici E, Creanga D, Caltun O (2011) Sunflower chlorophyll levels after magnetic nanoparticle supply. Afr J Biotechnol 10(2011):7092–7098

Vannini C, Domingo G, Onelli E, Mattia FD, Bruni I, Marsoni M, Bracale M (2014) Phytotoxic and genotoxic effects of silver nanoparticles exposure on germinating wheat seedlings. J Plant Physiol 171:1142–1148

Wang H, Kou X, Pei Z, Xiao JQ, Shan X, Xing B (2011) Physiological effects of magnetite (Fe_3O_4) nanoparticles on perennial ryegrass (*Lolium perenne* L.) and pumpkin (*Cucurbita mixta*) plants. Nanotoxicology 5:30–42

Wang Q, Ma X, Zhang W, Pei H, Chen Y (2012a) The impact of cerium oxide nanoparticles on tomato (*Solanum lycopersicum* L.) and its implications for food safety. Metallomics 4:1105–1112

Wang Z, Xie X, Zhao J, Liu X, Feng W, White JC, Xing B (2012b) Xylem- and phloem-based transport of CuO nanoparticles in maize (*Zea mays* L.). Environ Sci Technol 46:4434–4441

Wang J, Koo Y, Alexander A, Yang Y, Westerhof S, Zhang QB, Schnoor JL, Colvin VL, Braam J, Alvarez PJJ (2013a) Phytostimulation of poplars and Arabidopsis exposed to silver nanoparticles and Ag^+ at sublethal concentrations. Environ Sci Technol 47:5442–5449

Wang Q, Ebbs SD, Chen Y, Ma X (2013b) Trans-generational impact of cerium oxide nanoparticles on tomato plants. Metallomics 5:753–759

Xingmao M, Jane G-L, Yang D, Andrei K (2010) Interactions between engineered nanoparticles (ENPs) and plants: phytotoxicity, uptake and accumulation. Sci Total Environ 408:3053–3061

Yang L, Watts DJ (2005) Particle surface characteristics may play an important role in phytotoxicity of alumina nanoparticles. Toxicol Lett 158:122–132

Yin L, Cheng Y, Espinasse B, Colman BP, Auffan M, Wiesner M, Rose J, Liu J, Bernhardt ES (2011) More than the ions: the effects of silver nanoparticles on *Lolium multiflorum*. Environ Sci Technol 45:2360–2367

Yin L, Colman BP, McGill BM, Wright JP, Bernhardt ES (2012) Effects of silver nanoparticle exposure on germination and early growth of eleven wetland plants. PLoS ONE 7:e47674

Yoon SJ, Kwak JI, Lee WM, Holden PA, An YJ (2014) ZnO nanoparticles delay soybean development: a standard soil microcosm study. Ecotoxicol Environ Saf 100:131–137

Zhao L, Peng B, Hernandez-Viezcas JA, Rico C, Sun Y, Peralta-Videa JR, Tang X, Niu G, Jin L, Varela-Ramirez A, Zhang JY, Gardea-Torresdey JL (2012) Stress response and tolerance of *Zea mays* to CeO_2 nanoparticles: cross talk among H_2O_2, heat shock protein, and lipid peroxidation. ACS Nano 6:9615–9622

Zhao L, Sun Y, Hernandez-Viezcas JA, Servin AD, Hong J, Niu G, Peralta-Videa JR, Duarte-Gardea M, Gardea-Torresdey JL (2013) Influence of CeO_2 and ZnO nanoparticles on cucumber physiological markers and bioaccumulation of Ce and Zn: a life cycle study. J Agri Food Chem 61:11945–11951

Zhu H, Han J, Xiao JQ, Jin Y (2008) Uptake, translocation, and accumulation of manufactured iron oxide nanoparticles by pumpkin plants. J Environ Monit 10:713–717

Zhu Z-J, Wang H, Yan B, Zheng H, Jiang Y, Miranda OR, Rotello VM, Xing B, Vachet RW (2012) Effects of surface charge on the uptake and distribution of gold nanoparticles in four plant species. Environ Sci Technol 46:12391–12398

Chapter 6
Effect of Nanoparticles on Plants with Regard to Physiological Attributes

M. Sheikh Mohamed and D. Sakthi Kumar

Abstract The growth parameters of plants are influenced by various biotic and abiotic factors. The increased interference of humans with the environment has led to heightened concern over such activities on the living systems, including plants. With tremendous progress being made in the field of engineering, manufacturing, construction, etc., onus has shifted to the possible effects of such developments on the ecosystem. Nanotechnology has emerged as an indispensable tool for the future, with its reach spanning across diverse domains. Such a rapid advance has resulted in the exodus of various types of nanomaterials into the environment. Thus, it becomes essential to understand the imminent effects, either advantageous or deleterious, of these nanomaterials on the living subjects advertently or inadvertently exposed to them. Numerous studies have focused on the effects of such nanomaterials in the nanoparticulate form on the mammalian system, with increased studies on the plant system as well. Due to the complex nature of uptake and translocation mechanism present in plants, it has been relatively difficult to unanimously devise a general dataset of the effects that nanoparticles (NPs) have on them. Research over the past years has documented mostly toxic effects of the NPs, either during the germination stage or with respect to the shoot–root length, while few others have explored the possibilities of utilizing them as carriers for chemicals as herbicides, pesticides, fertilizers, or in some cases genes. There have been numerous contradictory findings with some reports suggesting growth enhancing effects and others observing retarding effects of similar NPs on similar or different plant species. Such contradictions and lack of conclusive observations has slowed down the impact of nanotechnology in the agriculture industry when compared with the medical scene. This scenario demands a comprehensive calibration of the analysis and interpretation of NP–plant interaction and effects thereof from the

M. Sheikh Mohamed · D. Sakthi Kumar (✉)
Bio-Nano Electronics Research Centre, Toyo University, 2100, Kujirai,
Kawagoe, Saitama 350-8585, Japan
e-mail: sakthi@toyo.jp

C. Kole et al. (eds.), *Plant Nanotechnology*,
DOI 10.1007/978-3-319-42154-4_6

physiological, biochemical, and photosynthetic level to the molecular level to decisively devise a verdict on the actual effects of nanoparticles on the plant system. This chapter summarizes the research conducted so far in this field and attempts at providing an outlook for the future.

Keywords Nanotechnology · Nanoparticles · Plants · Physiological · Biochemical · Photosynthesis · Toxicity

6.1 Introduction

Nanotechnology has established itself as one of the fastest and biggest research and development fields in recorded history. Although previous technological revolutions as space exploration, semiconductors, and biotechnology have made it big, they were and still are, severely confined to their respective domains and find fewer interest from other disciplines. Nanotechnology, on the other hand, though started off like the aforementioned fields of study and research, limited mostly to the electronics industry, has expanded its horizons of application by the amalgamation of nearly all subjects of science viz., biology, chemistry, physics, etc. The rapid strides of nanotechnology in electronics and manufacturing have recently been paralleled by advances in medical nanotechnology, with an ever-increasing list of scientific publications, patents, and products being commercially released.

Although, this introduction of nanotechnology is highly impressive, like many of its predecessor technologies, it also has been very slow to exert its influence on the agricultural domain. The main reasons include limited research and knowledge base available on the impact of nanomaterials (NMs) on the plant system. Still, researchers world over have tried to elucidate the effects of this technology in plants, both in vitro and in nature. Most of the works have focused on the deleterious effects while a few have shown promising applications in boosting the native functions as growth, yield, and biomass enhancement, improving the photosynthesis conversion efficiency, and as carriers for chemicals as herbicides, pesticides, fertilizers, or genes (Galbraith 2007; Torney et al. 2007; DeRosa et al. 2010; Nair et al. 2010; Lahiani et al. 2013; Kole et al. 2013; Cossins 2014; Giraldo et al. 2014; Siddiqui and Al-Whaibi 2014). Despite all the information available on the toxicity of nanoparticles (NPs) to plant system, the appropriate elucidation of physiological, biochemical, and molecular mechanisms is crucial for the future of this technology in wide-scale agricultural implications (Fig. 6.1).

In this chapter, we discuss on the positive as well as negative impacts of NPs on the plant system, with respect to physiological parameters.

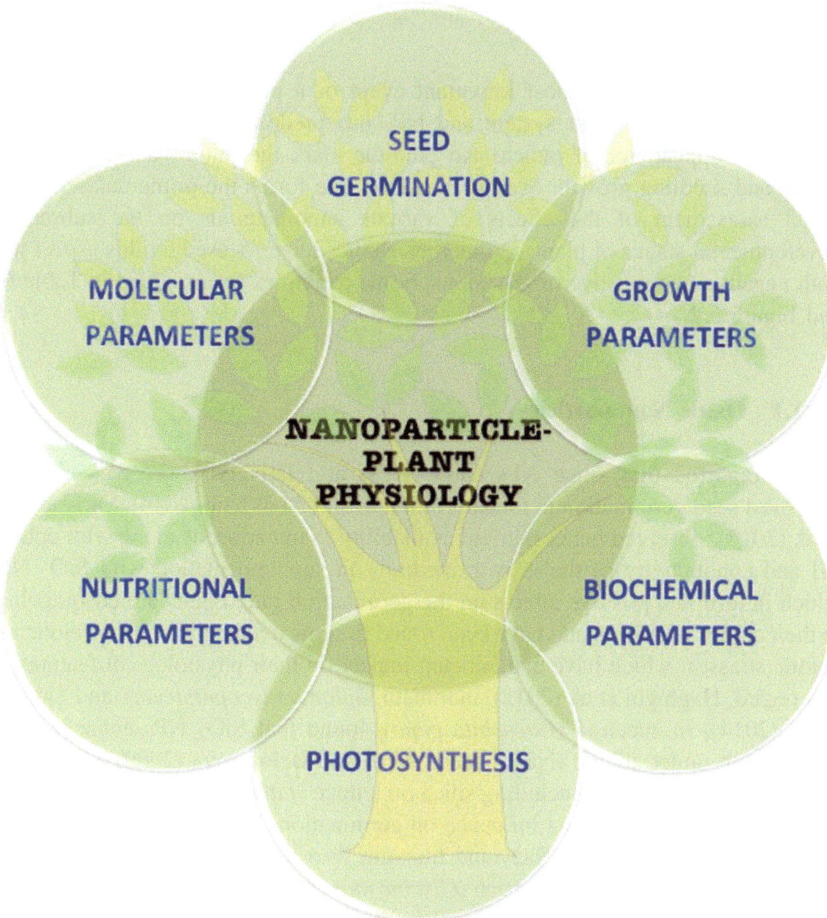

Fig. 6.1 Major parameters to be assessed while investigating the effects of nanoparticles on plant system

6.2 Effects of Nanoparticles on Germination, Growth, and Development

The interaction and subsequent effects of NPs on any biological system, including plants, depend primarily on the inherent physicochemical properties of the NPs, such as size, shape, charge, chemical composition, surface modifications, and reactivity (Ma et al. 2010; Khodakovskaya et al. 2012). Also, the NP–plant relation crucially depends on their concentration and varies from plant to plant. The primary focus of this chapter will be on the recorded role of NPs in seed germination, plant growth and biomass yield, biochemical parameters, and photosynthesis based on the type of NPs administered.

6.2.1 Germination

Seed germination is the most important event of a plant life. The first point of contact between the plant system and NPs has predominantly been through the seeds. The appearance of radicle and plumule mark the initiation of seed germination and seedling growth. Seed germination rate forms the prime dataset for the initial assessment of the effects of various nanomaterials on the subsequent developmental stages of plants. Numerous studies have focused on this aspect with both negative and positive observations. Some examples are provided in Table 6.1 and Figure 6.2.

6.2.1.1 Oxide Nanoparticles

It was reported that silica NPs (SiO_2 NPs) at relatively lower concentrations improved seed germination in tomato (Siddiqui and Al-Whaibi 2014). Suriyaprabha et al. (2012), observed better nutrient availability to maize seeds along with optimal pH and conductivity of the growth medium on supplementation with SiO_2 NPs, which in turn had positive effects on the germination rate. Plants are cosmopolitan in their choice of habitat and have been found to grow even under various biotic and abiotic stresses, which have a significant impact on their physiological features. In this regard, Haghighi et al. (2012), in tomato (*Solanum lycopersicum*), and Siddiqui et al. (2014), in zucchini (*Cucurbita pepo*), found that SiO_2 NPs enhanced seed germination under abiotic (NaCl) stress. Shah and Belozerova (2009) assessed the effects of a range of NPs including silica on lettuce (*Lactuca sativa*) and found that all the NPs had a significant influence on germination. Lu et al. (2002) and Zheng et al. (2005) observed that SiO_2 and titanium dioxide NPs (TiO_2 NPs) positively impact seed germination in soybean (*Glycine max*) and spinach (*Spinacea oleracea*) by increasing nitrate reductase and enhancing the uptake and utilization of water and nutrients.

Beneficial effects of lower concentrations of zinc oxide NPs (ZnO NPs) on seed germination have been observed in a variety of plant species as peanut (*Arachis hypogea*) (Prasad et al. 2012), soybean (Sedghi et al. 2013), wheat (*Triticum aestivum*) (Ramesh et al. 2014), and onion (*Allium cepa*) (Raskar and Laware 2014). ZnO NPs showed differential effects when tested on cucumber (*Cucumis sativus*), alfalfa (*Medicago sativa*), and tomato, with only the former exhibiting enhancement in germination (De la Rosa et al. 2013). The determination of phytotoxicity of metallic NPs and their oxides is relatively complex, primarily due to the potential dissolution of ions released from the NPs and their associated toxicity (Ma et al. 2010). Germination and root growth of zucchini seeds in hydroponic culture augmented with ZnO NPs presented no negative effects (Stampoulis et al. 2009) whereas in the case of rye grass (*Lolium perenne*) and maize (*Zea mays*), the germination was significantly inhibited by nano-Zn (35 nm) and zinc oxide (15–25 nm), respectively (Lin and Xing 2007).

Table 6.1 Effects of different nanoparticles on the germination of seeds [positive effect (+), negative effect (−), no effect (N)]

Nanoparticle	Plant	Effect on germination	Reference
Graphene oxide	Fava bean	+	Anjum et al. (2014)
CNTs	Tomato	+	Morla et al. (2011)
MWCNTs	Barley, soybean maize	+	Lahiani et al. (2013)
ZnO	Peanut	+	Prasad et al. (2012)
Au	Arabidopsis	+	Kumar et al. (2013)
Ag	*Boswellia ovalifoliolata*	+	Savithramma et al. (2012)
TiO_2	Fennel	+	Feizi et al. (2013b)
Se	Tobacco	+	Domokos-Szabolcsy et al. (2012)
TiO_2	Garden sage	+	Feizi et al. (2013a)
SiO_2	Maize	+	Suriyaprabha et al. (2012)
Au	Glory lily	+	Gopinath et al. (2014)
SiO_2	Tomato	+	Siddiqui et al. (2014)
CNTs	Onion, Indian mustard mungbean	+	Ghodake et al. (2010), Mondal et al. (2011)
CNTs	Rice	+	Nair et al. (2010)
TiO_2	Spinach	+	Zheng et al. (2005)
TiO_2	Wheat	+	Feizi et al. (2012)
Si, Pd, Au, Cu	Lettuce	+	Shah and Belozerova (2009)
SiO_2 and TiO_2	Soybean	+	Lu et al. (2002)
Zero-valent Fe	Flax, barley, rye	−	El-Temsah and Joner (2010)
Ag	Rye	−	El-Temsah and Joner (2010)
Ag	Barley	−	El-Temsah and Joner (2010)
Si	Zucchini	−	Stampoulis et al. (2009)
Al	Rye	−	Lin and Xing (2007)
ZnO	Maize	−	Lin and Xing (2007)
CeO_2	Alfalfa, tomato, cucumber, maize, soybean	−	Lopez-Moreno et al. (2010b)
Zero-valent Fe	Flax, red clover white, meadow fescue, barley, rye	N	El-Temsah and Joner (2010)
Al	Radish, rapeseed, lettuce, maize, cucumber	N	Lin and Xing (2007)
Ag	Flax	N	El-Temsah and Joner (2010)
Au	Cucumber, lettuce	+	Barrena et al. (2009)
Si	Zucchini	N	Stampoulis et al. (2009)
Cu	Lettuce	N	Shah and Belozerova (2009)

(continued)

Table 6.1 (continued)

Nanoparticle	Plant	Effect on germination	Reference
Au	Lettuce	N	Shah and Belozerova (2009)
Pd–Al(OH)$_2$	Lettuce	N	Shah and Belozerova (2009)
SiO_2	Lettuce	N	Shah and Belozerova (2009)
Al_2O_3	Radish, rapeseed, rye, lettuce, maize, cucumber	N	Lin and Xing (2007)
r-TiO_2	Spinach	+	Zheng et al. (2005)
SiO_2 + TiO_2	Soybean	+	Lu et al. (2002)
Au/Cu	Lettuce	N	Shah and Belozerova (2009)
MWCNTs	Radish, rapeseed, rye, lettuce, maize, cucumber	N	Lin and Xing (2007)
MWCNTs	Zucchini	N	Lin and Xing (2007)

Several examples exist depicting the positive aspects of TiO_2 NPs on plants (Zheng et al. 2005; Hong et al. 2005a; Yang et al. 2007; Gao et al. 2008). TiO_2 NPs have been observed to enhance seed germination and promoted radicle and plumule growth of canola (*Brassica napus*) seedlings (Mahmoodzadeh et al. 2013). Zheng et al. (2005) presented enhanced growth in spinach when TiO_2 NPs were administered to the seeds. Recently, seed priming has been found to increase the seed vigor and germination synchronization, which has resulted in growth enhancement of many crops under particularly stressful conditions (Carvalho et al. 2011). For example, anatase NPs treatment to parsley seeds increased the germination rate index of the test subject (Dehkourdi and Mosavi 2013). In another case, biogenic anatase NPs were used to increase the seedling vigor and germination percentage of tridax daisy (*Tridax procumbens*) by the possible triggering of antioxidative mechanism in germinating seeds under chilling (Bhati-Kushwaha et al. 2013). Feizi et al. (2013a) verified the germination rate enhancement of common sage (*Salvia officinalis*) while the seeds were exposed to 60 mg L^{-1} of bulk and TiO_2 NPs.

6.2.1.2 Carbon Materials

Carbon nanotubes (CNTs), owing to their unique mechanical, electrical, thermal, and chemical properties have found commendable presence in plant science as well. It has been prominently observed that CNTs are able to penetrate the cell wall and membrane of cells, facilitating the enhanced uptake of water and nutrients by forming additional transport channels and also act as delivery systems for certain specialized chemicals to cells. Various studies have demonstrated the ability of multiwalled CNTs (MWCNTs) to positively influence seed germination and plant growth. Villagarcia et al. (2012) and Tiwari et al. (2014) found that MWCNTs induced increased water, Ca, and Fe uptake, which enhanced the seed germination

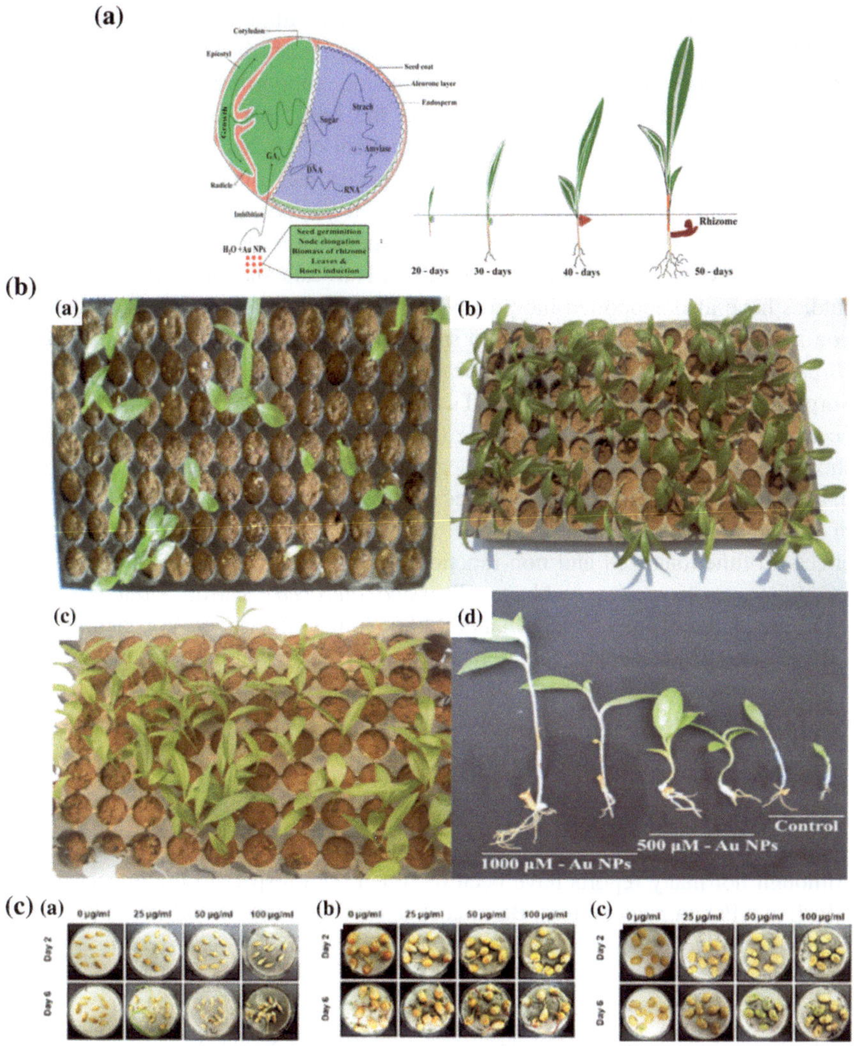

Fig. 6.2 **a** Schematic representation of the effect of Au NP-induced lily seed germination, node elongation, biomass of rhizome, leaf, and root initiation. **b** Effect of Au NPs on lily seed germination: *a* Control, *b* 500 µM Au NPs, *c* 1000 µM Au NPs for a duration of 30 day, *d* Induction of node elongation, biomass of rhizome, leaf, and root initiation of Au NP-treated samples for a duration of 40 day (Gopinath et al. 2014). **c** Germination of crop seeds (soybean, barley, corn) exposed to MWCNTs through the airspray technique. Phenotype of control and MWCNT-coated seeds of *a* barley, *b* corn, and *c* soybean are presented on the second and sixth day after MWCNT spray treatment (Adopted from Lahiani et al. 2013)

and plant growth. Raman spectroscopy and transmission electron microscopy (TEM) revealed the presence of MWCNTs aggregates inside the seed coats of barley (*Hordeum vulgare*), soybean, and maize, supporting the assumption that MWCNTs have the tendency to penetrate the seed coat to facilitate water and nutrient supply to the germinating seeds (Lahiani et al. 2013). Highly maximized germination rate was observed in crop species as tomato, hybrid Bt cotton (*Gossypium hirsutum*), Indian mustard (*Brassica juncea*), urdbean (*Vigna mungo*), and rice (*Oryza sativa*) with MWCNTs treatment (Nair et al. 2010; Gajanan et al. 2010; Morla et al. 2011; Mondal et al. 2011; Nalwade and Neharkar 2013). Other studies have also supported the positive influence of MWCNTs on seed germination and growth of six different crop species (radish (*Raphanus sativus*), rapeseed (*Brassica napus*), rye, lettuce, maize, and cucumber) (Lin and Xing 2007). Surprisingly, the effects of MWCNTs and single-walled CNTs (SWCNTs) have been observed to vary. For example, zucchini plants exposed to MWCNTs did not show any detrimental effects on seed germination and root elongation whereas a marked decrease in the biomass was recorded during further growth in the presence of SWCNTs (Stampoulis et al. 2009). Similarly, Cañas et al. (2008) studied the effects of functionalized and non-functionalized SWCNTs on root length parameters of six crop species (cabbage (*Brassica oleracea*), carrot (*Daucus carota*), cucumber, lettuce, onion, and tomato). Although the effects were concentration and species dependent, the non-functionalized CNTs exhibited higher degree of phytotoxicity when compared to its functionalized counterpart, signifying the important role of NP surface modifications.

6.2.1.3 Metal Nanoparticles

Although not many reports have been recorded with respect to the impact of gold NPs (Au NPs) on plant system, Barrena et al. (2009) in lettuce and cucumber, Arora et al. (2012) in Indian mustard; Savithramma et al. (2012) in *Boswellia ovalifoliolata,* and Gopinath et al. (2014) in lily have reported that Au NPs improve seed germination in the respective plant species.

Krishnaraj et al. (2012) observed that biologically synthesized silver NPs (Ag NPs) showed a significant effect on seed germination of hydroponically grown water hyssop (*Bacopa monneri*). Also, biosynthesized Ag NPs enhanced seed germination and seedling growth in *Boswellia ovaliofoliolata* (Savithramma et al. 2012). However, Yin et al. (2012) reported on the enhanced germination rate of trumpet weed (*E. fistulosum*) alone out of 11 wetland plant species (rye grass, switch grass (*Panicum virgatum*), sallow sedge (*Carex lurida*), broom sedge (*C. scoparia*), fox sedge (*C. vulpinoidea*), fringed sedge (*C. crinita*), trumpet weed, simple pokeweed (*Phytolaca americana*), wool grass (*Scirpus cyperinus*), scarlet lobelia (*Lobelia cardinalis*), and soft rush (*Juncus effusus*) on treatment with Ag NPs.

6.2.2 *Growth Parameters*

Plant growth is characterized by increase in biomass of the germinated seeds. The length of roots and shoot, number of laterals, number and size of leaves, total biomass, and yield represent the major growth parameters. Many studies on NP–plant interactions have focused on these factors and reports of both enhanced and retarded growth have been documented. Most of the NPs applied to seedlings or plants as such are through the roots, which causes a kind of bias in determining their effects. The main reason behind this speculative assessment is that the movement of NPs through the plant tissues (translocation) has not been clearly understood, although there are a few reports available. Overall, the observed effects are linked to the NP interaction with roots, either promoting or blocking nutrient supply and subsequent translocation to higher tissues. Table 6.2 provides a few examples of the various NPs influencing the growth parameters of plants.

6.2.2.1 Oxide Nanoparticles

SiO_2 NPs were found by Bao-shan et al. (2004) to improve seedling growth and quality, including mean height, root collar diameter, main root length, and the number of lateral roots while affecting the synthesis of chlorophyll in Changbai larch (*Larix olgensis*) seedlings. Wang et al. (2014) treated rice plants with bare quantum dots (QDs) and silica-coated QDs and found that the latter significantly promoted root growth.

Jaberzadeh et al. (2013) documented the growth enhancement in wheat plants on exposure to TiO_2 NPs under water-deficit stress. Zheng et al. (2005) described the enhanced growth performance of spinach when TiO_2 NPs were sprayed onto the leaves. Nano-anatase treatment to parsley (*Petroselinum crispum*) seeds had a positive effect on the root/shoot length, fresh weight, vigor index, and chlorophyll content (Dehkourdi and Mosavi 2013).

Juhel et al. (2011) evaluated the effects of alumina NPs on the growth, morphology, and photosynthesis of the aquatic plant duck weed (*Lemna gibba*) and concluded that aluminum oxide NPs (alumina and Al_2O_3 NPs) enhanced the growth of duck weed significantly with an evident increase in the biomass, which in turn was explained to be proportional to the morphological adjustments of the plant in response to alumina NPs exposure, such as increase in root length, number of fronds per colony, and photosynthetic efficiency.

Clément et al. (2013) characterized the phytotoxicity of the Ti NPs (crystal anatase or rutile) in daphnia and algae, rotifers, and plants as model organisms. TiO_2 NPs with anatase crystal structure were toxic in all of the tests at higher concentrations, but due to their antimicrobial properties, a significant growth of the roots was observed. As the rutile form exhibits lipophilicity, the TiO_2 NPs produce larger aggregates in aqueous medium, resulting in reduced effects on biological organisms and a lower toxicity compared with anatase form.

Table 6.2 Effect of nanoparticles on the overall growth parameters of plants [positive effect (+), negative effect (−), no effect (N)]

Nanoparticle	Plant	Growth parameters	Reference
CNTs	Alfalfa, wheat	Root elongation (+)	Miralles et al. (2012)
SWCNTs	Onion, cucumber	Root elongation (+)	Cañas et al. (2008)
MWCNTs	Wheat	Root growth (+)	Wang et al. (2012)
ZnO	Peanut	Root growth (+)	Prasad et al. (2012)
Au	Arabidopsis	Root length (+)	Kumar et al. (2013)
Ag	Common bean, maize	Root length (+)	Salama (2012)
TiO_2	Arabidopsis	Root length (+)	Lee et al. (2010)
Aluminum Oxide	Arabidopsis	Root length (+)	Lee et al. (2010)
Alumina	Duckweed	Root length root growth (+)	Juhel et al. (2011)
Zero-valent Iron Oxide	Arabidopsis	Root elongation (+)	Kim et al. (2014)
Co_3O_4	Radish	Root elongation (+)	Wu et al. (2012)
Ag	Tomato	Root growth (−)	Song et al. (2013)
TiO_2	Tomato	Root length (N)	Song et al. (2013)
Ni	Tomato	Root growth (−)	Faisal et al. (2013)
ZnO	Cluster bean	Root growth (+)	Raliya and Tarafdar (2013)
CNTs	Wheat	Root growth (+)	Wang et al. (2012)
TiO_2	Wheat	Root elongation (+)	Larue et al. (2012)
Al_2O_3	Arabidopsis	Root growth and elongation (N)	Lee et al. (2010)

(continued)

Table 6.2 (continued)

Nanoparticle	Plant	Growth parameters	Reference
Al	Radish, rapeseed	Root growth (+)	Lin and Xing (2007)
CeO_2	Maize, alfalfa, soybean	Root growth (+)	Lopez-Moreno et al. (2010b)
ZnO	Soybean	Root growth (+)	Lopez-Moreno et al. (2010a)
MWCNTs	Rye	Root length (+)	Lin and Xing (2007)
SWCNTs	Onion, cucumber	Root length (+)	Cañas et al. (2008)
Cu	Wheat	Root (−)	Lee et al. (2008)
Al	Rye	Root length (−)	Lin and Xing (2007)
Al	Maize, lettuce	Root length (−)	Lin and Xing (2007)
Zn	Radish, rapeseed, rye, lettuce, maize, cucumber	Root length (−)	Lin and Xing (2007)
ZnO	Rye	Root tips (−); root cap (−)	Lin and Xing (2007)
ZnO	Radish, rapeseed, rye, lettuce, maize, cucumber	Root growth (−)	Lin and Xing (2007)
ZnO	Maize	Root growth (−)	Stampoulis et al. (2009)
CeO_2	Maize, tomato, alfalfa	Root growth (−)	Lopez-Moreno et al. (2010b)
Al_2O_3	Maize, cucumber, carrot, cabbage	Root growth (−); root length (−)	Yang and Watts (2005); Lin and Xing (2007)
CNTs	Tomato	Root reduction	Cañas et al. (2008)
CNTs	Lettuce	Root length (−)	Cañas et al. (2008)
MWCNTs	Lettuce	Root length (−)	Lin and Xing (2007)
Au	Arabidopsis	Shoot length (+)	Kumar et al. (2013)
Ag	Bean, maize	Shoot length (+)	Salama (2012)
Ag	Rice	Shoot growth (−)	Mirzajani et al. (2013)
ZnO	Cluster bean	Shoot growth (+)	Raliya and Tarafdar (2013)

(continued)

Table 6.2 (continued)

Nanoparticle	Plant	Growth parameters	Reference
CeO_2	Maize, alfalfa, soybean	Shoot growth (+)	Lopez-Moreno et al. (2010b)
$SiO_2 + TiO_2$	Soybean	Shoot growth (+)	Lu et al. (2002)
Ag	Barley, flax, rye	Shoot length (−)	El-Temsah and Joner (2010)
Cu	Mungbean	Shoot growth (−)	Lee et al. (2008)
CeO_2	Alfalfa, tomato, cucumber, maize	Shoot growth (−)	Lopez-Moreno et al. (2010b)
CNTs	Tomato	Seedling growth (+)	Morla et al. (2011)
wsCNTs	Chickpea	Growth rate (+)	Tripathi et al. (2011)
MWCNTs	Tobacco	Growth rate (+)	Khodakovskaya et al. (2012)
Ag	*Boswellia ovaliofoliolata*	Seedling growth (+)	Savithramma et al. (2012)
SiO_2	Maize	Growth parameters (+)	Yuvakkumar et al. (2011), Suriyaprabha et al. (2012)
TiO_2	Duck weed	Plant growth (+)	Song et al. (2012)
Ag	Mungbean, Sorghum	Plant growth (−)	Lee et al. (2012)
GA-Ag	Rye, switch grass, sallow sedge, broom sedge, fox sedge, fringed sedge, simple pokeweed, wool grass, scarlet lobelia, soft rush	Plant growth (−)	Yin et al. (2012)
Se	Tobacco	Plant growth (+)	Domokos-Szabolcsy et al. (2012)
Alumina	Duck weed	Plant growth (+)	Juhel et al. (2011)
Ag	Duck weed	Plant growth (−)	Gubbins et al. (2011)
CNTs	Indian mustard, mungbean	Seedling growth (+)	Mondal et al. (2011)
TiO_2	Spinach	Plant growth (+)	Yang et al. (2006)
SiO_2 and TiO_2	Soybean	Plant growth (+)	Lu et al. (2002)
Cu	Mungbean, wheat	Seedling growth (−)	Lee et al. (2008)

ZnO NPs were instrumental in significantly improving the plant biomass, shoot and root growth, and root area in cluster bean (*Cyamopsis tetragonoloba*) rhizosphere (Raliya and Tarafdar 2013). Mahajan et al. (2011), employing correlative light and scanning microscope, and inductively coupled plasma/atomic emission spectroscopy, evidently revealed the presence of ZnO NPs in roots of mungbean (*Vigna radiata*) and chickpea (*Cicer arietinum*), which related to promotion of root/shoot length and biomass. Helaly et al. (2014) augmented MS media with nano-ZnO and found increased somatic embryogenesis, shooting, and subsequent regeneration of plantlets. Lin and Xing (2007) found retarded root growth of six higher plant species when treated with 2000 mg L^{-1} nano-Zn or ZnO NPs.

Faisal et al. (2013) investigated the nickel oxide NP (NiO NP)-induced phytotoxicity in the roots of tomato seedlings. Short duration treatment of tomato seeds to NiO NPs resulted in a significant repression of root growth. This anomaly was responsible for an oxidative imbalance, evidenced from the enhancement in antioxidant enzyme levels. An ultrastructure analysis of root cells revealed the translocation of the NiO NPs in the cell cytoplasm, characterizing changes in the structure of the organelles. Also, enhancement in activity of oxidative stress-related enzymes and mitochondrial dysfunction are related to the observed phytotoxicity.

6.2.2.2 Carbon Materials

Wang et al. (2012) recorded significantly enhanced root cell elongation and dehydrogenase activity with oxidized MWCNTs. The improved root and stem growth on MWCNTs exposure may be due to the uptake and accumulation of MWCNTs by roots with their subsequent translocation to leaves (Smirnova et al. 2012). The presence of water-soluble CNTs inside wheat plants was evidenced by Tripathi and Sarkar (2015) with scanning electron and fluorescence microscope. Furthermore, the authors linked this observation to the CNTs-induced root and shoot growth under both light and dark conditions. Interestingly, MWCNTs have been recognized to augment water retention, improve biomass, flowering, and fruit yield, and also enhance medicinal properties of plants (Khodakovskaya et al. 2013; Husen and Siddiqi 2014). A few examples of NPs influencing the yield and biomass of plants are compiled in Table 6.3. However, inhibitory effects of MWCNTs on plant growth have also been reported by many researchers (Begum and Fugetsu 2012; Ikhtiar et al. 2013; Tiwari et al. 2014; Begum et al. 2014). In another study, the uptake, accumulation, and transmission of natural organic matter (NOM)-suspended MWCNTs in rice were reported (Lin et al. 2009). The observations revealed a negative impact of the MWCNTs-NOM on the rice plants with delay in flowering and reduced seed set.

Carbon-based fullerol [$C_{60}(OH)_{20}$] NPs treatment resulted in increases of up to 54 % in biomass yield and 24 % in water content in bitter melon (*Momordica charantia*). A 20 % fruit length, 59 % fruit number, and 70 % fruit weight gain resulted in an overall improvement of up to 128 % in fruit yield (Kole et al. 2013).

Table 6.3 Effects of nanoparticles on the yield and biomass of plants [positive effect (+), negative effect (−), no effect (N)]

Nanoparticle	Plant	Yield/biomass	Reference
MWCNTs	Tomato	Number of flowers (+)	Khodakovskaya et al. (2013)
ZnO	Peanut	Yield (+)	Prasad et al. (2012)
Au	Arabidopsis	Early flowering and yield (+)	Kumar et al. (2013)
Iron oxide	Soybean	Yield (+)	Sheykhbaglou et al. (2010)
SWCNTs	Rice	Flowering (−); yield (−)	Burman et al. (2013)
MWCNTs	Wheat	Vegetative biomass (+)	Wang et al. (2012)
MWCNTs	Maize	biomass (+)	Tiwari et al. (2014)
ZnO	Chickpea	Shoot biomass (+)	Burman et al. (2013)
ZnO	Mungbean	Biomass (+)	Dhoke et al. (2013)
ZnO	Mungbean	Dry weight (+)	Patra et al. (2013)
Ag	Bean, maize	Dry vegetative weight (+)	Salama (2012)
S	Mungbean	Dry weight (+)	Patra et al. (2013)
Alumina	Duck weed	Biomass (+)	Juhel et al. (2011)
Iron oxide	Mungbean	Biomass (+)	Dhoke et al. (2013)
ZnFeCu oxide	Mungbean	Biomass (+)	Dhoke et al. (2013)
CeO_2	Arabidopsis	Biomass (+)	Ma et al. (2013)
CuO	Wheat	Biomass (+)	Dimkpa et al. (2012)
Ti and Ag	Tomato	Biomass (N)	Song et al. (2013)
Alumina	Duck weed	Biomass (+)	Juhel et al. (2011)
Zn	Maize	Biomass (N)	Zhao et al. (2013)
Ti	Bean, wheat, curly dock, pond weed	Biomass (N)	Jacob et al. (2013)
SiO_2	Maize	Biomass (+)	Suriyaprabha et al. (2012)
TiO_2	Spinach	Biomass (+)	Zheng et al. (2005)
TiO_2	Spinach	Dry weight (+)	Jacob et al. (2013)
Ag	Zucchini	Biomass (−)	Stampoulis et al. (2009)
Cu	Zucchini	Biomass (−)	Stampoulis et al. (2009)
ZnO	Rye	Biomass (−)	Lin and Xing (2007)
ZnO	Zucchini	Biomass (−)	Stampoulis et al. (2009)
CeO_2	Alfalfa	Biomass (−)	Lopez-Moreno et al. (2010b)
MWCNTs	Zucchini	Biomass (−)	Stampoulis et al. (2009)

6.2.2.3 Metal Nanoparticles

Au NPs have been seen to increase number of leaves, leaf area, plant height, and chlorophyll content, culminating in better crop yield (Arora et al. 2012; Gopinath et al. 2014).

Ag NPs were witnessed to increase the shoot/root length and leaf area of Indian mustard, common bean (*Phaseolus vulgaris*), and maize (Salama 2012; Sharma et al. 2012). Meanwhile, Gruyer et al. (2013) observed both positive and negative effects of Ag NPs on root elongation, with increment in barley roots, but inhibition in lettuce. Size and shape of the NPs under consideration play a major role in translating their effects on the morphological and physiological aspects of the host plant system. Syu et al. (2014) analyzed the effect of three varying morphologies of Ag NPs on the physiological and molecular response of Arabidopsis (*Arabidopsis thaliana*). They concluded that the decahedral Ag NPs exhibited the highest percentage of root promotion whereas the spherical particles had no effect. Lee et al. (2012) studied the phytotoxic aspects of Ag NPs (5–25 nm) on two edible crops, bean and sorghum in agar and soil cultures. In agar dispersed NPs experiment, bean and sorghum showed concentration-dependent growth inhibition whereas in the soil media fortified with Ag NPs, bean was not significantly affected and sorghum exhibited a slightly reduced growth rate. This study demonstrates the importance of media utilized for the dissolution of NPs on the toxicity of the plants. In the aquatic plant duck weed (used as an environmental toxicity plant indicator), Gubbins et al. (2011) indicated an inhibition of plant growth when plants were exposed to Ag NPs (5 mg L^{-1}) with a size ranging from 20 to 100 nm. Contrasting data obtained by Juhel et al. (2011), working with alumina NPs in duck weed, have shown an increase in the biomass accumulation. Ma et al. (2010) found that at lower concentrations of Ag NPs (1 mg L^{-1}), toxicity could be observed in seedlings of Arabidopsis plants. Studies on seed germination and root growth of hydroponically cultured zucchini plants in solution amended with Ag NPs showed no negative effects except for a decrease in total biomass and transpiration that was recorded on prolonged exposure to Ag NPs (Stampoulis et al. 2009).

Perchloric acid-coated magnetic NPs were tested on germinated maize seeds (Racuciu and Creanga 2009). Slight inhibitory effect was observed with brown spots on leaves at higher concentrations of the ferrofluid possibly due to generation of oxidative stress in leaf cells leading to altered photosynthesis rate and subsequent decreased metabolic activity.

Unmodified alumina NPs (13 nm) retarded the root elongation in maize, cucumber, soybean, carrot, and cabbage (Yang and Watts 2005). On the contrary, phenanthrene (a major constituent of polycyclic aromatic hydrocarbons)-loaded alumina NPs exhibited significantly decreased toxicity, implying on the relevance of appropriate surface modifications, which could facilitate in mitigating the phytotoxicity of NPs.

Copper NPs (Cu NPs) supplemented to agar culture media were tested for seedling growth of mungbean and wheat (Lee et al. 2008). Mungbean exhibited higher sensitivity to Cu NPs than wheat with noticeable inhibition in the growth of seedlings being observed. Similarly, Cu NPs were found to negatively influence the length of emerging roots of zucchini plants and had detrimental effects on the growth (Stampoulis et al. 2009). Jiang et al. (2014), with the wide medical applications of hydroxyapatite (HAP) in mind initiated a study on mungbean plants exposed to HAP NPs to understand the relationship between biocompatibility and biotoxicity of these NPs. The mungbean sprouts growth was inhibited, depending on the amount of HAP NPs that ruptured the cell wall and gained intracellular access, and also the Ca^{2+} concentrations were considered as the primary factors for cellular apoptosis and consequently for the observed inhibitory effect.

6.3 Biochemical Parameters

The biochemical evaluation of any system, including plants gives an approximate overview of the performance of that system. The biochemistry analysis of plants would shed light on the efficiency and extent of metabolic activity, actively taking place inside the subject, which is responsible for all the parameters of growth, development and reproduction and directly corresponds to the overall health of the plant. These parameters are greatly influenced by external biotic and abiotic factors, which include the NPs. Reports have been made on both the positive and negative influence of NPs on the biochemical features (Table 6.4).

6.3.1 Oxide Nanoparticles

It has been recorded that under high saline stress, SiO_2 NPs increase fresh and dry leaf weight, chlorophyll content, and proline accumulation. The reason for such a tolerance of plants to abiotic stress could be attributed to an increase in the accumulation of proline, free amino acids, nutrients, and enhanced activity of antioxidant enzymes due to the presence of SiO_2 NPs (Haghighi et al. 2012; Li et al. 2012; Siddiqui et al. 2012; Kalteh et al. 2014). Raliya and Tarafdar (2013) found that ZnO NPs were instrumental in significantly improving the chlorophyll content and protein synthesis, rhizospheric microbial population, acid phosphatase, alkaline phosphatase, and phytase activity in a cluster bean rhizosphere. ZnO NPs-supplemented MS media induced proline synthesis and increased activity of superoxide dismutase (SOD), catalase (CAT), and peroxidase resulting in heightened tolerance to biotic stress (Helaly et al. 2014). Hernandez-Viezcas et al. (2011) studied the effects of 10 nm ZnO NPs in hydroponic cultures of velvet mesquite at concentrations varying from 500 to 4000 mg L^{-1}. To evaluate NP-induced stress on the plant, specific activity of CAT and ascorbate peroxidase (APX) was performed. The NPs were recorded to increase the specific activity of CAT (in the root, stem, and leaves) and APX (only in the leaves), while no evidence of detrimental

Table 6.4 Biochemical aspects of nanoparticle plant interaction [positive effect (+), negative effect (-), no effect (N)]

Nanoparticle	Plant	Biochemical parameters	Reference
Al_2O_3	Tobacco	MicroRNA expression (+)	Burklew et al. (2012)
ZnO	Velvet mesquite	Levels of CAT and APOX (+)	Hernandez-Viezcas et al. (2011)
MWCNTs	Tomato	Uptake of nutrients (+)	Tiwari et al. (2013)
MWCNTs	Maize	Nutrient transport (+)	Tiwari et al. (2014)
ZnO	Cucumber	Micronutrients (+)	Zhao et al. (2014)
TiO_2	Wheat	Chlorophyll content (+)	Mahmoodzadeh et al. (2013)
TiO_2	Tomato	Net photosynthesis (+); transpiration and water conductance (+)	Qi et al. (2013)
TiO_2	Spinach	Enzymatic activities (+)	Yang et al. (2006)
Fe	Wheat, arabidopsis	Chlorophyll (−);	Larue et al. (2012), Marusenko et al. (2013)
ZnO	Cluster bean	Chlorophyll (+); protein content (+); P-nutrient-metabolizing enzymes (+)	Raliya and Tarafdar (2013)
SiO_2	Maize	Proteins (+); chlorophyll (+); phenols (+)	Suriyaprabha et al. (2012)
TiO_2	Spinach	Photosynthesis (+); nitrogen metabolism (+); oxidative stress (+)	Hong et al. (2005a), Lee et al. (2012), Lei et al. (2007)
SiO_2 and TiO_2	Soybean	Nitrate reductase activity (+)	Lu et al. (2002)
TiO_2	Spinach	RCA mRNA expression (+); protein levels (+);	Gao et al. (2006)
TiO_2	Spinach	N_2 fixation (+)	Linglan et al. (2008)
SiO_2 + TiO_2	Soybean	Nitrate reductase activity (+); water absorption (+); antioxidant potential (+)	Lu et al. (2002)
Ag	Zucchini	Transpiration (−)	Stampoulis et al. (2009)
TiO_2	Maize	Hydraulic conductivity (−); transpiration (−)	Asli and Neumann (2009)

aspects as chlorosis, necrosis, stunting, or wilting, even after 30 days of treatment, was observed, suggesting a significant tolerance level toward ZnO NPs. Kumari et al. (2011) during the evaluation of effects of ZnO NPs using root cells of onion showed that on increasing the ZnO NPs or the ZnO bulk concentrations, higher values for the thiobarbituric acid reactive species (TBARS) were observed. During reactive oxygen species (ROS) formation and release, fatty acid conversion to toxic lipid peroxides occurs, causing disruption of biological membranes (Gratão et al. 2005), facilitating the entry of and damage by NPs and metals, resulting in TBARS formation, which damages the membrane permeability and is predicted to be one the reasons for the observed phytotoxicity.

Effects of CuO NPs were studied in an economically important oil seed crop, Indian mustard. Significant increases in peroxidase enzyme activity and H_2O_2 formation were observed. The lipid peroxidation levels were found to have increased significantly in both the shoots and roots of seedlings. Gene expression studies revealed significant activation of CuZn SOD in roots and shoots while the MnSOD gene levels remained unchanged. Also, the CAT and APX expression levels were not observed to have changed in shoots. However, significant inhibition of CAT and APX was recorded in roots. The SOD enzyme activity also significantly increased in roots and shoots as a result of exposure to 50–500 mg L^{-1} of CuO NPs (Nair and Chung 2015).

Hydroponic cultures of cucumber, aerially treated with nano-ceria powder (CeO_2), displayed increased CAT activity in roots and decreased APX activity in leaves (Hong et al. 2014). TiO_2 NPs act as photocatalysts and are responsible for the induction of an oxidation–reduction reaction (Crabtree 1998). They regulate nitrogen metabolism-related enzymic activity such as glutamate dehydrogenase, nitrate reductase, glutamic–pyruvic transaminase, and glutamine synthase, which assist in the uptake of nitrates and facilitate the conversion of inorganic nitrogen to organic nitrogen in the form of protein and chlorophyll (Yang et al. 2006; Dehkourdi and Mosavi 2013; Mishra et al. 2014). According to the observations made by Hong et al. (2005b), TiO_2 NPs were seen to protect the chloroplasts from excess light by increasing the activity of antioxidant enzymes, such as CAT, peroxidase, and SOD. Nano-anatase TiO_2 has also been found to promote antioxidant stress by decreasing the production of superoxide radicals, malonyldialdehyde content, and hydrogen peroxide and enhancing the activities of SOD, APX, CAT, and guaiacol peroxidase resulting in the increased oxygen evolution rate in spinach chloroplasts under UV-B radiation (Lei et al. 2008). Phytotoxicity in tomato seedlings due to NiO NPs was partially related to the increase in caspase-3-like protease activity, which linked NiO NPs to trigger the intrinsic apoptotic pathway in tomato plants due to the release of the Ni ions (Faisal et al. 2013).

6.3.2 Carbon Materials

MWCNTs have been shown to improve the peroxidase and dehydrogenase activity (Smirnova et al. 2012). It was observed that graphene oxide (GO) exposure did not induce H_2O_2 production, formation of oxidative stress, increase in malondialdehyde content, or altered activities of antioxidant enzymes in Arabidopsis plants. These results along with other observations provided a physiological basis for the safety of GO (Zhao et al. 2015).

6.3.3 *Metal Nanoparticles*

Shah and Belozerova (2009) demonstrated Au NP-induced toxicity in plants due to markedly arrested aquaporin function, which help in the transportation of wide range of molecules including water. Au NPs treatment improved the chlorophyll and sugar content of test plants resulting in better crop yield (Arora et al. 2012; Gopinath et al. 2014).

Effects of biosynthesized Ag NPs on hydroponic cultures of water hyssop revealed the induction of protein, carbohydrate synthesis, and decreased total phenol contents in addition to reduced CAT and peroxidase activities (Krishnaraj et al. 2012). Ag NPs increased the biochemical attributes (chlorophyll, carbohydrate, and protein contents, antioxidant enzymes) of Indian mustard, common bean, and maize (Salama et al. 2012; Gruyer et al. 2013). Rezvani et al. (2012) found that Ag NP-induced root growth by blocking ethylene signaling in saffron (*Crocus sativus*). Syu et al. (2014) while analyzing the effect of three varying morphologies of Ag NPs on the physiological and molecular response of Arabidopsis concluded that the spherical particles triggered the highest levels of anthocyanin accumulation and Cu/Zn SOD in Arabidopsis seedlings when compared to decahedral NPs, which gave the lowest values. The Ag NPs were also responsible for the regulation of protein accumulations such as protochlorophyllide oxidoreductase, cell-division-cycle kinase 2, and fructose-1,6 bisphosphate aldolase along with activation of the aminocyclopropane-1-carboxylic acid-derived inhibition of root elongation. A proteomic approach (2-DE and NanoLC/FT-ICR MS identification) was employed to study the effects of colloidal suspension of spherical Ag NPs on rice. Results revealed an accumulation of protein precursors, indicative of the dissipation of a proton motive force upon Ag NP administration. The proteins were identified to be involved in oxidative stress tolerance, transcription and protein degradation, calcium ion regulation and signaling, cell division, apoptosis, and cell wall and DNA/RNA/protein direct damage (Mirzajani et al. 2013). The effects of Ag NPs and $AgNO_3$ on mustard (*Brassica nigra*) seed germination were investigated at physiological and molecular levels. Both nanoformulations inhibited lipase activity and soluble and reducing sugar contents along with increased transcription of heme oxygenase-1 (Amooaghaie et al. 2015).

6.4 Role of Nanoparticles in Photosynthesis

Photosynthesis is the most essential and vital physiological process in the plant kingdom. It involves the conversion of light energy to chemical energy in the chloroplasts, specifically using chlorophyll, and storing it in the bonds of sugar, which is later used as the energy currency to regulate various other processes. The only raw materials required for this are light energy, CO_2, and H_2O, which are abundantly available in nature. Still, the conversion efficiency of light to energy by plants remains only 2–4 % (Kirschbaum 2011). This significant deficiency has prompted a large number of researchers world over to either mimic the process of photosynthesis artificially or improve the existing efficiency in planta. NPs tend to interfere and alter the photosynthetic efficiency, photochemical fluorescence, and quantum yield in plants based on their inherent light interaction capabilities (Table 6.5).

Table 6.5 Interaction of nanoparticles with the photosynthetic machinery of plants [positive effect (+), negative effect (−), no effect (N)]

Nanoparticle	Plants	Mechanism involved	Reference
TiO_2	Spinach, tomato	Light absorption (+); quantum yield in PS-II (+)	Mingyu et al. (2007a), Lei et al. (2007)
TiO_2	Long raceme ulm	Light absorption (−)	Gao et al. (2013)
Au	Soybean	Light absorption (+)	Falco et al. (2011)
CeO_2	Alfalfa	Light absorption (−) and photochemical efficacy (−)	Gomez-Garay et al. (2014)
TiO_2	Fava bean	Quantum yield in PS-II (N)	Foltete et al. (2011)
CuO, TiO_2	Duck weed, long raceme ulm	Photochemical fluorescence (+)	Gao et al. (2013)
Ag	Indian mustard	Quantum yield in PS-I I (+)	Sharma et al. (2012)
Au	Soybean	Quantum yield (−)	Falco et al. (2011)
Mn, TiO_2	Spinach, mungbean	Splitting of water (+); evolution of oxygen (+)	Lei et al. (2007)
Mn, TiO_2	Spinach, mungbean	Photophosphorylation in ETC (+)	Lei et al. (2007)
CuO, TiO_2	Long raceme ulm	ETC activity (−)	Gao et al. (2013)
CeO_2	Moringa	ETC activity (+)	Gomez-Garay et al. (2014)
TiO_2	Spinach	Light absorption (+); energy conversion (+); CO_2 assimilation (+)	Baun et al. (2008)
TiO_2	Spinach	Chlorophyll Formation (+); Ribulosebisphosphate carboxylase/oxygenase activity (+); photosynthetic rate (+)	Zheng et al. (2005)
TiO_2	Spinach	Photosynthetic rate (+)	Hong et al. (2005a)
TiO_2	Spinach	Hill reaction and non-cyclic photophosphorylation (+)	Hong et al. (2005a)
TiO_2	Spinach	Rubisco activase expression (+)	Ma et al. (2008)
TiO_2	Spinach	Oxygen evolution (+); Rubisco carboxylation (+); Rubisco activase and photosynthesis (+)	Gao et al. (2006), Zheng et al. (2007), Gao et al. (2008)

6.4.1 Oxide Nanoparticles

A modified mesoporous silica (SBA)–photosystem II (PSII) complex demonstrated active light-driven electron transport from water to quinone molecules as a result of the stable activity of photosynthetic oxygen-evolving reaction (Noji et al. 2011). The PSII–SBA complex was proposed as a potential candidate for the development of photosensors and artificial photosynthetic systems. Siddiqui et al. (2014) and Xie et al. (2012) showed that SiO_2 NPs enhance the photosynthetic rate by improving activity of carbonic anhydrase (supplies CO_2 to Ribulose 1,5-bisphosphate carboxylase-RuBisCo) and synthesis of photosynthetic pigments. SiO_2 NP-enhanced gas exchange and chlorophyll fluorescence parameters, such as net photosynthetic/transpiration rate, stomatal conductance, PSII potential activity, effective photochemical efficiency, actual photochemical efficiency, electron transport rate, and photochemical quench, were reported.

The potential of nano-anatase TiO_2 in enhancing the light-harvesting complex content of plants can be readily compared with TiO_2-quantum dot (QD) solar energy conversion assemblies (Kongkanand et al. 2008). Nano-anatase TiO_2 possess photocatalyzing properties, which helps improve the light absorbance and subsequent conversion to chemical and electrical energy. It is also interesting to note that the TiO_2 NPs were found to protect chloroplasts from aging during long illumination regimes and promoted chlorophyll formation, in addition to stimulating RuBisCo activity and increasing photosynthesis (Hong et al. 2005b, c; Yang et al. 2006). These TiO_2 particles enhance the photosynthetic carbon assimilation by activating RuBisCo (Sharma et al. 2012). Ma et al. (2008) found enhancement of RuBisCo carboxylation with high rate of photosynthetic carbon reaction as a result of nano-anatase-induced marker gene for RuBisCo activase mRNA, enhanced protein levels, and activities of RuBisCo activase. Qi et al. (2013) investigated the exogenous application of TiO_2 NPs on plants and commented on the improved net photosynthetic rate, water conductance, and transpiration rate. Nano-anatase was observed to strongly promote electron transport chain reaction, photoreduction activity of PSII, O_2 evolution, and photophosphorylation of chlorophyll under both visible and ultraviolet light (Lei et al. 2007). Reports also suggest the nitrogen photoreduction to exercise positive effects on the improved growth of TiO_2-treated spinach plants (Yang et al. 2007; Mingyu et al. 2007b). An increase in the light-harvesting complex II (LHC II) content, which promotes energy transfer and oxygen evolution in PS-II, on thylakoid membranes of spinach was observed with the application of anatase NPs (Hong et al. 2005c; Lei et al. 2007). On the contrary, foliar applied anatase-TiO_2 NPs resulted in reduced PSII quantum yield, photochemical quenching and electron transfer rate, and chlorophyll fluorescence, but promoted higher non-photochemical quenching and water loss in long raceme ulm (Gao et al. 2013). Increased water loss due to decreased mesophyll activity and reduced electron transfer rate by blocking the electron transfer from quinone A (Q_A) to quinone B (Q_B) are suggested for the marked reduction in photosynthetic activity.

Fluorescence analysis of CeO_2 NPs (100–400 mg L^{-1})-treated alfalfa revealed a reduction in photochemical efficiency at 100 and 200 mg L^{-1} CeO_2 NP treatments. CeO_2 NPs at 200 mg L^{-1} were found to enhance the fluorescence levels of fully oxidized and completely reduced plastoquinone electron acceptor pool (Q_A), indicating the damage to PSII and the impairment of electron transport system (Gomez-Garay et al. 2014). These observations are contrary to the findings of Boghossian et al. (2013) and Giraldo et al. (2014) where isolated chloroplasts incubated with CeO_2 NPs displayed improved photosynthetic activity due partly to the ROS scavenging ability of CeO_2 NPs, which protected the chloroplasts from ROS damage. Chlorophyll *a* is known to be highly sensitive to photodegradation and many researchers have utilized this phenomenon as an indicator of NPs altering effects on the photosynthetic machinery. For example, Rico et al. (2013) found that CeO_2 NPs severely limited the chlorophyll content in rice. Fe_3O_4 and $CoFe_2O_4$ NPs also showed decreased chlorophyll content in sunflower (*Helianthus annus*) seedlings. When compared to the controls, about 50 % of chlorophyll reduction in Fe_3O_4 and 28 % in $CoFe_2O_4$ treatments were recorded (Ursache-Oprisan et al. 2011). In contrast, Ghafariyan et al. (2013) found superparamagnetic iron oxide NPs (SPIONs) to enhance the chlorophyll content in soybean. Another deleterious effect of NP on chlorophyll *a* fluorescence was observed in duck weed where CuO NPs markedly decreased the quantum yield and inhibited the photosynthetic process (Perreault et al. 2014). These NPs were also responsible for causing major modifications in PSII and decreased conversion of absorbed light energy via PSII e^- transport.

6.4.2 Carbon Materials

Recently, Giraldo et al. (2014) incorporated SWCNTs in isolated chloroplast and found that the photosynthetic activity was enhanced threefold along with relatively increased e^- transport rates. The researchers have augmented plants' ability to harvest more light energy by delivering carbon nanotubes into chloroplast, and also carbon nanotubes could serve as artificial antennae that allow chloroplast to capture wavelengths of light, which is not in their normal range, such as ultraviolet, green, and near infrared.

6.4.3 Metal Nanoparticles

According to Govorov and Carmeli (2007), metal NPs possess the tendency to influence the efficiency of energy conversions in photosynthetic systems. Chlorophyll would bind to Au and Ag NPs, forming a novel hybrid system that is projected to produce ten folds more excited electrons due to plasmon resonance and fast electron-hole separation. Such enhancement modules may assist in the design of future artificial light-harvesting systems. Electron transfer from excited fluorophore to Au or Ag NPs has been reported by numerous researchers (Barazzouk et al. 2005; Nieder et al. 2010; Beyer et al. 2011; Olejnik et al. 2013). The concentration-dependent effects of Au NPs (5–20 nm) on PSII chlorophyll *a* fluorescence quenching in soybean leaves were analyzed by Falco et al. (2011). The absorbance and fluorescence quenching were both found to be enhanced at higher concentration of Au, due to the light-absorbing tendency of Au and higher Au availability for electron transfer, respectively. On the contrary, lowest absorbance was recorded with larger Au NPs while greatest fluorescence quenching was registered with smallest Au NPs. This was due to the higher surface area of small NPs that absorb large amounts of chlorophyll molecules, facilitating better chlorophyll NPs energy transfer. In similar observations, 8-nm Au NPs enhanced quenching of chlorophyll fluorescence in a solution as a result of the enhanced e^- transfer from excited chlorophyll molecules to the metal NPs (Barazzouk et al. 2005). Falco et al. (2011) observed a shift in the fluorescence toward higher wavelength along with quenching of chlorophyll fluorescence in Au NP-treated soybean. Meanwhile, Sharma et al. (2012) documented the Ag NP-enhanced PSII quantum efficiency in Indian mustard.

Pradhan et al. (2013) analyzed the photoreduction activities in isolated chloroplasts of Mn NP-treated mungbean and revealed that the NPs alter the PSII by improving the photophosphorylation activity of electron transport chain (ETC) and by enhancing the H_2O splitting/O_2 evolution.

The influence of tetramethylammonium hydroxide (TMA-OH)-coated magnetic NPs on growth parameters of maize plants revealed an increased chlorophyll *a* level at low ferrofluid concentrations while an inhibited level was observed as the concentrations increased (Racuciu and Creanga 2009). Maize seeds germinated with magnetic fluid were exposed to an electromagnetic field. Post the electromagnetic field exposure, analysis revealed a decrease in essential assimilatory pigments with higher concentration of magnetic fluid solution. This could be attributed to the marginal localized heating due to the electromagnetic energy absorbed by magnetic NPs in plant tissues, which would have affected the redox reactions involved in photosynthesis process (Racuciu et al. 2009). It was also noted that alumina NPs can enhance the electron transfer efficiency of isolated photosynthetic reaction centers (Nadtochenko et al. 2008).

6.5 Other Physiological Parameters

Nano zerovalent iron (nZVI) triggered high plasma membrane H^+-ATPase activity in Arabidopsis, which resulted in a decrease in apoplastic pH, increase in leaf area, and wider stomatal aperture. Gene expression analysis revealed fivefold higher levels of

H^+-ATPase isoform responsible for stomatal opening, AHA2, in plants exposed to nZVI. The researchers demonstrated for the first time that nZVI enhances stomatal opening by inducing the activation of plasma membrane H^+-ATPase, leading to the possibility of increased CO_2 uptake (Kim et al. 2015). TiO_2-NPs' impact on wheat, rapeseed, and Arabidopsis evapotranspiration was evaluated (Larue et al. 2011). Similarly, TiO_2 NPs with diameters ranging from 14 to 655 nm did not impact wheat seed germination, biomass, and transpiration (Larue et al. 2012). The SiO_2 NPs apart from deleteriously affecting the plant height and shoot and root biomass also affected the contents of Cu, Mg in shoots, and Na in roots of transgenic cotton (Le et al. 2014). Nutritional analysis of pods from soybean plants cultivated in farm soil amended with CeO_2 NPs revealed that NPs at 1000 mg kg^{-1} had significantly less Ca but more

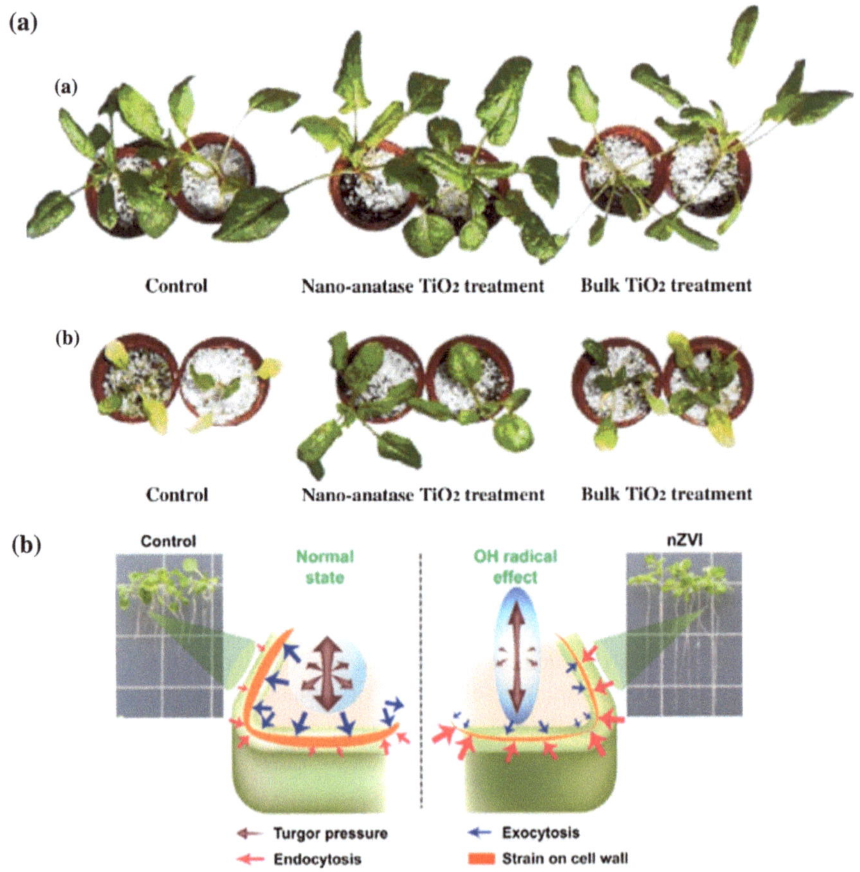

Fig. 6.3 **a** Effect of nano-anatase TiO_2 on growth of spinach. *a* Cultured by Hoagland solution. *b* Cultured by N-deficient Hoagland solution (Yang et al. 2007). **b** Conceptual model on how OH radical-induced cell wall loosening might alter endocytosis in root cells. Endocytosis (*red*) of the nZVI-treated seedlings (*right*) is enhanced compared to that of the control (*left*), and exocytosis (*blue*) is vice versa (Adopted from Kim et al. 2014)

P and Cu, while pods from 100 mg kg^{-1} nZnO had more Zn, Mn, and Cu. ZnO NP-treated plants shared significant correlations among Zn, P, and S in pods with Zn in roots. The data suggested the evident role of CeO_2 NPs and ZnO NPs in altering the nutritional value of soybean (Peralta-Videa et al. 2014). In a similar experiment, CeO_2 NPs effects on the nutritional aspects such as mineral, fatty acid, and amino acid content of wheat were studied. CeO_2 NPs modified S and Mn storage in grains and modified the amino acid composition and increased linolenic acid by up to 6.17 %. The linoleic acid content, however, was decreased by up to 1.63 %, compared to the other treatments (Rico et al. 2014). Khodakovskaya et al. (2012) demonstrated the ability of MWCNTs to enhance tobacco cell culture growth by upregulating the genes for cell divisions (*CycB*), cell wall formation (*NtLRX1*), and water transport (aquaporin, *NNtPIP1*). Lahiani et al. (2013) studied the MWCNTs-regulated gene expression for several water channel proteins in soybean, maize, and barley seed coats. Kumar et al. (2013) reported Au NPs have a significant role in altering microRNAs expression levels, which regulate various morphological, physiological, and metabolic processes in plants. Syu et al. (2014) documented the Ag NP-induced gene expressions such as indoleacetic acid protein 8 (IAA8), dehydration-responsive RD22 and 9-*cis*-epoxycarotenoid dioxygenase (NCED3), involved in various cellular events. Also, Ag NPs were found to negatively influence expression of ACC oxidase 2 and ACC synthase 7, underlining their role as inhibitors of ethylene perception and subsequently interfering with the ethylene biosynthesis in Arabidopsis seedlings (Fig. 6.3).

6.6 Discussion

As discussed earlier, nanotechnology is the only recent technological breakthrough that is revolutionizing every field it is introduced to. With trendsetting models in the electronics industry and unprecedented strides in medical arena, it seems to be only a matter of time when this technology breaks the bounds in the agro-industry. The major restricting force is the limited understanding available on the primary impacts of this technology on plants as such. Although numerous scientific articles have been published on NP–plant interactions, most revolving around the toxicological aspects of NPs, knowledge of their complex relationship with the crucial physiological and biochemical processes impacting a plant system is relatively novice. Plants in their native environment, domesticated or wild, face a number of challenges to survive. Majority of these are environmental factors such as water shortage, nutrient deficiency, alkalinity/acidity of soil, pollution, insects, and pests all of which encompass the biotic and abiotic agents influencing a plants growth and survival (Fig. 6.4).

(a)

Culture Time (day)		20	25	30	35	40	45
Glutamate dehydrogenase activity (μmol NADH·mg⁻¹ prot.·min)	Control	4.45 ±0.03a	3.33 ±0.02a	1.00 ±0.01a	1.37 ±0.01a	1.79 ±0.02a	2.65 ±0.02a
	Nano-anatase TiO_2	8.00 ±0.04b	4.67 ±0.03b	1.74 ±0.01b	2.16 ±0.01b	3.36 ±0.02b	3.88 ±0.03b
Glutamate synthase activity (u·mg⁻¹ prot.·min)	Control	2.83 ±0.03a	1.91 ±0.02a	1.01 ±0.01a	1.48 ±0.01a	1.92 ±0.02a	2.57 ±0.03a
	Nano-anatase TiO_2	6.19 ±0.04b	2.21 ±0.02b	1.88 ±0.01b	2.11 ±0.01b	2.13 ±0.02b	3.50 ±0.03b
Glutamic-pyruvic transaminase activity (μmol NADH·mg⁻¹ prot.·min)	Control	7.81 ±0.03a	6.82 ±0.03a	4.06 ±0.01a	4.46 ±0.02a	6.26 ±0.03a	7.66 ±0.04a
	Nano-anatase TiO_2	14.19 ±0.05b	8.06 ±0.04b	4.13 ±0.01b	6.76 ±0.03b	7.43 ±0.03b	8.63 ±0.04b

(b)

Treatment	Total soluble protein (mg kg^{-1})	Chlorophyll content (mg kg^{-1})	P uptake (mg kg^{-1})
Control	48.05	3.37	923.19
Ordinary ZnO	52.13	8.20	993.50
Nano ZnO	61.08	12.67	1023.27
LSD ($p = 0.05$)	0.02	0.07	0.50

(c)

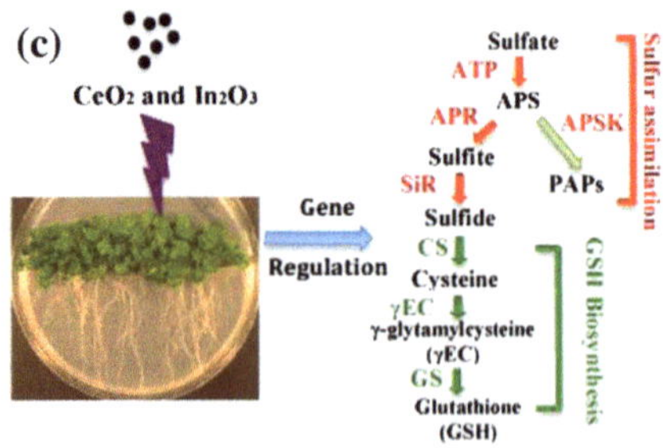

Fig. 6.4 **a** Effects of nano-anatase TiO_2 on the enzyme activities of NH^{4+} assimilation of spinach (Yang et al. 2006). **b** Total soluble protein, chlorophyll content, and P concentration in 6-week-old cluster bean plants (Raliya and Tarafdar 2013). The effects of cerium oxide (CeO_2) and indium oxide (In_2O_3) nanoparticles (NPs) exposure on *Arabidopsis thaliana* (L.) Heynh. were investigated and found to influence both the physiological and molecular level parameters (Adopted from Ma et al. 2013)

This chapter touched on the various major industrially relevant NPs currently under study for various different applications. The primary reason for initiating such studies, NP–plant interaction, was to assess the impact of the accidental release of such NPs into the environment and their potential toxicity to the environmental components including plants. Although much data suggested the toxicity of the NPs, evidenced by either germination inhibition or hindering in the growth and development, surprisingly in some cases, the effects were reverse, with noticeable enhancement of growth characteristics. Of course, the effects did vary according to the type, morphological, and chemical characteristics of the NPs and the type of plants being used (Fig. 6.5).

Initial studies were limited to basic germination and developmental features, which to date remains constant. Very few studies went beyond and analyzed the molecular and biochemical aspects of the effects. Therefore, it still remains an elusive task to concretely link a particular NPs' toxicity/beneficial attribute to a specific trait of the NP or the plant as similar NPs can show different effects on similar plant species in different experimental sets.

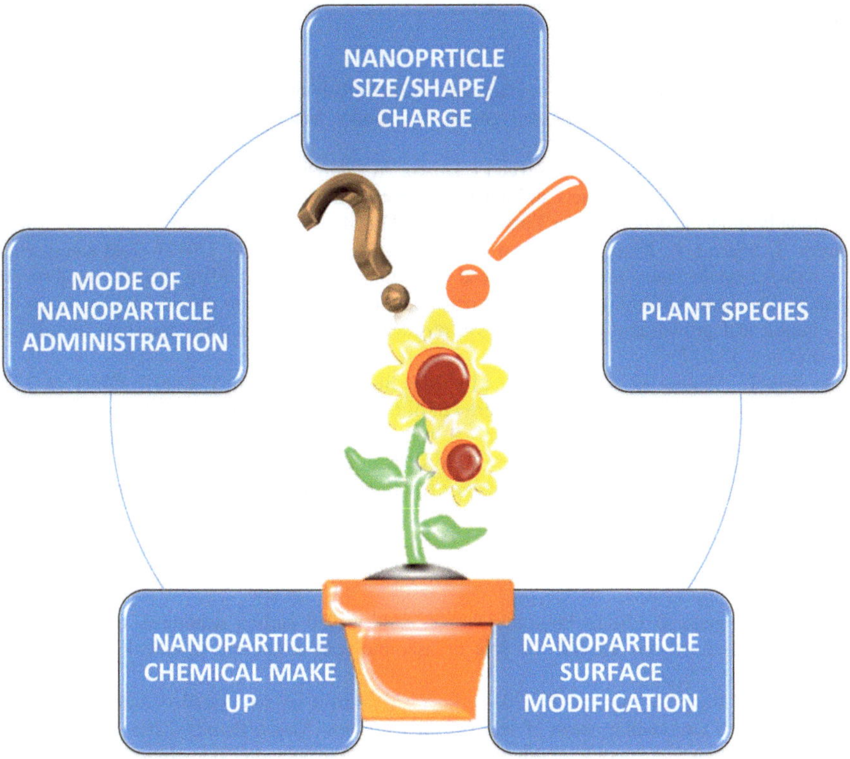

Fig. 6.5 There are multifarious determining factors while studying the effects of nanoparticles on plant system, which have to be cumulatively taken into consideration for a comprehensive understanding of the plant–nanoparticle interactions

6.7 Conclusion

The chapter discloses the use of some of the most important and commercially established NPs on the physiological parameters of various plant subjects. It is clear that the NPs exhibit both beneficial and negative influences. Understanding the chemical and physical processes of plants associated with their growth and development is critical in evaluating the role of NPs in either enhancing or retarding these features. From molecular interactions involved in photosynthesis diffusion of water, minerals, and nutrients to plant development, seasonality, and reproduction need to be thoroughly analyzed prior and post NP application. More comprehensive research needs to be performed to expand knowledge on the alterations induced by NPs on the physiological, biochemical, and molecular mechanisms of plants. Long-term studies need to be designed to assess the NPs role in regulating the physiological processes in plants to construct a database that would be helpful for current and future researchers to progress in the direction of setting a global nano-agro database accessible and useful for all.

References

Amooaghaie R, Tabatabaei F, Ali-mohammad A (2015) Role of hematin and sodium nitroprusside in regulating *Brassica nigra* seed germination under nanosilver and silver nitrate stresses. Ecotoxicol Environ Saf 113:259

Anjum NA, Singh N, Singh MK, Sayeed I, Duarte AC, Pereira E, Ahmad I (2014) Single-bilayer graphene oxide sheet impacts and underlying potential mechanism assessment in germinating faba bean (*Vicia faba* L.). Sci Total Environ 472:834

Arora S, Sharma P, Kumar S, Nayan R, Khanna PK, Zaidi MGH (2012) Gold-nanoparticle induced enhancement in growth and seed yield of *Brassica juncea*. Plant Growth Regul 66:303

Asli S, Neumann M (2009) Colloidal suspensions of clay or titanium dioxide nanoparticles can inhibit leaf growth and transpiration via physical effects on root water transport. Plant, Cell Environ 32:577

Bao-shan L, Shao-qi D, Chun-hui L, Li-jun F, Shu-chun Q, Min Y (2004) Effect of TMS (nano-structured silicon dioxide) on growth of Changbai larch seedlings. J Forest Res 15:138

Barazzouk S, Kamat PV, Hotchandani S (2005) Photoinduced electron transfer between chlorophyll a and gold nanoparticles. J Phys Chem B 109:716

Barrena R, Casals E, Colon J, Font X, Sanchez A, Puntes V (2009) Evaluation of the ecotoxicity of model nanoparticles. Chemosphere 75:850

Baun A, Hartmann NB, Grieger K, Kusk KO (2008) Ecotoxicity of engineered nanoparticles to aquatic invertebrates: a brief review and recommendations for future toxicity testing. Ecotoxicology 17:387

Begum P, Fugetsu B (2012) Phytotoxicity of multi-walled carbon nanotubes on red spinach (*Amaranthus tricolor* L) and the role of ascorbic acid as an antioxidant. J Hazard Mater 243:212

Begum P, Ikhtiari R, Fugetsu B (2014) Potential impact of multi-walled carbon nanotubes exposure to the seedling stage of selected plant species. Nanomaterials 4:203

Beyer SR, Ullrich S, Kudera S, Gardiner AT, Cogdell RJ, Kohler J (2011) Hybrid nanostructures for enhanced light-harvesting: plasmon induced increase in fluorescence from individual photosynthetic pigment-protein complexes. Nano Lett 11:4897

Bhati-Kushwaha H, Kaur A, Malik CP (2013) The synthesis and role of biogenic nanoparticles in overcoming chilling stress. Indian J Plant Sci 2:54

Boghossian AA, Sen F, Gibbons BM, Sen S, Faltermeier SM, Giraldo JP, Zhang CT, Zhang J, Heller DA, Strano MS (2013) Application of nanoparticle antioxidants to enable hyperstable chloroplasts for solar energy harvesting. Adv Energy Mater 3:881

Burklew CE, Ashlock J, Winfrey WB, Zhang B (2012) Effects of aluminum oxide nanoparticles on the growth, development, and microRNA expression of Tobacco (*Nicotiana tabacum*). PLoS ONE 7:1

Burman U, Saini M, Kumar P (2013) Effect of zinc oxide nanoparticles on growth and antioxidant system of chickpea seedlings. Toxicol Environ Chem 95:605

Cañas JE, Long M, Nations S, Vadan R, Dai L, Luo M, Ambikapathi R, Lee EH, Olszyk D (2008) Effects of functionalized and nonfunctionalized single walled carbon nanotubes on root elongation of select crop species. Environ Toxicol Chem 27:1922

Carvalho RF, Piotto FA, Schmidt D, Peters LP, Monteiro CC, Azevedo RA (2011) Seed priming with hormones does not alleviate induced oxidative stress in maize seedlings subjected to salt stress. Sci Agri 68:598

Clément L, Hurel C, Marmier N (2013) Toxicity of TiO_2 nanoparticles to cladocerans, algae, rotifers and plants—effects of size and crystalline structure. Chemosphere 90:1083

Cossins D (2014) Next generation: nanoparticles augment plant functions. The incorporation of synthetic nanoparticles into plants can enhance photosynthesis and transform leaves into biochemical sensors. The scientist, news & opinion. http://www.the-scientist.com/?articles.view/articleNo/39440/title/Next-Generation–Nanoparticles-Augment-Plant-Functions/

Crabtree RH (1998) A new type of hydrogen bond. Science 282:2000

De la Rosa G, Lopez-Moreno ML, de Haro D, Botez CE, Peralta-Videa JR, Gardea-Torresdey JL (2013) Effects of ZnO nanoparticles in alfalfa, tomato, and cucumber at the germination stage: root development and X-ray absorption spectroscopy studies. Pure Appl Chem 85:2161

Dehkourdi EH, Mosavi M (2013) Effect of anatase nanoparticles (TiO_2) on parsley seed germination (*Petroselinum crispum*) in vitro. Biol Trace Elem Res 155:283

DeRosa MC, Monreal C, Schnitzer M, Walsh R, Sultan Y (2010) Nanotechnology in fertilizers. Nat Nanotechnol 5:91

Dhoke SK, Mahajan P, Kamble R, Khanna A (2013) Effect of nanoparticles suspension on the growth of mung (*Vigna radiata*) seedlings by foliar spray method. Nanotechnol Dev 3:e1

Dimkpa CO, McLean JE, Latta DE, Manangón E, Britt DW, Johnson WP, Boyanov MI, Anderson AJ (2012) CuO and ZnO nanoparticles: phytotoxicity, metal speciation, and induction of oxidative stress in sand-grown wheat. J Nano Res 14:1

Domokos-Szabolcsy E, Marton L, Sztrik A, Babka B, Prokisch J, Fari M (2012) Accumulation of red elemental selenium nanoparticles and their biological effects in *Nicotinia tabacum*. Plant Growth Regul 68:525

El-Temsah YS, Joner EJ (2010) Impact of Fe and Ag nanoparticles on seed germination and differences in bioavailability during exposure in aqueous suspension and soil. Environ Toxicol 27:42

Faisal M, Saquib Q, Alatar AA, Al-Khedhairy AA, Hegazy AK, Musarrat J (2013) Phytotoxic hazards of NiO-nanoparticles in tomato: a study on mechanism of cell death. J Hazard Mater 250–251:318

Falco WF, Botero ER, Falcao EA, Santiago EF, Bagnato VS, Caires ARL (2011) In vivo observation of chlorophyll fluorescence quenching induced by gold nanoparticles. J Photochem Photobiol A 225:65

Feizi H, Moghaddam PR, Shahtahmassebi N, Fotovat A (2012) Impact of bulk and nano-sized titanium dioxide on wheat seed germination and seedling growth. Biol Trace Elem Res 146:101

Feizi H, Amirmoradi S, Abdollahi F, Jahedi Pour S (2013a) Comparative effects of nanosized and bulk titanium dioxide concentrations on medicinal plant *Salvia officinalis* L. Annu Rev Res Biol 3:814

Feizi H, Kamali M, Jafari L, Rezvani Moghaddam P (2013b) Phytotoxicity and stimulatory impacts of nanosized and bulk titanium dioxide on fennel (*Foeniculum vulgare* Mill). Chemosphere 91:506

Foltete AS, Masfaraud JF, Bigorgne E, Nahmani J, Chaurand P, Botta C, Labille J, Rose J, Férard JF, Cotelle S (2011) Environmental impact of sunscreen nanomaterials: ecotoxicity and genotoxicity of altered TiO_2 nanocomposites on *Vicia faba*. Environ Pollut 159:2515

Gajanan G, Deuk SY, Donghee P, Sung LD (2010) Phytotoxicity of carbon nanotubes assessed by *Brassica juncea* and *Phaseolus mungo*. J Nanoelectron Optoelectron 5:157

Galbraith DW (2007) Nanobiotechnology: silica breaks through in plants. Nat Nanotechnol 2:272

Gao FQ, Hong FS, Liu C, Zheng L, Su MY (2006) Mechanism of nano-anatase TiO_2 on promoting photosynthetic carbon reaction of spinach: inducing complex of Rubisco-Rubisco activase. Biol Trace Elem Res 111:286

Gao F, Liu C, Qu C, Zheng L, Yang F, Su M, Hong F (2008) Was improvement of spinach growth by nano-TiO_2 treatment related to the changes of Rubisco activase? Biometals 21:211

Gao J, Xu G, Qian H, Liu P, Zhao P, Hu Y (2013) Effects of nano-TiO_2 on photosynthetic characteristics of *Ulmus elongata* seedlings. Environ Pollut 176:63

Ghafariyan MH, Malakouti MJ, Dadpour MR, Stroeve P, Mahmoudi M (2013) Effect of magnetite nanoparticles on soybean chlorophyll. Environ Sci Technol 47:10645

Ghodake G, Seo YD, Lee DS (2010) Hazardous phototoxic nature of cobalt and zinc oxide nanoparticles assessed using *Allium cepa*. J Nanoelect Optoelect 5:157

Giraldo JP, Landry MP, Faltermeier SM, McNicholas TP, Iverson NM, Boghossian AA, Reuel NF, Hilmer AJ, Sen F, Brew JA, Strano MS (2014) Plant nanobionics approach to augment photosynthesis and biochemical sensing. Nat Mater 13:400

Gomez-Garay A, Pintos B, Manzanera JA, Lobo C, Villalobos N, Martin L (2014) Uptake of CeO_2 nanoparticles and its effect on growth of *Medicago arborea* In vitro plantlets. Biol Trace Elem Res 161:143

Gopinath K, Gowri S, Karthika V, Arumugam A (2014) Green synthesis of gold nanoparticles from fruit extract of *Terminalia arjuna*, for the enhanced seed germination activity of *Gloriosa superba*. J Nanostruct Chem 4:1

Govorov AO, Carmeli I (2007) Hybrid structures composed of photosynthetic system and metal nanoparticles: plasmon enhancement effect. Nano Lett 7:620

Gratão PL, Polle A, Lea PJ, Azevedo RA (2005) Making the life of heavy metal-stressed plants a little easier. Funct Plant Biol 32:481

Gruyer N, Dorais M, Bastien C, Dassylva N, Triffault-Bouchet G (2013) Interaction between sliver nanoparticles and plant growth. In: International symposium on new technologies for environment control, energy-saving and crop production in greenhouse and plant factory–greensys, Jeju, Korea, 6–11 Oct

Gubbins EJ, Batty LC, Lead JR (2011) Phytotoxicity of silver nanoparticles to *Lemna minor* L. Environ Pol 159:1551

Haghighi M, Afifipour Z, Mozafarian M (2012) The effect of N-Si on tomato seed germination under salinity levels. J Biol Environ Sci 6:87

Helaly MN, El-Metwally MA, El-Hoseiny H, Omar SA, El-Sheery NI (2014) Effect of nanoparticles on biological contamination of in vitro cultures and organogenic regeneration of banana. Aust J Crop Sci 8:612

Hernandez-Viezcas JA, Castillo-Michel H, Servin AD, Peralta-Videa JR, Gardea-Torresdey JL (2011) Spectroscopic verification of zinc absorption and distribution in the desert plant *Prosopis juliflora-velutina* (velvet mesquite) treated with ZnO nanoparticles. Chem Eng J 170:346

Hong F, Yang P, Gao FQ, Liu C, Zheng L (2005a) Effect of nano-TiO_2 on spectral characterization of photosystem particles from spinach. Chem Res Chin Univ 21:196

Hong F, Zhou J, Liu C, Yang F, Wu C, Zheng L, Yang P (2005b) Effect of nano-TiO_2 on photo-chemical reaction of chloroplasts of spinach. Biol Trace Elem Res 105:269

Hong F, Yang F, Ma ZN, Zhou J, Liu C, Wu C, Yang P (2005c) Influences of nano-TiO_2 on the chloroplast ageing of spinach under light. Biol Trace Elem Res 104:249

Hong J, Peralta-Videa Jose R, Rico C, Sahi S, Viveros MN, Bartonjo J, Zhao L, Gardea-Torresdey JL (2014) Evidence of translocation and physiological impacts of foliar applied CeO_2 nanoparticles on cucumber (*Cucumis sativus*) plants. Environ Sci Technol 48:4376

Husen A, Siddiqi KS (2014) Carbon and fullerene nanomaterials in plant system. J Nanotechnol 12:1

Ikhtiar R, Begum P, Watari F, Fugetsu B (2013) Toxic effect of multiwalled carbon nanotubes on lettuce (*Lactuca sativa*). Nano Biomed 5:18

Jaberzadeh A, Moaveni P, Moghadam HRT, Zahedi H (2013) Influence of bulk and nanoparticles titanium foliar application on some agronomic traits, seed gluten and starch contents of wheat subjected to water deficit stress. Not Bot Horti Agrobo 41:201

Jacob DL, Borchardt JD, Navaratnam L, Otte ML, Bezbaruah AN (2013) Uptake and translocation of Ti from nanoparticles in crops and wetland plants. Int J Phytoremed 15:142

Jiang H, Liu JK, Wang JD, Lu Y, Zhang M, Yang XH, Hong DJ (2014) The biotoxicity of hydroxyapatite nanoparticles to the plant growth. J Hazard Mater 270:71

Juhel G, Batisse E, Hugues Q, Daly D, van Pelt FN, O'Halloran J, Jansen MA (2011) Alumina nanoparticles enhance growth of *Lemna minor*. Aquat Toxicol 105:328

Kalteh M, Alipour ZT, Ashraf S, Aliabadi MM, Nosratabadi AF (2014) Effect of silica nanoparticles on basil (*Ocimum basilicum*) under salinity stress. J Chem Health Risks 4:49

Khodakovskaya MV, de Silva K, Biris AS, Dervishi E, Villagarcia H (2012) Carbon nanotubes induce growth enhancement of tobacco cells. ACS Nano 6:2128

Khodakovskaya MV, Kim BS, Kim JN, Alimohammadi M, Dervishi E, Mustafa T, Cernigla CE (2013) Carbon nanotubes as plant growth regulators: effects on tomato growth, reproductive system, and soil microbial community. Small 9:115

Kim JH, Lee Y, Kim EJ, Gu S, Sohn EJ, Seo YS, An HJ, Chang YS (2014) Exposure of iron nanoparticles to *Arabidopsis thaliana* enhances root elongation by triggering cell wall loosening. Environ Sci Technol 48:3477

Kim JH, Oh Y, Yoon H, Hwang I, Chang YS (2015) Iron nanoparticle-induced activation of plasma membrane H^+-ATPase promotes stomatal opening in *Arabidopsis thaliana*. Environ Sci Technol 49:1113

Kirschbaum MUF (2011) Does enhanced photosynthesis enhance growth? Lessons learned from CO_2 enrichment studies. Plant Physiol 155:117

Kole C, Kole P, Randunu RK, Choudhary P, Podila R, Ke PC, Rao AM, Marcus RK (2013) Nanobiotechnology can boost crop production and quality: first evidence from increased plant biomass, fruit yield and phytomedicine content in bitter melon (*Momordica charantia*). BMC Biotechnol 13:37

Kongkanand A, Tvrdy K, Takechi K, Kuno M, Kamat PV (2008) Quantum dot solar cells. Tuning photoresponse through size and shape of CdSe–TiO_2 architecture. J Amer Chem Soc 130:4007

Krishnaraj C, Jagan EG, Ramachandran R, Abirami SM, Mohan N, Kalaichelvan PT (2012) Effect of biologically synthesized silver nanoparticles on *Bacopa monnieri* (Linn.) Wettst. Plant growth metabolism. Process Biochem 47:51

Kumar V, Guleria P, Kumar V, Yadav SK (2013) Gold nanoparticle exposure induces growth and yield enhancement in *Arabidopsis thaliana*. Sci Total Environ 461:462

Kumari M, Khan SS, Pakrashi S, Mukherjee A, Chandrasekaran N (2011) Cytogenetic and genotoxic effects of zinc oxide nanoparticles on root cells of *Allium cepa*. J Hazard Mater 190:613

Lahiani MH, Dervishi E, Chen J, Nima Z, Gaume A, Biris AS, Khodakovskaya MV (2013) Impact of carbon nanotube exposure to seeds of valuable crops. ACS Appl Mater Interf 5:7965

Larue C, Khodja H, Herlin-Boime N, Brisset F, Flank AM, Fayard B, Chaillou S, Carriere M (2011) Investigation of titanium dioxide nanoparticles toxicity and uptake by plants. J Phys: Conf Ser 304:012057

Larue C, Laurette H. Herlin-Boime N, Khodja H, Fayard B, Flank AM, Brisset F, Carriere M (2012) Accumulation, translocation and impact of TiO_2 nanoparticles in wheat (*Triticum aestivum* spp.): influence of diameter and crystal phase. Sci Tot Ennviron 431:197

Le V, Rui Y, Gui X, Li X, Liu S, Han Y (2014) Uptake, transport, distribution and Bio-effects of SiO_2 nanoparticles in Bt-transgenic cotton. J Nanobiotechnol 12:50

Lee WM, An YJ, Yoon H, Kweon HS (2008) Toxicity and bioavailability of copper nanoparticles to terrestrial plants *Phaseolus radiatus* (Mung bean) and *Triticum aestivum* (Wheat); plant agar test for water-insoluble nanoparticles. Environ Toxicol Chem 27:1915

Lee CW, Mahendra S, Zodrow K, Li D, Tsai YC, Braam J, Alvarez PJJ (2010) Developmental phytotoxicity of metal oxide nanoparticles to *Arabidopsis thaliana*. Environ Toxicol Chem 29:669

Lee WM, Kwak JI, An YJ (2012) Effect of silver nanoparticles in crop plants *Phaseolus radiatus* and *Sorghum bicolor*: media effect on phytotoxicity. Chemosphere 86:491

Lei Z, Mingyu S, Chao L, Liang C, Hao H, Xiao W, Xiaoqing L, Fan Y, Fengqing G, Fashui H (2007) Effects of nanoanatase TiO_2 on photosynthesis of spinach chloroplasts under different light illumination. Biol Trace Elem Res 119:68

Lei Z, Mingyu S, Xiao W, Chao L, Chunxiang Q, Liang C, Hao H, Xiaoqing L, Fashui H (2008) Antioxidant stress is promoted by nano-anatase in spinach chloroplasts under UV-Beta radiation. Biol Trace Elem Res 121:69

Li B, Tao G, Xie Y, Cai X (2012) Physiological effects under the condition of spraying nano-SiO_2 onto the *Indocalamus barbatus* McClure leaves. J Nanjing For Univ (Natl Sci Edn) 36:161

Lin D, Xing B (2007) Phytotoxicity of nanoparticles: inhibition of seed germination and root growth. Environ Pollut 150:243

Lin S, Reppert J, Hu Q, Hudson JS, Reid ML, Ratnikova TA, Rao AM, Luo H, Ke PC (2009) Uptake, translocation, and transmission of carbon nanomaterials in rice plants. Small 5:1128

Linglan M, Chao L, Chunxiang Q, Sitao Y, Jie L, Fengqing G, Fashui H (2008) Rubisco activase mRNA expression in spinach: modulation by nanoanatase treatment. Biol Trace Elem Res 122:168

Lopez-Moreno ML, De La Rosa G, Hernandez-Viezcas JA, Castillo-Michel Botez CE, Peralta-Videa JR, Gardea- Torresdey JL (2010a) Evidence of the differential biotransformation and genotoxicity of ZnO and CeO_2 nanoparticles on soybean (*Glycine max*) plants. Environ Sci Technol 44:7315

Lopez-Moreno ML, De La Rosa G, Hernandez-Viezcas JA, Peralta-Videa JR, Gardea-Torresdey JL (2010b) X-ray absorption spectroscopy (XAS) corroboration of the uptake and storage of CeO_2 nanoparticles and assessment of their differential toxicity in four edible plant species. J Agric Food Chem 58:3689

Lu C, Zhang C, Wen J, Wu G, Tao M (2002) Research on effect of nanometer materials on germination and growth enhancement of *Glycine max* and its mechanism. Soybean Sci 21:168

Ma L, Liu C, Qu C, Yin S, Liu J, Gao F, Hong F (2008) Rubisco activase mRNA expression in spinach: modulation by nanoanatase treatment. Biol Trace Elem Res 122:168

Ma X, Geiser-Lee J, Deng Y, Kolmakov A (2010) Interactions between engineered nanoparticles (ENPs) and plants: phytotoxicity, uptake and accumulation. Sci Total Environ 408:3053

Ma C, Chhikara S, Xing B, Musante C, White JC, Dhankher OP (2013) Physiological and molecular response of *Arabidopsis thaliana* (L.) to nanoparticle cerium and indium oxide exposure. ACS Sustain Chem Eng 1:768

Mahajan P, Dhoke SK, Khanna AS (2011) Effect of nano-ZnO particle suspension on growth of mung (*Vigna radiata*) and gram (*Cicer arietinum*) seedlings using plant agar method. J Nanotechnol 1(696535). doi:10.1155/2011/696535

Mahmoodzadeh H, Nabavi M, Kashefi H (2013) Effect of nanoscale titanium dioxide particles on the germination and growth of canola (*Brassica napus*). J Ornam Hortic Plants 3:25

Marusenko Y, Shipp J, Hamilton GA, Morgan JLL, Keebaugh M, Hill H, Dutta A, Zhuo X, Upadhyay N, Hutchings J, Herckes P, Anbar AD, Shock E, Hartnett HE (2013) Bioavailability of nanoparticulate hematite to *Arabidopsis thaliana*. Environ Pollut 174:150

Mingyu S, Hong F, Liu C, Wu X, Liu X, Chen L, Gao F, Yang F, Li Z (2007a) Effects of nano-anatase TiO_2 on absorption, distribution of light and photoreduction activities of chloroplast membrane of spinach. Biol Trace Elem Res 118:120

Mingyu S, Wu X, Liu C, Qu C, Liu X, Chen L, Huang H, Hong F (2007b) Promotion of energy transfer and oxygen evolution in spinach photosystem II by nano-anatase TiO_2. Biol Trace Elem Res 119:183

Miralles P, Johnson E, Church TL, Harris AT (2012) Multiwalled carbon nanotubes in alfalfa and wheat: toxicology and uptake. J Roy Soc Interf 9:3514

Mirzajani F, Askari H, Hamzelou S, Farzaneh M, Ghassempour A (2013) Effect of silver nanoparticles on *Oryza sativa* L. and its rhizosphere bacteria. Ecotoxicol Environ Saf 88:48

Mishra V, Mishra RK, Dikshit A, Pandey AC (2014) Interactions of nanoparticles with plants: an emerging prospective in the agriculture industry. In: Ahmad P, Rasool S (eds) Emerging technologies and management of crop stress tolerance: biological techniques. Academic Press, Elsevier, Cambridge, USA, vol 1, p 159

Mondal A, Basu R, Das S, Nandy P (2011) Beneficial role of carbon nanotubes on mustard plant growth: an agricultural prospect. J Nanopart Res 13:4519

Morla S, Ramachandra Rao CSV, Chakrapani R (2011) Factors affecting seed germination and seedling growth of tomato plants cultured in vitro conditions. J Chem Bio Phys Sci B 1:328

Nadtochenko VA, Nikandrov VV, Gorenberg AA, Karlova MG, Lukashev EP, Yu A, Semenov Bukharina NS, Kostrov AN, Permenova EP, Sarkisov OM (2008) Nanophotobiocatalysts based on mesoporous titanium dioxide films conjugated with enzymes and photosynthetic reaction centers of bacteria. High Energy Chem 42:591

Nair PM, Chung IM (2015) Study on the correlation between copper oxide nanoparticles induced growth suppression and enhanced lignification in Indian mustard (*Brassica juncea* L.). Ecotoxicol Environ Saf 113:302

Nair R, Varghese SH, Nair BG, Maekawa T, Yoshida Y, Kumar DS (2010) Nanoparticulate material delivery to plants. Plant Sci 179:154

Nalwade AR, Neharkar SB (2013) Carbon nanotubes enhance the growth and yield of hybrid Bt cotton Var. ACH-177-2. Int J Adv Sci Tech Res 3:840

Nieder JB, Bittl R, Brecht M (2010) Fluorescence studies into the effect of plasmonic interactions on protein function. Angew Chem Int Edn 49:10217

Noji T, Kamidaki C, Kawakami K, Shen JR, Kajino T, Fukushima Y, Sekitoh T, Itoh S (2011) Photosynthetic oxygen evolution in mesoporous silica material: adsorption of photosystem II reaction center complex into 23 nm nanopores in SBA. Langmuir 27:705

Olejnik M, Krajnik B, Kowalska D, Twardowska M, Czechowski N, Hofmann E, Mackowski S (2013) Imaging of fluorescence enhancement in photosynthetic complexes coupled to silver nanowires. Appl Phys Lett 102:083703

Patra P, Choudhury SR, Mandal S, Basu A, Goswami A, Gogoi R, Srivastava C, Kumar R, Gopal M (2013) Effect sulfur and ZnO nanoparticles on stress physiology and plant (*Vigna radiata*) nutrition. In: Giri PK, Goswami DK, Perumal A (eds) Advanced nanomaterials and nanotechnology. Springer, Berlin, Heidelberg, Germany, p 301

Peralta-Videa JR, Hernandez-Viezcas JA, Zhao L, Diaz BC, Ge Y, Preister JH, Holden PA, Gardea-Torresday J (2014) Cerium dioxide and zinc oxide nanoparticles alter the nutritional value of soil cultivated soy bean plants. Plant Physiol Biochem 80:128

Perreault F, Samadani M, Dewez D (2014) Effect of soluble copper released from copper oxide nanoparticles solubilisation on growth and photosynthetic processes of *Lemna gibba* L. Nanotoxicology 8:374

Pradhan S, Patra P, Das S, Chandra S, Mitra S, Dey KK, Akbar S, Palit P, Goswami A (2013) Photochemical modulation of biosafe manganese nanoparticles on *Vigna radiata*: a detailed molecular, biochemical, and biophysical study. Environ Sci Technol 47:13122

Prasad TNVKV, Sudhakar P, Sreenivasulu Y, Latha P, Munaswamy V, Reddy KR, Sreeprasad TSP, Sajanlal R, Pradeep T (2012) Effect of nanoscale zinc oxide particles on the germination, growth and yield of peanut. J Plant Nutr 35:905

Qi M, Liu Y, Li T (2013) Nano-TiO_2 improve the photosynthesis of tomato leaves under mild heat stress. Biol Trace Elem Res 156:323

Racuciu M, Creanga D (2009) Biocompatible magnetic fluid nanoparticles internalized in vegetal tissues. Roman J Phys 54:115

Racuciu M, Miclaus S, Creanga D (2009) The response of plant tissues to magnetic fluid and electromagnetic exposure. Rom J Biophys 19:73

Raliya R, Tarafdar JC (2013) ZnO nanoparticle biosynthesis and its effect on phosphorous-mobilizing enzyme secretion and gum contents in clusterbean (*Cyamopsis tetragonoloba* L.). Agric Res 2:48

Ramesh M, Palanisamy K, Babu K, Sharma NK (2014) Effects of bulk & nano-titanium dioxide and zinc oxide on physio-morphological changes in *Triticum aestivum* Linn. J Glob Biosci 3:415

Raskar SV, Laware SL (2014) Effect of zinc oxide nanoparticles on cytology and seed germination in onion. Int J Curr Microbiol Appl Sci 3:467

Rezvani N, Sorooshzadeh A, Farhadi N (2012) Effect of nano-silver on growth of saffron in flooding stress. World Acad Sci Eng Technol 1:517

Rico CM, Morales MI, Barrios AC, McCreary R, Hong J, Lee WY, Nunez J, Peralta-Videa JR, Gardea-Torresday JL (2013) Effect of cerium oxide nanoparticles on the quality of rice (*Oryza sativa* L.) grains. J Agric Food Chem 61:11278

Rico CM, Lee SC, Rubenecia R, Mukerjee A, Hong J, Peralta-Videa JR, Gardea-Torresdey JL (2014) Cerium oxide nanoparticles impact yield and modify nutritional parameters in wheat (*Triticum aestivum* L.). J Agric Food Chem 62:9669

Salama HMH (2012) Effects of silver nanoparticles in some crop plants, common bean (*Phaseolus vulgaris* L.) and corn (*Zea mays* L.). Int Res. J Biotechnol 3:190

Savithramma N, Ankanna S, Bhumi G (2012) Effect of nanoparticles on seed germination and seedling growth of *Boswellia ovalifoliolata* an endemic and endangered medicinal tree taxon. Nano Vision 2:61

Sedghi M, Hadi M, Toluie SG (2013) Effect of nano zinc oxide on the germination of soybean seeds under drought stress. Ann West Uni Timisoara Ser Biol XVI:73

Shah V, Belozerova I (2009) Influence of metal nanoparticles on the soil microbial community and germination of lettuce seeds. Water Air Soil Pollut 197:143

Sharma P, Bhatt D, Zaidi MG, Saradhi PP, Khanna PK, Arora S (2012) Silver nanoparticle-mediated enhancement in growth and antioxidant status of *Brassica juncea*. Appl Biochem Biotechnol 167:2225

Sheykhbaglou R, Sedghi M, Shishevan MT, Sharifi RS (2010) Effects of nano-iron oxide particles on agronomic traits of soybean. Not Sci Biol 2:112

Siddiqui MH, Al-Whaibi MH (2014) Role of nano-SiO_2 in germination of tomato (*Lycopersicum esculentum* seeds Mill.). Saudi Biol Sci 21:13

Siddiqui MH, Mohammad F, Khan MMA, Al-Whaibi MH (2012) Cumulative effect of nitrogen and sulphur on *Brassica juncea* L. genotypes under NaCl stress. Protoplasma 249:139

Siddiqui MH, Al-Whaibi MH, Faisal M, Al Sahli AA (2014) Nano-silicon dioxide mitigates the adverse effects of salt stress on *Cucurbita pepo* L. Environ Toxicol Chem 33:2429

Smirnova E, Gusev A, Zaytseva O, Sheina O, Tkachev A, Kuznetsova E, Lazareva E, Onishchenko G, Feofanov A, Kirpichnikov M (2012) Uptake and accumulation of multiwalled carbon nanotubes change the morphometric and biochemical characteristics of *Onobrychis arenaria* seedlings. Front Chem Sci Eng 6:132

Song G, Gao Y, Wu H, Hou W, Zhang C, Ma H (2012) Physiological effect of anatase TiO_2 nanoparticles on *Lemna minor*. Environ Toxicol Chem 31:2147

Song U, Jun H, Waldman B, Roh J, Kim Y, Yi J, Lee EJ (2013) Functional analyses of nanoparticle toxicity: a comparative study of the effects of TiO_2 and Ag on tomatoes (*Lycopersicon esculentum*). Ecotoxicol Environ Saf 93:60

Stampoulis D, Sinha SK, White JC (2009) Assay-dependent phytotoxicity of nanoparticles to plants. Environ Sci Technol 43:9473

Suriyaprabha R, Karunakaran G, Yuvakkumar R, Rajendran V, Kannan N (2012) Silica nanoparticles for increased silica availability in maize (*Zea mays* L) seeds under hydroponic conditions. Curr Nanosci 8:902

Syu YY, Hung JH, Chen JC, Chuang HW (2014) Impacts of size and shape of silver nanoparticles on *Arabidopsis* plant growth and gene expression. Plant Physiol Biochem 83:57

Tiwari DK, Dasgupta-Schubert N, Villaseñor LM, Tripathi D, Villegas J (2013) Interaction of carbon nanotubes with mineral nutrients for the promotion of growth of tomato seedlings. Nano Stud 7:87

Tiwari DK, Dasgupta-Schubert N, Villaseñor-Cendejas LM, Villegas J, Carreto-Montoya L, Borjas-García SE (2014) Interfacing carbon nanotubes (CNT) with plants: Enhancement of growth, water and ionic nutrient uptake in maize (*Zea mays*) and implications for nanoagriculture. Appl Nanosci 4:577

Torney F, Trewyn BG, Lin VS-Y, Wang K (2007) Mesoporous silica nanoparticles deliver DNA and chemicals into plants. Nat Nanotechnol 2:295

Tripathi S, Sarkar S (2015) Influence of water soluble carbon dots on the growth of wheat plant. Appl Nanosci 5:609

Tripathi S, Sonkar SK, Sarkar S (2011) Growth stimulation of gram (*Cicer arietinum*) plant by water soluble carbon nanotubes. Nanoscale 3:1176

Ursache-Oprisan M, Focanici E, Creanga D, Caltun O (2011) Sunflower chlorophyll levels after magnetic nanoparticle supply. Afr J Biotechnol 10:7092

Villagarcia H, Dervishi E, Silva K, Biris AS, Khodakovskaya MV (2012) Surface chemistry of carbon nanotubes impacts the growth and expression of water channel protein in tomato plants. Small 8:2328

Wang X, Han H, Liu X, Gu X, Chen K, Lu D (2012) Multi-walled carbon nanotubes can enhance root elongation of wheat (*Triticum aestivum*) plants. J Nanopart Res 14:1

Wang A, Zheng Y, Peng F (2014) Thickness-controllable silica coating of CdTe QDs by reverse microemulsion method for the application in the growth of rice. J Spectrosc (169245). doi:10.1155/2014/169245

Wu SG, Huang L, Head J, Chen DR, Kong IC, Tang YJ (2012) Phytotoxicity of metal oxide nanoparticles is related to both dissolved metals ions and adsorption of particles on seed surfaces. J Pet Environ Biotechnol 3:126

Xie Y, Li B, Zhang Q, Zhang C (2012) Effects of nano-silicon dioxide on photosynthetic fluorescence characteristics of *Indocalamus barbatus* McClure. J Nanjing Forest Univ (Natl Sci Ed) 2:59

Yang L, Watts DJ (2005) Particle surface characteristics may play an important role in phytotoxicity of alumina nanoparticles. Toxicol Lett 158:122

Yang F, Hong F, You W, Liu C, Gao F, Wu C, Yang P (2006) Influence of nano-anatase TiO_2 on the nitrogen metabolism of growing spinach. Biol Trace Elem Res 110:179

Yang F, Liu C, Gao F, Su M, Wu X, Zheng L, Hong F, Yang P (2007) The improvement of spinach growth by nano-anatase TiO_2 treatment is related to nitrogen photoreduction. Biol Trace Elem Res 119:77

Yin L, Colman BP, McGill BM, Wright JP, Bernhardt ES (2012) Effects of silver nanoparticle exposure on germination and early growth of eleven wetland plants. PLoS ONE 7:1

Yuvakkumar R, Elango V, Rajendran V, Kannan NS, Prabu P (2011) Influence of nanosilica powder on the growth of maize crop (*Zea mays* L.). Int J Green Nanotechnol 3:80

Zhao L, Hernandez-Viezcas JA, Peralta-Videa JR, Bandyopadhyay S, Peng B, Munoz B, Kellerce AA, Gardea-Torresdey JL (2013) ZnO nanoparticle fate in soil and zinc bioaccumulation in corn plants (*Zea mays*) influenced by alginate. Environ Sci: Processes Impacts 15:260

Zhao L, Peralta-Videa JR, Rico CM, Hernandez-Viezcas JA, Sun Y, Niu G, Duarte-Gardea M, Gardea-Torresdey JL (2014) CeO_2 and ZnO nanoparticles change the nutritional qualities of cucumber (*Cucumis sativus*). J Agric Food Chem 62:2752

Zhao S, Wang Q, Zhao Y, Rui Q, Wang D (2015) Toxicity and translocation of graphene oxide in *Arabidopsis thaliana*. Environ Toxicol Pharmacol 39:145

Zheng L, Hong F, Lu S, Liu C (2005) Effect of nano-TiO_2 on strength of naturally aged seeds and growth of spinach. Biol Trace Elem Res 104:83

Zheng L, Su M, Liu C, Chen L, Huang H, Wu X, Liu X, Yang F, Gao F, Hong F (2007) Effects of nanoanatase TiO_2 on photosynthesis of spinach chloroplasts under different light illumination. Biol Trace Elem Res 119:68

Chapter 7
Molecular Mechanism of Plant–Nanoparticle Interactions

Shweta Jha and Ramesh Namdeo Pudake

Abstract Research and development in the field of nanotechnology are rapidly progressing in all aspects of human life. Recently, the use of engineered nanomaterial (ENM) is being conceptualized in the field of agriculture and food industry. These ENMs are often released into the environment and pose toxicity risk due to potential uptake by crop plants. Standard developmental and physiological methods to measure phytotoxicity including seed germination, root elongation, and enzymatic assays are not sensitive enough while evaluating nanoparticle toxicity to terrestrial plant species. Also, unique properties of nanomaterials allow them to interact with biological systems. Understanding the nature of interactions between nanoparticles and plants is crucial for assessing their uptake, distribution, and toxicity associated with exposure of plants to nanoparticles. However, little progress has been made toward understanding the impact of nanomaterials at molecular level, which is an important step in evaluation of the possible mechanisms of observed effects *in planta*. Analysis of changes in gene expression through transcriptomics constitutes a powerful approach toward understanding the mechanism of phytotoxicity and molecular responses of plants exposed to nanoparticles. Also, global protein profiling, emerging as a new field "nanotoxicoproteomics," can be used for understanding plant responses to toxic nanomaterials. The present chapter reviews the current knowledge on phytotoxicity assessment and interactions of nanoparticles with plants at the cellular level and discusses the future aspects to improve our knowledge of this field.

Keywords Nanoparticles · Nano-ecotoxicology · Nano-toxicogenomics · Nano-toxicoproteomics · Phytotoxicity · Risk assessment

---

S. Jha (✉)
Biotechnology Lab, Department of Botany (UGC-Centre of Advanced Study), Jai Narain Vyas University, Jodhpur 342001, Rajasthan, India
e-mail: jha.shweta80@gmail.com; sj.bo@jnvu.edu.in

R.N. Pudake
Amity Institute of Nanotechnology, Amity University Uttar Pradesh, Sector 125, Noida 201313, Uttar Pradesh, India
e-mail: rnpudake@amity.edu

C. Kole et al. (eds.), *Plant Nanotechnology*,
DOI 10.1007/978-3-319-42154-4_7

7.1 Introduction

Nanoparticles (NPs) are considered as building block of nanotechnology. They have size in the range of nanoscale (1–100 nm) and possess a high surface-to-volume ratio, resulting in higher solubility and surface reactivity (Remédios et al. 2012). Nanoparticulate matter can be produced by natural processes or by anthropogenic manufacturing. Any material that is intentionally produced in the nanoscale to have novel properties is called a manufactured/engineered nanoparticle (MNP/ENP). The properties of nanomaterials such as size, shape, structure, and surface characteristics result in novel physicochemical and biological properties different from their corresponding bulk counterparts (Handy et al. 2008). They can be spherical, tubular, irregularly shaped and can also exist in aggregated forms. United States Environmental Protection Agency (US EPA) has classified engineered nanoparticles in four different categories based on their composition, viz. (1) carbon-based materials, e.g., carbon nanotubes (CNTs) and fullerenes, (2) metal-based nanoparticles, e.g., Ag NP, metal oxides such as TiO_2, alumina (Al_2O_3), CuO, ZnO, iron oxide nanoparticles, such as magnetite (Fe_3O_4) and maghemite (γ-Fe_2O_3), quantum dots (QDs), (3) dendrimers, and (4) bio-inorganic complexes (US EPA 2007).

Nanoparticles have enormous functions and almost infinite applications in different fields including biotechnology, pharmaceuticals (e.g., antibacterial agent), biomedicine and drug delivery, cosmetics, electronics, optics, material science, textiles, energy sectors, water treatment technology, environmental remediation, and food and agriculture industry (Remédios et al. 2012). Nanotechnology is one of the most important tools for crop improvement in modern agriculture. The key applications of nanotechnology in agriculture include nano-agrochemicals (nanopesticides, nanofertilizers), agri-food production, nanobiosensors, agri-environment, nanobiocomposites, biofuels, organic agriculture, particle farming, nanofoods, postharvest management, and nanoparticle-mediated gene transfer for genetic improvement of plants (e.g., for enhancing nutritional quality/shelf-life, developing abiotic/biotic stress resistance) (Sekhon 2014). Presently, the number of patents filed or granted for manufacturing processes of nanopesticides, nanofertilizers, and nanosensors has also increased.

Plant species have evolved in the presence of various natural metal and organic nanomaterials. However, there is many-fold increase in the chances of intentional and nonintentional exposure of plants to nanomaterials with the ongoing increasing production and utilization of ENMs in a variety of fields. The increased applications could pose significant threats to the environment and human health (Hood 2004). Although the toxicity of released nanomaterials to human cell lines, mammalian models, microbes, and other ecological indicator organisms has been investigated intensely, their effects on plants have received less attention. As plants constitute the first trophic level of the terrestrial food chain, their exposure to nanomaterials may have significant implications for human health (Klaine et al. 2008; Ma et al. 2010; Rico et al. 2011). To characterize the risk of phytotoxicity, studies have been performed in plants using standard morphological and physiological assays, such as seed germination, root elongation, biomass, lipid peroxidation, and enzymatic activity (Remédios et al. 2012). But to achieve promised benefits offered by agri-nanotechnology at the fullest, the comprehensive studies on effects of ENMs on growth and development of valuable agricultural plant species are needed at whole-genome and whole-proteome level. However, to our knowledge, there are only few reports describing the plant–nanoparticle interaction at the molecular level, and almost nothing is known about the molecular mechanism of nanoparticle-mediated phytotoxicity.

In this chapter, recently published studies on transcriptomics and proteomics in plants exposed to different types of nanoparticles have been reviewed, which necessitates on further research to understand the molecular response triggered by nanoparticles in plants.

7.2 Aspects of Nanoparticles' Association with Plants

The increasing production and widespread utilization of nanoparticles entails the risk of their indiscriminate release to the environment (Gottschalk and Nowack 2011), raising concerns for environmental hazards and adverse health effects (Bhatt and Tripathi 2011). Nanoparticles can enter the environment due to unintentional human activities through accidental release, such as combustion, atmospheric emissions, domestic wastewater, transport and chemical manufacturing; or through intentional releases such as during environmental remediation efforts, and crop improvement. The release of engineered nanoparticles in the environment may cause adverse effects on edible plants by entering into water and soil, thus affecting the whole food chain. Thus, nanoparticles should be treated as a "new" group of contaminants that may pose a serious threat to the environment. It is necessary to evaluate their environmental fate and potential toxicity by developing appropriate risk assessment methods.

Nanotoxicological studies have been performed on a few model organisms, with a focus on mammalian cytotoxicity and impacts on animals and bacteria (Roco 2005). However, little is known about the potential effects of nanomaterials on plants. Like other living organisms, number of nanoscale materials has been shown to be absorbed by plant cells and translocated to various tissues and plant organs (Liu et al. 2009; Nair et al. 2010; Khodakovskaya et al. 2011; Aken 2015; Chen et al. 2015; Mattiello et al. 2015). This can provide a route for their bioaccumulation into the food chain (Zhu et al. 2008), constituting a significant link in eco-toxicological studies. Therefore, assessment of potential phytotoxicity of engineered nanoparticles is particularly crucial.

Several recent studies have evaluated nanoparticle-mediated phytotoxicity as well as their ecotoxicity (Barrena et al. 2009; Lee et al. 2010, 2012, 2013). The effect of nanoparticles on different plants varies greatly, and both positive and negative effects have been reported, depending on the plant species and nature/composition of nanoparticles, size, concentration, and exposure time. For instance, various types of nanoparticles were found to negatively affect seed germination rate, root elongation, and biomass in plants. Phytotoxicity of silver nanoparticles (Ag NPs) was studied on rice (*Oryza sativa*) (Mazumdar and Ahmed 2011), cucumber (*Cucumis sativus*) and lettuce (*Lactuca sativa*) seeds (Barrena et al. 2009), ryegrass (*Lolium perenne*), barley (*Hordeum vulgare*) (El-Temsah and Joner 2012), common duckweed (*Lemna minor*) (Gubbins et al. 2011), gibbous duckweed (*Lemna gibba*) (Farrag 2015), zucchini (*Cucurbita pepo*) (Stampoulis et al. 2009; Craig and Jason 2011), tomato (*Solanum lycopersicum*), radish (*Raphanus sativus*) (Shiny et al. 2013), mung bean (*Vigna radiata*), and sorghum (*Sorghum bicolor*) (Lee et al. 2012). Similarly, CuO and NiO NPs showed deleterious impacts on growth of lettuce, radish, and cucumber (Wu et al. 2012). TiO_2 NPs can inhibit leaf growth and transpiration via impairing root water transport (Asli and Neumann 2009), whereas both TiO_2 and ZnO NPs negatively affect rice and wheat (*Triticum aestivum*) growth (Boonyanitipong et al. 2011; Du et al. 2011). ZnO showed maximum toxic effects on seed germinations and root elongation, followed by Fe_3O_4 and SiO_2 in Arabidopsis (*Arabidopsis thaliana*) (Lee et al. 2010). Another study on ryegrass and Indian mustard (*Brassica juncea*) indicated that plant growth was inhibited by ZnO NPs with gradual increase in proline content, antioxidant enzyme activities, and lipid peroxidation (Lin and Xing 2008; Rao and Shekhawat 2014). Phytotoxicity of Co_3O_4 and ZnO NPs was investigated using the roots of onion (*Allium cepa*) as an indicator organism (Ghodake et al. 2011), and CuO and ZnO NPs in wheat and cucumber (Dimkpa et al. 2012; Kim et al. 2012) showing increased lipid peroxidation, oxidized glutathione, increased production of reactive oxygen species (ROS), higher peroxidase and catalase activities, and decreased chlorophyll content. Similarly, phytotoxicity of nano-CuO was assessed in soybean (*Glycine max*), chickpea (*Cicer arietinum*) (Adhikari et al. 2012), and in rice (Shaw and Hossain 2013). In another study, the developmental phytotoxicity exerted by five types of nanoparticles—multi-walled carbon nanotubes (MWCNTs), nAl, nAl_2O_3, nZn, and nZnO—was investigated on six different plant species—radish, oilseed rape (*Brassica napus*), ryegrass, lettuce, maize (*Zea mays*), and cucumber (Lin and Xing 2007). This study reported significant inhibition of ryegrass germination by nZn and maize germination by nZnO or nAl_2O_3, whereas no inhibition was observed for MWCNT. Interestingly, nAl caused both positive and negative effects on root elongation, depending on the plant species. Slomberg and Schoenfisch (2012) evaluated the phytotoxicity of silica nanoparticles as reduced development and chlorosis in Arabidopsis plants. Alumina nanoparticles showed inhibition of root elongation and growth of five crop species: maize, cucumber, soybean, cabbage (*Brassica oleracea*), and carrot (*Daucus carota*) in hydroponic culture medium (Yang and Watts 2005). Nanoparticle-mediated phytotoxicity of rare earth elements such as nanoceria was determined on alfalfa (*Medicago sativa*), maize, cucumber, tomato (López-Moreno et al. 2010a), and Arabidopsis (Ma et al. 2013).

In contrast to these negative effects, several studies demonstrated that nanoparticles can have positive or neutral effects on plants. In one study with MWCNTs, it was found that they can even penetrate thick seed coat of tomato plants. When the growth medium was supplemented with different concentrations of CNTs, seed germination was quick and vegetative biomass was also increased (Khodakovskaya and Biris 2015). In another study, MWCNTs have enhanced the growth of tobacco (*Nicotiana tabacum*) cell culture by 55–64 % over control (Khodakovskaya et al. 2012). MWCNTs also exerted positive effects on root growth of Indian mustard and urd bean (*Vigna mungo*) (Ghodake et al. 2010). Although Ag NPs are often detrimental to plant growth, several studies have demonstrated the growth-stimulatory effects of Ag NPs. For example, the small size of Ag NPs at lower concentration has induced growth of Arabidopsis (Syu et al. 2014), shallow sedge (*Carex lurida*)—a wetland plant (Yin et al. 2012), kidney bean (*Phaseolus vulgaris*), and maize (Salama 2012). Nano-TiO_2 was also found to enhance spinach growth (Zheng et al. 2004; Yang et al. 2007), or did not affect germination and root elongation of wheat, oilseed rape and Arabidopsis (Larue et al. 2011), lettuce, radish, and cucumber (Wu et al. 2012). Similarly, Hernandez-Viezcas et al. (2011) reported no significant toxic effect of ZnO NPs in a wild desert velvet mesquite (*Prosopis juliflora-velutina*) plant. Exposure to SiO_2 NPs caused no significant effect in zucchini (Stampoulis et al. 2009). Platinum nanoparticles also exhibit excellent germination index and did not result in harmful effects in tomato and radish (Shiny et al. 2013). Likewise, in a recent report, Cu NPs did not impart any toxicity to the plant system including morphological or physiological alterations in mung bean (Pradhan et al. 2015).

The positive impact of nano-SiO_2 and nano-TiO_2 mixture in soybean was shown to be due to increased nitrate reductase activity, absorption potential, and antioxidant system activity (Lu et al. 2002). Likewise, enhancement of many physiological parameters such as increased photosynthetic activity, nitrogen metabolism, and activity of antioxidant enzymes by metal-based nanomaterials was reported in Arabidopsis, spinach, peanut (*Arachis hypogea*), and duckweed (Yang et al. 2007; Gao et al. 2008; Ze et al. 2011; Giraldo et al. 2014; Farrag 2015).

7.3 Plant–Nanoparticle Interactions at Molecular Level

Nanoparticles can interact with biological systems such as plants chemically or mechanically; and these specific interactions originate mainly from their small size, large surface area, and intrinsic catalytic reactivity (Dietz and Herth 2011). Understanding the nature of interactions between nanoparticles and plants is critical to assess their positive and negative environmental and agricultural impact. However, information on the interactions of nanoparticles with these organisms is rather limited. To achieve the necessary comprehensive and mechanistic understanding of the risks posed by nanoparticles in the environment, and for the development of biomarkers for nanoparticle toxicity, the underlying biochemical and molecular mechanisms of plant–nanoparticle interactions must be evaluated.

Most studies on the phytotoxicity of nanomaterials have been based on phenotypic parameters such as seed germination and root elongation as suggested by

US Environmental Protection Agency (US EPA 1996), which evaluate the acute effects of nanoparticles on plant metabolism. However, these standard phytotoxicity tests are highly dependent on plant type and specific nanoparticle properties and may not be sensitive enough while evaluating nanoparticles' toxicity to terrestrial plant species.

The mode of action of nanoparticles on cellular structures is multiple, complex, and still poorly understood. To date, several biochemical markers have been used to study the effect of environment stress on plants, such as metabolites composition, membrane integrity, and activity of enzymes. Physiological studies have demonstrated that smaller nanoparticles can travel through symplast (like plasmodesmata), whereas larger nanoparticles accumulate in the apoplastic space. The most commonly described mechanisms of nanoparticle toxicity in plants include cell surface coating causing mechanical damage or clogging of pores (Asli and Neumann 2009; Dietz and Herth 2011), increased production of ROS causing oxidative stress (Miralles et al. 2012), and release of toxic metal ions (Dietz and Herth 2011; Miralles et al. 2012). However, there are only few studies describing nanoparticle–plant interaction and NP-phytotoxicity at molecular level.

7.3.1 Nanotoxicogenomics

The term "toxicogenomics" describes the application of genomics tools to toxicology aspects (Nuwaysir et al. 1999). Analysis of changes in gene expression through high-throughput methods such as cDNA microarrays or quantitative real-time PCR (qRT-PCR) constitutes a powerful approach for understanding the mechanisms of toxicity and molecular responses in cells exposed to nanomaterials (Fig. 7.1) (Xu et al. 2011). The role of toxicogenomics in regulatory ecotoxicology and risk assessment was reviewed by Boverhof and Zacharewski (2005), Ankley et al. (2006), and Magdolenova et al. (2014). Gene expression analyses are typically conducted to complement morphological and/or physiological investigations and provide unique information on specific toxicity pathways and modes of action that cannot be obtained directly from other approaches (Ankley et al. 2006; Dietz and Herth 2011). In addition, unique gene expression patterns may help in development and validation of promising biomarkers suitable for high-throughput screening methods, and for better understanding of the toxicity of nanoparticles (Merrick and Bruno 2004; Thomas et al. 2011). Changes in gene expression have been shown to be induced by very low dose of contaminants and may help investigate the cellular impact of chronic toxicity associated with nanoparticles (Poma and Di Giorgio 2008; Poynton and Vulpe 2009).

Although transcriptional analyses have been widely used to study the molecular basis of nanoparticle toxicity in a variety of organisms including microbes, humans, mammalian cell lines, and other model organisms (Asharani et al. 2009), only limited investigations have been conducted to assess the molecular mechanism of nanoparticle–plant interactions and nanoparticle-mediated phytotoxicity. Nanoparticle-

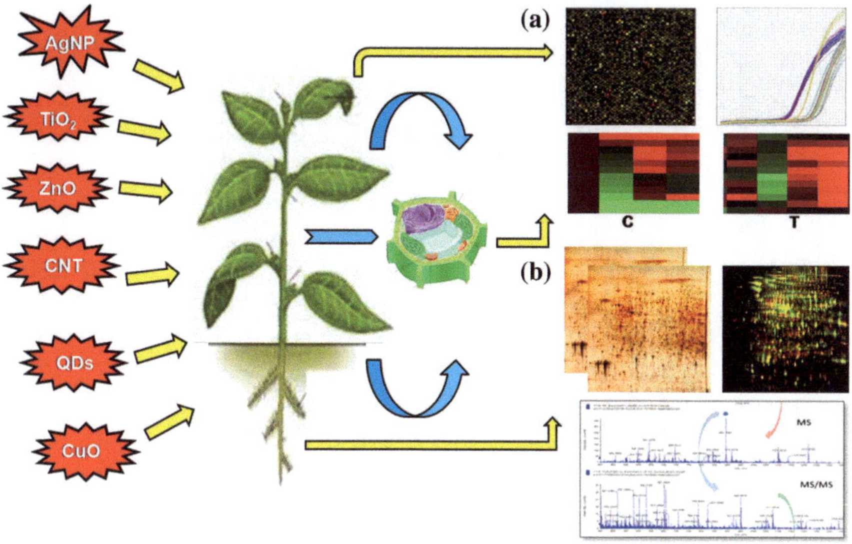

Fig. 7.1 Molecular mechanisms to evaluate engineered nanoparticles (ENPs)–plant interactions and nanoparticle-mediated phytotoxicity. Cellular responses due to exposure of plants to different types of nanoparticles can be measured through (**a**) studies of changes in gene expression at whole-genome level through cDNA microarray or qRT-PCR and (**b**) global proteome profiling to identify differentially expressed proteins and to measure their relative abundances through qualitative/quantitative 2-DE coupled with MS. Abbreviations: *C,* control; *T,* treatment; *qRT-PCR,* quantitative real-time PCR; *2-DE,* two-dimensional polyacrylamide gel electrophoresis; *MS,* mass spectrometry

induced gene expression changes in plants and microorganisms, and genotoxic aspects of nanoparticle interactions with plants have been reviewed recently (Remédios et al. 2012; Aken 2015). Differential transcriptional regulation of the genes involved in cell cycle regulation and DNA repair has been reported in human cells and male medaka (*Oryzias latipes*) upon exposure to Ag NPs (Asharani et al. 2012; Pham et al. 2012). Enhanced genotoxicity caused by Ag NPs has been reported from DNA repair-deficient mammalian cells (Lim et al. 2012). A comprehensive review of gene expression analyses in plants exposed to different types of nanomaterials has been presented here (Table 7.1). These findings may provide a link between the plant phenotype and the associated genes affected by nanoparticles, and thus open up possibility to identify the genetic mechanisms behind toxic effects of nanoparticles on plants.

7.3.1.1 Silver Nanoparticles (Ag NPs)

The Ag NPs alter the physiochemical properties and metabolic pathways of the plant, although the underlying mechanisms and specific pathways involved remain unclear. Ag NPs have been shown to be genotoxic in plant cells as exposure to Ag

Table 7.1 Molecular analysis for nanoparticle-mediated phytotoxicity: studies on effect of different NPs on plants

S. No.	Type of NP	Size of NP	Concentration of NP (dose)	Exposure time	Model plant	Toxicity assay/method	Reference
1	Ag NPs	60 nm	12.5, 25, 50, 100 mg l^{-1}	4 h	Faba bean	Cytogenetic analysis	Patlolla et al. (2012)
2	Ag NPs	10 nm	2.5 mg kg^{-1}	14 days	Wheat	Gene expression analysis (semi-qRT-PCR)	Dimkpa et al. (2013)
3	Ag NPs	10 nm	10 mg l^{-1}	5 days	Eruca	Proteomics (2-DE and nano-LC–ESI–MS/MS)	Vannini et al. (2013)
4	Ag NPs	20 nm	5 mg l^{-1}	10 days	Arabidopsis	Transcriptomics (expression microarrays)	Kaveh et al. (2013)
5	Ag NPs	20 nm	0.2, 0.5, 1 mg l^{-1}	14 days	Arabidopsis	Gene expression analysis (qRT-PCR)	Nair and Chung (2014a)
6	Ag NPs	20 nm	0.2, 0.5, 1 mg l^{-1}	24, 48,72 h	Arabidopsis	Gene expression analysis (qRT-PCR)	Nair and Chung (2014b)
7	Ag NPs	18.34 nm (0.1–1000 nm)	30, 60 μg ml^{-1}	20 days	Rice	Proteomics (2-DE and nano-LC/FT-ICR MS)	Mirzajani et al. (2014)
8	Ag NPs	20 nm	215 μg l^{-1}	72 h	Chlamydomonas	Proteomics (2-DE and LC–ESI–MS/MS)	Lindgren (2014)
9	Ag NPs	13.2 nm (5–15.5 nm)	10 mg l^{-1}	5 days	Wheat	Proteomics (2-DE and LC–ESI–MS/MS), genetic analysis (AFLP)	Vannini et al. (2014)
10	Ag NPs	15 nm	2 mg l^{-1}	0, 2, 4 days	Soybean	Gel-free proteomics (nano-LC–ESI–MS/MS), gene expression analysis (qRT-PCR)	Mustafa et al. (2015a)

(continued)

Table 7.1 (continued)

S. No.	Type of NP	Size of NP	Concentration of NP (dose)	Exposure time	Model plant	Toxicity assay/method	Reference
11	Ag NPs	20 nm	5 mg l^{-1}	10 days	Arabidopsis	Gene expression analysis (Microarray)	Kohan-Baghkheirati and Geisler-Lee (2015)
12	Ag NPs	Decahedral AgNPs (45 ± 5 nm), triangular (47 ± 7 nm) and spherical (8 ± 2 nm)	–	–	Arabidopsis	Gene expression analysis (qRT-PCR)	Syu et al. (2014)
13	Ag NPs, ZnO and TiO_2	–	–	–	Medicago	Gene expression analysis (microarray and qRT-PCR)	Chen et al. (2015)
14	Ag–silica hybrid complex	30 nm	10 mg l^{-1}	4 weeks	Arabidopsis	Gene expression analysis (semi-qRT-PCR)	Chu et al. (2012)
15	Al_2O_3	Not specified	0.1, 0.5, 1 %	3 weeks	Tobacco	miRNA expression analysis (qRT-PCR)	Burklew et al. (2012)
16	CuO NPs		0, 20, 50, 100, 200, 400 and 500 mg/L	14 days	Indian mustard	Gene expression analysis (qRT-PCR)	Nair and Chung (2015)
17	MWCNTs	Length 500 nm–1 μm, OD 20 nm	100 μg ml^{-1}	6 h, 24 h, 4d, 8d, 16d, 25 d	Tobacco cells	Gene expression analysis (qRT-PCR), Western blotting	Khodakovskaya et al. (2012)
18	MWCNTs	Length >1 μm, OD 15–40 nm	50, 100, 200 μg ml^{-1}	1 day	Soybean, barley, maize	Gene expression analysis (semi-qRT-PCR)	Lahiani et al. (2013)
19	MWCNTs, SWCNTs, Graphene	MWCNT: length 6 μm, OD 10–35 nm SWCNT: length >1 μm, OD 0.86–2.22 nm, graphene: diameter 100–120 nm thickness of 2–5 nm	50,100, 200 μg ml^{-1}	10 days	Tomato	Transcriptomics (microarray), gene expression analysis (qRT-PCR)	Khodakovskaya et al. (2011)

(continued)

Table 7.1 (continued)

S. No.	Type of NP	Size of NP	Concentration of NP (dose)	Exposure time	Model plant	Toxicity assay/method	Reference
20	SWCNTs	Length 5–30 μm, OD 1–2 nm	250 μg ml^{-1}	24 h, 36 h, 48 h, 72 h	Arabidopsis, rice	Gene expression analysis (semi-qRT-PCR, qRT-PCR)	Shen et al. (2010)
21	SWCNTs	Length ≈ 30 μm, OD 1–2 nm	20 mg l^{-1}	36, 48, 60, 72 h	Maize	Gene expression analysis (qRT-PCR), immunostaining	Yan et al. (2013)
22	SWCNTs	50–100 nm	25, 50, and 100 μg/ml	10–20 days	Barley, maize, rice, soybean, switchgrass, tomato and tobacco cell culture	Gene expression analysis (microarray)	Lahiani et al. (2015)
23	TiO_2, bulk TiO_2	TiO_2 5 nm, bulk TiO_2 155 nm	0.25 %	35 days	Arabidopsis	Gene expression analysis (qRT-PCR), SDS-PAGE	Ze et al. (2011)
24	TiO_2	25 nm	0.1, 1, 2.5, 5 %	3 weeks	Tobacco	Gene expression analysis (qRT-PCR)	Frazier et al. (2014)
25	TiO_2, CeO_2	–	–	12 days	Arabidopsis	Gene expression analysis (microarrays and qRT-PCR)	Tumburu et al. (2015)
26	TiO_2, Ag NP, MWCNTs	Ag NPs 10–80 nm, TiO_2 NPs 10–40 nm, MWCNTs 4–12 nm	Ag NPs 0.2 μg ml^{-1}, TiO_2 NPs 20 μg ml^{-1}, MWCNTs 25 μg ml^{-1}	2 days	Arabidopsis	Transcriptomics (microarray), gene expression analysis (qRT-PCR)	García-Sánchez et al. (2015)
27	CdS QDs	5 nm	0, 40, 80 mg l^{-1}	21 days	Arabidopsis	Transcriptomics (microarray), gene expression analysis (qRT-PCR)	Marmiroli et al. (2014)
28	GO	40–50 nm	10–1000 μg l^{-1}	4 weeks	Arabidopsis	Gene expression analysis (qRT-PCR)	Wang et al. (2014)

(continued)

Table 7.1 (continued)

S. No.	Type of NP	Size of NP	Concentration of NP (dose)	Exposure time	Model plant	Toxicity assay/method	Reference
29	CeO_2	10 ± 1 nm	400 or 800 mg/kg seed	10, 15, and 20 days	Maize	Proteomics (SDS-PAGE)	Zhao et al. (2012)
30	CeO_2, In_2O_3	CeO_2 10–30 nm, In_2O_3 20–70 nm	$0 - 2000$ mg l^{-1}	25 days	Arabidopsis	Gene expression analysis (qRT-PCR)	Ma et al. (2013)
31	ZnO	20 nm	4 mg l^{-1}	7 days	Arabidopsis	Gene expression analysis (microarray)	Landa et al. (2015)
32	ZnO, CeO_2	ZnO NPs 8 nm, CeO_2 NPs 7 nm	2000–4000 mg l^{-1}	7 days	Soybean	Genetic analysis (random amplified polymorphic DNA assay)	López-Moreno et al. (2010b)
33	ZnO, TiO_2, FS	ZnO <100 nm, TiO_2 <150 nm, FS: C60 76 %, C70 22 %	100 mg l^{-1}	7 days	Arabidopsis	Transcriptomics (microarray)	Landa et al. (2012)

NPs has resulted in chromosomal aberrations and disruption of cell division in onion (Kumari et al. 2009). Ag NPs can induce DNA damage in plants, causing the formation of chromatin bridges, disturbed metaphase, and multiple chromosomal breaks (Panda et al. 2011). In addition, genotoxicity of engineered Ag NPs to plant cells was inferred through exposure on root tip cells of faba bean (*Vicia faba*) that significantly enhanced the number of chromosomal aberrations and micronuclei, and decreased the mitotic index, due to penetration of nanoparticles into the plant system causing impaired mitosis (Patlolla et al. 2012). Recently, an increase in DNA damage has been reported by DNA laddering, comet, and TUNEL assays in turnip—*Brassica rapa* subsp. *rapa*, when plants were treated with higher concentrations of Ag NPs (Thiruvengadam et al. 2015).

Gene expression analyses of the model plant Arabidopsis by RT-PCR have provided new insights into the molecular mechanisms of plant responses to Ag NPs. Dimkpa et al. (2012) investigated that exposure of commercial Ag NPs to wheat in a sand growth matrix causes oxidative stress, as indicated by the accumulation of oxidized glutathione, and induced expression of a metallothionein (*MT*) gene involved in detoxification by metal ion sequestration (Fig. 7.2). Interestingly, both nanosized Ag–silica hybrid (NSS) complex and reduced Ag NPs induced plant defense response in Arabidopsis, as revealed by significant upregulation of pathogenesis-related (*PR1*, *PR2,* and *PR5*) genes involved in systemic acquired resistance (SAR) (Fig. 7.2), and pretreatment with the NSS complex induced more pathogen resistance to the virulent pathogen *Pseudomonas syringe* pv. Tomato DC3000 (Chu et al. 2012).

The transcriptional response of Arabidopsis plants exposed to Ag NPs was analyzed using whole-genome cDNA expression microarrays (Kaveh et al. 2013). This has been resulted in upregulation of 286 genes, including the genes primarily associated with metal and oxidative stress (e.g., vacuolar cation/proton exchanger, superoxide dismutase, cytochrome P450-dependent oxidase, and peroxidase), and down regulation of 81 genes, including the genes involved in plant defense system and hormonal stimuli (e.g., auxin-regulated gene involved in organ size-*ARGOS*, ethylene signaling pathway, and SAR against pathogens) (Fig. 7.2). Similar molecular studies were performed to understand the toxic effects of exposure to Ag NPs in Arabidopsis (Nair and Chung 2014a). Real-time PCR analysis showed significant transcriptional modulation of genes involved in sulfur assimilation and glutathione biosynthesis, viz. ATP sulfurylase (*ATPS*), 3′-phosphoadenosine 5′-phosphosulfate reductase (*APR*), sulfite reductase (*SiR*), cysteine synthase (*CS*), glutamate–cysteine ligase (*GCL*), glutathione synthetase (*GS2*) with upregulation of glutathione S-transferase (*GSTU12*), glutathione reductase (*GR*), and phytochelatin synthase (*PCS1*) genes (Fig. 7.2). In another study by the same group, the expression of cell cycle genes proliferating cell nuclear antigen (*PCNA*) and DNA mismatch repair (*MMR*) was found to be modulated as a result of oxidative stress caused by Ag NPs exposure in Arabidopsis (Nair and Chung 2014b) (Fig. 7.2). Another study in Arabidopsis revealed that around 110 genes were uniquely expressed in Ag NP stress, mainly involved in three biological functions, viz. genetic response to infection by fungal pathogens, anion transport, and the processes related to cell wall/plasma membrane (Kohan-Baghkheirati and Geisler-Lee 2015).

In recent studies with other plants such as turnip, Ag NPs at higher concentrations had induced genes involved in glucosinolates and phenolics biosynthesis. These biomolecules are considered to be an integral part of defense mechanism of Brassicaceae and related plant families (Thiruvengadam et al. 2015). Also, the model legume—*Medicago truncatula*—was recently used for toxicogenomic studies with Ag NPs by mixing these ENPs in soil and allowed them to age for six months in the field. The results suggested that root nodulation was hampered by increased phytotoxicity of Ag NPs as compared to bulk metal (Chen et al. 2015). Gene expression profiling in roots by microarray revealed that around 239 genes were downregulated in *M. truncatula*, related to nitrogen metabolism, nodulation, metal homeostasis, and various stress responses (Chen et al. 2015).

7.3.1.2 Titanium Oxide Nanoparticles (TiO_2 NPs)

TiO_2 nanoparticles may exert a negative impact on plant growth and development. A dose-dependent alteration in mitotic activity and chromosomal aberrations was observed as an effect of nano-TiO_2 in purple broad bean (*Vicia narbonensis*) and maize, indicating genotoxic effects of these nanoparticles (Castiglione et al. 2011).

MicroRNAs (miRNAs) are newly discovered post-transcriptional gene regulators, which belong to the small endogenous class of noncoding RNAs ($\sim$20–22 nt). MicroRNAs function to alter gene expression by either targeting mRNAs for degradation or inhibiting translation (Zhang et al. 2006), and mediate plant development as well as abiotic stress responses such as drought and salinity (Sunkar and Zhu 2004). Interestingly, miRNAs have also been shown to play a significant role in plant response to nanoparticles by regulating gene expression. In a study by Frazier et al. (2014), nano-TiO_2 exposure to tobacco plants significantly affected the expression profiles of miRNAs. Low concentrations of TiO_2 significantly induced *miR395* and *miR399* expression with 285-fold and 143-fold increase, respectively, that might be responsible for reduction in growth of tobacco seedlings upon exposure to TiO_2 NPs (Fig. 7.2).

Recently, García-Sánchez et al. (2015) evaluated the transcriptome changes of Arabidopsis in response to different types (metallic TiO_2 and Ag; carbonaceous MWCNTs) and sizes (4–80 nm) of nanoparticles, and compared them with those under biotic (necrotizing fungus or hemibiotrophic bacterium infection) or abiotic (saline, drought, or wounding) stress inducers. A set of 16 comparable transcriptome profiles were produced to monitor early changes in gene expression upon nanoparticle and stress exposure. Nanoparticle exposure downregulated a significant number of genes involved in response to microbial pathogens, resulting in increased bacterial survival and colonization. Also, nanoparticle-induced suppression involved genes related to phosphate starvation and conditions related to promotion of root hair development. The root-hair-less phenotype was recovered by exogenous supply of salicylic acid (SA). In this study, the effect of stress-responsive phytohormone abscisic acid (ABA) was also tested on the gene expressions and phenotypes of nanoparticle-exposed plants, revealing a basic common molecular mechanism of the early response of Arabidopsis to different nanoparticles, regardless of the composition (García-Sánchez et al. 2015).

Some studies have shown that TiO_2 NPs could significantly promote photosynthesis and plant growth, but its mechanism is still unclear. In a study by Ze et al. (2011), a significant increase in expression of light-harvesting complex II (*LHCII*) b gene was observed in Arabidopsis upon TiO_2 NP exposure (Fig. 7.2), resulting in overall improvement in photosynthesis, which might be due to promotion of energy transport from Chl b and carotenoid to Chl a, distribution of light energy from PSI to PSII, and increase in fluorescence quantum yield and water photolysis.

7.3.1.3 Zinc Oxide Nanoparticles (ZnO NPs)

The mechanisms of phytotoxicity may be highly nanoparticle-specific. Landa et al. (2012) reported gene expression changes upon long-term exposure to TiO_2 NPs, ZnO NPs, and fullerene soot (FS) using oligonucleotide microarrays. Nano-ZnO was found to be most toxic and resulted in 660 upregulated and 826 downregulated genes, whereas FS caused differential gene expression with 232 upregulated and 189 downregulated genes. Abiotic (oxidative, salt, water deficit) and biotic (wounding and defense to pathogens) stress-responsive genes were induced by ZnO NP and FS exposure, whereas the downregulated genes by ZNO NP exposure were mainly involved in cell organization and biogenesis, including translation, nucleosome assembly, and microtubule-based process (Fig. 7.2). Fullerene soot largely repressed the genes involved in electron transport and energy pathways. Only mild changes in gene expression were observed upon TiO_2 NP exposure, which resulted in only 80 upregulated and 74 downregulated genes, mainly involved in response to biotic and abiotic stimuli, indicating minimal toxicity.

7.3.1.4 Carbon Nanotubes (CNTs)

Single-walled carbon nanotubes (SWCNTs) have been shown to exert adverse effects on Arabidopsis and rice leaf protoplasts through oxidative stress, leading to a certain amount of programmed cell death (PCD)/apoptosis, DNA damage, and chromatin condensation, as revealed by terminal deoxynucleotidyl transferase-mediated dUTP nick end-labeling (TUNEL)-positive reaction (Shen et al. 2010). Conversely, the exposure of SWCNTs to maize seedlings promotes growth of seminal roots (Yan et al. 2013). These effects are related to the differential expression and upregulation of the associated genes encoding epigenetic modification enzymes, leading to global deacetylation of histone H3, similar to other abiotic stress response mechanism.

Khodakovskaya et al. (2012) have demonstrated that the growth of tobacco cell culture (callus) can be highly enhanced by the introduction of MWCNTs in the

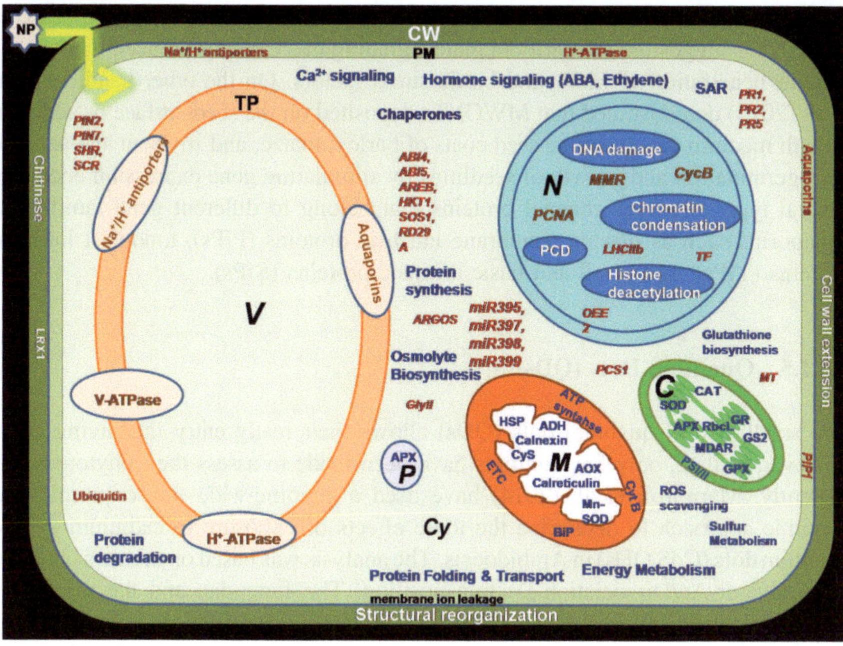

Fig. 7.2 Nanoparticles (NPs)-induced modulation of cellular transcriptome and proteome in plants. The scheme is based on the various published reports on nanoparticle-mediated changes in gene expression and protein abundance. Major metabolic and signaling pathways affected by nanoparticle exposure are also highlighted. Abbreviations: *CW*, cell wall; *PM*, plasma membrane; *N*, nucleus; *V*, vacuole; *M*, mitochondria; *C*, chloroplast; *P*, peroxisome; *Cy*, cytosol; *TF*, transcription factors; *ROS*, reactive oxygen species; *HSP*, heat shock protein; *BiP*, binding immunoglobulin proteins; *SOD*, superoxide dismutase; *APX*, ascorbate peroxidase; *CAT*, catalase; *MDHAR*, monodehydroascorbate reductase; *GR*, glutathione reductase; *GPX*, glutathione peroxidase; *CyS*, cysteine synthase; *LHC, IIb* light-harvesting complex IIB; *CycB*, cyclin B; *MMR*, mismatch repair gene; *PCNA*, proliferating cell nuclear antigen; *OEE2*, oxygen-evolving complex 2; *RbcL*, large subunit of Rubisco; *AOX*, alternate oxidase; *ETC*, electron transport chain; *ADH*, alcohol dehydrogenase; *miR*, microRNA; *PCS1*, phytochelatin synthase 1; *MT*, metallothionein; *PIP1*, plasma membrane intrinsic protein 1; *GlyII*, glyoxalase II; *PR*, pathogenesis related; *SAR*, systemic acquired resistance; *ARGOS*, auxin-regulated gene involved in organ size; *PIN*, peptidyl-prolyl cis-trans isomerase NIMA-interacting; *SHR*, short root; *SCR*, scarecrow; *ABI*, ABA-insensitive; *AREB*, ABA-responsive elements-binding proteins; *HKT1*, high-affinity K^+ transporter 1; *SOS1*, salt overly sensitive 1; *RD29A*, responsive to desiccation 29A

growth medium and is directly correlated to the overexpression of marker genes for cell division (*CycB*), cell wall extension (*NtLRX1*), and water transport (aquaporin gene *NtPIP1*) in tobacco cells exposed to MWCNTs (Fig. 7.2). Similarly, the observed physiological responses of tomato plants exposed to MWCNTs were linked with the complex sets of information provided by microarray analysis along with photothermal and photoacoustic imaging of nanoparticles (Khodakovskaya et al. 2011). The expression of tomato aquaporin gene *LeAqp2* and a number of other genes related to plant responses to environmental stress and pathogen infection were found to be upregulated in tomato leaves and roots by exposure to MWCNTs, suggesting similarity in plant signaling mechanisms induced by both—a cellular penetration of ENPs and biotic stress factors. On the other hand, Lahiani et al. (2013) demonstrated that MWCNTs deposited on the seed surface or added in growth medium can penetrate seed coats of barley, maize, and soybean and activate seed germination and growth of seedlings by stimulating gene expression encoding several types of water channel proteins that belong to different gene families of aquaporins such as plasma membrane intrinsic proteins (PIPs), tonoplast intrinsic proteins (TIPs), and small and basic intrinsic proteins (SIPs).

7.3.1.5 Quantum Dots (QDs)

The small size of quantum dots (QDs) allows their ready entry into living cells; however, to date, only few attempts have been made to assess their phytotoxicity. Recently, Marmiroli et al. (2014) have used a genome-wide transcriptomic and genomic approach to determine the toxic effects of exposure to cadmium sulfide quantum dots (CdS QDs) in Arabidopsis. The analysis was based on mutants obtained by Activator-Ac/Dissociation-Ds transposition. The functions and the regulatory networks of the genes involved in CdS QDs uptake, translocation, detoxification, and accumulation were determined using both a genome-wide top-down and a bottom-up approach. A number of genes were identified, whose transcript abundance was correlated with the CdS QDs tolerance like an *MYB* containing gene *ELM2*, and a gene involved in photosynthesis of high chlorophyll fluorescence 101 (*HCF101*). *MYB* gene family encodes for transcription factors that play an important role in response to biotic and abiotic stress (Dubos et al. 2010), and universally conserved *HCF101* gene is involved in metal transition during photosynthesis (Yruela 2013).

7.3.1.6 Other Nanoparticles

The graphene oxide (GO) exposure combined with drought/salt stress induced severe alterations in expression patterns of the genes required for root development (*PIN2*, *PIN7*, *SHR*, and *SCR*) and abiotic stress (*ABI4*, *ABI5*, *AREB1*, *HKT1*, *SOS1*, and *RD29A*) in Arabidopsis seedlings (Wang et al. 2014). This has been resulted in a loss of morphology along with increase in ROS production or membrane ion leakage and decrease in activity of superoxide dismutase (SOD) or catalase (CAT) enzymes (Fig. 7.2). In another study, copper oxide nanoparticles (CuO NPs) were also shown to mediate DNA damage in terrestrial plants (Atha et al. 2012).

As discussed earlier, miRNAs have been found to mediate plant stress responses to nanoparticle stress, such as alumina (Al_2O_3) NPs (Burklew et al. 2012). Expression profile of certain miRNAs (*miR395*, *miR397*, *miR398*, and *miR399*) was significantly upregulated upon exposure of tobacco plants to 1 % Al_2O_3 nanoparticles (Fig. 7.2), providing the ability of plants to withstand stress to alumina nanoparticles in the environment.

The genotoxic effects from the exposure of soybean plants to rare earth element (RER) nanoparticles of cerium oxide (CeO_2) were demonstrated by López-Moreno et al. (2010b) using random amplified polymorphic DNA (RAPD) assay to detect DNA damage and mutations caused by nanoparticles. Results of this study showed the appearance of four new bands at 2000 mg l^{-1} and three new bands at 4000 mg l^{-1} treatment of CeO_2 NPs. In addition, Ma et al. (2013) evaluated physiological and molecular responses triggered by exposure of CeO_2 and indium oxide (In_2O_3) NPs on Arabidopsis. Results showed alteration of gene expression related to stress response such as the sulfur metabolism, glutathione (GSH) biosynthesis pathway, and a series of genes involved in the detoxification of metal toxicity in plants upon exposure to both types of elements.

7.3.2 Nanotoxicoproteomics

"Toxicoproteomics" is a branch of toxicology, which includes the proteomic study of toxicity caused by environmental stress inducers, released chemicals, nanoparticles, or any other toxic substance that may cause significant cellular responses (Wetmore and Merrick 2004; Gao et al. 2009). Nanotoxicoproteomics combines with principles and methods of toxicology and proteomics to unravel molecular mechanism of nanoparticle toxicity. Comparative proteomics can be used as a powerful technique for abiotic stress-related research in crop plants through identification of novel stress-responsive proteins (Barkla et al. 2013). Understanding the dynamics of expression of these proteins may provide direct insights into their function and interactions. Alteration in protein accumulation under nanoparticle stress is closely interrelated to plant phenotypic response, as changes at transcript level do not always match with alteration at protein level. Therefore, investigation of changes in plant proteome is highly important since proteins are direct effectors of plant stress response. Studies of cellular reactions at protein level upon nanoparticle exposure can significantly contribute to our understanding of physiological mechanisms underlying plant–nanoparticle interaction. Moreover, the power of proteomic technologies is demonstrated in analyzing post-translational modifications, protein–protein/toxicant interactions, function, and subcellular localization.

Differential protein expression profiling represents the core of proteomic approaches, and several techniques are available for the differential analysis of protein expression. Two-dimensional polyacrylamide gel electrophoresis (2-DE) remains the primary method for this, and densitometric analysis of 2-DE gels along with mass spectrometry (MS) has enabled the researchers to identify differentially expressed protein spots (Fig. 7.1). Protein chip or microarray is also emerging as a robust tool for identification of protein interaction with other biomolecules. Thus, proteomic studies could lead to identification and characterization of key proteins under nanoparticle exposure, which can be used as potential biomarkers for nanoparticle phytotoxicity.

Nanotechnology-based proteomic biomarker development is still in its infancy, and as yet only a few proteomic studies have examined the effects of nanoparticles on different organisms. It has been used to determine cytotoxicity of different nanoparticles for bacteria, fungi, rats, mice, Daphnia, and human cell lines (as reviewed by Abdelhamid and Wu 2015). However, in plants, the reports describing proteomic studies in response to nanoparticle exposure are very rare and mostly deal with investigation of Ag NPs phytotoxicity (Table 7.1). Some studies had been done with other nanoparticles such as Al_2O_3 (Mustafa et al. 2015b), CdS quantum dots (Marmiroli et al. 2015), and CeO_2 (Zhao et al. 2012).

Proteomic responses of *Eruca sativa* roots exposed to Ag NPs and $AgNO_3$ revealed insights into the mode of action of Ag NPs or $AgNO_3$ (Vannini et al. 2013). Ag NP and $AgNO_3$ appear to share some common mechanisms of action such as accumulation of proteins related to sulfur metabolism, e.g., Jacalin lectin family (JAC) proteins, proteins involved in stress/defense response, cell cycle, protein folding, transport, and activation of ROS detoxification pathways, such as type 2 peroxiredoxin (PRX) and superoxide dismutase (SOD) (Fig. 7.2). While Ag NPs' treatment alone was shown to disturb ER functions, as revealed by down-regulation of two ER chaperones, binding protein 1 (BiP1) and the heat shock protein 70–2, and two vacuolar-type proton ATPase (VATPase) subunits (Fig. 7.2). The data from these proteomic studies strongly indicate that the effects of Ag NPs are not solely due to the release of Ag^+ into the surrounding environment. Similarly, gel-based proteomic analysis of rice identified a total of 28 responsive proteins that change in abundance upon exposure to different concentrations of Ag NPs colloidal suspension (Mirzajani et al. 2014). The identified proteins were involved in normal cell metabolic processes such as transcription, protein synthesis/degradation, cell division, and apoptosis, along with proteins related to Ca^{2+} regulation and signaling, oxidative stress tolerance, and direct damage to cell walls and DNA/RNA/proteins. It was hypothesized that the Ag NPs enter into the cell and condense the DNA/protein, thus inhibiting the normal reproduction. Furthermore, the increase in detoxification enzymes implies a toxic effect due to production of ROS and metal toxicity in the presence of Ag NPs (Fig. 7.2). Therefore, the mechanism of Ag NP toxicity may result in parallel action of the silver nanoform and released silver ion.

In another study, genomic and proteomic changes induced by Ag NPs were analyzed in wheat seedlings using DNA fingerprinting technique and 2-DE coupled with LC–ESI–MS/MS (Vannini et al. 2014). No significant DNA polymorphism was observed by nanoparticle treatment at the genomic level (4.6 and 3.7 % for treated roots and shoots, respectively), whereas 2-DE profiling of roots and shoots treated with 10 mg l^{-1} of Ag NPs revealed an altered expression of several proteins mainly involved in primary metabolism, protein synthesis/folding, cell defense, and stress responses in multiple cellular compartments, suggesting that metabolic adaptation of plants plays an important role in mitigating unfavorable changes in the environment (Fig. 7.2). Similarly, Lindgren (2014) investigated the toxic mechanisms of Ag NPs and $AgNO_3$ in the green alga *Chlamydomonas reinhardtii* through proteomic approach using 2-DE and LC–ESI–MS/MS. Both Ag NPs and $AgNO_3$ tend to regulate in similar ways as oxygen-evolving enhancer protein 2 (OEE2) of PSII was downregulated by both treatments (Fig. 7.2). The main drawback of this study was that a vast number of proteins were matched for each spot and it was not possible to identify the exact proteins that contribute to the difference in protein regulation. However, the results suggested that the Ag NP toxicity is mainly due to release to free silver ions.

Recently, Ag NPs-induced changes in the proteome profiles of roots and cotyledons of soybean exposed to flooding stress were evaluated using a gel-free proteomic technique to elucidate the underlying mechanism of nanoparticle-mediated growth promotion of soybean under flooding stress (Mustafa et al. 2015a). Ag NPs primarily affected the abundances of 107 root proteins predominantly associated with stress, signaling, and cell metabolism (Fig. 7.2). Comparative proteomic analysis revealed significant increase in abundances of the glyoxalase II and fermentation-related proteins (pyruvate decarboxylase 2 and alcohol dehydrogenase 1) under flooding stress, while a decrease was observed by Ag NPs treatment, implying Ag NPs-mediated metabolic shift from fermentative pathways toward normal cellular processes and formation of comparatively low cytotoxic by-products, which might be responsible for better growth performance of Ag NPs-treated soybeans under flooding stress (Mustafa et al. 2015a). The same group had later studied the effects of Al_2O_3 nanoparticles on soybean. Al_2O_3 nanoparticles (50 mg l^{-1}) were found to enhance the seedling growth after NP flooding, as compared to ZnO and Ag nanoparticles. Proteins that were significantly altered during exposure to alumina NPs belonged to a group of proteins involved in energy metabolism, glycolysis, and lipid metabolism. A gene involved in nitrogen metabolism—*NmrA* like negative transcriptional regulator—was five times upregulated, while flavoprotein that protects organisms from oxidative stress (flavodoxin-like quinone reductase) was downregulated. Moreover, there was less cell death in roots with Al_2O_3 nanoparticles flooding when compared to flooding-treated soybean. Thus, the growth of soybean promoted by Al_2O_3 nanoparticles might be through regulation of energy metabolism and cell death (Mustafa et al. 2015b).

For a better understanding of the interaction of cadmium sulfide quantum dots (CdS QDs) and Arabidopsis, Marmiroli et al. (2015) used two independent mutants which were showing tolerance to CdS QDs. The proteomic analysis of tolerance response in these two mutants confirmed the results of the earlier transcriptomic analysis (Marmiroli et al. 2014), which implies that a significant level of translational and post-translational regulation must have been taking place, presumably triggered by the CdS QD treatment. Similarly, the potential toxicity of CeO_2 NPs in maize was evaluated using proteomic approaches, and it was found that proteins such as catalase (CAT), ascorbate peroxidase (APX), and heat shock protein 70 (HSP 70) might help the plants to defend against oxidative injury and to survive NP exposure (Zhao et al. 2012).

7.4 Conclusion and Prospects

The growing public debate on the toxicity and environmental impact of released nanoparticles has not been yet fully established. There is currently no regulation specific to utilization, release, and maximum acceptable levels of nanoparticles that may pose a serious threat to the environment and human health. In the United States, the utilization and potential release of nanoparticles are regulated by the US EPA. However, studies on the potential toxicity of nanoparticles to ecological terrestrial test species are still lacking (US EPA 2007). The reports from few recent studies have advanced our knowledge of the toxicological impact of several types of nanomaterials. There are still many unresolved issues and challenges concerning the positive and negative biological effects of nanoparticles. There is a critical need to collect more experimental data about the ecotoxicity of different kinds of nanoparticles to support further regulatory efforts by federal agencies.

High-throughput "omics" techniques such as transcriptomics, proteomics, and metabolomics may provide new insights into the biochemical and molecular responses of an organism and play a fundamental role in understanding the mechanisms of cellular toxicity of nanoparticles, as well as other environmental contaminants. These profiling techniques could be used to support aspects of regulatory decision-making in ecotoxicology. However, these techniques have many parallel challenges about data collection, integration, and interpretation and mostly rely on advanced expertise and expensive resources. Furthermore, because of the vast amount of information generated, the analysis of transcriptomic data requires sophisticated bioinformatic approaches. As increasing numbers of transcriptomic datasets are published, there is a critical need for uniformizing gene expression analysis platforms and more integrated data processing (Ankley et al. 2006). In addition, current infrastructure and expertise need to be expanded to enable meaningful analysis of genomic data with increased resource investment.

Despite its great promises for understanding the modes of action of nanoparticles on biological systems, toxicogenomics can lead to unclear results. Genome-wide expression analyses typically reveal cascades of regulated pathways involving set of genes, which do not necessarily provide evidence of causative relationships between toxic stimuli and transcriptional responses. Gene expression analyses are therefore most meaningful when integrated with relevant morphological, physiological, and/or proteomic investigations (Ideker et al. 2001; Waters and Fostel 2004). Proteomics is a valid choice of method to further understand the toxicant effects and to make better toxicity predictions by finding new biomarkers. Comparative proteomic studies may represent an important step to understand nanoparticle–plant interactions as a whole (Matysiak et al. 2015). Although tools exist to detect alterations in transcriptome or proteome profile in various organisms, the genomes of most plant species have not been fully sequenced and annotated till date. Ongoing genome sequencing projects, in long term, would obviate issues related to the global identification of gene products/proteins in key test species used for ecotoxicological risk assessments.

Nevertheless, future perspectives on nanoparticle–plant interaction will depend on a thorough understanding of the molecular mechanisms responsible for the particular response triggered by engineered nanomaterials. Along with a boost of new methodological approaches, it could be expected that next-generation sequencing (NGS) techniques, as well as quantitative proteomic approaches along with post-translational proteomics (glyco-/phosphoproteomics) and interactomics, would contribute to a detailed characterization of genes/protein toward a better understanding of nanoparticle-mediated phytotoxicity.

Acknowledgments **SJ** gratefully acknowledges Science and Engineering Research Board, Department of Science & Technology, Government of India for DST-SERB Young Scientist grant (SB/YS/LS-39/2014), University Grants Commission, India, for UGC start-up grant (F.30-50/2014-BSR), and Special Assistance Program (UGC-SAP-CAS) in the Centre for Advanced Studies in Botany, J.N.V. University, Jodhpur.

RNP gratefully acknowledges the funding under Start-up Research Grant (Life Sciences) by Science and Engineering Research Board, Department of Science & Technology, Government of India (SB/FT/LS-104/2012).

The authors declare no financial or commercial conflict of interest.

References

Abdelhamid HN, Wu H-F (2015) Proteomics analysis of the mode of antibacterial action of nanoparticles and their interactions with proteins. Trends Anal Chem 65:30–46

Adhikari T, Kundu S, Biswas AK, Tarafdar JC, Rao AS (2012) Effect of copper oxide nano particle on seed germination of selected crops. J Agri Sci Technol 2:815–823

Aken BV (2015) Gene expression changes in plants and microorganisms exposed to nanomaterials. Curr Opin Biotechnol 33:206–219

Ankley GT, Daston GP, Degitz SJ, Denslow ND, Hoke RA, Kennedy SW, Miracle AL, Perkins EJ, Snape J, Tillitt DE et al (2006) Toxicogenomics in regulatory ecotoxicology. Environ Sci Technol 40:4055–4065

Asharani PV, Mun GLK, Hande MP, Valiyaveettil S (2009) Cytotoxicity and genotoxicity of silver nanoparticles in human cells. ACS Nano 3:279–290

Asharani PV, Sethu S, Lim HK, Balaji G, Valiyaveettil S, Hande MP (2012) Differential regulation of intracellular factors mediating cell cycle, DNA repair and inflammation following exposure to silver nanoparticles in human cells. Genome Integr 3:1–14

Asli S, Neumann PM (2009) Colloidal suspensions of clay or titanium dioxide nanoparticles can inhibit leaf growth and transpiration via physical effects on root water transport. Plant, Cell Environ 32:577–584

Atha DH, Wang H, Petersen EJ, Cleveland D, Holbrook RD, Jaruga P, Dizdaroglu M, Xing B, Nelson BC (2012) Copper oxide nanoparticle mediated DNA damage in terrestrial plant models. Environ Sci Technol 46:1819–1827

Barkla BJ, Vera-Estrella R, Pantoja O (2013) Progress and challenges for abiotic stress proteomics of crop plants. Proteomics 13:1801–1815

Barrena R, Casals E, Colón J, Font X, Sánchez A, Puntes V (2009) Evaluation of the ecotoxicity of model nanoparticles. Chemosphere 75:850–857

Bhatt I, Tripathi BN (2011) Interaction of engineered nanoparticles with various components of the environment and possible strategies for their risk assessment. Chemosphere 82:308–317

Boonyanitipong P, Kositsup B, Kumar P, Baruah S, Dutta J (2011) Toxicity of ZnO and TiO_2 nanoparticles on germinating rice seed *Oryza sativa* L. Int J Biosci Biochem Bioinf 1(4):282–285

Boverhof DR, Zacharewski TR (2005) Toxicogenomics in risk assessment: applications and needs. Toxicol Sci 89:352–360

Burklew CE, Ashlock J, Winfrey WB, Zhang BH (2012) Effects of aluminum oxide nanoparticles on the growth, development, and microRNA expression of tobacco (*Nicotiana tabacum*). PLoS ONE 7

Castiglione MR, Giorgetti L, Geri C, Cremonini R (2011) The effects of nano-TiO_2 on seed germination, development and mitosis of root tip cells of *Vicia narbonensis* L. and *Zea mays* L. J Nanopart Res 13:2443–2449

Chen C, Unrine JM, Judy JD, Lewis RW, Guo J, McNear DH Jr, Tsyusko OV (2015) Toxicogenomic responses of the model legume *Medicago truncatula* to aged biosolids containing a mixture of nanomaterials (TiO_2, Ag, and ZnO) from a pilot wastewater treatment plant. Environ Sci Technol 49:8759–8768

Chu H, Kim HJ, Kim JS, Kim MS, Yoon BD, Park HJ, Kim CY (2012) A nanosized Ag-silica hybrid complex prepared by gamma-irradiation activates the defense response in Arabidopsis. Radiat Phys Chem 81:180–184

Craig M, Jason CW (2011) Toxicity of silver and copper to *Cucurbita pepo*: differential effects of nano and bulk-size particles. Environ Toxicol 26: n/a. doi:10.1002/tox.20667

Dietz KJ, Herth S (2011) Plant nanotoxicology. Trends Plant Sci 16(11):582–589

Dimkpa CO, McLean JE, Latta DE, Manangón E, Britt DW, Johnson WP, Boyanov MI, Anderson AJ (2012) CuO and ZnO nanoparticles: phytotoxicity, metal speciation, and induction of oxidative stress in sand-grown wheat. J Nanopart Res 14:1125

Dimkpa CO, McLean JE, Martineau N, Britt DW, Haverkamp R, Anderson AJ (2013) Silver nanoparticles disrupt wheat (*Triticum aestivum* L.) growth in a sand matrix. Environ Sci Technol 47:1082–1090

Du W, Sun Y, Ji R, Zhu J, Wu J, Guo H (2011) TiO_2 and ZnO nanoparticles negatively affect wheat growth and soil enzyme activities in agricultural soil. J Environ Monit 13:822–828

Dubos C, Stracke R, Grotewold E, Weisshaar B, Martin C, Lepiniec L (2010) MYB transcription factors in *Arabidopsis*. Trends Plant Sci 15(10):573–581

El-Temsah YS, Joner EJ (2012) Impact of Fe and Ag nanoparticles on seed germination and differences in bioavailability during exposure in aqueous suspension and soil. Environ Toxicol 27:42–49

Farrag HF (2015) Evaluation of the growth responses of *Lemna gibba* L. (duckweed) exposed to silver and zinc oxide nanoparticles. World Appl Sci J 33(2):190–202

Frazier TP, Burklew CE, Zhang BH (2014) Titanium dioxide nanoparticles affect the growth and microRNA expression of tobacco (*Nicotiana tabacum*). Funct Integr Genom 14:75–83

Gao F, Liu C, Qu C, Zheng L, Yang F, Su M et al (2008) Was improvement of spinach growth by nano-TiO_2 treatment related to the changes of Rubisco activase? Bio Metals 21:211–217

Gao Y, Holland RD, Yu LR (2009) Quantitative proteomics for drug toxicity. Brief Funct Genom Proteom 8:158–166

García-Sánchez S, Bernales I, Cristobal S (2015) Early response to nanoparticles in the *Arabidopsis* transcriptome compromises plant defence and root-hair development through salicylic acid signalling. BMC Genom 16:341

Ghodake G, Seo YD, Park DH, Lee DS (2010) Phytotoxicity of carbon nanotubes assessed by *Brassica juncea* and *Phaseolus mungo*. J Nanoelect Optoelect 5:157–160

Ghodake G, Seo YD, Lee DS (2011) Hazardous phytotoxic nature of cobalt and zinc oxide nanoparticles assessed using *Allium cepa*. J Hazard Mat 186:952–955

Giraldo JP, Landry MP, Faltermeier SM et al (2014) Plant nanobionics approach to augment photosynthesis and biochemical sensing. Nat Mater 13(4):400–408

Gottschalk F, Nowack B (2011) The release of engineered nanomaterials to the environment. J Environ Monitor 13:1145–1155

Gubbins EJ, Batty LC, Lead JR (2011) Phytotoxicity of silver nanoparticles to *Lemna minor* L. Environ Pollut 159:1551–1559

Handy RD, Kamme F, Lead JR, Hasselov M, Owen R, Crane M (2008) The ecotoxicology and chemistry of manufactured nanoparticles. Ecotoxicology 17:287–314

Hernandez-Viezcas JA, Castillo-Michel H, Servin AD, Peralta-Videa JR, Gardea-Torresdey JL (2011) Spectroscopic verification of zinc absorption and distribution in the desert plant *Prosopis juliflora-velutina* (velvet mesquite) treated with ZnO nanoparticles. Chem Eng J 170 (1–3):346–352

Hood E (2004) Nanotechnology: looking as we leap. Environ Health Perspect 112:a740–a749

Ideker T, Thorsson V, Ranish JA, Christmas R, Buhler J et al (2001) Integrated genomic and proteomic analyses of a systematically perturbed metabolic network. Science 292:929–934

Kaveh R, Li YS, Ranjbar S, Tehrani R, Brueck CL, Van Aken B (2013) Changes in *Arabidopsis thaliana* gene expression in response to silver nanoparticles and silver ions. Environ Sci Technol 47:10637–10644

Khodakovskaya MV, Biris AS (2015) Method of using carbon nanotubes to affect seed germination and plant growth. US Patent Application 14/716, 117

Khodakovskaya MV, de Silva K, Nedosekin DA, Dervishi E, Biris AS, Shashkov EV, Galanzha EI, Zharov VP (2011) Complex genetic, photothermal, and photoacoustic analysis of nanoparticle-plant interactions. Proc Natl Acad Sci USA 108(3):1028–1033

Khodakovskaya MV, de Silva K, Biris AS, Dervishi E, Villagarcia H (2012) Carbon nanotubes induce growth enhancement of tobacco cells. ACS Nano 6:2128–2135

Kim S, Lee S, Lee I (2012) Alteration of phytotoxicity and oxidant stress potential by metal oxide nanoparticles in *Cucumis sativus*. Water Air Soil Pollut 223:2799–2806

Klaine SJ, Alvarez PJJ, Batley GE, Fernandes TF, Handy RD, Lyon DY, Mahendra S, McLaughlin MJ, Lead JR (2008) Nanomaterials in the environment: behavior, fate, bioavailability and effects. Environ Toxicol Chem 27:1825–1851

Kohan-Baghkheirati E, Geisler-Lee J (2015) Gene expression, protein function and pathways of *Arabidopsis thaliana* responding to silver nanoparticles in comparison to silver ions, cold, salt, drought, and heat. Nanomaterials 5:436–467

Kumari M, Mukherjee A, Chandrasekaran N (2009) Genotoxicity of silver nanoparticles in *Allium cepa*. Sci Total Environ 407:5243–5246

Lahiani MH, Dervishi E, Chen J, Nima Z, Gaume A, Biris AS, Khodakovskaya MV (2013) Impact of carbon nanotube exposure to seeds of valuable crops. ACS Appl Mater Interf 5:7965–7973

Lahiani MH, Chen J, Irin F, Puretzky AA, Green MJ, Khodakovskaya MV (2015) Interaction of carbon nanohorns with plants: uptake and biological effects. Carbon 81:607–619

Landa P, Vankova R, Andrlova J, Hodek J, Marsik P, Storchova H, White JC, Vanek T (2012) Nanoparticle-specific changes in *Arabidopsis thaliana* gene expression after exposure to ZnO, TiO_2, and fullerene soot. J Hazard Mater 241:55–62

Landa P, Prerostova S, Petrova S, Knirsch V, Vankova R, Vanek T (2015) The transcriptomic response of *Arabidopsis thaliana* to Zinc Oxide: a comparison of the impact of nanoparticle, bulk, and ionic zinc. Environ Sci Technol 49(24):14537–14545

Larue C, Khodja H, Herlin-Boime N, Brisset F, Flank AM, Fayard B, Chaillou S, Carrière M (2011) Investigation of titanium dioxide nanoparticles toxicity and uptake by plants. J Physics 304(1): Article ID 012057

Lee CW, Mahendra S, Zodrow K, Li D, Tsai YC, Braam J, Alvarez PJ (2010) Developmental phytotoxicity of metal oxide nanoparticles to *Arabidopsis thaliana*. Environ Toxicol Chem 29:669–675

Lee W-M, Kwak JI, An Y-J (2012) Effect of silver nanoparticles in crop plants *Phaseolus radiatus* and *Sorghum bicolor*: media effect on phytotoxicity. Chemosphere 86:491–499

Lee S, Kim S, Kim S, Lee I (2013) Assessment of phytotoxicity of ZnO NPs on a medicinal plant, *Fagopyrum esculentum*. Environ Sci Pollut Res 20:848–854

Lim HK, Asharani PV, Hande MP (2012) Enhanced genotoxicity of silver nanoparticles in DNA repair deficient mammalian cells. Front Genet 3:104–116

Lin D, Xing B (2007) Phytotoxicity of nanoparticles: inhibition of seed germination and root growth. Environ Pollut 150:243–250

Lin D, Xing B (2008) Root uptake and phytotoxicity of ZnO nanoparticles. Environ Sci Technol 42:5580–5585

Lindgren AL (2014) The effects of silver nitrate and silver nanoparticles on *Chlamydomonas reinhardtii*: a proteomic approach. Degree Project, Department of Biology and Environmental Sciences, University of Gothenburg, Germany

Liu Q, Chen B, Wang Q, Shi X, Xiao Z, Lin J, Fang X (2009) Carbon nanotubes as molecular transporters for walled plant cells. Nano Lett 9:1007–1010

López-Moreno M, de la Rosa G, Hernandez-Viezcas J, Peralta-Videa J, Gardea-Torresdey J (2010a) X-ray absorption spectroscopy (XAS) corroboration of the uptake and storage of CeO_2 nanoparticles and assessment of their differential toxicity in four edible plant species. J Agric Food Chem 58:3689–3693

López-Moreno M, de la Rosa G, Hernandez-Viezcas J, Castillo-Michel H, Botez C, Peralta-Videa J, Gardea-Torresdey J (2010b) Evidence of the differential biotransformation and genotoxicity of ZnO and CeO_2 nanoparticles on soybean (*Glycine max*) plants. Environ Sci Technol 44:7315–7320

Lu CM, Zhang CY, Wen JQ, Wu GR, Tao MX (2002) Research of the effect of nanometer materials on germination and growth enhancement of *Glycine max* and its mechanism. Soybean Sci 21:168–172

Ma XM, Geiser-Lee J, Deng Y, Kolmakov A (2010) Interactions between engineered nanoparticles (ENPs) and plants: phytotoxicity, uptake and accumulation. Sci Total Environ 408:3053–3061

Ma CX, Chhikara S, Xing BS, Musante C, White JC, Dhankher OP (2013) Physiological and molecular response of *Arabidopsis thaliana* (L.) to nanoparticle cerium and indium oxide exposure. ACS Sustain Chem Eng 1:768–778

Magdolenova Z, Collins A, Kumar A, Dhawan A, Stone V, Dusinska M (2014) Mechanisms of genotoxicity. A review of in vitro and in vivo studies with engineered nanoparticles. Nanotoxicology 8:233–278

Marmiroli M, Pagano L, Sardaro MLS, Villani M, Marmiroli N (2014) Genome-wide approach in *Arabidopsis thaliana* to assess the toxicity of cadmium sulfide quantum dots. Environ Sci Technol 48:5902–5909

Marmiroli M, Imperiale D, Pagano L, Villani M, Zappettini A, Marmiroli N (2015) The proteomic response of *Arabidopsis thaliana* to cadmium sulfide quantum dots, and its correlation with the transcriptomic response. Front Plant Sci 6

Mattiello A, Filippi A, Pošćić F, Musetti R, Salvatici MC, Giordano C, Vischi M, Bertolini A, Marchiol L (2015) Evidence of phytotoxicity and genotoxicity in *Hordeum vulgare* L. exposed to CeO_2 and TiO_2 nanoparticles. Front Plant Sci 6

Matysiak M, Kapka-Skrzypczak L, Brzóska K, Gutleb AC, Kruszewski M (2015) Proteomic approach to nanotoxicity. J Proteom. doi:10.1016/j.jprot.2015.10.025

Mazumdar H, Ahmed GU (2011) Phytotoxicity effect of Silver nanoparticles on *Oryza sativa*. Int J Chem Technol Res 3(3):1494–1500

Merrick BA, Bruno ME (2004) Genomic and proteomic profiling for biomarkers and signature profiles of toxicity. Curr Opin Mol Ther 6:600–607

Miralles P, Church TL, Harris AT (2012) Toxicity, uptake, and translocation of engineered nanomaterials in vascular plants. Environ Sci Technol 46:9224–9239

Mirzajani F, Askari H, Hamzelou S, Schober Y, Römpp A, Ghassempour A, Spengler B (2014) Proteomics study of silver nano particles toxicity on *Oryza sativa* L. Ecotoxicol Environ Safe 108:335–339

Mustafa G, Sakata K, Hossain Z, Komatsu S (2015a) Proteomic study on the effects of silver nanoparticles on soybean under flooding stress. J Proteom 122:100–118

Mustafa G, Sakata K, Komatsu S (2015b) Proteomic analysis of flooded soybean root exposed to aluminum oxide nanoparticles. J Proteom 128:280–297

Nair PMG, Chung IM (2014a) Assessment of silver nanoparticle-induced physiological and molecular changes in *Arabidopsis thaliana*. Environ Sci Pollut Res 21:8858–8869

Nair PMG, Chung IM (2014b) Cell cycle and mismatch repair genes as potential biomarkers in *Arabidopsis thaliana* seedlings exposed to silver nanoparticles. Bull Environ Contam Toxicol 92:719–725

Nair PM, Chung IM (2015) Study on the correlation between copper oxide nanoparticles induced growth suppression and enhanced lignification in Indian mustard (*Brassica juncea* L.). Ecotoxicol Environ Safe 113:302–313

Nair R, Varghese SH, Nair BG, Maekawa T, Yoshida Y, Kumar DS (2010) Nanoparticulate material delivery to plants. Plant Sci 179:154–163

Nuwaysir EF, Bittner M, Trent J, Barrett JC, Afshari CA (1999) Microarrays and toxicology: the advent of toxicogenomics. Mol Carcinogen 24:153–159

Panda KK, Achary MM, Krishnaveni R, Padhi BK, Sarangi SN, Sahu SN et al (2011) In vitro biosynthesis and genotoxicity bioassay of silver nanoparticles using plants. Toxicol In Vitro 25:1097–1105

Patlolla AK, Berry A, May L, Tchounwou PB (2012) Genotoxicity of silver nanoparticles in *Vicia faba*: a pilot study on the environmental monitoring of nanoparticles. Int J Environ Res Public Health 9:1649–1662

Pham CH, Yi J, Gu MB (2012) Biomarker gene response in male Medaka (*Oryzias latipes*) chronically exposed to silver nanoparticle. Ecotoxicol Environ Safe 78:239–245

Poma A, Di Giorgio ML (2008) Toxicogenomics to improve comprehension of the mechanisms underlying responses of in vitro and in vivo systems to nanomaterials: a review. Curr Genom 9:571–585

Poynton H, Vulpe C (2009) Ecotoxicogenomics: emerging technologies for emerging contaminants. J Am Water Resour Assoc 45:83–96

Pradhan S, Patra P, Mitra S, Dey KK, Basu S, Chandra S, Palit P, Goswami A (2015) Copper nanoparticle (CuNP) nanochain arrays with a reduced toxicity response: a biophysical and biochemical outlook on *Vigna radiata*. J Agric Food Chem 63:2606–2617

Rao S, Shekhawat GS (2014) Toxicity of ZnO engineered nanoparticles and evaluation of their effect on growth, metabolism and tissue specific accumulation in *Brassica juncea*. J Environ Chem Eng 2:105–114

Remédios C, Rosário F, Bastos V (2012) Environmental nanoparticles interactions with plants: morphological, physiological, and genotoxic aspects. J Bot 2012:1–8. doi:10.1155/2012/751686

Rico CM, Majumdar S, Duarte-Gardea M, Peralta-Videa JR, Gardea-Torresdey JL (2011) Interaction of nanoparticles with edible plants and their possible implications in the food chain. J Agric Food Chem 59:3485–3498

Roco MC (2005) Environmentally responsible development of nanotechnology. Environ Sci Technol 39:106A–112A

Salama HMH (2012) Effects of silver nanoparticles in some crop plants, common bean (*Phaseolus vulgaris* L.) and corn (*Zea mays* L.). Int Res J Biotechnol 3:190–197

Sekhon BS (2014) Nanotechnology in agri-food production: an overview. Nanotechnol Sci Appl 7:31–53

Shaw AK, Hossain Z (2013) Impact of nano-CuO stress on rice (*Oryza sativa* L.) seedlings. Chemosphere 93:906–915

Shen C-X, Zhang Q-F, Li J, Bi F-C, Yao N (2010) Induction of programmed cell death in *Arabidopsis* and rice by single-wall carbon nanotubes. Am J Bot 97:1602–1609

Shiny PJ, Mukerjee A, Chandrasekaran N (2013) Comparative assessment of the phytotoxicity of silver and platinum nanoparticles. Proceedings of the international conference on advanced nanomaterials & emerging engineering technologies. Sathyabama University, Chennai, pp 391–393

Slomberg DL, Schoenfisch MH (2012) Silica nanoparticle phytotoxicity to *Arabidopsis thaliana*. Environ Sci Technol 46:10247–10254

Stampoulis D, Sinha SK, White JC (2009) Assay-dependent phytotoxicity of nanoparticles to plants. Environ Sci Technol 43:9473–9479

Sunkar R, Zhu JK (2004) Novel and stress-regulated microRNAs and other small RNAs from Arabidopsis. Plant Cell 16(8):2001–2019

Syu YY, Hung JH, Chen JC, Chuang HW (2014) Impact of size and shape of silver nanoparticles on *Arabidopsis* plant growth and gene expression. Plant Physiol Biochem 83:57–64

Thiruvengadam M, Gurunathan S, Chung IM (2015) Physiological, metabolic, and transcriptional effects of biologically-synthesized silver nanoparticles in turnip (*Brassica rapa* ssp. *rapa* L.). Protoplasma 252:1031–1046

Thomas CR, George S, Horst AM, Ji ZX, Miller RJ, Peralta-Videa JR, Xia TA, Pokhrel S, Madler L, Gardea-Torresdey JL et al (2011) Nanomaterials in the environment: from materials to high-throughput screening to organisms. ACS Nano 5:13–20

Tumburu L, Andersen CP, Rygiewicz PT, Reichman JR (2015) Phenotypic and genomic responses to titanium dioxide and cerium oxide nanoparticles in Arabidopsis germinants. Environ Toxicol Chem 34:70–83

United States Environmental Protection Agency (1996) Ecological test guidelines (OPPTS 850, 4200): seed germination/root elongation toxicity test. US EPA, Washington, DC

United States Environmental Protection Agency (2007) Nanotechnology White Paper: Tech. Rep. EPA 100/B-07/001, Science Policy Council, US EPA, Washington, DC

Vannini C, Domingo G, Onelli E, Prinsi B, Marsoni M, Espen L, Bracale M (2013) Morphological and proteomic responses of *Eruca sativa* exposed to silver nanoparticles or silver nitrate. PLoS ONE 8(7):e68752

Vannini C, Domingo G, Onelli E, De Mattia F, Bruni I, Marsoni M, Bracale M (2014) Phytotoxic and genotoxic effects of silver nanoparticles exposure on germinating wheat seedlings. J Plant Physiol 171:1142–1148

Wang Q, Zhao S, Zhao Y, Rui Q, Wang D (2014) Toxicity and translocation of graphene oxide in Arabidopsis plants under stress conditions. RSC Adv 4:60891–60901

Waters MD, Fostel JM (2004) Toxicogenomics and systems toxicology: aims and prospects. Nat Rev Genet 5:936–948

Wetmore BA, Merrick BA (2004) Toxicoproteomics: proteomics applied to toxicology and pathology. Toxicol Pathol 32(6):619–642

Wu SG, Huang L, Head J, Chen DR, Kong IC, Tang YJ (2012) Phytotoxicity of metal oxide nanoparticles is related to both dissolved metals ions and adsorption of particles on seed surfaces. J Pet Environ Biotechnol 3:4

Xu LM, Takemura T, Xu MS, Hanagata N (2011) Toxicity of silver nanoparticles as assessed by global gene expression analysis. Mater Express 1:74–79

Yan S, Zhao L, Li H, Zhang Q, Tan J, Huang M, He S, Li L (2013) Single-walled carbon nanotubes selectively influence maize root tissue development accompanied by the change in the related gene expression. J Hazard Mater 246:110–118

Yang L, Watts DJ (2005) Particle surface characteristics may play an important role in phytotoxicity of alumina nanoparticles. Toxicol Lett 158:122

Yang F, Liu C, Gao F, Su M, Wu X, Zheng L, Hong F, Yang P (2007) The improvement of spinach growth by nano-anatase TiO_2 treatment is related to nitrogen photo reduction. Biol Trace Elem Res 119:77–88

Yin L, Colman BP, McGill BM, Wright JP, Bernhardt ES (2012) Effects of silver nanoparticle exposure on germination and early growth of eleven wetland plants. PLoS ONE 7:e47674

Yruela I (2013) Transition metals in plant photosynthesis. Metallomics 5(9):1090–1109

Ze YG, Liu C, Wang L, Hong MM, Hong FS (2011) The regulation of TiO_2 nanoparticles on the expression of light-harvesting complex II and photosynthesis of chloroplasts of *Arabidopsis thaliana*. Biol Trace Elem Res 143:1131–1141

Zhang BH, Pan XP, Cobb GP, Anderson TA (2006) Plant microRNA: a small regulatory molecule with big impact. Dev Biol 289(1):3–16

Zhao L, Peng B, Hernandez-Viezcas JA, Rico C, Sun Y, Peralta-Videa JR, Tang X, Niu G, Jin L, Varela-Ramirez A, Zhang JY, Gardea-Torresdey JL (2012) Stress response and tolerance of

Zea mays to CeO_2 nanoparticles: cross talk among H_2O_2, heat shock protein, and lipid peroxidation. ACS Nano 6:9615–9622

Zheng L, Hong F, Lv S, Liu C (2004) Effect of nano-TiO_2 on strength of naturally aged seeds and growth of spinach. Biol Trace Elem Res 101:1–9

Zhu H, Han J, Xiao JQ, Jin Y (2008) Uptake, translocation, and accumulation of manufactured iron oxide nanoparticles by pumpkin plants. J Environ Monit 10:713–717

Chapter 8
Uptake, Translocation, Accumulation, Transformation, and Generational Transmission of Nanoparticles in Plants

Pradeep Kumar Shukla, Pragati Misra and Chittaranjan Kole

Abstract The field of plant nanotechnology has recently been up-surged into a new epoch of discovery to dissect the intricate processes and mechanisms for better understanding of plant's functional biology in response to nanoparticle exposure. This chapter reviews the current scenario of pathways, mechanisms, and patterns of uptake, translocation, accumulation, transformation, and generational transmission of nanoparticles in plants. Experimental data support that symplastic route is the dominant and highly regulated pathway for transporting NPs within plants and facilitated by a vast array of carrier proteins, aquaporins, interconnected ion channels, endocytosed pathway, or novel pores for the entry of nanoparticles. Xylem being the most preferred plant tissue along with phloem and stomatal opening for absorption and transportation of nanoparticles. Engineered and carbon-based nanoparticles have shown different responses for their transport and utilization in different plants. Engineered nanomaterials are translocated and accumulated differentially within stems, leaves, trichomes, petioles, and fruits of different plants. At subcellular locations, engineered nanomaterials are accumulated in cell walls, cytoplasm, seldom plastids, nuclei, and small vesicles. Carbon-based

P.K. Shukla (✉)
Department of Biological Sciences, School of Basic Sciences,
Sam Higginbottom Institute of Agriculture, Technology and Sciences,
Allahabad 211007, UP, India
e-mail: pradeepshuklak@yahoo.co.in

P. Misra
Department of Molecular and Cellular Engineering, Jacob School of Biotechnology and Bioengineering, Sam Higginbottom Institute of Agriculture, Technology and Sciences,
Allahabad 211007, UP, India
e-mail: pragati.misra@shiats.edu.in

C. Kole
Bidhan Chandra Agricultural University, Mohanpur 741252, West Bengal, India
e-mail: ckole2012@gmail.com

Present Address:
C. Kole
Jacob School of Biotechnology and Bioengineering, Sam Higginbottom Institute of Agriculture, Technology and Sciences, Allahabad 211007, India

C. Kole et al. (eds.), *Plant Nanotechnology*,
DOI 10.1007/978-3-319-42154-4_8

nanomaterials have shown superior prospective for internalization. Uptake, accumulation, and generational transmission of NOM-suspended carbon nanopartcles in rice plants have been reported. Uptake and biodistribution of fullerol was confirmed almost in all plant organs including petioles, leaves, flowers, and fruits in bitter melon. Carbon nanotubes have shown the possibilities for effective penetration into seed coat. Single-walled carbon nanotubes have shown their capability to penetrate chloroplasts and accumulate on thylakoids and stroma in spinach, whereas, multi-walled carbon nanotubes were observed in the seeds and root systems of the developed tomato seedlings. It is certain that not a single transportation mechanism, but a diverse array of multiple mechanisms at physiological, biochemical, and molecular levels are involved for penetration, acquisition, and in planta trafficking of nanoparticles. The goal of this chapter is to put individual experimental efforts back together to unveil the possible enigmas of mechanisms of internalization of nanoparticles, pathways of their movement, and patterns of accumulation and their generational transmission.

Keywords Nanoparticles · Engineered nanomaterials · Carbon-based nanomaterials · Uptake-mechanism · Translocation-pattern · Generational transmission

8.1 Introduction

The chemistry of the Earth is unimaginably complex mostly due to a splendid tangled web of interdependencies of living and lifeless components that include a vast, diverse, and global array of naturally occurring nanomaterials (Wiesner et al. 2011).

To understand the potential benefits of applying nanotechnology to agriculture, the primary step should be to analyze penetration, transport, interaction, and possible significant roles of nanoparticles (NPs) in plants (Lee et al. 2008). Uptake, translocation, and accumulation of NPs may depend on the plant species and the size, kinds, chemical composition, and stability of the NPs (Rico et al. 2011).

The impact of natural, engineered, and incidental nanomaterials (NMs) on higher plants and their beneficial and harmful effects in different plant systems at the physiological, biochemical, and genetic levels has recently been examined and documented in the literatures (Yang and Watts 2005; Zheng et al. 2005; Lin and Xing 2007; Torney et al. 2007; Lei et al. 2008; Zhu et al. 2008; Lin et al. 2009; Ma et al. 2010; Rico et al. 2011; Miralles et al. 2012a; Bhattacharya et al. 2012; Prasad et al. 2012; Remedios et al. 2012; Kole et al. 2013; Azimi et al. 2014; Rad et al. 2014; Shyla and Natarajan 2014; Chutipaijit 2015; Cicek and Nadaroglu 2015; Roohizadeh et al. 2015; Ebbs et al. 2016). These reports explain the effect of different nanomaterials, alone or in combination, on diverse types of plants/vegetation at different growth and developmental stages, but the vital questions regarding the uptake, accumulation, translocation, and transmission of nanomaterials in plant cells and tissues are still unsolved (Navarro et al. 2008). The cell wall of plants, algae, and fungi is the primary site for the interaction and a barrier for the entrance of engineered nanoparticles (ENPs). Mechanisms allowing ENPs to penetrate through cell walls and membranes are yet to be well understood. Inside cells, ENPs might directly elicit alterations of membranes and other cell structures and molecules, as well as protective mechanisms (Navarro et al. 2008). The cell wall of plants prevents the entrance of different kinds of elements into cells, and the NPs having a lesser diameter than the pores of cell wall can, therefore, easily cross the pores and can penetrate inside the cell. Nanoparticles can also utilize stomata and/or also the base of hairs for entry into the leaves' surface, and are then transported to different organs of the plant (Nair et al. 2010). Among the carbon-based NPs, only the fullerene C_{70} and fullerols have been reported to get readily accumulated in plants. Conversely, most of the metal-based NPs were found to be taken up and accumulated in plants, although some conflicting data exists (Rico et al. 2011).

Engineered nanomaterials (ENMs) can play pivotal roles to regulate photosynthetic processes, oxidative stress, antioxidative enzyme activity, radical scavenging ability, gene expression, and macromolecular (DNA, protein, carbohydrates, fatty acid, lignin) modification within edible plants (Rico et al. 2011). The absorption of minerals by the plant is nonselective; some of these metal ions (in conjunction with anions) may be toxic beyond the tolerance limit of the plant. After the absorption, NPs are subsequently translocated and finally accumulated in different parts of the plants establishing complex with carrier proteins. Selection criterion of a particular NP by a specific plant species while rejecting other NPs remains unclear. If NPs are larger than the diameter of pores present in the root hairs, they tend to accumulate at the surface, and if NPs are smaller, they get absorbed and transported to other parts of the plants. Some NPs are accumulated in extracellular space, while others are inside the cell (Husen and Siddiqui 2014). The understanding of ionic metal transport in plants may not accurately predict ENPs' transport mechanism (Ebbs et al. 2016). The present review provides a basic platform to understand the possible mechanisms of NP uptake, transport, internalization, and their generational and transgenerational transmission.

8.2 Physiological Aspects of Possible Mechanisms of Uptake, Transport, and Accumulation of Nanoparticles in Plants

Uptake and transport of NPs are integral to their successful functioning in the plant systems. Experimental data are very limited, and many proposed mechanisms are under intense debate to explain uptake, transport, and accumulation of NPs. Accumulation and transformation of NPs in plant cells and tissues suggest a possible mechanism for NP penetration (Lin et al. 2009). Proper understanding of mobilization/remobilization mechanisms of nutrient elements in particulate forms and their conversion into plant operational forms in planta could provide a promising pathway for micronutrient transport as NPs or the packaging of nutrients in general, in NP encapsulations, which are also capable of being taken up intact by plants (DeRosa et al. 2010; Gogos et al. 2012; Zhang et al. 2012). The uptake and distribution of metal ion/metal itself in the plant is a matter of debate and challenge to the scientific community. It is not clear whether nanocrystals are formed outside the plants and then transported through the membrane into various parts or the NPs are formed within the plant by the reduction of the metal salt (Husen and Siddiqui 2014).

Particulate forms of mineral nutrients could be mobilized and remobilized via the xylem and phloem, respectively (Wang et al. 2012a, b, 2013a, b, c). It is now well-known that plasma membrane (via apoplast) and plasmodesmal (symplastic) transport mechanisms both play central roles in nutrient internalization along with the water. Water, from the soil, can be absorbed by the roots and then can move radially across into xylem tracheary elements. Subsequently, xylem structures are important determinants to regulate the speed of water transport, and different xylem structures may validate diverse uptake kinetics of NPs (Fig. 8.1) (Ma et al. 2010; Rico et al. 2011; Mishra et al. 2014). Root also absorbs water-dissolved minerals, and these metal salts ascend in ionic form and subsequently are reduced to elemental form as NPs. Water movement through the root apoplast is driven only by pressure gradients, while transport across a membrane-delimited pathway implicates capillary

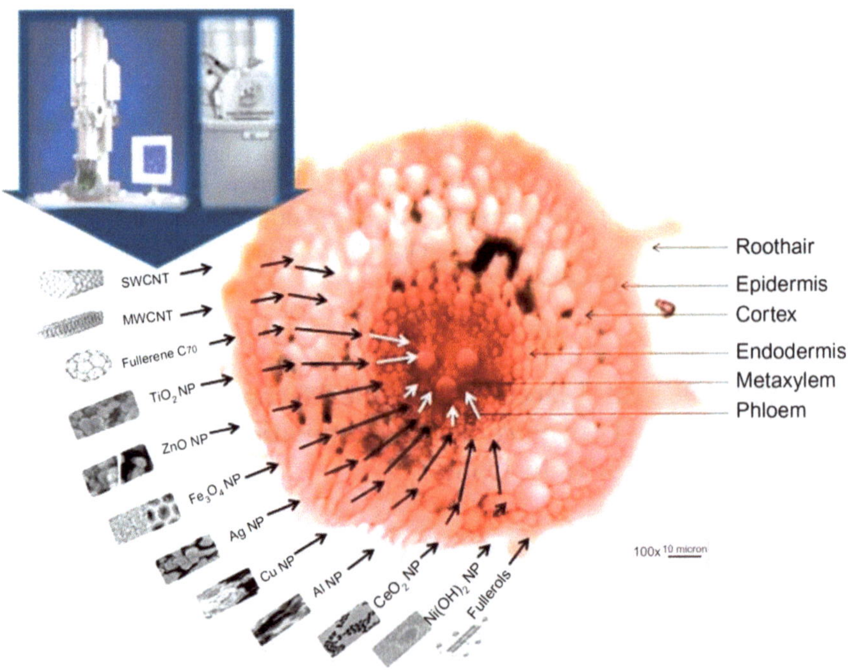

Fig. 8.1 Schematic representation showing uptake of metal-based NPs and carbon-based NPs by root system. Root anatomical structures represent internalization, cellular translocation, and cellular localization of different NPs in root zones [adapted from Mishra et al. (2014)]

action, osmotic pressure, and osmotic gradients (Gardea-Torresdey et al. 2005; Patrick et al. 2015). The transportation of NPs is supposed to pass through the epidermis and cortex and finally to stele of the plant (Shankar et al. 2003).

The stele is the central part of the root containing the pith (if present), vascular tissue, and pericycle and occurs on the inside of the endodermis. At strategic locations (endodermis and sometimes exodermis), the root apoplast is blocked by casparian bands composed of lignin deposited in cell walls. Therefore, it is prerequisite for nutrients to penetrate plasma membranes of each endodermal cell and transport to the stele through the symplasm or transcellularly, or after effluxed from endodermal cells via the apoplast. Some endodermal and exodermal cells have a conspicuous absence of lignin and suberin lamellae and are referred to as passage cells (Fig. 8.2) (Patrick et al. 2015). Also, mobilization of NPs is known to be very prompt, ensuring participation of phloem transport and confirming the nutrient availability to all parts of the plant. Further, the presence of NPs was confirmed in extracellular space and within some cells in the *Cucurbita* plants (Gonzalez-Melendi et al. 2008). The results obtained from both lower and higher plants demonstrate that ion uptake is characterized by (White 2012): (1) Selectivity: Certain mineral elements are taken up preferentially, while others are discriminated against or almost excluded; (2) accumulation: The concentration of elements can be much higher in cell sap than in the external solution; and (3) genotype: There are distinct differences between plant species in their ion uptake characteristics.

ENMs may diffuse in the space between the cell wall and plasma membrane (through porous cell walls): a route well-known as the apoplastic pathway (Lin et al. 2009). Transmission electron microscopy (TEM) images of root cross sections confirmed the presence of NPs in the apoplast, cytoplasm, and nuclei of the endodermal cells in ryegrass (*Lolium perenne*) (Lin and Xing 2008). Through the apoplast, particles may directly reach the endodermis without crossing the edge of epidermal and cortical cells. However, aggregates often accumulate in the endodermis as a result of the significant apoplast barrier imposed by the waxy casparian strip (Larue et al. 2012a; Zhao et al. 2012b; Patrick et al. 2015). For efficient translocation, ENMs in apoplastic flow must eventually unify into the symplast so as to penetrate into vascular system (Deng et al. 2014).

The symplastic route is hypothesized to be the more important and highly regulated pathway for transporting ENMs within crops. It has been hypothesized that cellular penetration and trafficking of ENMs could be accomplished by binding to a vast array of carrier proteins, through appropriate aquaporins, via interconnected ion channels, via endocytosed pathway, or by crafting novel pores (carbon nanotubes) (Rico et al. 2011; Patrick et al. 2015). Depending upon genotype and environmental conditions, 30–80 % of water flow across roots occurs via the cell-to-cell pathway (symplasmic plus transcellular) (Patrick et al. 2015). Superior expression of aquaporin proteins and upregulation of water channel genes were found to support possible passive uptake mechanisms (Khodakovskaya et al. 2012). Endocytosis has been proved, through the use of temperature control and the addition of wortmannin (an endocytosis inhibiting agent), as one of the possible transport mechanisms (Onelli et al. 2008; Liu et al. 2009; Iversen et al. 2012; Miralles et al. 2012b).

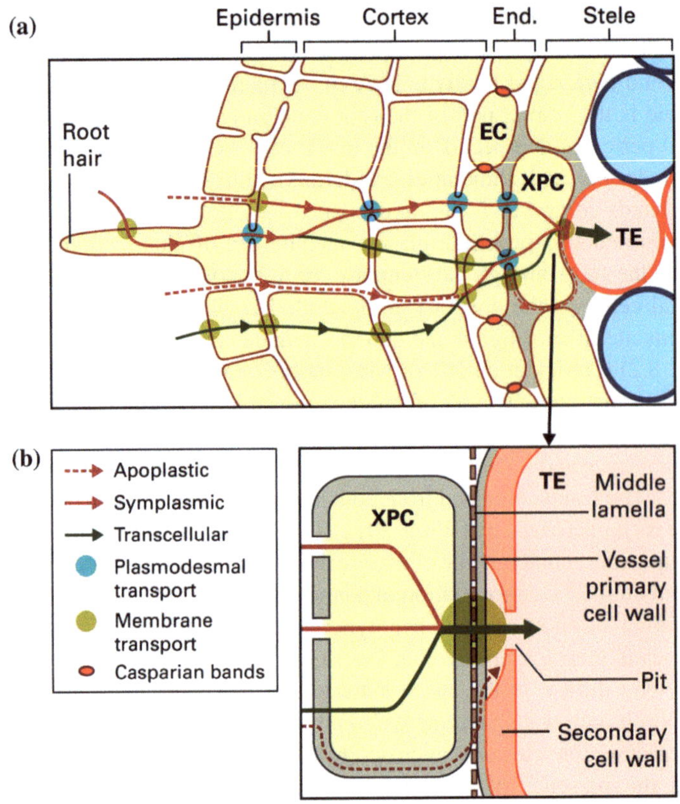

Fig. 8.2 Possible uptake routes of nanoparticle transport in plant system through xylem (including apoplastic, symplasmic, transcellular, plasmodesmal transport, membrane transport, and casparian band). **a** Diagrammatic representation of transverse cross section of a root showing arrangement of root hair, epidermis, cortex, endodermis (End.), and stele. Nutrient and water transport routes across the root are indicated as well as key transport steps through plasmodesmata and plasma membranes. At strategic locations, the root apoplast is blocked by casparian bands composed of lignin deposited in cell walls. Some endodermal cells (ECs) have a conspicuous absence of lignin and suberin lamellae and are referred to as passage cells. **b** Diagram showing possible clues for loading xylem tracheary elements (TE) by Xylem parenchyma cells (XPCs) through half-bordered pits; this process is facilitated by specific membrane transporters [adapted from Patrick et al. (2015)]

Chang and colleagues advocated the involvement of an energy-independent route for mesoporous silica nanoparticles (MSNs) uptake and further proposed possible routes of MSNs uptake by Arabidopsis (*Arabidopsis thaliana*) roots. The MSNs, which are in contact with the cell membrane, can be internalized via endocytosis (scheme A) and remain in internal vesicles inside the cells. However, most MSNs cross the plasma membrane directly (scheme B), and then, the particles endure in the cytoplasm or translocate to other organelles (e.g., plastids or nuclei) (scheme C), which is a specific benefit for cargo delivery (Fig. 8.3) (Chang et al. 2013).

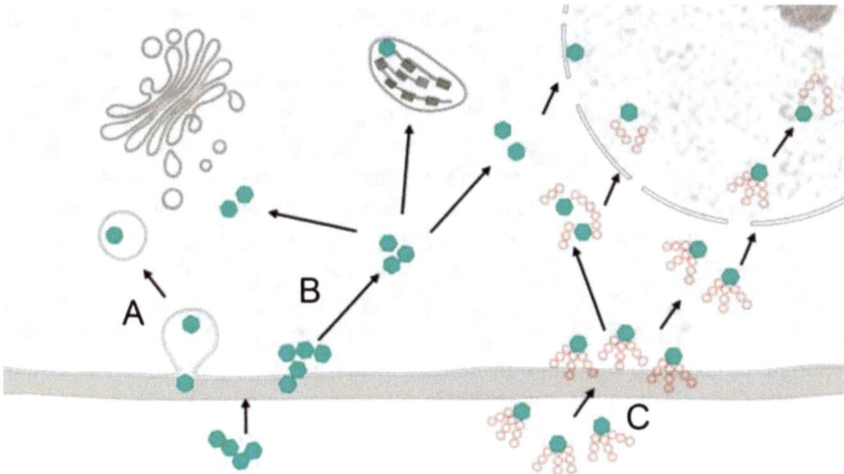

Fig. 8.3 Schematic representation showing possible pathways and localizations of TMAPS/F-MSNs after penetration into the Arabidopsis root system. *Scheme A*—internalization of TMAPS/F-MSNs through endocytosis; *Scheme B*—direct penetration of TMAPS/F-MSNs through plasma membrane as a primary route (after crossing plasma membrane, TMAPS/F-MSNs may localize into cytoplasm or may penetrate organelles, e.g., plastids and the nucleus); *Scheme C*—DNA complexed with TMAPS/F-MSN may internalize into the plant cell and then transported to the nucleus [adapted from Chang et al. (2013)]

Interaction of ENMs with plant cell membranes could alter depending on hydrophobic/hydrophilic nature. Hydrophilic nanomaterials tend to adsorb on bilayer membrane surface, and further, they can bind to intracellular vesicles, while hydrophobic NMs favors to embed into the hydrophobic core of the membrane without resulting in any membrane injury and leakage (Li et al. 2008; Stark 2011). ENMs in the cytoplasm may be embedded by certain proteins or with some specific biomolecules that form a corona (Nel et al. 2009). ENM-containing endosomes or ENM–protein complex (typically with a diameter of 20–50 nm) could undergo effective mobilization to neighboring cells via plasmodesmata. Ultrasmall TiO_2 NPs have been found to disorder structural integrity of microtubular networks of plasmodesmata in Arabidopsis (Wang et al. 2011; Larue et al. 2012a). Moreover, Rab proteins were hypothesized to serve as master regulators for intracellular trafficking of ENMs to specific zones near plasmodesmatal connections (Cifuentes et al. 2010). As a result, ENMs transportation may be regulated through the customary and harmonized action of membranous organelles, diverse array of transport proteins, and interconnected complex trans-walled channels. Because the symplastic flow is inhabited with a diversity and high volume of materials, this pathway shows the possibilities to be highly stringent and well-organized for inter- and intracellular transportation of ENMs through endodermis and into stele and subsequent vascular tissues (Deng et al. 2014).

NPs can also adapt some novel mechanisms to enter plant cells. NPs can penetrate plant root system and/or other plant tissues through selective ion channels, binding to carrier proteins, via aquaporins (water conducting channels), through endocytosed pathway, sometimes also forming new apertures (preferentially for carbon nanotubes, CNTs) or by attaching to organic compounds in the environmental media. NPs exhibit higher surface area to mass ratio in comparison with the bulk metals; therefore, they possess superior reactivity with living and nonliving surroundings. The NPs may be complexed with large numbers of specific and nonspecific membrane transporter proteins or wide varieties of chemicals in root exudates and, subsequently, be transported to the plants (Watanabe et al. 2008; Kurepa et al. 2010). Most of the metal-based NPs (MB NPs) that have been reported as taken up by plants include elements for which ion transporters have been identified (Hall and Williams 2003). NPs can accumulate inside the plant root and/or shoot tissues as intact particles (Gardea-Torresdey et al. 2014; Antisari et al. 2015). Some organic molecules released from root tips may transform metal salts into NPs via reduction, and then, these NPs are transported into plant system (Gardea-Torresdey et al. 2002; Sharma et al. 2007). Size exclusion limits and lateral heterogeneity studies of stomatal foliar uptake pathway for aqueous solutes and water-suspended NPs have suggested that the stomatal pathway differs fundamentally from the cuticular foliar uptake pathway (Eichert et al. 2008). Entry of NPs into the plant was confirmed through leaf surface. Special structural features such as trichrome and hypodermis in a leaf of murici (*Byrsonima sericea*) and araçá (*Psidium guineense*) probably formed a barrier, reducing the penetration of metal ions into the mesophyll as observed by the lower iron leaf content and iron accumulation in trichomes (Da silva et al. 2006). Ion transporters specific to cell membrane have been reported for efficient uptake of NPs in the plants (Hall and Willams 2003). Adsorption and aggregation of the NPs were confirmed by scanning electron microscopy (SEM) analysis on the root surface of ryegrass (Lin and Xing 2008).

The genetic response of plants in the presence of NPs is also a topic of discussion. Differences in xylem anatomical structures may lead to different internalization route of NPs into vasculature of plant systems. Solutes may follow either apoplastic or symplastic mode of internalization or sometimes through plasmodesmatal connection for entry into vascular tissues. More studies are required in the field to confirm this hypothetical view (Singh et al. 2015). Wheat (*Triticum aestivum*), maize (*Zea mays*), spinach (*Spinacia oleracea*), zucchini (*Cucurbita pepo*), rapeseed (*Brassica napus*), and some desert plants showed their differential ability for metal-based NPs to penetrate seeds without affecting germination (Răcuciu and Creangă 2009; Stampoulis et al. 2009; De la rosa et al. 2011; Pokhrel and Dubey 2013; Kouhi et al. 2014; Taran et al. 2014; Chichiriccò and Poma 2015).

Till now, symplastic transport is the most accepted pathway for NP uptake. Some studies also explained the apoplastic mode of transport, whereas some other studies have explained a diverse array of involvement of plasmodesmata, carrier proteins, aquaporins, ion channels, and endocytosis. While xylem being the most preferred plant tissue along with the phloem and stomatal opening plays a significant role in absorption and transport processes, a well-defined mechanism of NP uptake and transport is still under question and need to be explored.

8.3 Uptake, Translocation, Accumulation, and Transformation of Engineered Nanomaterials in Plants

The in planta uptake and internalization of ENMs is a dynamic phenomenon and may depend on exposure conditions, chemical properties of ENMs (including surface charge, particle size, hydrophobic/hydrophilic nature, aggregation state, and protein/biomolecule adsorption), and crop species (Nedosekin et al. 2011). Concentration and charge of the NP are important determinants for uptake of NP in roots and translocation to above ground plant tissue such as leaves and stems (Burke et al. 2014). All NPs, once released into the environment, undergo dramatic and complex transformations through interactions with various chemicals and other factors (e.g., UV light, interaction with (in) organic ligands, redox reactions, bio-transformations, and aggregation) (Wiesner et al. 2011). ENMs have been shown to translocate and accumulate differentially within stems, leaves, petioles, and fruits of different crops. Direct imaging or whole-plant mapping confirms possible evidences for uptake via roots, translocation through vasculature, and aggregation of ENMs in different plant parts. Based on the experimental observations, the following patterns for ENMs uptake and translocation are evident (Deng et al. 2014): (1) In shoots, transpiration flow pattern and/or the leaf architecture plausibly regulate translocation of ENMs and their accumulation near or within vasculature (Ghafariyan et al. 2013); (2) long-distance transport of ENMs is size/dimension reliant, i.e., smaller aggregates or individual ENPs showing selective advantage and greater efficiency for long-distance transportation (viz. root system to subapical tissues), as compared to larger aggregates from ENMs of same type; (3) accumulation of ENMs (expressed as amount per dry weight tissue) in leaf is higher than that of stems; and (4) some particular sites of distribution of ENMs (away from vascular transport), e.g., leaf periphery and trichomes, may be associated with detoxifying pathways (Cifuentes et al. 2010).

8.3.1 Metal-Based Nanoparticles

8.3.1.1 Silica-Based Nanoparticles

Chang and colleagues showed the delivery of DNA using 100-nm mesoporous silica nanoparticles (MSNs) to the cortical cells and endodermis of intact roots of Arabidopsis. The localization and subcellular distribution of MSNs in the roots were examined through transmission electron microscopy (TEM) and confocal laser scanning microscopy (CLSM). The results showed that TMAPS/F-MSNs (*N*-trimethoxysilylpropyl-*N*,*N*,*N*-trimethylammonium chloride labeled MSNs) were present in cortical cells, endodermal cells, pericycle, and vasculature of the root. Subcellular distribution studies of MSNs in root cells showed that MSNs were accumulated in cell walls, cytoplasm, seldom plastids, nuclei, and small vesicles. In addition, the accumulation of some MSNs in vesicles advocates that endocytosis might be one of the uptake routes. Most importantly, the occurrence of MSNs at the cell nucleus notifies that the nanoparticles could penetrate the nuclear envelope or nuclear pore (Fig. 8.4) (Chang et al. 2013).

Uptake and cellular distribution of fluorescently labeled MSNs, with size of 20-nm harboring integrated pores with an estimated diameter of 2.58 nm, were investigated in four plant species, viz. lupin (*Lupinus albus*), wheat (*Triticum aestivum*), maize (*Zea mays*), and Arabidopsis. The results obtained from the study revealed that MSNs transported into the roots via symplastic and apoplastic pathways and further destined to the aerial parts of the plants including the stems and leaves through the conducting tissues of the xylem. The results also confirmed that MSNs sufficiently penetrated the cell wall, entered the endodermis and intercellular spaces and to the vascular tissue, and finally transported to the aerial parts of the plants. Moreover, when MSNs were taken up by the protoplasts, the accumulation of MSNs was also observed in the chloroplast. It was also hypothesized that the translocation and broader internalization of MSNs in plants will facilitate them to be utilized as a novel delivery means for the transportation of different sized biomolecules into plants (Sun et al. 2014). Uptake, transport, and distribution of SiO_2 NPs were also examined in Bt-transgenic cotton (*Gossypium* spp.). Results, as revealed by TEM analysis, confirmed the presence of SiO_2 NPs in the xylem sap. Also, SiO_2 NPs were transported from roots to shoots via xylem sap in both nontransgenic and Bt-transgenic cotton. The presence of dark dots (particles) in the endodermal region and vascular cylinder (under 2000 mg L^{-1} SiO_2 NPs treatment) was confirmed in both Bt-transgenic cotton and nontransgenic cotton. The presence of SiO_2 NPs was more prominent on the root outer epidermis, whereas only a few were located in intercellular spaces. These results exemplified that most of SiO_2 NPs were adhered on root surface and only a very small amount of NPs could succeeded to penetrate roots. Moreover, Si content in the Bt-transgenic roots was higher than nontransgenic when treated with 2000 mg L^{-1} SiO_2 NPs, suggesting that SiO_2 NPs have great potential to penetrate into the root of Bt-transgenic cotton as compared to nontransgenic cotton (Le et al. 2014). An attempt was made to use MSNs as carriers to deliver Cre recombinase protein (immobilized on gold-plated MSNs) into maize cells (Martin-Ortigosa et al. 2014). SiO_2 NPs, at concentration range 10–1000 mg/l, showed accumulation and aggregation on the root surface of pear plant. Aggregation was very prominent at higher concentrations (500 and 1000 mg/l) of NPs, whereas an insignificant amount of particles were found to be attached to the roots for $NSiO_2$ at 10 and 100 mg/l (Zarafshar et al. 2015).

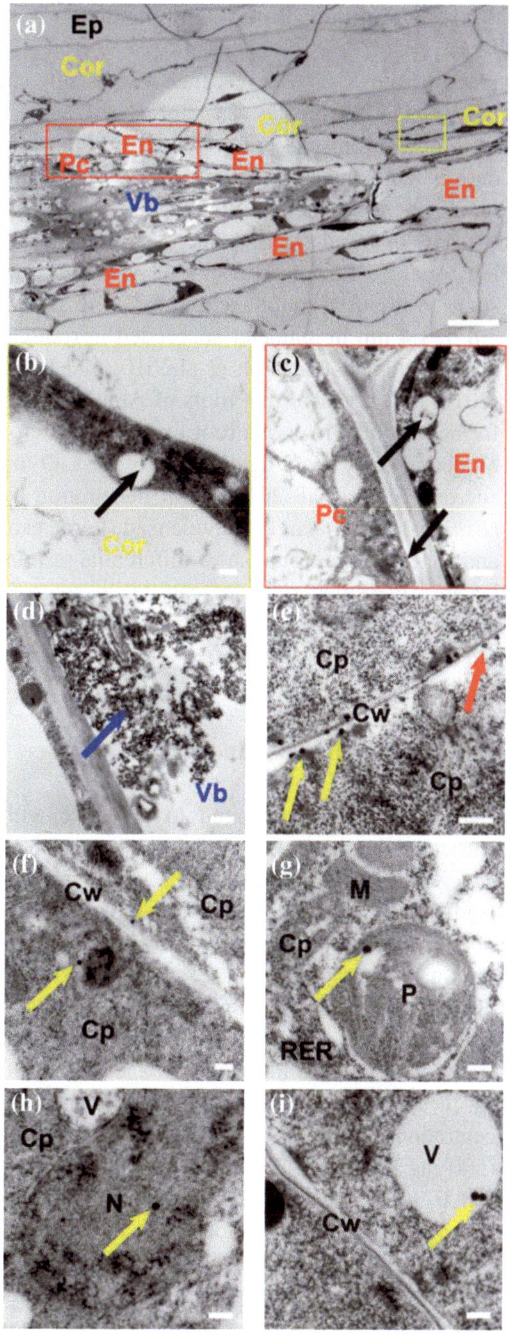

Fig. 8.4 Transmission electron microscopy of Arabidopsis root tissue confirming distribution of MSN. **a** Root section showing organization of tissues from the epidermal cells to the vascular bundle; **b The** presence of MSNs (*black arrow*) in the cortical cell (Cor) as seen in an enlarged view of the *yellow box* in **a**; **c The** presence of MSNs (*black arrow*) in the endodermal (En) and pericycle (Pc) cells in an enlarged view of the *red box* in **a**; **d The** presence of MSNs (*blue arrow*) in the vascular bundle (Vb); **e** localization of MSNs is in the cell wall (Cw) (*red arrow*) or penetration through the plasma membrane (entered the cell) (*yellow arrows*); **f–h** MSNs accumulation (*yellow arrow*) in the cytoplasm (Cp) (**f**) or in the plastid (P) (**g**); (**h**) nucleus (N) after penetrating cell wall; **i** MSNs accumulation in vesicles (V) (*yellow arrows*). *Scale bars* **a** 20 mm; **b** 200 nm; **c** and **d** 500 nm; **e–i** 200 nm. *Ep* epidermis; *Cor* cortex; *En* endodermis; *Pc* pericycle; *Vb* vascular bundle; *Cp* cytoplasm; *Cw* cell wall; *P* plastid; *M* mitochondrion; *N* nucleus; *V* vacuole; *RER* rough endoplasmic reticulum [adapted from Chang et al. (2013)]

8.3.1.2 Titanium-Based Nanoparticles

Electron and X-ray fluorescence microscopy studies established penetrating ability of TiO_2 nanoconjugates (2.8 ± 1.4 nm) on seedlings of Arabidopsis grown on agar medium. The results confirmed distribution of TiO_2 nanoconjugates into the epidermis and underlying palisade tissue, signifying stomatal contribution and involvement of endocytotic vesicles in the internalization process. Further, mass spectroscopy and electron microscopy analysis evidenced the foliar uptake following aerial treatments. However, TiO_2 nanoconjugates smaller than 5 nm remained stuck to the seed mucilage and failed to penetrate, while TiO_2 nanoconjugates of 2.8 ± 1.4 nm in diameter succeeded in root cell penetration up to inside vacuoles and the nucleus (Kurepa et al.2010; Chichiriccò and Poma 2015). Kurepa et al. (2010) also revealed that roots of Arabidopsis bounded by pectin hydrogel capsule formed by mucilaginous root exudates, which may play miraculous role either by hindering or by enabling the entry of the TiO_2 nanoconjugates with Alizarin red S or sucrose (Rico et al. 2011). Leaf penetration by TiO_2 NPs in wheat and rapeseed (*Brassica napus*) was also evidenced (Larue et al. 2012b). TiO_2 NPs differing in size and concentration could show differential responses for seed growth and germination because the small particles can easily enter the cell wall pores of the plant and transport to various other parts (Lu et al. 2002). Doping TiO_2 NPs with N could affect plant translocation of NPs to above ground plant tissue (Burke et al. 2014).

8.3.1.3 Zinc-Based Nanoparticles

ZnO NPs have been found to associate with highly vacuolated and collapsed cortical cells along with the shrinking and partial death of the vascular cells (Lin and Xing 2008). The ZnO NPs were absorbed by the plant roots and circulated equivalently throughout the plant tissues. But All ENPs may not be similarly operative for all crops. Unlike CeO_2 NPs, ZnO NPs were found to be translocated into above ground plant tissue, suggesting that uptake and translocation are dependent on NP type (Priestera et al. 2012).

Uptake and accumulation of ZnO NPs (8 nm) were investigated in soybean (*Glycine max*) seedlings at the range of 500–4000 mg L^{-1}. The uptake of Zn NPs by the soybean seedlings was significantly higher at 500 mg L^{-1} than the concentrations at 1000 mg L^{-1} and above. This may be because at lower concentration (500 mg L^{-1}), the NPs have lesser aggregation, whereas at high concentrations (1000–4000 mg L^{-1}), the probability of agglomerates formation is proposed. Passage of oversized agglomerates through the cell pore walls, therefore, becomes problematic. This ultimately reduces uptake and accumulation in case of ZnO NPs as understood from the results (Lopez-Moreno et al. 2010a). ZnO NPs were absorbed as Zn^{2+} oxidation state by hydroponically grown soybean plants. Later, it was hypothesized that ZnO NPs transformed in Zn^{2+} oxidation state at the root surface (Lopez-Moreno et al. 2010a). Similar results were also obtained by Dimkpa et al. (2013) and Wang et al. (2013a, b, c).

Scanning electron microscopy and energy dispersive analysis of X-rays (SEM-EDAX) showed Zn uptake by the peanut (*Arachis hypogea*) seeds treated with nanoscale ZnO. Thin sections of the peanut embryo were analyzed by SEM. Although, an expected, low Zn concentration in peanut seeds was observed in EDAX spectra, EDAX images confirmed that the regions showing higher C and N concentrations also exhibited high accumulation of Zn in the seeds treated with nanoscale ZnO. The postharvest leaf and kernel samples were analyzed using Atomic Absorption Spectrophotometer (AAS) to estimate the zinc content (Prasad et al. 2012).

8.3.1.4 Copper-Based Nanoparticles

CuO NPs were transported to the shoots and translocated back to the roots via phloem (Shankar et al. 2003). CuO NPs were taken up by maize and wheat in the particulate form (Dimkpa et al. 2012, 2013; Wang et al. 2012a, b). Uptake and translocation of Cu NPs in mung bean (*Vigna radiata*) and wheat in agar growth medium were evaluated. The results showed that the Cu NPs were able to cross the cell membrane and agglomerate in the cells. A significant relationship between the bioaccumulated NPs in plant tissues and growth media was also established. It was also noticed that mung bean was more sensitive than wheat to toxicity of Cu NPs probably due to root anatomical differences (Lee et al. 2008; Rico et al. 2011).

Copper NPs exhibited greater ability for uptake in shoots than copper bulk particles (BPs). Results revealed that total uptake into the shoots was approximately three times greater for the NPs. Scanning transmission electron microscopy (STEM) images of radish (*Raphanus sativus*) shoot samples did not reveal any significant evidence of electron-dense deposits, and energy dispersive spectroscopy (EDS) analysis did not reveal specific elemental signals for Cu in either control samples or samples exposed to 500 mg/L NPs (Atha et al. 2012).

8.3.1.5 Silver-Based Nanoparticles

Uptake and distribution of silver nanoparticles (SNPs) were investigated in Indian mustard (*Brassica juncea*) and alfalfa (*Medicago sativa*). Alfalfa, in contrast to Indian mustard, showed better uptake with a parallel upsurge in the metal concentration and exposure time (Harris and Bali 2008). In another study, Ag NPs did not seem to accumulate Ag in any form in Indian mustard plants (Haverkamp and Marshall 2009). The silver NPs were found to be located in the nucleus (Monica and Cremonini 2009). The seeds of *Boswellia ovalifoliolata*—an endemic and globally threatened medicinal tree species placed in MS medium containing SNPs, showed 90 % germination, in contrast to 70 % germination without SNPs. It was proposed that SNPs can penetrate through seed coat and may stimulate the embryo for germination (Savithramma et al. 2012). Uptake and the internalization of silver NPs and its bulk counterpart were first time compared in zucchini (*Cucurbita pepo*) plants. Plants exposed to 10–1000 mg L^{-1} Ag NPs exhibited 4.7 times higher Ag concentration in the shoots than those treated with bulk Ag powder at similar concentrations (Stampoulis et al. 2009).

The leaves of lettuce (*Lactuca sativa*) plant, when sprayed with the salt $AgNO_3$ and with Ag–NPs, which were both round (38.6 nm in diameter) and nonround (38.2 nm × 57.8 nm), and had hydrodynamic diameters of 47.9 nm ± 29.2 nm, evidenced by the cuticular and stomatal uptake of NPs and translocation into the vascular tissue. Translocation pathways seemed to be both apoplastic and symplastic. Transformation cycles within the plant involving the binding of Ag^+ ions to thiol groups and the conversion of Ag^+ ions in Ag–NPs, starting from the dissolution of both the salt $AgNO_3$ and Ag–NPs were also proposed (Larue et al. 2014). Accumulation of Ag NPs was found to be accumulated in vacuoles of root cell. Deposition of both individual and the aggregate particle was observed inside the cell wall, indicating the penetration of Ag particle inside the cells. Spherical Ag NPs with a diameter of 20 nm were observed inside the plant cell. Regarding transportation of smaller particles inside the cells, it was hypothesized that cell wall thickness (about 5–20 nm) may respond as natural molecular sieves, allowing transport of smaller nanoparticles through larger pores to enter in the protoplasm (Figs. 8.5 and 8.6) (Mazumdar 2014).

8.3.1.6 Cerium-Based Nanoparticles

Seedlings of soybean, alfalfa, maize, and tomato (*Solanum lycopersicum*) exhibited Ce accumulation in tissues with the increased external concentration of CeO_2 NPs (7 nm) (Lopez-Moreno et al. 2010a, b). This differential accumulation could be the result of variances in root microstructures and the physical and chemical interfaces between the NPs and diverse variety of the root exudates in the rhizosphere.

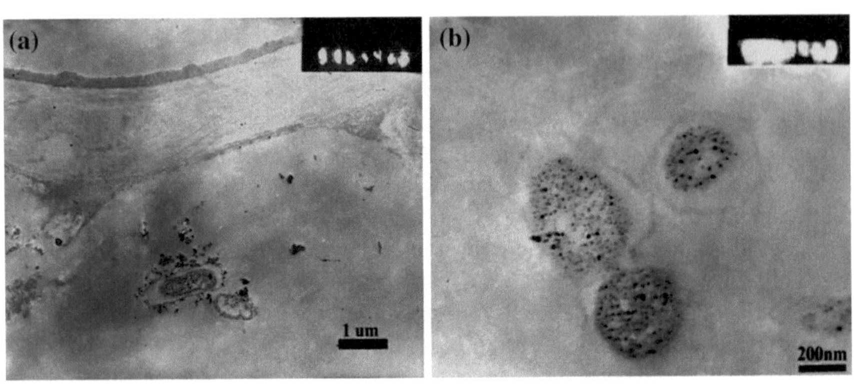

Fig. 8.5 TEM images showing ultrastructures of *V. radiata* roots treated with Ag NPs at a concentration of 1000 μg/mL. **a** Accumulation of AgNPs inside the cell and **b** accumulation of AgNPs inside vacuoles [adapted from Mazumdar (2014)]

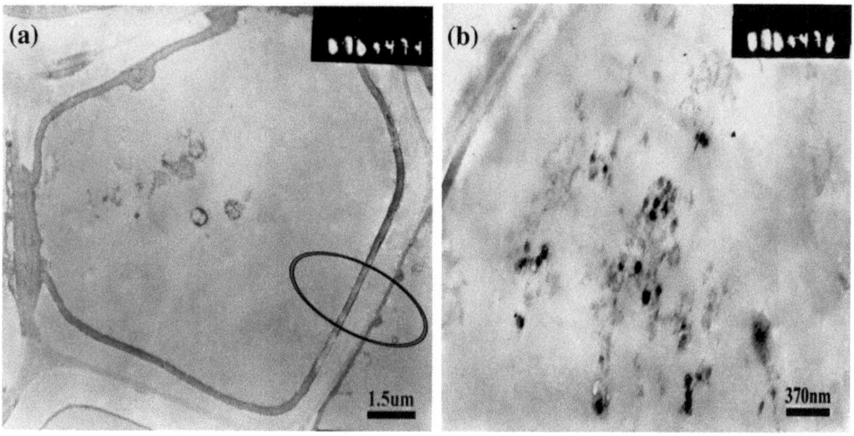

Fig. 8.6 TEM images showing ultrastructures of *B. campestris* roots treated with Ag NPs at a concentration of 1000 μg/mL. **a** Accumulation of AgNPs inside whole cell, **b** enlarged view (*encircled*) of image **a** showing accumulation of AgNPs in plasmodesmata and cell wall [adapted from Mazumdar (2014)]

Aerosol or suspension of CeO_2 NPs was absorbed by the corn leaves but did not translocate to new leaves. Application of NPs along with the irrigation water did not evidence for any detectable translocation of the NPs within the plant (Birbaum et al. 2010). Soybean plants also exhibited uptake and accumulation of CeO_2 NPs and did not show biotransformation (Lopez-Moreno et al. 2010a). CeO_2–NPs exhibited primary diameter of 8 nm ± 1 nm and hydrodynamic diameter of 1373 nm ± 32 nm, internalized by the roots and translocated to the shoots when added to the soil where maize plants were growing. The translocation pathway was proposed to be apoplastic. The studies also pointed out that the mobility of NPs and NP accumulation in the roots and translocation to shoots were favorably influenced by an organic substance in the soil and alginates, respectively (Zhao et al. 2012a, b, 2014). Unexpectedly, soybean plants, treated with nano-CeO_2, showed reduced leaf counts irrespective of its concentration. Even the lowest concentration of nano-CeO_2 showed growth retardation in the harvested plant (Priestera et al. 2012; Husen and Siddiqi 2014). Tomato plants were treated with low concentrations of CeO_2 NPs (10 mg/L) to investigate its effect on seed quality and the development of second-generation seedlings. These NPs in fact slightly improved the growth of the plant (first-generation seedlings) but, at the same time, weakened the capacity to respond to the fertilization effect of the CeO_2 NPs. The accumulation of CeO_2 NPs in plant seeds and fruit tissues suggested that they have a high impact that can influence subsequent generations. These results demonstrate that although the instant results are positive, there is the need to evaluate the long-term, multigenerational impact of NPs on plants. The results showed that the benefits obtained in the first generation, as illustrated in the previous works, were not persistent in seedlings of the second generation. This study provided, probably, the first evidence

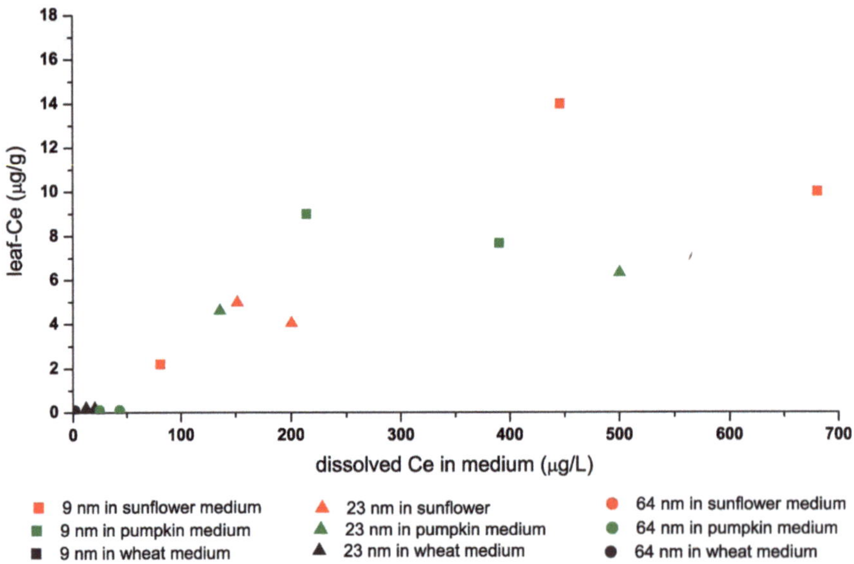

Fig. 8.7 Graphical representation of concentration (µg/L) of dissolved Ce in plant-growth medium (x-axis) against Ce concentration (µg/g) inside the leaves of the plants (sunflower, pumpkin, and wheat) grown on that medium (y-axis) at the end of the experiment [adapted from Schwabe et al. (2015)]

of the transgenerational impact of CeO_2 NPs on the development and growth of tomato plants (Wang et al. 2013d).

An investigation regarding uptake of differently sized CeO_2 NPs by three crop plants including pumpkin (*Cucurbita maxima*), wheat, and sunflower (*Helianthus annuus*) revealed that Ce NPs larger than 20 nm did not translocate from roots to shoots. Ce uptake was particularly high for particles smaller than 10 nm due to their greater dissolution rates (Fig. 8.7). Experiments with Zr/CeOx NP revealed that Ce NP was not the solitary, but to a significant degree, dissolved Ce(III) ions, were also adequate forms of NPs for uptake. The study highlighted that dissolution of CeO_2 NPs in soil solution was significantly influenced by plant root activity and that uptake of dissolved Ce(III) trailed by reprecipitation needs to be considered as an important pathway to explain CeO_2 NPs uptake by plants. Further, NP-root-exposure studies confirmed that translocation of Ce was species-dependent. Sunflower had a high affinity for Ce-ion accumulation inside the leaves when Ce was supplied as dissolved ions (Fig. 8.8), while no significant difference between pumpkin and wheat were observed (Schwabe et al. 2015).

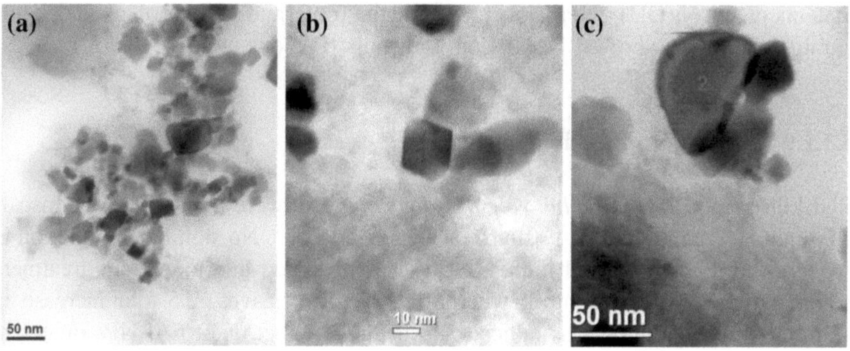

Fig. 8.8 STEM-EDX analysis of NPs extracted from Zr/CeOx-treated sunflower leaves. **a** Bright-field STEM image showing occurrence of NPs in sunflower leaves; **b** STEM image confirming the presence of rhombic NP in higher magnification. **c** STEM image of round-shaped NPs [adapted from Schwabe et al. (2015)]

8.3.1.7 Iron-Based Nanoparticles

The influence of magnetic nanoparticles coated with tetramethylammonium hydroxide was analyzed on the growth of maize plants in early ontogenetic stages (Răcuciu and Creangă 2007). Magnetite (Fe_3O_4 NPs, with 20 nm diameter) NPs uptake was analyzed using a vibrating sample magnetometer by pumpkin seedlings grown under hydroponic conditions. The results confirmed that signals of magnetic NPs were detected in roots, stems, and leaves of edible pumpkin plants but no uptake occurred in Fe_3O_4 NPs-treated lima bean (*Phaseolus limensis*) plants. It was, therefore, proposed that uptake of Fe_3O_4 NPs also depends on the plant species (Zhu et al. 2008). Epidermal cells of leaf petioles of living pumpkin plants accumulated carbon-coated Fe NPs. Results also showed that accumulation site (epidermal cells) was closer to the application site, whereas no NPs were noticed in the cells located distant from the application points or near the xylem (Corredor et al. 2009). ENMs were detected in shoots within a period of 24 h, when sunflower, tomato, pea (*Pisum sativum*), and wheat plant were exposed to the carbon-coated magnetic NPs (Cifuentes et al. 2010). Application of Fe_3O_4 NPs increased the translocation of Fe to leaf tissue, and positively charged Fe_3O_4 NPs caused a reduction in root colonizing rhizobia (Burke et al. 2015). Rapid accumulation of engineered iron NPs in leaves of aquatic plant, Brazilian waterweed (*Egeria densa*), was confirmed using electron spin resonance, two photon, and confocal microscopy (Spori et al. 2014).

8.3.1.8 Nickel-Based Nanoparticles

Uptake and translocation of $Ni(OH)_2$ NPs (8.7 nm) in mesquite (*Prosopis* sp.) were investigated. The X-ray absorption near edge structure (XANES) spectra confirmed that uncoated $Ni(OH)_2$ NPs were observed in roots and shoots of plants, while citrate-coated NPs showed Ni NPs only in roots (Parsons et al. 2010).

8.3.1.9 Aluminum-Based Nanoparticles

Red kidney beans (*Phaseolus vulgaris*) and ryegrass were treated with nanoscale aluminum (Al) particles (1–100 nm) for uptake analysis. No significant variation in Al concentration in the red kidney beans was observed due to Al NPs treatment compared to untreated control, whereas in ryegrass leaves, 2.5-fold increase in aluminum concentration was noticed. No negative effect due to Al NPs treatment was observed on the growth of red kidney beans and ryegrass in the tested concentration range (Doshi et al. 2008).

8.3.1.10 Other Metal-Based Nanoparticles

Plants when exposed to Fe and Mn also exhibited incidence of particulate Fe oxide and Mn (Ghafariyan et al. 2013; Pradhan et al. 2013). In a similar fashion, MgO NPs were observed in roots, when applied via foliar application (Wang et al. 2013a, b, c). Notably, the same crop showed differential absorption pattern for different nutrient elements provided in particulate form through the root, and it was also evident where wheat showed differential pattern for CuO versus ZnO NPs, confirming Cu existence in wheat shoot mainly as CuO particles and a lower amount of dissolved forms, and Zn as Zn phosphate (Dimkpa et al. 2012, 2013). Development and growth processes of the mung bean plant were prominently affected by foliar spray of the NP suspensions of ZnO, FeO, and ZnFeCu-oxide. Enhancements in root and shoot length as well as accumulation of biomass were recorded for NPs-treated plant as compared to the nontreated plants. The maximum enhancement was found at 50 ppm ZnFeCu-oxide followed by 50 ppm FeO and least for 20 ppm ZnO depending on their chemical composition, size, and surface energy (Dhoke et al. 2013).

Alfalfa seedlings, when exposed to Au(III) and Ag(I) ions through agar solid growth media, got reduced and accumulated as Au and Ag NPs (Gardea-Torresdey et al. 2002, 2003). Similar observations, regarding accretion and biotransformation of Ag(I) and Pt(II) ions into Ag and Pt NPs, were also recorded in alfalfa and Indian mustard (*Brassica juncea*) seedlings. TEM images confirmed accumulation of Pt NPs ranging between 3 and 100 nm with different morphologies in roots of alfalfa (Harris and Bali 2008; Bali et al. 2010). Experimental evidence suggested that gold NPs were able to translocate and accumulate in the soybean plants after seed inoculation (Falco et al. 2011; Maharramov et al. 2015).

8.3.2 Carbon-Based Nanoparticles

Carbon-based nanomaterials (CNMs) have shown superior prospective for internalization through leaves' surface and further translocation to the root system of the plant. However, their foliar uptake is not well recognized. Further, CNMs are not considered potential contaminants in the liquid phase (Ke and Qiao 2007; Deng et al. 2014). Conversely, the hydrophobicity of NMs can be obviated through their interaction with natural organic matter (NOM), when discharged into the environment (Hyung et al. 2007). Uptake and translocation of CNMs to aerial parts were provided by many researchers (Lin and Xing 2007; Cañas et al. 2008; Khodakovskaya and Biris 2009; Lin et al. 2009; Nedosekin et al. 2011; Smirnova et al. 2011; Bhattacharya et al. 2012; Kole et al. 2013; Cicek and Nadaroglu 2015). The first evidence on the uptake, accumulation, and generational transmission of NOM-suspended carbon NPs in rice plants was provided by Lin et al. (2009). The potential impact of nanomaterial exposure on plant development and genetic consequences through plant–nanomaterial interactions was documented by these authors. The abundance of NOM (a heterogeneous mixture of proteins, lipids, amino acids, and peptides that are derived from decomposed animals and plants) in natural soil and water sources permits its interaction with NPs to provoke water solubility and kinesis in the environment (Davies et al. 1997; Ke and Lamm 2011). An in vivo flow cytometry analysis in tomato stems showed that the average velocity of quantum dot–carbon nanotube conjugates was approximately 0.2 mm/s (Nedosekin et al. 2011). Kole et al. (2013) investigated uptake and biodistribution of a fullerene derivative $C_{60}(OH)_{20}$, or "fullerol" in bitter melon (*Momordica charantia*). Uptake, translocation, accumulation, transformation, and generational transmission of carbon-based NPs are scantly examined; some of them are discussed below.

8.3.2.1 Fullerene Nanoparticles

Probably for the first time, Lin et al. (2009) investigated the uptake, accumulation, and generational transmission of NOM-suspended fullerene in rice plants. Suspensions of fullerene C_{70} and multiwalled carbon nanotubes (MWNTs) in NOM solution at a concentration of 100 mg L^{-1} in Milli-Q water were prepared. Dynamic uptake, compartment distribution, and transformation of fullerene C_{70} in rice plants were characterized, and transgenerational transmission of C_{70} particles to the next progeny through seeds was detected. Results showed that distribution of C_{70} particle was not reliant on concentration. The prevalent C_{70} particles were dominant in the roots as well as on the stems and leaves of the 2-week-old plants (Fig. 8.9). However, C_{70} particle was predominantly present in or near the stems' vasculature systems, lesser in the leaves in the mature (6-month-old) plants, and least in the seeds due to the multiplied uptake rates, therefore reducing the amount of translocated NPs. However, no C_{70} aggregates were found in the epidermis, plausibly due to a greater distance from the vascular system. Furthermore, no C_{70}

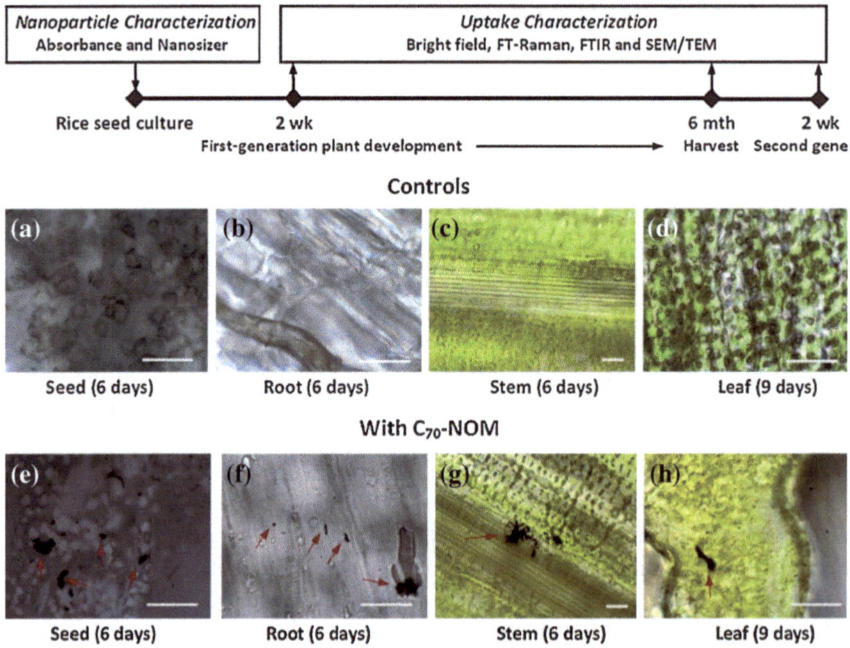

Fig. 8.9 Schematic representations of experimental details and bright-field imaging of C_{70} uptake by 1-week-old rice plants. *First row*—experiment scheme. *Second row*—bright-field images of controls seeds (**a**), root (**b**), stem (**c**), leaf (**d**). *Third row*—aggregates of NPs (as shown by *arrows*) observed in seeds (**e**), root (**f**), stem (**g**), leaf (**h**) treated with C_{70}–NOM. The scale bars are 20 mm for all images [adapted from Lin et al. (2009)]

was found left in the roots of the mature plants, indicating robust transport of NMs from the roots to the aerial parts of the plant. It was hypothesized in the previous studies that penetration of C_{70} nanoparticles may ensue via osmotic pressure, capillary forces and pores on cell walls (≈3.5–5 nm) (Carpita et al. 1979) or through intercellular plasmodesmata (≈50–60 nm at midpoint) (Smith 1978) or via the highly regulated symplastic route. NPs' small dimension and self-assembly and from the NP interactions with plant organelles and the NOM are important factors for the integration of NPs by plant species. Interestingly, though much less frequently, C_{70} NPs were also marked in the leaf tissues of the second-generation plants grown without the addition of NMs (Fig. 8.10) (Lin et al. 2009).

Uptake and accumulation of two fullerene derivatives (i.e., a $C_{60}(OH)_{20}$ molecule a supramolecular assembly of C_{70}–NOM) were investigated in onion (*Allium cepa*) plants. To avoid the structural complexity of NOM, a Temple Northeastern–Birmingham (TNB) model (Davies et al. 1997), expressing a monomer of the humic substance in NOM, was used in present exploration. TNB monomers were attached to the surfaces of the C_{70} molecules through hydrophobic interaction. Interaction of NPs with onion plant cells was dependent on particle size and surface properties. When plant cells were exposed a higher concentration of $C_{60}(OH)_{20}$ NPs

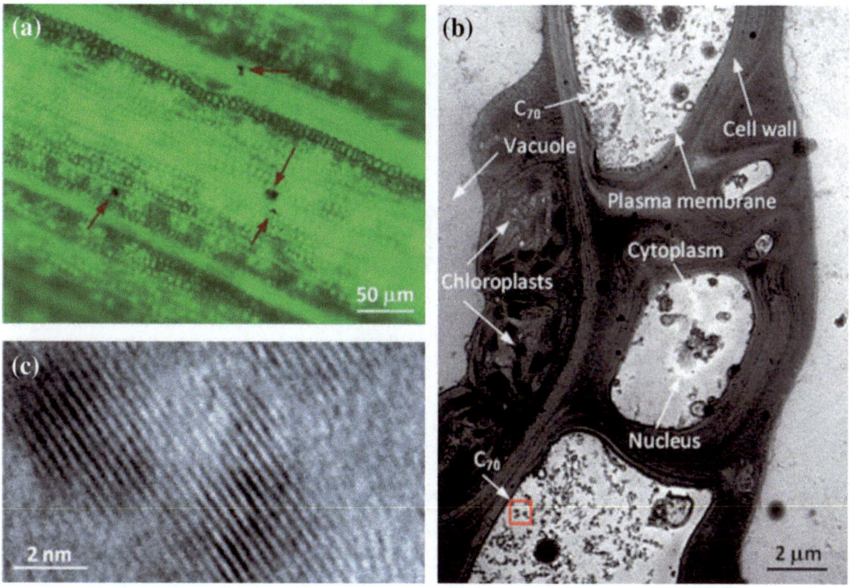

Fig. 8.10 Experimental evidences of generational transmission of C_{70} NPs. **a** Bright-field image showing aggregation of C_{70} (indicated by *arrows*) appeared mostly in or near the vascular system of the leaf in second-generation rice plant. **b** TEM image of C_{70} particles in the leaf cells (plant cell walls and other organelles) of a 2-week-old rice plant. **c** Enlarged view of TEM image of the C_{70} particles in (**b**) (*red square*). FFT analysis of the TEM image confirmed the lattice spacing of the C_{70} particles to be 0.257 nm [adapted from Lin et al. (2009)]

(i.e., 70 mg L^{-1}), a steady upsurge in cellular damage was observed. It was hypothesized that the presence of a thick, rigid, and porous cell wall acts as a barrier for large and hydrophobic NPs and their aggregates while imposing slight interference to the translocation of hydrophilic NPs (Fig. 8.11) (Chen et al. 2010; Ke and Lamm 2011).

8.3.2.2 Fullerol Nanoparticles

Uptake, biodistribution, and accumulation of fullerol (a fullerene derivative) were examined in bitter melon (*Momordica charantia*) through bright-field imaging (BFI) and Fourier-transformed infrared (FTIR) spectroscopy by Kole et al. (2013). Seeds were treated with five stock concentrations (0.943, 4.72, 9.43, 10.88, and 47.2 nM) of fullerol, $C_{60}(OH)_{20}$, nanoparticles, referred to as C_1, C_2, C_3, C_4, and C_5, respectively, and C_0 was controlled without fullerol, $C_{60}(OH)_{20}$, nanoparticles. Black aggregates observed through FTIR analysis confirmed biodistribution of fullerols almost in all plant organs including petioles, leaves, flowers, and fruits (Fig. 8.12). The result had confirmed that most of the stem and fruit samples (excluding C_0 and C_1) exhibited distinct FTIR peaks common to fullerols across the 1500–1700 cm^{-1}

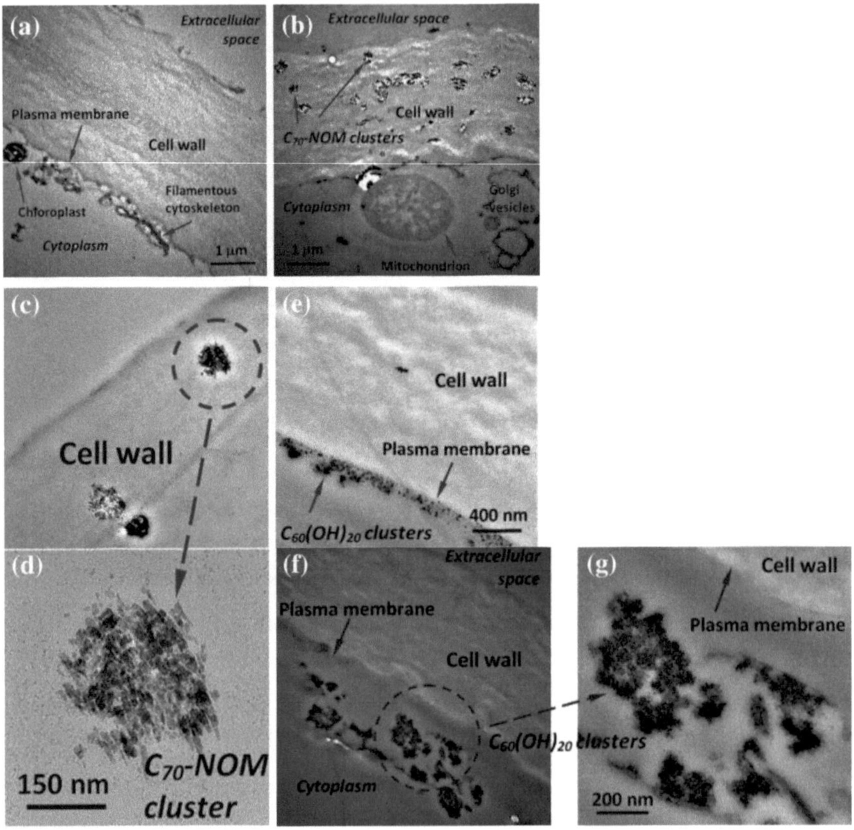

Fig. 8.11 Transmission electron microscopy of *Allium cepa* plant cells showing uptake and translocation of carbon nanoparticle. **a** TEM imaging of control plants showing plant cell wall and plasma membrane. **b–d** Aggregation of C_{70}–NOM at plant cell wall showing C_{70}–NOM clusters ranging 50–400 nm, when exposed to C_{70}–NOM concentration at 50 mg L^{-1}. **d** Enlarged image of a C_{70}–NOM cluster as *encircled* in **c**. **e–g** Translocation of $C_{60}(OH)_{20}$ across plant cell walls. **e** Accumulation of $C_{60}(OH)_{20}$ clusters at cell wall and a plasma membrane interface. **f** Localization of C_{70}–NOM clusters in intracellular space. **g** Magnified view of C_{70}–NOM clusters in intracellular space as encircled in **f** [adapted from Chen et al. (2010)]

spectral region, suggesting the presence of fullerols in the samples, whereas fullerol-like IR peaks were found absent in sample C_0, obviously reflecting the absence of the NM. Intense FTIR signal for fullerols was observed only in the fruits from C3 and C5 samples. This result was predictable as the C5 seeds were treated at the highest fullerol concentration. Authors proposed that the foremost mechanisms for the fullerol uptake could be through the diverse array of (1) transpirational stream generated through rapid water evaporation from the aerial plant parts especially leaves, (2) in planta NPs concentration gradient, or (3) hydrophobic interface between the NPs and the waxy coatings among plant cells (Fig. 8.13) (Kole et al. 2013).

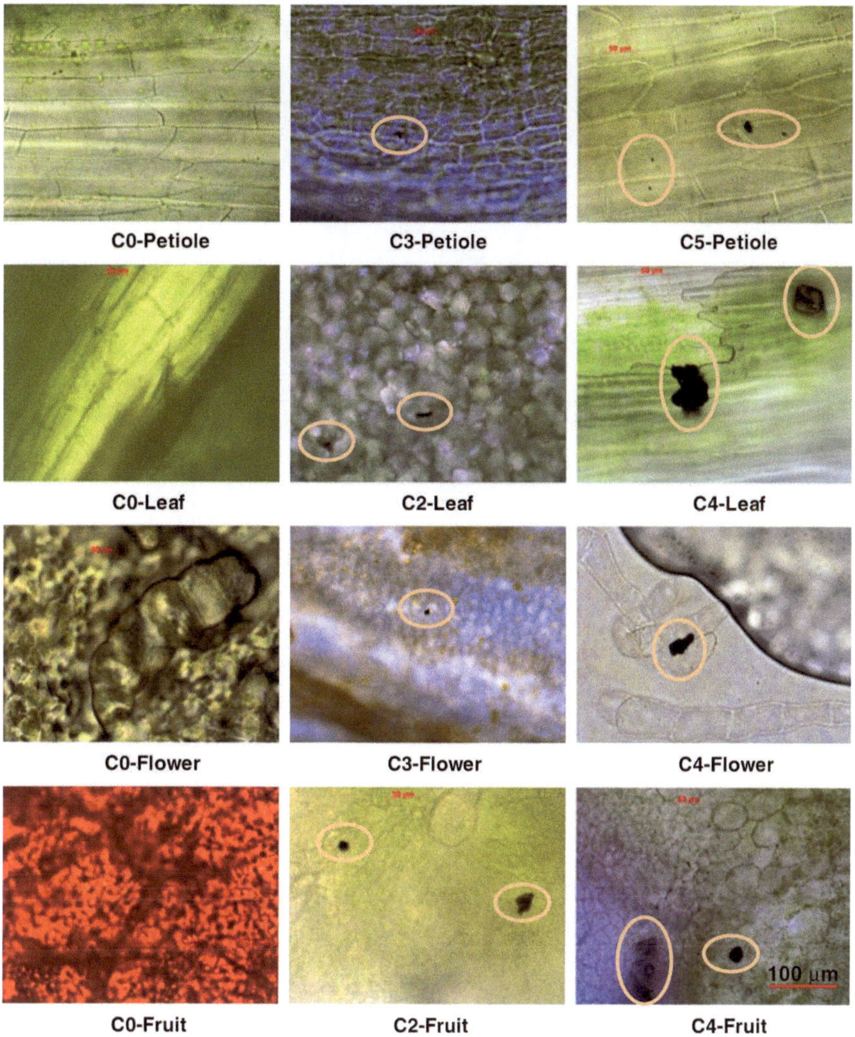

Fig. 8.12 Fullerol biodistribution in different plant organs including petioles, leaves, flowers, and fruits of bitter melon. The *circles* highlight black aggregates which were later confirmed by FTIR as fullerols. Fullerol, $C_{60}(OH)_{20}$, and nanoparticles (BuckyUSA) were dissolved in Milli-Q water (pH 6.5) to prepare five stock concentrations (0.943, 4.72, 9.43, 10.88, and 47.2 nM), referred to C1, C2, C3, C4, and C5, respectively, whereas C0 is control without fullerol nanoparticles [adapted from Kole et al. (2013)]

8.3.2.3 Single-Walled Carbon Nanotubes

Single-walled carbon nanotubes (SWNTs) are capable to transverse across the plant cell wall and cell membrane as well (Liu et al. 2009). NPs act as smart treatment delivery systems for plant systems (Gonzalez-Melendi et al. 2008). Due to a

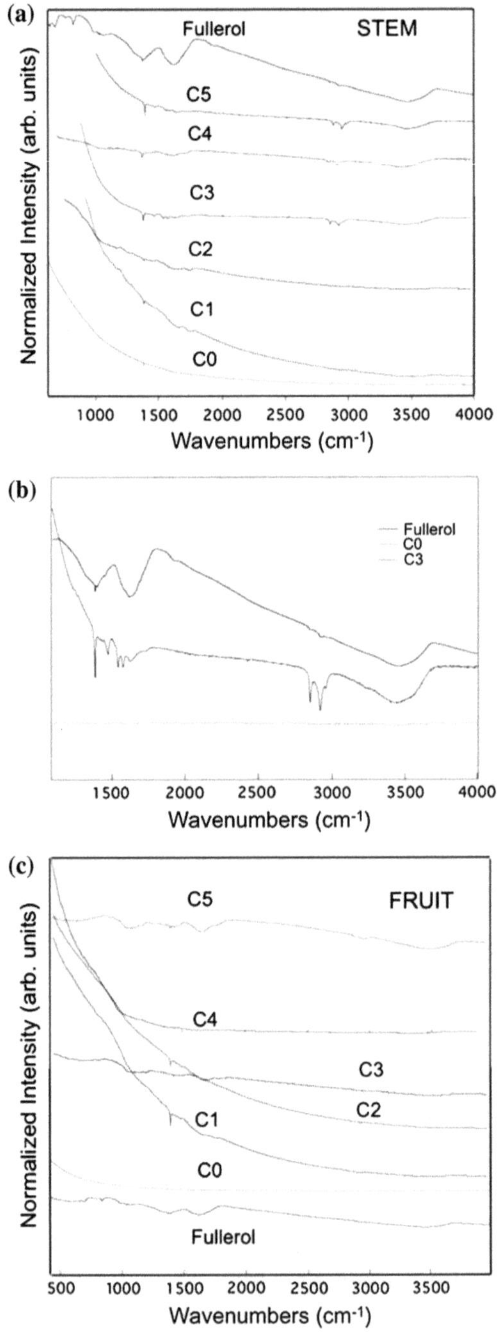
(a)
Fullerol
STEM
C5
C4
C3
C2
C1
C0
Normalized Intensity (arb. units)
1000 1500 2000 2500 3000 3500 4000
Wavenumbers (cm-1)
(b)
Fullerol
C0
C3
1500 2000 2500 3000 3500 4000
Wavenumbers (cm-1)
(c)
C5
FRUIT
C4
C3
C2
C1
C0
Fullerol
Normalized Intensity (arb. units)
500 1000 1500 2000 2500 3000 3500
Wavenumbers (cm-1)

◀ **Fig. 8.13** Fourier-transformed infrared spectroscopy of fullerols in different plant organs of bitter melon. **a** FTIR data for stem samples. C1–C5 samples exhibit clear FTIR signatures for fullerol. All the spectra were counterpoise for precision. **b** Fullerol peaks (~1580–1640 cm^{-1}) of scaled and expanded view of C3 sample. **c** FTIR data for fruit samples. C1–C5 samples display precise fullerol signatures. All the spectra were counterpoise for clarity. Sample C5 shows very distinct features similar to fullerols due to preliminary incubation of seeds in highest fullerol concentration [adapted from Kole et al. (2013)]

thickness of seed coats, penetration of NPs into seeds could be difficult compared to plant cell walls and membranes (Srinivasan and Saraswathi 2010). But carbon nanotubes may effectively penetrate seed coat possibly due to an enlarged water uptake (Khodakovskaya and Biris 2009; Ganguly et al. 2014). The low surface friction of CNTs facilitates the flow of organic substances into the cytoplasm (Whitby and Quirke 2007).

Confocal microscopy studies confirmed that SWCNTs (length <500 nm) bound noncovalently to fluorescein isothiocyanate (FITC) dye, followed a penetration pathway via endocytosis in suspension cultures of intact tobacco (*Nicotiana tabacum*) cells line BY-2 (Liu et al. 2009). A similar approach of membrane penetration via endocytosis was also reported in in vitro cultures of rice and Arabidopsis cells (enzymatically treated for removing walls) (Shen et al. 2010).

By applying advanced methods, SWCNTs have shown their capability to penetrate chloroplasts and accumulate on thylakoids and stroma in spinach (*Spinacea oleracea*) leaves. Further, it was also found that once penetrated in the membranes of spinach chloroplasts, SWCNTs improved the flow of electrons and photosynthetic activity, thereby exciting action on the uptake of light with near-infrared wavelengths. Moreover, SWCNTs were noticed to be sensitive to nitric oxides (NOx), and therefore, plants-harboring nanotubes could be used as indicators for NOx (Giraldo et al. 2014). Apical meristem at the base of tomato roots (where root elongation occurs) showed a high accumulation of CNTs (Cañas et al. 2008). SWCNTs were found accumulated on the peripheral surface of the main root and also in secondary roots in the form of nanotube sheets (Tan and Fugetsu 2007).

Tomato seeds inoculated in CNTs-complemented agar medium showed the presence of CNTs inside the seeds escorted with higher moisture percentage (Srinivasan and Saraswathi 2010). Nanotubes also served as potential nanotransporters to deliver DNA and small dye molecules into intact plant cells (Lin et al. 2009; Savithramma et al. 2012). Further, endocytotic method of the nanotubes penetration was reported in the nucleus, plastids, and vacuoles, and nanotubes also induced organelle recycling in tobacco and periwinkle (*Catharanthus roseus*) plants (Serag et al. 2011a, b, 2012a, b; Chichiriccò and Poma 2015).

8.3.2.4 Multi-walled Carbon Nanotubes

Multi-walled carbon nanotubes (MWCNTs) were observed in the seeds and root systems of the developed tomato seedlings (Khodakovskaya and Biris 2009), whereas the cell walls of rice cell suspension limited the access of the MWCNTs into the cellular cytoplasm (Tan and Fugetsu 2007). At high concentrations, MWNTs showed higher affinity for the epidermis and the waxy casparian strips of the roots, and consequently, MWNTs adsorbed to the plant root surfaces. Interestingly, at higher concentration, MWNTs aggregated at root surface caused a blockage at the plant roots and root hairs, thereby impeding the uptake of water, nutrients, and NOM, as well as plant development. Delayed flowering and reduction in seed setting was observed in the rice plants nurtured with MWNT–NOM (400 mg/L), compared to the controls or the NOM-fed plants (Bhattacharya et al. 2012). MWCNTs showed insignificant particle uptake and translocation, whereas it enhanced germination and root elongation in alfalfa and wheat (Miralles et al. 2012a). Investigation of potential effects of oxidized multi-walled carbon nanotubes (o-MWCNTs) (differing in length ranging from 50 and 630 nm) on wheat physiology and development revealed that enhanced root growth and higher plant biomass were observed in the plants exposed to o-MWCNT compared to the control (Han et al. 2012; Cicek and Nadaroglu 2015).

8.4 Prospects of Transgenerational Transmission of Nanoparticles: Possible Clues

Nanoparticles, being miniature in size, can penetrate easily into plant cells, interrelate with biomolecules, and may not hold as the promise for transgenerational transmission. Plant cell–NP interaction could regulate plant gene expression and related metabolic pathways. However, the transgenerational transmission of NP and their associated genetic, physiological, biochemical, and molecular avenues still have a huge gap. Following are some studies, showing some clues for transgenerational transmission of the NP.

A mesoporous silica nanoparticle system was pragmatic to transport DNA and chemicals into isolated plant cells (protoplasts from tobacco culture) and intact leaves (young maize embryos), (Torney et al. 2007). Although biocompatible, ZnO NPs can influence the genetic material of terrestrial plants. The presence of new bands may reveal a change in the priming sites leading to new annealing events. Also, large deletions and homologous recombination could lead to the appearance of new bands. High concentrations of CeO_2 NPs and Zn ions released from NPs can alter redox chemistry, thereby increasing oxidative stress leading to DNA damage that affects random amplified polymorphic DNA (RAPD) profiles. RAPD profile of soybean DNA revealed the presence of four new bands at 2000 mg L^{-1} CeO_2 NPs and three new bands at 4000 mg L^{-1} CeO_2 NPs. RAPD profiles confirmed that both ZnO and CeO_2 NPs influence the integrity of the DNA, but CeO_2 NPs caused the highest effect on the genetic stability of soybean plants (Atienzar and Jha 2006;

Singh et al. 2009; Lopez-Moreno et al. 2010a). ZnO NPs interfered with the development of mitosis and inhibited mitotic division in onion (*Allium cepa*). It was hypothesized that inhibition of DNA synthesis at S-phase or an arrest at the G_2 phase of the cell cycle could be the reasons for this cytotoxic effect (Duan and Wang 1995; Borboa and De la Torre 1996; Sudhakar et al. 2001). ZnO NPs also led to an upsurge of chromosomal aberrations (Shaymurat et al. 2011). High-resolution gas chromatography/isotope dilution mass spectrometry (GC/IDMS) confirmed CuO NP-induced accumulation of multiple DNA lesions in three plant systems including radish, perennial ryegrass (*Lolium perenne*), and annual ryegrass (*Lolium rigidum*) (Dizdaroglu 1985; Jaruga et al. 2008). CuO NPs can significantly affect formation and accumulation of DNA base lesions in radish seedlings compared to DNA damage in grassland plants (perennial and annual ryegrass). Moreover, DNA damage in all three plant systems was dependent on exposure time and dose of NP (Atha et al. 2012).

Chromosomal aberrations, micronuclei, and DNA damage were observed in root-tip meristematic cells of onion and maize when exposed to silver NPs, zinc oxide NPs, and coated magnetic NPs of ferrofluid. It was further recommended that NMs could penetrate plant system and may interact with intracellular components causing destruction to cell division (Răcuciu and Creangă 2007; Kumari et al. 2009, 2011, 2012; Patlolla 2013). Effect of NPs (ZnO and TiO_2) on the plant regeneration frequency was investigated for the study of plant line improvement and genetic transformation in the future (Chutipaijit 2015). Nano-CeO_2 may impact the second-generation seedlings growth establishing transgenerational effect. The experimental finding revealed higher Ce accumulation in the fruit from 200 mg/L $nCeO_2$ treatment, indicating potential transgenerational effects (Wang et al. 2013d; Hong et al. 2015). It was hypothesized that the ferrophase might penetrate the nuclear membrane, and magnetic fluids can target the extra nuclear DNA preferably the plastome. The magnetic NPs can influence chromosomal aberrations and perturbation of the proliferative capacity (Răcuciu and Creangă 2009).

Cellular "injection" with carbon nanofibers containing foreign DNA has been used to modify genetically golden rice (AZoNano.com 2014). Nanoparticles, nanofibers, and nanocapsules are capable of carrying foreign DNA and gene-modifying chemicals. This virtue has established nanobiotechnology as a novel industry with new tools to modify genes and even produce new organisms (Torney et al. 2007). It is easier for coated nanoparticles to penetrate cell wall, where the genes might be inserted at factual target site and after that activated in a precise and controlled manner, without any toxic side or after effects. This procedure has already been realistic to introduce DNA successfully to plants, including tobacco and maize plants (Galbraith 2007; Park et al. 2008; Kovalchuk et al. 2012; Sekhon 2014).

8.5 Conclusion and Future Perspectives

In almost all studies, patterns of in planta uptake, translocation, accumulation, and transformation of NPs are poorly understood and only limited data and experimental evidences are available to explain uptake, translocation, and transmission of NPs. However, it is clear that the size of NPs appears to be the critical factor for uptake, accumulation, and further translocation and transgenerational transmission. As the concentration of metal-based NPs or carbon-based NPs increases, the growth increases and reaches an optimum value after which constant or retardation in growth occurs, indicating consequences of NPs accumulation. A NP may either adopt some common routes such as symplastic, apoplastic, or plasmodesmata pathway for transmission to a different part of plant system or may travel through some novel mechanism such as by binding to a diverse array of carrier proteins, through appropriate aquaporins, through interconnected ion channels, via endocytosed pathway, by creating new pores, or by binding to organic chemicals. But, the exact mechanism regarding uptake, translocation, and accumulation of NPs is yet not confirmed. Followings are some possible criteria which could outline "nature of mechanism" for NP uptake, translocation, and accumulation.

1. **Selectivity**: Structure and biochemistry of plasma membrane are important determining features for NP internalization. Membranes harbor specific ion channels for selective ions; therefore, it would be interesting to know that how NPs reacts with the plasma membrane.
2. **Size and charge of NPs**: Nanosized particles have greater degree of freedom for movement; hence, absorption and trafficking of NPs may be reliant on size of the particles, and in some cases, particle charge and hydrophobic/hydrophilic nature of NPs may also determine absorption and trafficking of NPs.
3. **Aquaporins**: Since water plays the crucial role in nutrient internalization. Therefore, NPs-water channel (aquaporins) interaction would of a greater concern and can play a central role in the transmission of the NPs.
4. **Essentiality versus nonessentiality**: Nature and physiology of plant system for essential mineral NPs and nonessential mineral NPs may be important determinants to regulate uptake and trafficking of NPs.
5. **Xylem anatomy**: Variations in xylem structures may authenticate different uptake kinetics of NPs and therefore, influence translocation of NPs in plant system.
6. **NP–plant interaction**: NPs uptake and transmission may also be dependent on plant species; i.e, the same NP may behave differentially with different plants having different genetic backgrounds.

Further research needs to address questions about the mobilization/remobilization mechanisms of NPs and their conversion into operational forms in planta, thereby providing a promising pathway for NPs transmission. It is, therefore, a challenge for scientific community to solve physiological, molecular, and genetic mechanisms for NPs' internalization and trafficking. Furthermore, a holistic view of mitigating the adverse effects of NMs on plant development also needs to be answered.

References

Antisari LV, Carbone S, Gatti A, Vianello G, Nannipieri P (2015) Uptake and translocation of metals and nutrients in tomato grown in soil polluted with metal oxide (CeO_2, Fe_3O_4, SnO_2, TiO_2) or metallic (Ag, Co, Ni) engineered nano particles. Environ Sci Pollut Res 22:1841–1853

Atha DH, Wang H, Petersen EJ, Cleveland D, Holbrook RD, Jaruga P, Dizdaroglu M, Xing B, Nelson BC (2012) Copper oxide nano particle-mediated DNA damage in terrestrial plant models. Environ Sci Technol 46:1819–1827

Atienzar FA, Jha AN (2006) The random amplified polymorphic DNA (RAPD) assay and related techniques applied to genotoxicity and carcinogenesis studies: a critical review. Mutat Res 613:76–102

Azimi R, Borzelabad MJ, Feizi H, Azimi A (2014) Interaction of SiO_2 nano particles with seed pre chilling on germination and early seedling growth of tall wheatgrass (*Agropyron elongatum* L.). Polish J Chem Technol 16(3):25–29

AZoNano.com (2013) Nanofibers to be used in drug delivery, gene therapy, crop engineering and environmental monitoring [webpage on the Internet]. AZoM.com Pty. Ltd, Manchester, UK. Updated 11 June 2013. Available from: http://www.azonano.com/article.aspx?ArticleID=114. Accessed 19 April 2014

Bali R, Siegel R, Harris AT (2010) Biogenic Pt uptake and nano particle formation in *Medicago sativa* and *Brassica juncea*. J Nanopart Res 12:3087–3095

Bhattacharya P, Salonen E, Ke PC (2012) Transformation of engineered nanostructures in the natural environment. In: Barnard AS, Guo H (eds) Nature's Nano Structures. Pan Stanford Publishing Pte. Ltd, Temasek Boulevard, Singapore, pp 509–536

Birbaum K, Brogiolli R, Schellenberg M, Martinoia E, Stark WJ, Gunther D, Limbach L (2010) No evidence for cerium dioxide nano particle translocation in maize plants. Environ Sci Technol 44(22):8418–8423

Borboa L, De la Torre C (1996) The genotoxicity of Zn(II) and Cd(II) in *Allium cepa* root meristematic cells. New Phytol 134:481–486

Burke DJ, Zhu S, Pablico-Lansigan MP, Hewins CR, Samia ACS (2014) Titanium oxide nano particle effects on the composition of soil microbial communities and plant performance. Biol Fertil Soils 50:1169–1173

Burke DJ, Nicole PN, Shu F, Situ SF, Abenojar EC, Porche M, Kraj P, Lakliang Y, Samia ACS (2015) Iron oxide and titanium dioxide nano particle effects on plant performance and root associated microbes. Int J Mol Sci 16:23630–23650

Cañas JE, Long M, Nations S, Vadan R, Dai L, Luo M, Ambikapathi R, Lee EH, Olszyk D (2008) Effects of functionalized and non-functionalized single-walled carbon-nano tubes on root elongation of select crop species. Nanomat Environ 27:1922–1931

Carpita N, Sabularse D, Montezinos D, Delmer DP (1979) Determination of the pore size of cell walls of living plant cells. Science 205(4411):1144–1147

Chang F-P, Kuang L-Y, Huang C-A, Jane W-N, Hung Y, Yue-ie CH, Mou C-Y (2013) A simple plant gene delivery system using mesoporous silica nanoparticles as carriers. J Mater Chem B 1:5279–5287

Chen R, Ratnikova TA, Stone MB, Lin S, Lard M, Huang G, Hudson JS, Ke PC (2010) Differential uptake of carbon nano particles by plant and mammalian cells. Small 6:612–617

Chichiriccò G, Poma A (2015) Penetration and toxicity of nano materials in higher plants. Nanomaterials 5:851–873

Chutipaijit S (2015) Establishment of condition and nano particle factors influencing plant regeneration from aromatic rice (*Oryza sativa*). Int J Agric Biol 17:1049–1054

Cicek S, Nadaroglu H (2015) The use of nanotechnology in the agriculture. Adv Nano Res 3 (4):207–223

Cifuentes Z, Custardoy L, de la Fuente JM, Marquina C, Ibarra MR, Rubiales D, Pérez-de-Luque A (2010) Absorption and translocation to the aerial part of magnetic carbon-coated nano particles through the root of different crop plants. J Nanobiotechnol 8(26):1–8

Corredor E, Testillano PS, Coronado MJ, Gozalez-Melendi P, Fernandez-Pacheco R, Marquina C, Ibarra MR, de la Fuente JM, Rubiales D, Perez de Luque A, Risueno MC (2009) Nano particle penetration and transport in living pumpkin plants: in situ subcellular identification. BMC Plant Biol. doi:10.1186/1471-2229-9-45

Da silva LC, Oliva MA, Azevedo AA, De Araujo MJ (2006) Response of resting a plant species to pollution from an iron palletization factory. Water Air Soil Pollut 75:241–256

Davies G, Fataftah A, Cherkasskiy A, Ghabbour EA, Radwan A, Jansen SA, Kolla S, Paciolla MD, Buermann W, Balasubramanian M, Budnick J, Xing B (1997) Tight metal binding by humic acids and its role in biomineralization. J Chem Soc-Dalton Transact 21:4047–4060

De la Rosa G, Lopez-Moreno ML, Hernandez-Viescaz J, Montes MO, Peralta-Videa JR, Gardea-Torresdey JL (2011) Toxicity and biotransformation of ZnO nano particles in the desert plants *Prosopis juliflora-velutina*, *Salsola tragus* and *Parkinsonia florida*. Int J Nanotechnol 8:492–506

Deng Y, White JC, Xing B (2014) Interactions between engineered nano materials and agricultural crops: implications for food safety. J Zhejiang Univ SCI A (Appl Phys Eng) 15(8):552–572

DeRosa MC, Monreal C, Schnitzer M, Walsh R, Sultan Y (2010) Nanotechnology in fertilizers. Nat Nanotechnol 5:91–94

Dhoke SK, Mahajan P, Kamble R, Khanna A (2013) Effect of nano particles suspension on the growth of mung (*Vigna radiata*) seedlings by foliar spray method. Nanotechnol Dev (3)1:1–5

Dimkpa CO, McLean JE, Latta DE, Manangón E, Britt DW, Johnson WP, Boyanov MI, Anderson AJ (2012) CuO and ZnO nano particles: phytotoxicity, metal speciation and induction of oxidative stress in sand-grown wheat. J Nanopart Res 14:1125

Dimkpa CO, Latta DE, McLean JE, Britt DW, Boyanov MI, Anderson AJ (2013) Fate of CuO and ZnO nano and micro particles in the plant environment. Environ Sci Technol 47:4734–4742

Dizdaroglu M (1985) Application of capillary gas-chromatography mass-spectrometry to chemical characterization of radiation-induced base damage of DNA: implications for assessing DNA-repair processes. Ann Biochem 144(2):593–603

Doshi R, Braida W, Christodoulatos C, Wazne M, O'Connor G (2008) Nano-aluminum: transport through sand columns and environmental effects on plants and soil communities. Environ Res 106:296–303

Duan CQ, Wang HX (1995) Cytogenetical toxical effects of heavy metals on *Vicia faba* and inquires into the *Vicia*-micronucleus. Acta Bot Sin 37:14–24

Ebbs SD, Bradfield SJ, Kumar P, White JC, Musante C, Mac X (2016) Accumulation of zinc, copper, or cerium in carrot (*Daucus carota*) exposed to metal oxide nano particles and metal ions. Environ Sci Nano (Advance Article). doi:10.1039/C5EN00161G

Eichert T, Kurtz A, Steiner U, Goldbach HE (2008) Size exclusion limits and lateral heterogeneity of the stomatal foliar uptake pathway for aqueous solutes and water suspended nano particles. Physiol Planta 134:151–160

Falco WF, Botero ER, Falcão EA, Santiag EF, Bagnato VS, Caires ARL (2011) In vivo observation of chlorophyll fluorescence quenching induced by gold nano particles. J Photochem Photobiol A Chem 225:65–71

Galbraith DW (2007) Nano biotechnology: silica breaks through in plants. Nat Nanotechnol 5:272–273

Ganguly S, Das S, Dastidar SG (2014) Effect of zinc sulphide nano particles on germination of seeds of *Vigna radiata* and their subsequent acceleration of growth in presence of the nano particles. Euro J Biomed Pharma Sci 1(2):273–280

Gardea-Torresdey JL, Parsons JG, Gomez E, Peralta-Videa J, Troiani HE, Santiago P, Yacaman MJ (2002) Formation and growth of Au nano particles inside live alfalfa plants. Nano lett 2:397–401

Gardea-Torresdey JL, Gomez E, Peralta-Videa J, Parsons JG, Troiani HE, Yacaman MJ (2003) Alfalfa sprouts: a natural source for the synthesis of silver nanoparticles. Langmuir 19 (4):1357–1361

Gardea-Torresdey J, Rodriguez E, Parsons JG, Peralta-Videa JR, Meitzner G, Cruz-Jimenez G (2005) Use of ICP and XAS to determine the enhancement of gold phyto extraction by *Chilopsis linearis* using thiocyanate as a complexing agent. Ann Bioanal Chem 382:347–352

Gardea-Torresdey JL, Rico CM, White JC (2014) Trophic transfer, transformation, and impact of engineered nanomaterials in terrestrial environments. Environ Sci Technol 48:2526–2540

Ghafariyan MH, Malakouti MJ, Dadpour MR, Stroeve P, Mahmoudi M (2013) Effects of magnetite nano particles on soybean chlorophyll. Environ Sci Technol 47:10645–10652

Giraldo JP, Landry MP, Faltermeier SM, Mc Nicholas TP, Iverson NM, Boghossian AA, Reuel NF, Hilmer AJ, Sen F, Brew JA (2014) Plant nano bionics approach to augment photosynthesis and biochemical sensing. Nat Mater 13:400–408

Gogos A, Knauer K, Bucheli TD (2012) Nanomaterials in plant protection and fertilization: current state, foreseen applications, and research priorities. J Agric Food Chem 60:9781–9792

Gonzalez-Melendi P, Fernandez Pacheco R, Coronado MJ, Corredor E, Testillano PS, Risueno MC, Marquina C, Ibarra MR, Rubiales D, Perez-De-Luque A (2008) Nanoparticles as smart treatment-delivery systems in plants: assessment of different techniques of microscopy for their visualization in plant tissues. Ann Bot 101:187–195

Hall JL, Williams LE (2003) Transition metal transporters in plants. J Exp Bot 54:2601–2613

Han H, Wang X, Liu X, Gu X, Chen K, Lu D (2012) Multi-walled carbon nano tubes can enhance root elongation of wheat (*Triticum aestivum*) plants. J Nanopart Res 14:841–851

Harris AT, Bali R (2008) On the formation and extent of uptake of silver nano particles by live plants. J Nanopart Res 10:691–695

Haverkamp RG, Marshall AT (2009) The mechanism of metal nano particle formation in plants: limits on accumulation. J Nanopart Res 11:1453–1463

Hong J, Wang L, Sun Y, Zhao L, Niu G, Tan W, Rico CM, Peralta-Videa JR, Gardea-Torresdey JL (2015) Foliar applied nanoscale and microscale CeO_2 and CuO alter cucumber (*Cucumis sativus*) fruit quality. Sci Total Environ. doi:10.1016/j.scitotenv.2015.08.029

Husen A, Siddiqi KS (2014) Phytosynthesis of nanoparticles: concept, controversy and application. Nanoscale Res Lett 9:229

Hyung H, Fortner JD, Hughes JB, Kim JH (2007) Natural organic matter stabilizes carbon nanotubes in the aqueous phase. Environ Sci Technol 41(1):179–184

Iversen TG, Frerker N, Sandvig K (2012) Uptake of ricin B-quantum dot nanoparticles by a micro pinocytosis like mechanism. J Nanobiotechnol 10:33

Jaruga P, Kirkali G, Dizdaroglu M (2008) Measurement of formamido pyrimidines in DNA. Free Radical Biol Med 45:1601–1609

Ke PC, Lamm MH (2011) A biophysical perspective of understanding nano particles at large. Phys Chem Chem Phys 13:7273–7283

Ke PC, Qiao R (2007) Carbon nano materials in biological systems. J Phys Conden Matt 19 (37):373101. doi:10.1088/0953-8984/19/37/373101

Khodakovskaya MV, Biris AS (2009) Method of using carbon nanotubes to affect seed germination and plant growth. WO 2011059507 A1—patent application

Khodakovskaya MV, de Silva K, Biris AS, Dervishi E, Villagarcia H (2012) Carbon nano tubes induce growth enhancement of tobacco cells. ACS Nano 6:2128–2135

Kole C, Kole P, Randunu KM, Choudhary P, Podila R, Ke PC, Rao AM, Marcus RK (2013) Nanobiotechnology can boost crop production and quality: first evidence from increased plant biomass, fruit yield and phytomedicine content in bitter melon (*Momordica charantia*). BMC Biotechnol 13:37

Kouhi SM, Lahouti M, Ganjeali A, Entezari MH (2014) Comparative phytotoxicity of ZnO nano particles, ZnO micro particles, and Zn^{2+} on rapeseed (*Brassica napus* L.): investigating a wide range of concentrations. Toxicol Environ Chem 96:861–868

Kovalchuk I, Ziemienowicz A, Eudes F, inventors Plantbiosis Ltd., assignee (2012) T-DNA/protein nano-complexes for plant transformation. United States patent US 20120070900 A1, 22 Mar 2012

Kumari M, Mukherjee A, Chandrasekaran N (2009) Genotoxicity of silver nano particles in *Allium cepa*. Sci Total Environ 407:5243–5246

Kumari M, Khan SS, Pakrashi S, Mukherjee A, Chandrasekaran N (2011) Cytogenetic and genotoxic effects of zinc oxide nano particles on root cells of *Allium cepa*. J Hazard Mater 190:613–621

Kumari M, Ernest V, Mukherjee A, Chandrasekaran N (2012) In vivo nano toxicity assays in plant models. Meth Mol Biol 926:399–410

Kurepa J, Paunesku T, Vogt S, Arora H, Rabatic BM, Lu J, Wanzer MB, Woloschak GE, Smalle JA (2010) Uptake and distribution of ultrasmall anatase TiO_2 alizarin red S nano conjugates in *Arabidopsis thaliana*. Nano Lett 10:2296–2302

Larue C, Laurette J, Herlin-Boime N, Khodja H, Fayard B, Flank AM, Brisset F, Carriere M (2012a) Accumulation, translocation and impact of TiO_2 nanoparticles in wheat (*Triticum aestivum* spp.) Influence of diameter and crystal phase. Sci Total Environ 431:197–208

Larue C, Veronesi G, Flank AM, Surble S, Herlin-Boime N, Carriere M (2012b) Comparative uptake and impact of TiO_2 nano particles in wheat and rapeseed. J Toxicol Environ Health A 75:722–734

Larue C, Castillo-Michel H, Sobanska S, Cécillon L, Bureau S, Barthès V, Ouerdane L, Carrière M, Sarret G (2014) Foliar exposure of the crop *Lactuca sativa* to silver nano particles: evidence for internalization and changes in Ag speciation. J Hazard Mater 261:98–106

Le VN, Rui Y, Gui X, Li X, Liu S, Han Y (2014) Uptake, transport, distribution and Bio-effects of SiO2 nanoparticles in Bt-transgenic cotton. J Nanobiotechnol 12:50

Lee WM, An YJ, Yoon H, Kwbon HS (2008) Toxicity and bioavailability of copper nano particles to the terrestrial plants mung bean (*Phaseolus radiatus*) and wheat (*Triticum aestivum*): plant agar test for water-insoluble nanoparticles. Environ Toxicol chem 27:1915–1921

Lei Z, Mingyu S, Xiao W (2008) Antioxidant stress is promoted by nano-anatase in spinach chloroplasts under UV-B radiation. Biol Trace Elem Res 121:69–79

Li Y, Chen X, Gu N (2008) Computational investigation of interaction between nano particles and membranes: hydrophobic/hydrophilic effect. J Phys Chem B 112(51):16647–16653

Lin D, Xing B (2007) Phytotoxicity of nano particles: inhibition of seed germination and root growth. Environ Pollut 150:243–250

Lin D, Xing B (2008) Root uptake and phytotoxicity of ZnO nano particles. Environ Sci Technol 42:5580–5585

Lin S, Reppert J, Hu Q, Hudson JS, Reid ML, Ratnikova TA, Rao AM, Luo H, Ke PC (2009) Uptake, translocation, and transmission of carbon nano materials in rice plants. Small 5:1128–1132

Liu Q, Chen B, Wang Q, Shi X, Xiao Z, Lin J, Fang X (2009) Carbon nano tubes as molecular transporters for walled plant cells. Nano Lett 9:1007–1010

Lopez-Moreno ML, De La Rosa G, Hernandez-Viezcas JA, Castillo-Michel H, Botez CE, Peralta-Videa JR, Gardea-Torresdey JL (2010a) Evidence of the differential biotransformation and genotoxicity of ZnO and CeO_2 nanoparticles on soybean (*Glycine max*) plants. Environ Sci Technol 44:7315–7320

Lopez-Moreno ML, De La Rosa G, Hernandez-Viezcas JA, Peralta-Videa JR, Gardea-Torresdey JL (2010b) X-ray absorption spectroscopy (XAS) corroboration of the uptake and storage of CeO_2 nanoparticles and assessment of their differential toxicity in four edible plant species. J Agric Food Chem 58:3689–3693

Lu CM, Zhang CY, Wen JQ, Wu GR, Tao MX (2002) Research of the effect of nanometer materials on germination and growth enhancement of *Glycine max* and its mechanism. Soybean Sci 21:168–172

Ma X, Geiser-Lee J, Deng Y, Kolmakov A (2010) Interactions between engineered nano particles (ENPs) and plants: phytotoxicity, uptake and accumulation. Sci Total Environ 408:3053–3061

Maharramov AM, Ahmadov IS, Ramazanov MA, Aliyeva SQ, Ramazanli VN (2015) Fluorescence emission spectrum of elodea leaves exposed to nano particles. J Biomater Nanobiotechnol 6:135–143

Martin-Ortigosa S, Peterson DJ, Valenstein JS, Lin VS-Y, Trewyn BG, Lyznik LA, Wang K (2014) Mesoporous silica nanoparticle-mediated intracellular Cre protein delivery for maize genome editing via *loxP* site excision. Plant Physiol 164:537–547

Mazumdar H (2014) The impact of silver nano particles on plant biomass and chlorophyll content. Research inventy Int J Eng Sci 4(7):12–20

Miralles P, Johnson E, Church TL, Harris AT (2012a) Multiwalled carbon nanotubes in alfalfa and wheat: toxicology and uptake. J Roy Soc Interf 9(77):3514–3527

Miralles P, Church TL, Harris AT (2012b) Toxicity, uptake, and translocation of engineered nano materials in vascular plants. Environ Sci Technol 46(17):9224–9239

Mishra V, Mishra RK, Dikshit A, Pandey AC (2014) Interactions of nanoparticles with plants: an emerging prospective in the agriculture industry. In: Ahmad P, Rasool S (eds) Emerging technologies and management of crop stress tolerance: biological techniques, vol 1. Elsevier Academic Press, New York, pp 159–180

Monica RC, Cremonini R (2009) Nanoparticles and higher plants. Caryologia 62:161–165

Nair R, Varghese SH, Nair BG, Maekawa T, Yoshida Y, Kumar DS (2010) Nano particulate material delivery to plants. Plant Sci 179:154–163

Navarro E, Baun A, Behra R, Hartmann NB, Filser J, Miao AJ, Quigg A, Santschi PH, Sigg L (2008) Environmental behavior and ecotoxicity of engineered nano particles to algae, plants, and fungi. Ecotoxicology 17:372–386

Nedosekin DA, Khodakovskaya MV, Biris AS, Wang D, Xu Y, Villagarcia H, Galanzha EI, Zharov VP (2011) In vivo plant flow cytometry: a first proof-of concept. Cytometry A 79 (10):855–865

Nel AE, Madler L, Velegol D, Xia T, Hoek EMV, Somasundaran P, Klaessig F, Castranova V, Thompson M (2009) Understanding biophysicochemical interactions at the nano-bio interface. Nat Mater 8(7):543–557

Onelli E, Prescianotto-Baschong C, Caccianiga M, Alessandra M (2008) Clathrin-dependent and independent endocytic pathways in tobacco protoplasts revealed by labelling with charged nanogold. J Exp Bot 59(11):3051–3068

Park IY, Kim IY, Yoo MK, Choi YJ, Cho MH, Cho CS (2008) Mannosylated polyethylenimine coupled mesoporous silica nanoparticles for receptor-mediated gene delivery. Int J Pharm 359:280–287

Parsons JG, Lopez ML, Gonzalez CM, Peralta-Videa JR, Gardea-Torresdey JL (2010) Toxicity and biotransformation of uncoated and coated nickel hydroxide nanoparticles on mesquite plants. Environ Toxicol Chem 29:1146–1154

Patlolla AK (2013) Environmental toxicity monitoring of nanomaterials using *Vicia faba* GENE-TOX assay. J Nanomed Nanotechnol 4:e129. doi:10.4172/2157-7439.1000e129

Patrick JW, Tyerman SD, Bel AJE (2015) Long-distance transport. In: Buchanan BB, Gruissem W, Jones RL (eds) Biochemistry and molecular biology of plants, 2nd edn. Wiley, West Sussex, pp 658–710

Pokhrel LR, Dubey B (2013) Evaluation of developmental responses of two crop plants exposed to silver and zinc oxide nanoparticles. Sci Total Environ 452–453:321–332

Pradhan S, Patra P, Das S, Chandra S, Mitra S, Dey KK, Akbar S, Palit P, Goswami A (2013) Photochemical modulation of biosafe manganese nano particles on *Vigna radiata*: a detailed molecular, biochemical, and biophysical study. Environ Sci Technol 47:13122–13131

Prasad TNVKV, Sudhakar P, Sreenivasulu Y, Latha P, Munaswamy V, Reddy KR, Sreeprasad TS, Sajanlal PR, Pradeep T (2012) Effect of nano scale zinc oxide particles on the germination, growth and yield of peanut. J Plant Nutri 35(6):905–927

Priestera JH, Gea Y, Mielkea RE, Horsta AM, Moritzb SC, Espinosae K, Gelbf J, Walkerg SL, Nisbetb RM, Ani YJ, Schimelb JP, Palmere RG, Hernandez-Viezcasc JA, Zhaoc L, Gardea-Torresdey JL, Holdena PA (2012) Soybean susceptibility to manufactured nano materials with evidence for food quality and oil fertility interruption. Proc Natl Acad Sci USA 109:14734–14735

Răcuciu M, Creangă D (2007) TMA-OH coated magnetic nanoparticles internalized in vegetal tissues. Rom J Phys 52:395–402

Răcuciu M, Creangă D (2009) Cytogenetical changes induced by β-cyclodextrin coated nanoparticles in plant seeds. Rom J Phys 54:125–131

Rad JS, Karimi J, Mohsenzadeh S, Rad MS, Moradgholi J (2014) Evaluating SiO_2 Nano particles effects on developmental characteristic and photosynthetic pigment contents of *Zea mays* L. Bull Environ Pharmaco Life Sci 3:194–201

Remedios C, Rosario F, Bastos V (2012) Environmental nano particles interactions with plants: morphological, physiological, and genotoxic aspects. J Bot 1–8 doi:10.1155/2012/751686

Rico CM, Majumdar S, Duarte-Gardea M, Peralta-Videa JR, Gardea-Torresdey JL (2011) Interaction of nano particles with edible plants and their possible implications in the food chain. J Agric Food Chem 59:3485–3498

Roohizadeh G, Majd A, Arbabian S (2015) The effect of sodium silicate and silica nano particles on seed germination and growth in the *Vicia faba* L. Trop Plant Res 2(2):85–89

Savithramma N, Ankanna S, Bhumi G (2012) Effect of nano particles on seed germination and seedling growth of *Boswellia ovalifoliolata*—an endemic and endangered medicinal tree. Taxon Nano Vision (1, 2 & 3):61–68

Schwabe F, Tanner S, Schulin R, Rotzetter A, Stark W, Quadt A, Nowack B (2015) Dissolved cerium contributes to uptake of Ce in the presence of differently sized CeO_2-nanoparticles by three crop plants. Metallomics 7:466–477

Sekhon BS (2014) Nanotechnology in agri-food production: an overview. Nanotechnol Sci Appl 7:31–53

Serag MF, Kaji N, Gaillard C, Okamoto Y, Terasaka K, Jabasini M, Tokeshi M, Misukami H, Bianco A, Baba Y (2011a) A functional platform for controlled subcellular distribution of carbon nanotubes. ACS Nano 5:9264–9270

Serag MF, Kaji N, Venturelli E, Okamoto Y, Terasaka K, Tokeshi M, Mizukami H, Ugent KB, Bianco Baba Y (2011b) Trafficking and subcellular localization of multiwalled carbon nanotubes in plant cells. ACS Nano 5:493–499

Serag MF, Braeckmans K, Habuchi S, Kaji N, Bianco A, Baba Y (2012a) Spatiotemporal visualization of subcellular dynamics of carbon nano tubes. Nano Lett 12:6145–6151

Serag MF, Kaji N, Tokeshi M, Baba Y (2012b) Introducing carbon nano tubes into living walled plant cells through cellulase-induced nanoholes. RSC Adv 2:398–400

Shankar SS, Ahmad A, Sastry M (2003) Geranium leaf assisted biosynthesis of silver nano particles. Biotechnol Prog 19(1627–1631):44

Sharma NC, Sahi SV, Nath S, Parsons JG, Gardea-Torresdey JL, Pal T (2007) Synthesis of plant-mediated gold nano particles and catalytic role of biomatrix-embedded nano materials. Environ Sci Technol 41:5137–5142

Shaymurat T, Gu J, Xu C, Yang Z, Zhao Q, Liu Y, Liu Y (2011) Phytotoxic and genotoxic effects of ZnO nano particles on garlic (*Allium sativum* L.): a morphological study. Nanotoxicology 6 (3):241–248

Shen CX, Zhang QF, Li J, Bi FC, Yao N (2010) Induction of programmed cell death in *Arabidopsis* and rice by single wall carbon nanotubes. Am J Bot 97:1602–1609

Shyla KK, Natarajan N (2014) Customizing Zinc oxide, silver and titanium dioxide nano particles for enhancing groundnut seed quality. Ind J Sci Technol 17:1376–1381

Singh N, Manshian B, Jenkins GJ, Griffiths SM, Williams PM, Maffeis TG, Wright CJ, Doak SH (2009) Nano geno toxicology: the DNA damaging potential of engineered nano materials. Biomaterials 30(23–24):3891–3914

Singh A, Singh NB, Hussain I, Singh H, Singh SC (2015) Plant-nanoparticle interaction: an approach to improve agricultural practices and plant productivity. Int J Pharmaceut Sci Inven 4 (8):25–40

Smirnova EA, Gusev AA, Zaitseva ON, Lazareva EM, Onishchenko GE, Kuznetsova EV, Tkachev AG, Feofanov AV, Kirpichnikov MP (2011) Multi-walled carbon nano tubes penetrate into plant cells and affect the growth of *Onobrychis arenaria* seedlings. Acta Nat 3:99–106

Smith H (ed) (1978) The molecular biology of plant cells. University of California Press, Berkeley

Spori CL, Prigent G, Schaer M, Crittin M, Matus P, Laroche T, Sikora B, Kaminska I, Fronc K, Elbaum D, Digigow R, Fink A, Ahmadov I, Khalilov R, Ramazanov M, Forró L, Sienkiewicz A (2014) Uptake and biomagnification of multifunctional magnetic and NIR sensitive nano particles by aquatic plants: electron spin resonance, two photon and confocal microscopy studies. In: Proceedings of the Nano-Tera Annual Plenary Meeting, Lausanne, Switzerland, 19–20 May 2014, p 82

Srinivasan C, Saraswathi R (2010) Nano-agriculture-carbon nano tubes enhance tomato seed germination and plant growth. Curr Sci 99:274–275

Stampoulis D, Sinha SK, White JC (2009) Assay-dependent phytotoxicity of nano particles to plants. Environ Sci Technol 43:9473–9479

Stark WJ (2011) Nano particles in biological systems. Angew Chem Int Ed Engl 50(6):1242–1258

Sudhakar R, Gowda N, Venu G (2001) Mitotic abnormalities induced by silk dyeing industry effluents in the cells of *Allium cepa*. Cytologia 66:235–239

Sun D, Hussain HI, Yi Z, Siegele R, Cresswell T, Kong L, Cahill DM (2014) Uptake and cellular distribution, in four plant species, of fluorescently labeled mesoporous silica nanoparticles. Plant Cell Rep 33:1389–1402

Tan XM, Fugetsu B (2007) Multi-walled carbon nano tubes interact with cultured rice cells: evidence of a self-defense response. J Biomed Nanotechnol 3:285–288

Taran N, Batsmanova L, Konotop Y, Okanenko A (2014) A redistribution of elements of metals in plant tissues under treatment by non-ionic colloidal solution of biogenic metal nano particles. Nanoscale Res 9:354–357

Torney F, Trewyn BG, Lin VS, Wang K (2007) Mesoporous silica nano particles deliver DNA and chemicals into plants. Nat Nanotechnol 2:295–300

Wang H, Kou X, Pei Z, Xiao JQ, Shan X, Xing B (2011) Physiological effects of magnetite (Fe_3O_4) nanoparticles on perennial ryegrass (*Lolium perenne* L.) and pumpkin (*Cucurbita mixta*) plants. Nanotoxicology 5(1):30–42

Wang J, Mao H, Zhao H, Huang D, Wang Z (2012a) Different increases in maize and wheat grain zinc concentrations caused by soil and foliar applications of zinc in Loess Plateau China. Field Crops Res 135:89–96

Wang Z, Xie X, Zhao J, Liu X, Feng W, White JC, Xing B (2012b) Xylem- and phloem-based transport of CuO nano particles in maize (*Zea mays* L). Environ Sci Technol 46:4434–4441

Wang J, Koo Y, Alexander A, Yang Y, Westerhof S, Zhang QB, Schnoor JL, Colvin VL, Braam J, Alvarez PJJ (2013a) Phytostimulation of poplars and *Arabidopsis* exposed to silver nanoparticles and Ag^+ at sublethal concentrations. Environ Sci Technol 47:5442–5449

Wang P, Menzies NW, Lombi E, McKenna BA, Johannessen B, Glover CJ, Kappen P, Kopittke PM (2013b) Fate of ZnO nano particles in soils and Cowpea (*Vigna unguiculata*). Environ Sci Technol 47:13822–13830

Wang Q, Ebbs SD, Chen Y, Ma X (2013c) Trans-generational impact of cerium oxide nano particles on tomato plants. Metallomics 5(6):753–759

Wang WN, Tarafdar JC, Biswas P (2013d) Nano particle synthesis and delivery by an aerosol route for watermelon plant foliar uptake. J Nanopart Res 15:1417

Watanabe T, Misawa S, Hiradate S, Osaki M (2008) Root mucilage enhances aluminum accumulation in *Melastoma malabathricum*, an aluminum accumulator. Plant Signal Behav 3:603–605

Whitby M, Quirke N (2007) Fluid flow in carbon nano tubes and nano pipes. Nat Nanotechnol 2:87–94

White PJ (2012) Ion uptake mechanisms of individual cells and roots: short-distance transport. In: Marschner P (ed) Marschner's mineral nutrition of higher plants. Elsevier, London, pp 7–47

Wiesner MR, Lowry GV, Casman E, Bertsch PM, Matson CW, Di Giulio RT, Liu J, Hochella MF Jr (2011) Meditations on the ubiquity and mutability of nano-sized materials in the environment. ACS Nano 5(11):8466–8470

Yang L, Watts DJ (2005) Particle surface characteristics may play an important role in phytotoxicity of alumina nano particles. Toxicol Lett 158:122–132

Zarafshar M, Akbarinia M, Askari H, Hosseini SM, Rahaie M, Struve D (2015) Toxicity assessment of SiO_2 nanoparticles to pear seedlings. Int J Nanosci Nanotechnol 11(1):13–22

Zhang M, Ellis EA, Cisneros-Zevallos L, Akbulut M (2012) Uptake and translocation of polymeric nano particulate drug delivery systems into ryegrass. RSC Advances 2:9679–9686

Zhao L, Peralta-Videa JR, Varela-Ramirez A, Castillo-Michel H, Li C, Zhang J, Aguilera RJ, Keller AA, Gardea-Torresdey JL (2012a) Effect of surface coating and organic matter on the uptake of CeO_2–NPs by corn plants grown in soil: insight into the uptake mechanism. J Hazard Mater 225–226:131–138

Zhao L, Peralta-Videa JR, Ren M, Varela-Ramirez A, Li C, Hernandez-Viezcas JA, Aguilera RJ, Gardea-Torresdey JL (2012b) Transport of Zn in a sandy loam soil treated with ZnO NPs and uptake by corn plants: electron microprobe and confocal microscopy studies. Chem Eng J 184:1–8

Zhao L, Peralta-Videa JR, Peng B, Bandyopadhyay S, Corral-Diaz B, Osuna-Avila P, Montes MO, Keller AA, Gardea-Torresdey JL (2014) Alginate modifies the physiological impact of CeO_2 nano particles in corn seedlings cultivated in soil. J Environ Sci 26:382–389

Zheng L, Hong F, Lu S, Liu C (2005) Effect of nano-TiO(2) on strength of naturally aged seeds and growth of spinach. Biol Trace Elem Res 104:83–92

Zhu H, Han J, Xiao JQ, Jin Y (2008) Uptake, translocation, and accumulation of manufactured iron oxide by pumpkin plants. J Environ Monit 10:713–717

Chapter 9
Nanotechnology for Crop Improvement

Pragati Misra, Pradeep Kumar Shukla, Krishnendu Pramanik, Sanghdeep Gautam and Chittaranjan Kole

Abstract Nanotechnology has the potential to reinforce the mission toward evergreen revolution by enhancing agricultural productivity with limited inputs. It is emerging as a paradigm shift and evolving as a promising tool to begin a new era of precise farming techniques and therefore may provide a possible solution for crop improvement, even in challenging environments. Employment of engineered nanoparticles (ENPs), whether carbon- or metal-based, may be the future solution to increase crop production for feeding the fast-growing world population. This chapter provides an overview of the current knowledge on the effects of nanoparticles for

P. Misra (✉) · S. Gautam
Department of Molecular and Cellular Engineering,
Jacob School of Biotechnology and Bioengineering,
Sam Higginbottom Institute of Agriculture, Technology and Sciences,
Allahabad 211007, India
e-mail: pragatimisra3@rediffmail.com; pragati.misra@shiats.edu.in

S. Gautam
e-mail: sanghdeep007@gmail.com

P.K. Shukla
Department of Biological Sciences, School of Basic Sciences,
Sam Higginbottom Institute of Agriculture, Technology and Sciences,
Allahabad 211007, India
e-mail: pradeepshuklak@yahoo.co.in

K. Pramanik
Department of Agricultural Biotechnology, BC Agricultural University,
Mohanpur 741252, India
e-mail: kpramanik7@gmail.com

C. Kole
Bidhan Chandra Agricultural University, Mohanpur 741252, West Bengal, India
e-mail: ckole2012@gmail.com

Present Address:
C. Kole
Jacob School of Biotechnology and Bioengineering,
Sam Higginbottom Institute of Agriculture, Technology and Sciences,
Allahabad 211007, India

C. Kole et al. (eds.), *Plant Nanotechnology*,
DOI 10.1007/978-3-319-42154-4_9

crop improvement. Overriding influences of different carbon, metal-based and metal oxide nanoparticles on different growth parameters (number of seminal root initiation, root elongation, shoot length, number of seeds, flowers and its quality), ultimately leading to increased plant biomass and yield have been presented. Throughout this chapter, the beneficial role of nanoparticles through enhanced seed germination, increased root and shoot length, fruit and crop yield, and substantial increase in vegetative biomass of seedlings and plants in many crops including maize, wheat, alfalfa, soybean, mustard, mung bean, tomato, potato, lettuce, spinach, onion, peanut, borage, Arabidopsis, cluster bean, and bitter melon is highlighted. The experimental evidences for enhancement of secondary metabolites through nanoparticle treatment under in vivo and in vitro conditions are presented. Although implementation of nanotechnology for agriculture sustainability via enhanced yield, biomass, and secondary metabolite is at juvenile stage, world will witness exceptional and unparalleled prospective of nanoparticles for invigorating agriculture in many ways. It is evident that more investigations are urgently required to know the type of nanoparticle, size, concentration, and mode of application to enable its application on large scale for crop improvement.

Keywords Nanoparticles · Agriculture · Yield · Biomass · Secondary metabolite

9.1 Introduction

Agriculture in the twenty-first century is facing manifold challenges for producing more food by addressing the problems of rapidly growing global population, unpredictable climate change, decreasing agricultural productivity, variable labor force, and increased urbanization. These problems seem to intensify ferociously by 2050 when we have to feed the population of over 9 billion. Agriculture as a source of food, feed, and fiber has always been increasingly important in a world of diminishing resources and with an ever-increasing global population (Brennan 2012).

To counteract this scenario, the agriculture-dependent countries have to adopt more advanced technologies, labor-saving practices, and methods. Nanotechnology is a promising tool and has the potential to foster a new era of precise farming techniques and therefore may emerge as a possible solution for these problems. Nanotechnology may increase agricultural potential to harvest higher yields in an ecofriendly way even in the challenging environments (Sugunan and Dutta 2008). This is expected to be a potential complement to plant molecular breeding and genetic engineering besides traditional plant breeding in the near future.

The potential of nanotechnologies in agricultural practices is still unrevealed and needs to be explored to a large extent. Nanotechnologies can benefit agriculture in multiple dimensions. Introduction of nanomaterial in agriculture aims particularly to increase the yield through optimized nutrient management, minimal loss of nutrient in fertilization, and reduced application of plant protection chemicals (Chen et al. 2013). Engineered nanomaterials (ENMs) can alter agronomic traits including plant growth, biomass production, physiological parameters that directly influence yield, and quality of produce of plants grown to full maturity (Gardea-Torresdey et al. 2014) (Fig. 9.1). The use of nanoparticles in plant science is attracting attention of the researchers due to its beneficial effects (Zheng et al. 2005). Nanoagrotechnology currently focuses target farming involving the use of nanoparticles (NPs) with unique properties to boost crop and livestock productivity (Scott and Chen 2002; Batsmanova et al. 2013).

Nanotechnology has the potential to improve global food production and food quality through increased plant protection, detection of diseases, monitoring plant growth, and reduced waste for strengthening agriculture sustainability (Frewer et al. 2011; Gruère et al. 2011; Biswal et al. 2012; Ditta 2012; Prasad et al. 2012; Sonkaria et al. 2012; Pérez-de-Luque and Hermosín 2013).

The application of nanotechnology in agriculture also involves precise delivery of fertilizers to increase plant growth and yield (Liu et al. 2006; Naderi and Danesh-Shahraki 2013), sensors for monitoring soil quality, and pesticides for pest and disease management (Liu et al. 2008).

In recent years, scientists have been trying to reveal the potential of the nanobiotechnology as a promising tool in the field of crop improvement through an array of experiments in different directions. The role of NPs, either metal-based (MBNPs) or carbon-based (CBNPs), has been documented in many research articles in relation to their uptake, internalization, translocation, persistence, and effect on growth and overall development in many plant species of different commercial importance. Some of these studies have shown beneficial role on plant growth and

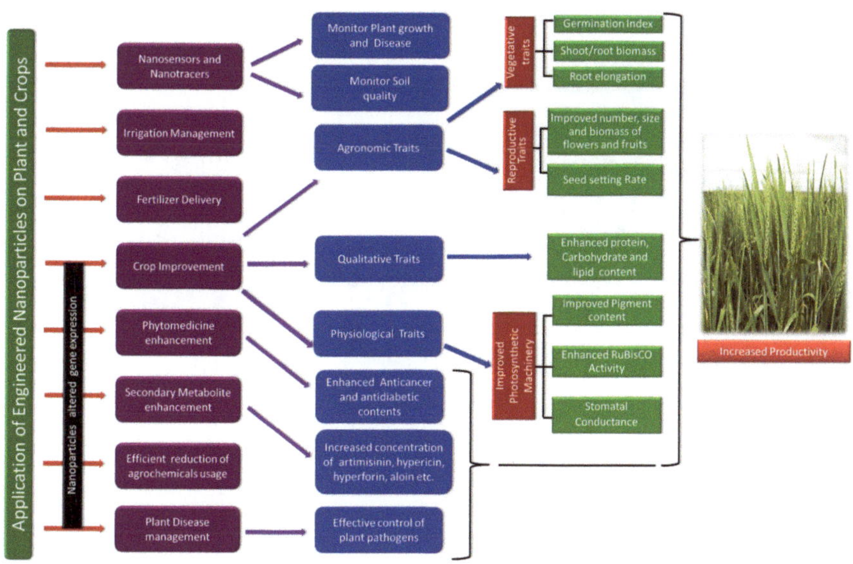

Fig. 9.1 Positive effects of engineered nanoparticles on plants and crops finally leading to increased productivity

development upon exposure to NPs (Lu et al. 2002; Shah and Belozerova 2009; Sharon et al. 2010; Sheykhbaglou et al. 2010; Kole et al. 2013; Razzaq et al. 2016), while others show negative effects (Lee et al. 2008, 2010, 2012, 2013; Zhu et al. 2008; Barrena et al. 2009; Kumari et al. 2009; Stampoulis et al. 2009; Yin et al. 2011).

The beneficial role of NPs has been evidenced through the successful demonstration of enhanced percentage in seed germination (Lu et al. 2002; Khodakovskaya et al. 2009; Nair et al. 2010; Gopinath et al. 2014), increased root and shoot length (Fig. 9.2) (Liu et al. 2005; Hafeez et al. 2015), increased fruit yield (Kole et al. 2013), enhanced phytomedicine content (Kole et al. 2013), and a substantial increase in vegetative biomass of seedlings and plants in many crops including wheat (*Triticum aestivum*), maize (*Zea mays*), ryegrass (*Lolium perenne*), alfalfa (*Medicago sativa*), soybean (*Glycine max*), rapeseed (*Brassica napus*), tomato (*Solanum lycopersicum*), radish (*Raphanus sativus*), lettuce (*Lactuca sativa*), spinach (*Spinacia oleracea*), onion (*Allium cepa*), pumpkin (*Cucurbita maxima*), cucumber (*Cucumis sativus*), and bitter melon (*Momordica charantia*). Augmentation in many biochemical parameters related to plant growth and development has also been reported that facilitates enhanced photosynthetic activity and nitrogen-use efficiency in many crops including soybean (Ngo et al. 2014), spinach (Hong et al. 2005; Zheng et al. 2005; Yang et al. 2006; Gao et al. 2008; Klaine et al. 2008; Linglan et al. 2008), and peanut (*Arachis hypogea*) (Liu et al. 2005; Prasad et al. 2012). However, the molecular mechanism underlying overall

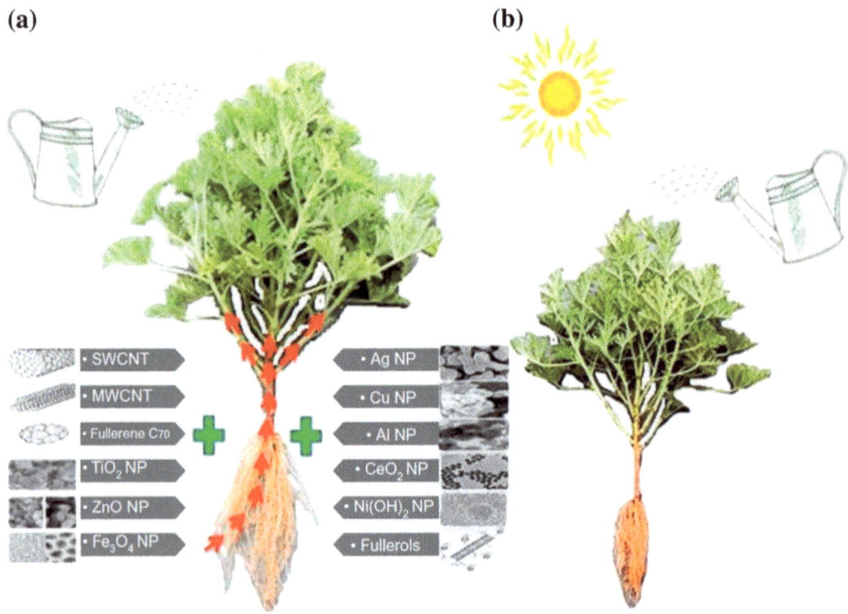

Fig. 9.2 Changes in plant growth on exposure to different NPs **a** plant treated with NPs in addition to other basic requirements, showing enhanced growth in comparison with **b** showing plant growth in the absence of NPs (adapted from Mishra et al. 2014)

development documented in various plant species on the application of different NP formulations is yet to be deciphered through rigorous experiments. Although the use of NPs in crop improvement is still under investigation, we can expect to see its use on a regular basis in farmers' fields in the near future.

Particular types of NPs in low concentrations have not displayed any harmful effect to plants but instead are capable of activating specific physiological and molecular responses. For example, TiO_2 nanoparticles (0.25–4 %) are able to promote photosynthesis and nitrogen metabolism in spinach and, therefore, improve the growth of the plants (Zheng et al. 2005; Klaine et al. 2008). Khodakovskaya et al. (2009) demonstrated that relatively low doses (10–40 μg/mL) of multiwalled carbon nanotubes (MWCNTs) were able to penetrate thick seed coats, increase germination, and stimulate growth in tomato plants (Khodakovskaya et al. 2009, 2012). However, the effects of NPs are influenced by the media and the mode of application. Zhu et al. (2008) studied the uptake of 20-nm-sized iron oxide NPs (Fe_3O_4 NPs) in pumpkin and lima beans (*Phaseolus lunatus*). Under hydroponic conditions, indications of magnetic NPs were found in roots, stems, and leaves, while the plants growing in soil or in sand did not show any signs of magnetic NPs confirming no particle uptake.

During the past few years, there has been extensive interest in applying NPs to plants for agricultural management (Nanotechnology in Agriculture and Food 2006; Torney et al. 2007; Khodakovskaya et al. 2009, 2012; Ashrafi et al. 2010; Serag et al. 2011b, 2012a; Husen and Siddiqi 2014; Razzaq et al. 2016). The genetic implications of such NP-induced positive changes have been validated through investigations on enhanced mRNA expression and protein level in spinach (Gao et al. 2008) by nano-TiO_2, generational transmission of fullerol through seeds in rice (Lin et al. 2009), and changes in gene expression at plant and cellular levels in tomato and tobacco (Khodakovskaya et al. 2009, 2012; Villagarcia et al. 2012) by MWCNTs.

9.2 Demonstration of Nanoparticle-Mediated Enhancement of Plant Biomass and Yield

Despite the high potential of NPs in enhancing plant growth and development, only few reports are available which document the improvement in agronomic traits in terms of an increased leaf and pod dry weight and improved grain yield in soybean when exposed to nanoiron oxide (Sheykhbaglou et al. 2010), borage (*Borago officinalis*) by silver nanoparticles (SNPs) (Seif et al. 2011), bitter melon by fullerols (Kole et al. 2013), mung bean by silver and $PbNO_3$ (Najafi and Jamei 2014), and wheat by SNPs (Razzaq et al. 2016). The role of ENPs of varying size and concentration on plants is given in Table 9.1. ENPs may be classified into the metal (or nonmetal) and metal oxide nanoparticles. Most widely used ENPs examined in the field of crop improvement include nanoferrous/ferric oxides (Liu et al. 2005; Sheykhbaglou et al. 2010; Alidoust and Isoda 2013; Bakhtiari et al. 2015), nanosilver (Vakhrouchev and Golubchikov 2007; Sharma et al. 2012; Razzaq et al.

Table 9.1 Enhanced biomass, productivity, and yield of different plants through nanoparticle treatment

NPs	Optimum concentration	Plant	Effects	Reference
CNT_S, $MWCNT_S$, fullerols	50 and 200 μg mL^{-1}	Tomato	Plant height and number of flowers	Khodakovskaya et al. (2013)
	47.2 nM	Bitter melon	Fruit yield	Kole et al. (2013)
Ag NPs	50 ppm	Potato	Weight and yield of potato mini-tubers	Tahmasbi et al. (2011)
	60 ppm	Common bean, maize	Dry weight of root and shoot	Salama (2012)
	60 ppm	Borage	Seed yield	Seif et al. (2011)
	–	Basil	Seed yield	Nejatzadeh-Barandozi et al. (2014)
	25–50 ppm	Wheat	Growth and yield	Razzaq et al. (2016)
Au NPs	10 ppm	Indian mustard	Growth and seed yield	Arora et al. (2012)
	10 μg mL^{-1}	Arabidopsis	Root and shoot length, early flowering	Kumar et al. (2013)
	1000 μM	Flame lily	Vegetative growth	Gopinath et al. (2014)
Ti NPs	0.25 % w/v	Spinach	Fresh and dry weights	Yang et al. (2007)
	20 g L^{-1}	Wheat	Biomass and yield	Jaberzadeh et al. (2013)
Si, Pd, Au, and Cu NPs	0.013 and 0.066 % w/w	Lettuce	Shoot–root ratio	Shah and Belozerova (2009)
Nanocrystalline powders (Fe, Co, and Cu)		Soybean	Growth and crop yield	Ngo et al. (2014)
Iron oxide NPs	0.5–75 g L^{-1}	Soybean	Yield and quality	Sheykhbaglou et al. (2010)
	50 ppm	Mung bean	Biomass yield	Dhoke et al. (2013)
	0.04 % w/v	Wheat	Grain yield, spike weight, protein content	Bakhtiari et al. (2015)

(continued)

Table 9.1 (continued)

NPs	Optimum concentration	Plant	Effects	Reference
	300 ppm and He, Xe irradiation 10 min	Pea	Growth and yield	Al Sherbini et al. (2015)
Nanoanatase-TiO_2 NPs Nano-TiO_2 Rutile (TiO_2)	0.25–4 %	Spinach (naturally aged)	Plant dry weight	Zheng et al. (2005)
	0.01 and 0.03 %	Maize	Content of carotenoids and anthocyanin	Morteza et al. (2013)
ZnO NPs	20 ppm 1 ppm	Mung bean Gram	Root and shoot biomass	Mahajan et al. (2011)
	20 mg L^{-1}	Tomato	Growth and biomass production	Panwar et al. (2012)
	1000 ppm	Peanut	Stem and root growth, high yield	Prasad et al. (2012)
	500, 1000, 2000, 4000 ppm	Mung bean	Dry weight	Patra et al. (2013)
	1.5 ppm	Chick pea	Shoot and dry weights	Burman et al. (2013)
	10–40 μg ml^{-1}	Onion	Seed yield	Laware and Raskar (2014)
	50 mg	Mung bean	Biomass weight	Jayarambabu et al. (2015)
	10 mg L^{-1}	Cluster bean	Shoot length, root area, and plant biomass	Raliya and Tarafdar (2013)
Silicon dioxide NPs	15 kg ha^{-1}	Maize	Growth and growth parameters	Yuvakumar et al. (2011)
	–	Tomato	Antioxidant system	Haghighi et al. (2012)
	–	Squash	Antioxidant system under salt stress condition	Siddiqui et al. (2014)
CuO NPs	500 mg Kg^{-1}	Wheat	Biomass	Dimkpa et al. (2012)
	30 ppm	Wheat	Growth and yield	Hafeez et al. (2015)

(continued)

Table 9.1 (continued)

NPs	Optimum concentration	Plant	Effects	Reference
CeO_2 NPs	2000 mg L^{-1} 4000 mg L^{-1}	Maize, alfalfa, soybean	Shoot growth and biomass	López-Moreno et al. (2010)
	125, 250, 500 mg Kg^{-1} soil	Wheat	Yield and nutritional parameter	Rico et al. (2014)
$CaCO_3$ NPs	–	Mung bean	Seedling growth and biomass	Yugandhar and Savithramma (2013)

2016), nanogold (Arora et al. 2012; Kumar et al. 2013), nanocopper (Ngo et al. 2014), nano zinc oxide (Prasad et al. 2012; Burman et al. 2013; Raliya and Tarafdar 2013), nanotitanium oxide (Zheng et al. 2005; Morteza et al. 2013; Feizi et al. 2013), nanocerium oxide (Rico et al. 2014, 2015), carbon nanotubes, and fullerols (Khodakovskaya et al. 2009, 2013; Villagarcia et al. 2012; Kole et al. 2013).

9.2.1 Carbon Nanomaterials

Among the NPs, carbon nanomaterials (CNMs) have acquired a significant place due to their unique mechanical, electrical, chemical, and thermal properties. Moreover, the information regarding the effect of nanomaterial such as carbon nanotubes (CNTs) on plant physiology and development is very limited and needs to be explored.

To achieve the goals of "nanoagriculture," exhaustive research on the effects of nanotubes on seed germination and development of seedlings of valuable agricultural plant species is required. Various studies have been reported showing contradictory results depending on the size and concentration of NPs and the species of plants. Canas et al. (2008) reported that CNTs enhanced root elongation in onion and cucumber, whereas it significantly reduced the root length in tomato. The tomato seeds, exposed to multiwalled CNTs (MWCNTs), showed significant enhancement in seed germination and increase in vegetative biomass (Khodakovskaya et al. 2009, 2011). Studies have proposed that the enhanced water uptake efficiency (Khodakovskaya et al. 2009) due to the surface chemistry of carbon nanotubes (Villagarcia et al. 2012) and activation of water channel proteins (aquaporins) (Khodakovskaya et al. 2009) resulted in increase in seed germination and plant growth.

Serag et al. (2013) reported that the diameter and length of single-walled CNTs (SWCNTs) are the major restraining features for their effective penetration into the plant cell wall. Many researchers have shown the penetration of chemically shortened SWCNTs into both the cell wall and the cell membrane of tobacco

(*Nicotiana tabacum*) and periwinkle (*Catharanthus roseus*) (Liu et al. 2009; Serag et al. 2011a, b, 2012a, b).

Some studies conducted on CNMs evidently indicated their potential to enhance plant growth, nutrient uptake, seed germination, and fruit yield/quality. Khodakovskaya et al. (2009), for the first time, demonstrated that CNTs can penetrate thick seed coat and support water uptake inside tomato seeds. Molecular mechanisms of CNT-induced water uptake inside plant seeds are not clear and necessitate further investigation. Researchers also found that the nanotubes migrate through the vascular tissues in the plant. However, such positive effects of CNTs on seed germination and biomass could have noteworthy economic importance for agriculture, horticulture, and the energy sector, such as for production of biofuels (Mondal et al. 2011). In another report, Srinivasan and Saraswathi (2010) showed enhanced seed germination and growth rate in tomato seeds when exposed to CNTs. Gonzalez-melendi et al. (2008) reported the use of carbon nanoparticles as smart treatment delivery system in plants.

Khodakovskaya et al. (2012) investigated the potential of CNTs as regulators of seed germination and growth of tobacco cell culture. Varying concentrations (5–500 μg/ml) of MWCNTs were used for enhanced growth of tobacco cell culture, and results confirmed 55–64 % increase over control. Improved cell growth (16 % increase) was observed under the effect of activated carbon (AC) at low concentrations (5 μg/mL), whereas intense inhibition in the cellular growth was recorded at higher concentrations (100–500 μg/mL). Correlation between the stimulation of growth of cells exposed to MWCNTs, the upregulatory genes participating in cell division/cell wall formation, and water transport was established. Unique molecular mechanism is involved in the regulation of cell division and plant growth by CNTs and is associated with the activation of water channels (aquaporins) and specific genes indulged in the regulation of cell division and extension (Khodakovskaya et al. 2012).

Recently, Tiwari et al. (2014) observed the beneficial role of pristine MWCNTs in enhanced growth and biomass of maize seedlings at low concentrations by enhancing water absorption and concentrations of the essential nutrients Ca and Fe, but their effectiveness could be diminished by high concentrations of ions/polar species in the medium. They proposed a plausible utilization of CNTs for optimizing water transport in arid zone agriculture and for improving crop biomass yields.

Kole et al. (2013) reported the effects of a carbon-based nanoparticle, fullerol, on agroeconomic traits in bitter melon. The uptake, translocation, and accumulation of fullerol were confirmed through bright-field imaging and Fourier transform infrared spectroscopy. Varied effects (positive and non-consequential) were recorded on yield, plant biomass, fruit yield, and component characters, when seeds were treated with five varying concentrations (0.943, 4.72, 9.43, 9.88, and 47.2 nM) of fullerols (Fig. 9.3). Increase in biomass yield by 54 % and water content in plants by 24 % over control was observed after treatment with fullerol, whereas fruit length, fruit number, and fruit weight increased up to 20, 59, and 70 %, respectively, that resulted in the improvement of up to 128 % in fruit yield. The accumulation of fullerol in tissues of root, stem, petiole, leaf, flower, and fruit at particular concentrations was stated as the causal factor for increase in biomass and fruit yield.

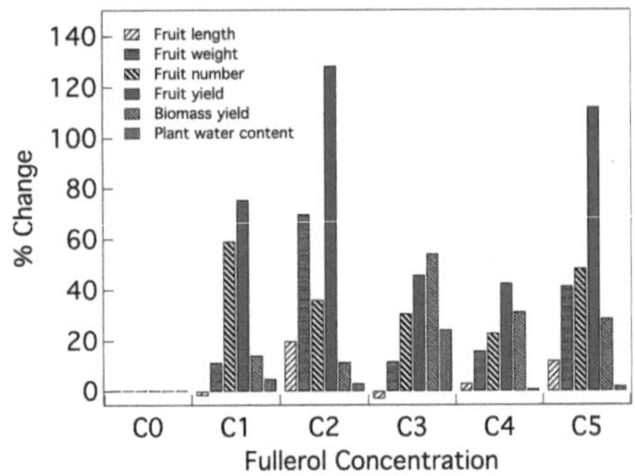

Fig. 9.3 Effect of fullerol at five concentrations (C_1–C_5) in comparison with control on changes (in %) in six plant characters. C_0 denotes control (without fullerol), and C_1–C_5 denote five fullerol concentrations (0.943, 4.72, 9.43, 9.88, and 47.2 nM), respectively (adapted from Kole et al. 2013)

Husen and Siddiqi (2014) proposed the potential of fullerene, C_{60}, and CNTs to improve the water retention capacity, biomass, and fruit yield in plants up to ~118 %. These findings can be taken into account as a remarkable achievement of agri-nanotechnology in the field of crop improvement.

Research-based evidences can confirm that CNMs may be considered as a promising nanoscale amendment for significantly suppressing microbial pathogens, improving plant growth, and promoting crop quality/yield. However, it is a challenge for research community to unravel the exact mechanism and dose-dependent pattern of CNMs on plant growth and development in physiological, metabolic, and molecular perspectives.

9.2.2 Metal-Based Nanoparticles

As already mentioned, nanoparticles can have both growth-promoting and harmful effects on crops. Toward this effort, different studies have been conducted to analyze the effect of metal-based nanoparticles (MBNP) such as silver (Ag), gold (Au), aluminum (Al), and copper (Cu) on plants. Application of MBNPs has been found to improve germination (Barrena et al. 2009; Yuvakumar et al. 2011), enhance growth and physiological activities (Shah and Belozerova 2009; López-Moreno et al. 2010; Salama 2012; Razzaq et al. 2016), increase water and fertilizer-use efficiency (Yuvakumar et al. 2011), inhibit abscission of reproductive organs of plant (Seif et al. 2011), and stimulate nodule formation (Taran et al. 2014).

Effects of super-dispersive iron, cobalt, and copper nanocrystalline powders on germination rate, plant growth, crop yield, and quality of soybean (Vietnamese species DT-51) were examined (Ngo et al. 2014). The soybean seeds treated with an extra low nanocrystalline dose (not more than 300 mg of each metal per hectare) were sowed on experimental landfill plot farming area of 180 m^2. Cobalt nanopowder exhibited a better germination effect than nanoscaled iron and copper wherein all the growth parameters exceeded the control ones with crop yield surpassing the control by 16 %.

SNPs have remarkable uses in crop production. Variable responses of SNPs have been reported by different researchers in various plants. SNPs affect plant growth by significantly inducing changes at physiological and molecular levels. Soaking of cotton seeds in SNPs produced favorable effects and reduced the amount of fertilizers applied through roots by half (Vakhrouchev and Golubchikov 2007). SNPs have catalytic effects (Ma et al. 2010), decreasing the abscission of reproductive organs of plants (Seif et al. 2011), and are known to increase chlorophyll contents (Sharma et al. 2012). Razzaq et al. (2016) reported positive effect of SNPs on wheat growth and yield when applied to soil. Exposure to 25–50 ppm SNPs significantly increased plant height and fresh and dry weights over the control. Application of SNPs at low concentrations (25 and 50 ppm) positively affected the number of seminal roots (Fig. 9.4). Favorable effects of soil-applied SNPs on growth might be due to more bioavailability and accumulation in plants, thereby stimulating growth. The highest grain number per spike, 100-grain weight, and yield per pot were recorded at 25 ppm SNPs (Fig. 9.5) (Razzaq et al. 2016). Sensible use of SNPs to soil can, therefore, improve the yield of wheat. However, further investigations are needed to explore precise concentration, suitable mode, and time of application to realize growth- and yield-enhancing potential of SNPs for wheat and other crops in an ecofriendly manner. Similarly, SNPs at 50 ppm was observed to increase remarkably total chlorophyll, chl-a, chl-b, root fresh weight in mung bean plants (Najafi and Jamei 2014). Similar effects of SNPs including increased fresh weight, root and shoot length, vigor index, and chlorophyll contents of seedlings of Indian mustard (*Brassica juncea*) were also reported (Sharma et al. 2012). Increased weight and yield of potato mini-tubers were reported by the

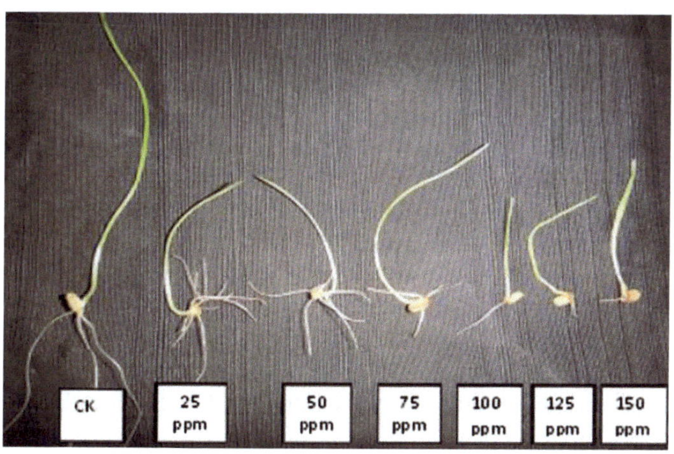

Fig. 9.4 Effect of varying concentrations of Ag NPs on the number of seminal roots in wheat (adapted from Razzaq et al. 2016)

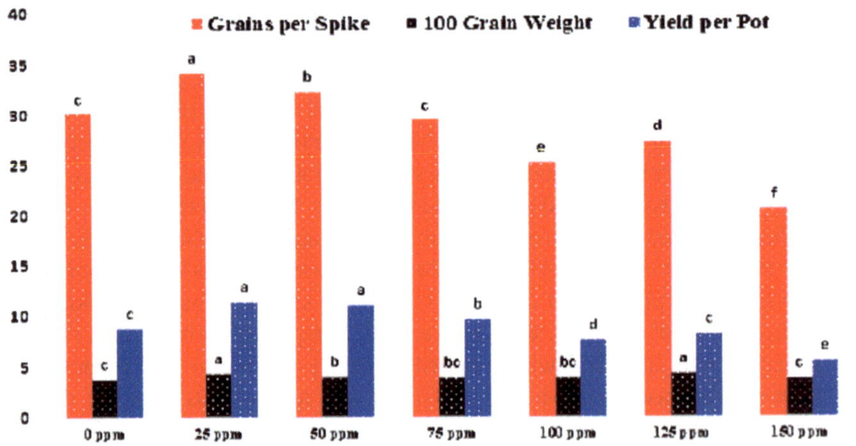

Fig. 9.5 Impact of different concentrations of soil-applied Ag NPs nanoparticles on crop yield in wheat (adapted from Razzaq et al. 2016)

application of 50 ppm nanosilver in combination with nitrogen and nitroxin (Tahmasbi et al. 2011).

Another study by Seif et al. (2011) studied the effect of nanosilver and silver nitrate on abscission and seed yield in borage plants. Varying concentrations (20, 40, and 60 ppm) of SNPs were used. Improvement in the seed yield was observed on raising the concentration of nanosilver from 20 to 60 ppm. In contrast, increasing the concentration of silver nitrate from 100 to 300 ppm led to the reduction in seed yield. Similar results were reported by Salama (2012) in common bean (*Phaseolus vulgaris* L.) and maize when exposed to five levels of SNPs (20, 40, 60, 80, and 100 ppm). Effects of SNPs on plant growth parameters such as shoot and root lengths, leaf surface area, chlorophyll, carbohydrate, and protein contents were investigated. The results revealed that lower concentrations (20, 40, and 60 ppm) of SNPs had positive impact on the growth of the plantlets, and in contrast, higher concentrations (80 and 100 ppm) of SNPs showed inhibitory effect. Enhancement in shoot and root lengths, leaf area, chlorophyll, carbohydrate, and protein contents was reported in both common bean and maize on exposure to increasing concentration of SNPs from 20 to 60 ppm.

SNPs also have strong antimicrobial effects and therefore effectively control and avoid plant diseases. SNPs at concentrations of 0.5–1000 ppm cause faster growth of plants and control pathogens (Ashrafi et al. 2010). Increased germination and enhanced seedling growth were reported by SNPs (Lu et al. 2002) that act as growth simulators (Sharon et al. 2010). SNPs can facilitate the plant to delay the senescence provoked by reactive oxygen species (ROS) formed due to oxidative stress. Oxidative stress induced senescence followed by 2,4-D triggered ROS generation in mung bean was suppressed by the application of 100 μL of SNPs (Karuppanapandian et al. 2011). The studies suggest that MBNPs enhance plant growth and development. Ag NPs increased root length in maize and cabbage (*Brassica oleracea* var. *capitata*) plants in comparison with $AgNO_3$ (Pokhrel and Dubey 2013).

The potential of Au nanoparticle as a promising tool to enhance seed yield was reported by Kumar et al. (2013). Total seed yield increased by threefold over the control when Arabidopsis (*Arabidopsis thaliana*) seeds were exposed to 10 μg/mL of Au NPs (24 nm size). Au NP treatment at both 10 and 80 μg/mL concentrations significantly improved seed germination rate, vegetative growth, and free radical scavenging activity. A significant correlation was found between expression of key plant regulatory molecules, microRNAs (miRNAs), seed germination, growth, and antioxidant potential of Arabidopsis upon Au NP exposure (Kumar et al. 2013). In another report, the role of gold NP on yield of Indian mustard was investigated. The positive effect of Au NPs was visible on various growth- and yield-related parameters including plant height, number of branches, stem diameter, number of pods, and seed yield. The average leaf area was not affected; however, there was an increase in the number of leaves per plant. At 10 ppm concentration of Au NP, the seed yield increased optimally, whereas the reducing sugar and total sugar content increased up to 25 ppm concentration of Au NP treatment (Arora et al. 2012).

Au NPs showed positive impact on seed germination and vegetative growth in flame lily (*Gloriosa superba*), an endangered medicinal plant (Gopinath et al. 2014). Two concentrations of Au NPs (500 and 1000 μM) were used for seed treatment. Seed germination rate (enhanced by 39.67 % than control) and vegetative growth of flame lily were significantly affected at 1000 μM concentration. Seed coat of Au NP-treated seeds showed increased permeability facilitating the entry of H_2O and O_2 into the cells and uptake of gold ions, which interacts with embryo cells. It stimulates the GA_3 activity resulting in the expression of α-amylase enzyme in the aleurone cell layer. α-amylase breaks down starch into simple sugar and accelerates the germination process (Fig. 9.6). More prominent effect on number of leaves, root initiation, and node elongation was observed in seeds exposed to

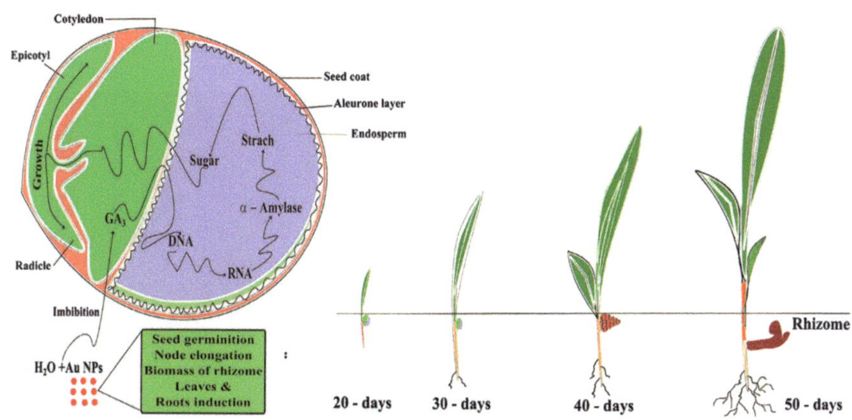

Fig. 9.6 Schematic representation of positive impact of Au NPs on *Gloriosa superba* resulting in enhanced germination, leaf and root initiation, node elongation, and biomass of rhizome (adapted from Gopinath et al. 2014)

1000 μM Au NPs (Fig. 9.7). Total biomass and fresh weight of flame lily rhizome were also increased by 2.40- and 5.18-fold after treatment with 500 and 1000 μM Au NPs respectively as compared to control.

Shah and Belozerova (2009) reported the influence of Si, Cu, Au, and Pd NPs on growth of lettuce plants after 15 days of incubation, which resulted in an increased shoot/root ratio compared to control.

Jaberzadeh et al. (2013) reported increased stem elongation, biomass, ear mass, seed number, yield, gluten and starch content, and early flowering in wheat when 20 gL^{-1} Ti NPs were applied through foliar spray as compared to their bulk material. Similarly, Yang et al. (2007) reported augmented fresh and dry weights as well as contents of total *N*, chlorophyll, and protein in leaves when spinach seeds were soaked in a solution of Ti NPs.

Iron is one of the essential elements for plant growth and plays an important role in the photosynthetic reactions. Iron is known to activate several enzymes contributing to RNA synthesis and enhance photosystem performance (Malakouti and Tehrani 2005). Soybean is sensitive to iron deficiency though different genotypes differ in iron consumption efficiency. Grain yield in soybean was increased by the application of iron in low-iron soils (Sheykhbaglou et al. 2010). Almeelbi and Bezbaruah (2012) reported the significant enhancement in plant growth and biomass of spinach by Fe NPs in hydroponic solution. Remarkably, Fe content in spinach leaves, stem, and root was increased by 11 to 21-fold. Amuamuha et al. (2012) also investigated the effect of different concentrations of Fe NPs (1, 2, and 3 gL^{-1}) on pot marigold (*Calendula officinalis*) at three growth stages, viz. stem elongation, flowering, and after harvest. Results revealed that the highest flower yield and essential oil percentage were achieved when 1 gL^{-1} Fe NPs were applied at the stem elongation stage.

Hafeez et al. (2015) examined the potential of copper NPs to increase growth and yield of wheat. The growth and yield were significantly increased in comparison with control when Cu NPs (at 10, 20, 30, 40, and 50 ppm) were applied to soil

Fig. 9.7 Effect of Au NPs at two concentrations (1000 and 500 μM) as compared to control on *G. superba* leaf and root initiation, node elongation, and biomass of rhizome for duration of 40 days (adapted from Gopinath et al. 2014)

in pots. However, the significant increase in the chlorophyll content, leaf area, number of spikes/pot, number of grains/spike, 100-grain weight, and grain yield was observed at 30 ppm Cu NPs. Results revealed that the enhanced growth and yield in wheat due to Cu NPs are concentration-dependent and further experimentation is required for the dose optimization and mode of application to maximize the yield of wheat.

In a recent study, Khan et al. (2016) investigated the effect of nine types of metal nanoparticles including monometallic and bimetallic alloy nanoparticles [Ag, Au, Cu, AgCu (1:3), AgCu (3:1), AuCu (1:3), AuCu (3:1), AgAu (1:3), AgAu (3:1)] on seed germination and biochemical profile of milk thistle (*Silybum marianum*) plant. Significant increase in seed germination was reported upon treatment with all the NPs suspensions as compared to control and was recorded the highest for Ag NPs suspension. Significant effect was also observed on the biochemical profile of milk thistle on exposure to metal NPs. Enhancement in total protein content, DPPH, peroxidase, and superoxide dismutase activity was recorded for the first week and then declined as the time progressed. Among all the NPs being used, the maximum enhancement was observed with Ag NPs. Significant potential of different monometallic and bimetallic NPs on medicinal plant species was proposed.

Taran et al. (2014) studied the effect of colloidal solution of molybdenum nanoparticles (Mo NPs) on the microbial composition in the rhizosphere of chick pea (*Cicer arietinum*). It was reported that exposure of chick pea seeds with the combination of colloidal solution of Mo NPs (8 mg/L) and microbial preparation stimulated the development of "agronomically valuable" microflora. Combined treatment resulted in increase in number of nodules per plant by four times, while single treatment with colloidal solution of Mo NPs increased the number of nodules twofold as compared to control.

9.2.3 Metal Oxide Nanoparticles

Enormous studies on the effect of metal oxide NPs on varying parameters such as germination, growth, and yield of plants have been documented (Hong et al. 2005; Liu et al. 2005; Zheng et al. 2005; Sheykhbaglou et al. 2010; Mahajan et al. 2011; Alidoust and Isoda 2013; Laware and Raskar 2014; Bakhtiari et al. 2015; Razzaq et al. 2016). In spinach, enhanced chlorophyll formation, photosynthesis, and plant dry weight was observed when exposed to TiO_2 NPs (Hong et al. 2005; Zheng et al. 2005). In another study, increase of 58.2 and 69.8 % in fresh and dry weights, respectively, and substantial rise in chlorophyll content, Rubisco activity, and photosynthetic rate were recorded in spinach when treated with anatase TiO_2 NPs (Linglan et al. 2008). Low dosage of nanosized TiO_2 enhanced seed germination indices of fennel (Feizi et al. 2013). Germination percent was highly improved following exposure to 60 ppm nanosized TiO_2. Nano-TiO_2 was suggested to be used for improvement of seed germination of fennel. Morteza et al. (2013) reported that nano-TiO_2 plays a significant role in increasing pigments in maize—higher amounts of pigments were obtained when sprayed with nano-TiO_2 at reproductive stages of plant, which finally led to increase in yield. Thus, an application of nano-TiO_2 can foster an increase in crop yield, especially in maize.

Chutipaijit (2015) investigated, for the first time, on the effect of TiO_2 NPs on regeneration frequency in aromatic rice (*Oryza sativa*) (cultivar KDML105). Application of TiO_2 NPs at an optimum concentration, i.e., 25 mg L^{-1} showed elevated green spots, plant regeneration, and the ratio of seedling number to the number of regenerated calli. Therefore, application of TiO_2 NPs showed positive response on regeneration efficiency in rice, probably due to improved plant metabolism (Fig. 9.8) (Chutipaijit 2015). Improved nitrate reductase activity and stimulated antioxidant system by mixture of TiO_2 and SiO_2 NPs were reported in soybean (Lu et al. 2002).

Many researchers have reported the positive effect of iron NPs on photosynthetic potential, growth, biomass, and yield in crop plants (Liu et al. 2005; Sheykhbaglou et al. 2010; Amuamuha et al. 2012; Alidoust and Isoda 2013; Bakhtiari et al. 2015). Researchers have shown that the application of nanoferric oxide (nano-Fe_2O_3) significantly affected nutrient absorption in peanut and resulted in increased growth and photosynthesis. The photosynthate and iron transfer rate to the leaves of peanut was promoted by nano-Fe_2O_3 as compared to other organic materials and iron citrate treatments (Liu et al. 2005). In another study, the effect of Fe_2O_3 NPs when applied to soybean via foliar and soil route was investigated. The enhancement in root elongation and photosynthetic potential were significantly higher when Fe_2O_3 NPs were administered to plants by foliar spray as compared to soil route, which may be due to precipitation of Fe ions (Alidoust and Isoda 2013).

A study was conducted to investigate the effect of nanoiron oxide particles on soybean yield and agronomic traits (Sheykhbaglou et al. 2010). Nanoiron oxide was applied at 5 levels (0, 0.25, 0.5, 0.75, and 1 gL^{-1}). It was observed that exposure to

Fig. 9.8 Regeneration frequency of rice callus on medium **a** without treatment **b** after treatment with 25 mg L^{-1} TiO_2 nanoparticles (adapted from Chutipaijit 2015)

nanoiron oxide at 0.75 gL^{-1} concentration caused improvement in leaf and pod dry weight. Grain yield was augmented to 48 % as compared to control, by the treatment of 0.5 gL^{-1} nanoiron oxide. Thus, it can be suggested that 0.5 gL^{-1} nanoiron oxide treatment resulted in increased total yield by an increase in leaf and pod dry weight (Table 9.2). Bakhtiari et al. (2015) reported that spraying of iron oxide NPs solution in varying concentrations (0, 0.01, 0.02, 0.03, and 0.04 %) significantly affected the wheat growth and yield. The measured trait included: spike weight, 1000-grain weight, biological yield, grain yield, and protein content of wheat. The highest values of spike weight (666.96 g), 1000-grain weight (37.96 g), biological yield (8895.0 kg/ha), grain yield (3776.5 kg/ha), and protein content (16.44 %) were achieved at 0.04 % iron oxide concentration. It was evident that spraying time and concentration are the major factors affecting all the measured traits. Racuciu and Creanga (2007) observed similar results on plant growth in maize at early ontogenetic stages after application of magnetic NPs coated with tetramethyl ammonium hydroxide (TMA-OH).

In most of the studies, the effect of ZnO NPs on plant growth depends on concentration. Root elongation in soybean was reported at 500 mg L^{-1}, whereas higher concentrations resulted in the reduction of root length. No effect on seed germination in soybean was observed at even higher concentration (4000 mg L^{-1}) (López-Moreno et al. 2010). Mahajan et al. (2011) demonstrated the effect of nano-ZnO particles on the growth of plant seedlings of mung bean and chick pea (*C. arietinum*). ZnO NPs showed concentration-dependent growth pattern in mung bean and chick pea seedlings. The maximum growth was found at 20 ppm for mung bean and 1 ppm for chick pea seedlings, and beyond this concentration, the growth was inhibited (Mahajan et al. 2011).

The effect of ZnO NPs on growth, flowering, and seed productivity of onion was studied (Laware and Raskar 2014). Six-month-aged onion bulbs (cut in half portions) were subjected to pot plantation and sprayed three times with varying concentrations (0, 10, 20, 30, and 40 μg ml^{-1}) of ZnO NPs at the interval of 15 days. The growth parameters including plant height and number of leaves per plant were assessed at the time of flowering, and the seed yield parameters such as number of seeded fruits per umbel, seed yield per umbel, and 1000-seed weight were determined at the time of harvest. Seed samples obtained from treated plants along with control were tested for germination and early seedling growth. Results revealed that

Table 9.2 Effect of different concentrations of nanoiron oxide on some agronomic traits in soybean (adapted from Sheykhbaglou et al. 2010)

Nanoiron oxide (g/l)	Pod dry weight (g)	Leaf + pod dry weight (g)	Yield (g m^{-2})
0	0.41^{b}	32.35^{b}	60.94^{b}
0.25	0.42^{ab}	42.35^{ab}	76.78^{ab}
0.5	0.44^{ab}	42.45^{ab}	90.22^{a}
0.75	0.48^{a}	45.84^{a}	88.33^{a}
1	0.45^{ab}	42.32^{ab}	80.39^{ab}

Means with different letters at each column have statistically difference at 5 % level the plants treated with ZnO NPs at the concentration of 20 and 30 μg ml^{-1} showed better growth and flowered 12–14 days earlier in comparison with control. Treated plants showed significantly higher values for seeded fruits per umbel, seed weight per umbel, and 1000-seed weight over control plants. It was confirmed that high-quality seed along with all other inputs (size, number, etc.) was responsible for enhancement in final yield. These results indicated that ZnO NPs can reduce flowering period in onion by 12–14 days and produce high-quality healthy seeds (Laware and Raskar 2014). The increase in vegetative growth in onion might be related to the fundamental role of ZnO in maintenance and protection of structural stability of cell membranes (Welch et al. 1982) and involvement in protein synthesis, functioning of membrane, cell elongation, as well as tolerance to various environmental stresses (Cakmak 2000). Prasad et al. (2012) suggested variable response of peanut seeds toward the treatment at various concentrations of both bulk $ZnSO_4$ and nanoscale ZnO particles. Absorption of ZnO NPs by plants was more as compared to $ZnSO_4$ bulk. Results also revealed the beneficial effects of NPs in enhancing plant growth, development, and yield in peanut at lower doses (1000 ppm), but at higher concentrations (2000 ppm), ZnO NPs were detrimental just as the bulk nutrients. Pod yield per plant was 34 % higher in plants treated with ZnO as compared to chelated bulk ZnSO4. Similar findings were reported by Raliya and Tarafdar (2013) on shoot length, root length, root area, and plant biomass in cluster bean (*Cymopsis tetragonoloba*), when 10 ppm ZnO NPs were foliar-sprayed on leaf of 14-day-old plant. Significant improvement was observed in shoot length (31.5 %), root length (66.3 %), root area (73.5 %), and plant biomass (27.1 %) over control in 6weekold plants because of the treatment with ZnO NPs (Fig. 9.9; Table 9.3).

Fig. 9.9 Effect of ZnO NPs on growth of cluster bean (6 weeks old). Plant treated with *n* ZnO-nano zinc oxide at 10 ppm concentration exhibited maximum growth as compared to O ZnO-ordinary zinc oxide and control (adapted from Raliya and Tarafdar 2013)

Table 9.3 Effect of nano-ZnO and ordinary- ZnO on some phenological parameters of 6 weeks aged cluster bean plants (adapted from Raliya and Tarafdar 2013)

Treatment	Shoot length (cm)	Root length (mm)	Root area (mm^{2})	Dry biomass (g^{-1})
Control	44.53	720.23	809.30	10.47
Ordinary ZnO	47.73	835.20	1241.47	11.60
Nano-ZnO	58.57	1197.70	1404.30	25.33
LSD (p = 0.05)	0.10	0.09	0.03	0.15

In another study, Kisan et al. (2015) examined the effect of nano-ZnO on the leaf physical and nutritional quality of spinach. The spinach plants were sprayed with varying concentrations (0, 100, 500, and 1000 ppm) of ZnO NPs after 14 days of sowing. At the time of maturity (45–50 days), the leaf physical parameters such as leaf length, leaf width, and leaf surface area were noted and nutritional parameters such as protein, carbohydrate, fat, and dietary fiber contents in leaf samples were determined. When 500 and 1000 ppm concentration of ZnO NPs were sprayed, increase in leaf length, width, surface area, and color of spinach leaves were recorded with respect to control. Similarly, elevated levels of protein and dietary fiber contents were observed in plants treated with ZnO NPs at the concentration of 500 and 1000 ppm in comparison with control leaf samples of spinach. It was proposed that the nanozinc oxide has a potential to be used as a biofortification agent for the improvement of protein and dietary fiber contents of spinach leaves and thereby reduces malnutrition.

Morales et al. (2013) assessed the impact of cerium oxide nanoparticles (CeO_2 NPs) on cilantro (*Coriandrum sativum*) plants grown in organic soil. Cilantro seeds were germinated, and plants were grown in organic soil treated with 0–500 mg kg^{-1} CeO_2 NPs for 30 days and analyzed by biochemical assays and spectroscopic techniques to determine the CeO_2 uptake, variations in macromolecules, and catalase (CAT) and ascorbate peroxidase (APX) activity. At 125 mg kg^{-1} concentration of CeO_2 NPs, plants produced longer roots and shoots, had a higher biomass production, and significantly increased in catalase activity in shoots and ascorbate peroxidase activity in roots. Furthermore, CeO_2 NPs downregulated the production of these defensive enzymes and altered the carbohydrates in shoots, signifying its role in changing the nutritional properties of cilantro. Thus, this study demonstrated the fertilizing effects of CeO_2 NPs, which helped plants to grow better. Although CeO_2 NPs produce stress in cilantro plants, at the same time they have antioxidant activity. In fact, CeO_2 NPs induce conformational changes within the plant (in the components of roots), visible in the spectra by vibrational shifting, but do not induce chemical reactions and substantial changes. From all these findings, positive effects of CeO_2 NPs on the development of plants were depicted.

In another study, Rico et al. (2014) examined the impact of CeO_2 NPs on agronomic traits, yield, and nutritional parameters in wheat. Wheat was grown in soil administered with 0, 125, 250, and 500 mg of $nCeO_2$ kg^{-1} (control, $n$$CeO_2$-L,

nCeO$_2$-M, and nCeO$_2$-H, respectively). The cultivated grains and tissues were studied for contents of minerals, fatty acids, and amino acids. Results revealed that nCeO$_2$-H improved plant growth, shoot biomass, and grain yield by 9.0, 12.7, and 36.6 %, respectively, relative to control Fig. 9.10). nCeO$_2$ modified S and Mn storage in grains. nCeO$_2$-L modified the amino acid composition and increased linolenic acid by up to 6.17 % but decreased linoleic acid by up to 1.63 %, compared to control. These findings evidenced the potential of CeO_2 NPs to modify crop physiology and food quality. nCeO$_2$ treatment caused a 6-day delay in spike formation and physiological maturity in wheat compared to control. Li et al. (2011) suggested that the extended period for spike formation and physiological maturity might be the reason for improved yield in wheat. More recently, it has been reported by Marchiol et al. (2016) that plants treated with nCeO$_2$ and nTiO$_2$ had a longer vegetative period than the control. This fact as such may not be undesirable. In fact, a longer vegetative phase may support higher biomass and grain yield as plants have comparatively more time to produce more photosynthetically active leaves and therefore more photosynthates (Dofing 1995).

The impact of nano-SiO$_2$ on the characteristics of seed germination was studied in tomato. Results revealed that the treatment with nSiO$_2$ significantly enhanced seed germination potential. Exposure of nSiO$_2$ increased seed germination percentage, mean germination time, seed germination index, seed vigor index, seedling fresh weight, and dry weight. An increase in germination parameters by the use of nSiO$_2$ may be inductive for the growth and yield of plants. Nonetheless, the present findings offer a scope to search out the mechanism of interaction between nanosilica and plants, since nSiO$_2$ could be used as a fertilizer for the crop improvement (Siddiqui and Al-Whaibi 2014).

The use of nanofertilizers in agriculture is an important approach to enhance agronomic production and ensure global food and nutritional security (Liu and Lal 2015; Servin et al. 2015). In the context of applicability of nanoparticles as nanotoxicants or nanonutrients, Liu et al. (2016) stated that manufactured NPs were not

Fig. 9.10 Yield response of wheat to different concentrations of nCeO$_2$. Wheat was cultivated to grain production in soil amended with 0, 125, 250, and 500 mg of nCeO$_2$ kg^{-1} (control, nCeO$_2$-L, nCeO$_2$-M, and nCeO$_2$-H, respectively). NCeO$_2$-H resulted in improved plant growth, shoot biomass, and grain yield by 9.0, 12.7, and 36.6 %, respectively, as compared to control. Ce content in roots increased with increased nCeO$_2$ concentration (adapted from Rico et al. 2014)

at all times more toxic than other chemical species comprising the same elements. MnOx NPs and FeOx NPs stimulated the growth of lettuce seedlings by 12–54 % and were found less toxic than their ionic counterparts. Fe or Mn NPs can significantly improve plant growth and has promising role as nanofertilizers for increasing agronomic productivity (Liu et al. 2016).

9.2.4 Iron Oxide Nanoparticle in Combination with Irradiation

The effect of the presowing laser irradiation (He–Ne) combined with different concentrations of iron NPs on growth and yield of pea was investigated (Al Sherbini et al. 2015). Leaf area, dry weight per plant, chlorophyll content, Fe and Mn concentration, pod protein, pod number, and yield per feddan (0.42 ha) were determined. Research findings indicated that treatment of seeds with He–Ne laser irradiation for 10-min exposure time combined with 300 ppm iron oxide NPs improved all the tested parameters significantly. It was concluded that separate or combined He–Ne laser irradiation at 10 min and 300 ppm of iron oxide NPs gave the best growth parameters and the highest yield, compared to the control.

9.3 Nanoparticle-Mediated Enhancement of Secondary Metabolites

A plant cell produces two types of metabolites: Primary metabolites are involved directly in growth and metabolism, viz. proteins, carbohydrates, and lipids, whereas secondary metabolites are considered as the end products of primary metabolism, viz. phenolics, flavonoids, alkaloids, resins, quinones, essential oils, lignins, tannins, steroids, terpenoids, etc.

Plant secondary metabolites are organic substances that are not directly involved in the normal plant growth, development, or reproduction; rather, they play some vital role in various signaling cascades, defense mechanism against microorganisms, etc. Secondary plant products are considered for their vital role in the survival of the plant in its ecosystem, time and again protecting plants against pathogen attack, insect attack, mechanical injury, and other types of biotic and abiotic stresses (Hartmann 2007). It has been documented in various research articles that most of these plant secondary metabolites have some beneficial role in the human body, so these are considered as phytomedicines. Secondary metabolites, also known as natural products or phytochemicals, are responsible for medicinal properties of plants to which they belong. Classification of secondary metabolites is based on the chemical structure, composition, their solubility in various solvents, or their biosynthetic pathway. They are mainly classified into three major groups: terpenoids, alkaloids, and phenolics (Kabera et al. 2014).

Plants offer a great diversity of bioactive small molecular metabolites that are potentially valuable as pharmaceuticals, nutraceuticals, and agrochemicals. Plant crude extract contains various novel bioactive constituents such as phenolics, flavonoids, alkaloids, resins, quinones, steroids, and terpenoids, which are responsible for the reduction of ionic compounds to bulk metallic NPs (Aswathy Aromal and Philip 2012). Primary and secondary metabolites are also reported to be involved in the synthesis of ecofriendly nanosized particles.

The complete level of secondary metabolites is generally low in many medicinally important plants. In the search for alternatives to enhance the production of desirable medicinal compounds in plants, nanotechnological approach, specifically ENPs are found to have great potential as a supplement to traditional agriculture.

Nowadays, researchers are developing novel techniques which facilitate the plants in the improvement of its innate functions. Nanoparticles are empowered with unique physicochemical properties and have the potential to boost the plant metabolism (Giraldo et al. 2014). Galbraith (2007) and Torney et al. (2007) reported the use of engineered NPs to deliver DNA and chemicals into plant cells. This research area offers new possibilities in plant biotechnology to target-specific genes' manipulation and expression in the specific cells of the plants.

9.3.1 Enhancement of Secondary Metabolites Through Nanotreatment In Vivo

Several strategies have been conducted to improve the yields of secondary metabolites also known as natural products or phytochemicals in plants. Only few studies reported the enhancement of secondary metabolite on treatment with NPs under in vivo condition, whereas the effects of different NPs have been reported on plant growth and metabolic function (Nair et al. 2010; Krishnaraj et al. 2012). The same concentration of individual NP may cause effects in diverse directions and ranges on different variables. Hence, selection of appropriate concentration of nanoparticle is essential for recognizing higher benefits for a target agroeconomic trait.

9.3.1.1 Enhancement of Phytomedicines

Kole et al. (2013) observed varied effects of seed treatment with five concentrations of fullerol on the content of five phytomedicines in bitter melon fruits. Contents of two anticancer phytomedicines, namely cucurbitacin B and lycopene, were enhanced by 74 and 82 %, at 9.88 and 47.2 nM fullerol, respectively. Antidiabetic phytomedicines, charantin, and insulin contents were augmented up to 20 and 91 %, when the seeds were treated with 4.72 and 9.88 nM fullerol, respectively (Fig. 9.11).

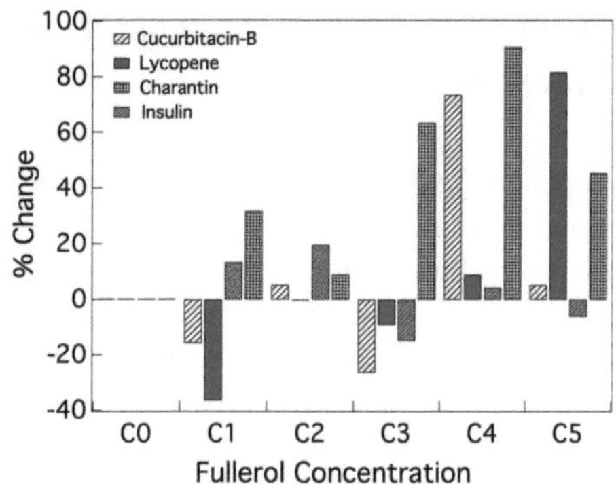

Fig. 9.11 Effect of fullerol at five concentrations (C_1–C_5) in comparison with control (C_0) on changes (in %) in the contents of four phytomedicines in bitter gourd. C_0 denotes control (without fullerol), C_1–C_5 denote five fullerol concentrations (0.943, 4.72, 9.43, 9.88, and 47.2 nM, respectively) (adapted from Kole et al. 2013)

9.3.1.2 Enhancement of Gum and Resins

Significant improvement in the gum content and its viscosity was reported in cluster bean seeds at crop harvest when the leaf of 14-day-old plant was foliar-sprayed with 10 mg L^{-1} ZnO NPs. Improved growth parameters and gum content might be due to adsorption of NPs on plant surface and taken up by the plants through natural nano- or microscale openings (Raliya and Tarafdar 2013).

9.3.1.3 Enhancement of Essential Oil

Amuamuha et al. (2012) recorded the effect of varying concentrations and time of nanoiron foliar application on the essential oil of pot marigold. Four concentrations (0, 1, 2, and 3 gL^{-1}) of iron NPs were used for spraying at different stages (foliar application at stem initialize, flowering, and after the first and second harvest). Significant influence of spraying time (growth stage) on the essential oil percent was observed at the first harvest and the essential oil yield at the third harvest. Similarly, nanoiron concentrations showed significant effect on the yield of essential oil at the first harvest. The highest percentage (1.573 %) of essential oil was reported when nanoiron was applied at the early stage (stem initialized) led to the maximum yield of essential oil (2.397 kg ha^{-1}) in the flower. The lowest essential oil percentage (0.981 %) was recorded when nanoiron was applied at later stages (after the second harvest).

9.3.2 *Enhancement of Secondary Metabolites Through Nanotreatment in Vitro*

Plant kingdom is the potential source of agrochemicals, flavors, and pharmaceuticals, known as secondary metabolites that have several economic advantages. In this regard, plants can be considered to be the best, non-polluting chemical factories. The chemical industries are making enormous efforts to synthesize these products, but the success rate is still limited. Usually, secondary metabolites, a rich source of pharmaceuticals with defensive properties, are synthesized by plants when exposed to different elicitors and/or inducer molecules (Zhao et al. 2005a, b). An "elicitor" can be defined as chemicals or bioagents from various sources which initiates or progresses biosynthesis of specific compounds responsible for physiological and morphological changes in the target living organism, when provided in very low concentrations to a living cell system. In plants, the elicitors can trigger the physiological and morphological changes and phytoalexin accumulation (Zhao et al. 2005a, b). Nowadays, various biotic and abiotic elicitors are practiced to trigger and concentrate the secondary metabolites and cell volume in suspension culture (Rao and Ravishankar 2002).

Among the various strategies available to increase the levels of metabolite of interest, application of elicitors in suspension culture is mostly trusted and practiced strategy. Elicitors in a precise concentration can be administered at desirable time to the suspension culture, resulting in achieving the highest levels of metabolite in a short span of time (Mulabagal and Tsay 2004).

The phenomenal surface characteristics of NP attribute to its extraordinary and unique properties. By increasing the number of atoms on surface, there is an increase in total free energy, resulted in the alteration of material characteristics. Nanoparticles have the potential to be used as novel effective elicitors in plant biotechnology for the elicitation of secondary metabolite production (Fakruddin et al. 2012). Many researchers have studied the role of NPs as elicitors (Aditya et al. 2010; Asghari et al. 2012; Sharafi et al. 2013; Zhang et al. 2013; Ghanati and Bakhtiarian 2014; Raei et al. 2014; Ghasemi et al. 2015; Yarizade and Hosseini 2015). Effect of NPs on enhancement of secondary metabolites is furnished in Table 9.4. A number of studies have supported the possible role of NPs as elicitors for enhancing the expression level of genes related to the production of secondary metabolite (Ghasemi et al. 2015; Yarizade and Hosseini 2015).

Nanoparticles have successfully offered a new strategy in enhancing the secondary metabolite production. But still an in-depth and consolidate insight in research is required to elucidate the effects of NPs in production mechanisms of secondary metabolite production in medicinal plants.

Table 9.4 Effect of nanoparticles on enhancement of secondary metabolites

Elicitors	Plant cell culture	Elicited product	References
AgNPs	*Taxus chinensis* (Chinese yew)	Paclitaxel	Choi et al. (2001)
	Salvia miltirorrhiza (Chinese sage)	Tanshinone	Zhang et al. (2004), Zhao et al. (2010)
	Saussurea medusa (Saw-wort)	Flavonoid jaceosidin Hispidulin	Zhao et al. (2005a, b)
	Bacop amonnieri (Brahmi)	Total phenol content	Krishnaraj et al. (2012)
	Artemisia annua (Sweet sagewort)	Artemisinin	Zhang et al. (2013)
	Datura metel (Datura)	Tropane alkaloids atropine	Shakeran et al. (2015)
CoNPs	*Artemisia annua (Sweet sagewort)*	Artemisinin	Ghasemi et al. (2015)
	Calendula officinalis (Marigold)	Saponin	Ghanati and Bakhtiarian (2014)
Fullerol	*Momordica charantia* (Bitter melon)	Cucurbitacin B, lycopene, charantin, and insulin	Kole et al. (2013)
TiO_2NPs	*Aloe vera*	Aloin	Raei et al. (2014)
	Cicer arietinum (chick pea)	Phenolic and flavonoid compounds	AL-Oubaidi and Kasid (2015)
ZnONPs	*Hypericum perforatum* (St John's wort)	Hypericin Hyperforin	Sharafi et al. (2013)
FeO_xNPs	*Hypericum perforatum* (St John's wort)	Hypericin Hyperforin	Sharafi et al. (2013)

9.3.2.1 Enhancement of Terpenoids

Artemisia annua is a medicinal plant that produces artemisinin as one of the secondary metabolites, which is a sesquiterpene lactone. Artemisinin is used against malaria parasite (*Plasmodium falciparum* and *P. vivax*) (Snow et al. 2005), for treating different types of cancers such as leukemia, colon cancer, breast cancer, and small carcinomas in lungs (Lei et al. 2011). Artemisinin is produced in very low quantity in *A. annua*. Thus, there is a hike in the price of medicines made from artemisinin, particularly for people in developing countries, where malaria is widely prevalent. Being very expensive, it is not economical to synthesize it chemically. Till now, scientists have not achieved a commercial method to enhance artemisinin content in spite of its known valuable medicinal properties (Ferreira et al. 1995). The study by Zhang et al. (2013) highlighted the potential of nanosilver particles as a novel and effective elicitor in plant biotechnology for the production of plant secondary metabolites. Exposure of Ag-SiO_2 core–shell nanoparticles (Ag NPs) resulted in increased artemisinin content in the hairy root culture of *A annua*. Recent investigations have reported the potential of lipid nanoparticles for parenteral delivery and the augmentation of antimalarial potential of artemether, a derivative of artemisinin (Aditya et al. 2010). Influence of nanocobalt on the expression level of involved genes and content in *Artemisia* was examined (Ghasemi et al. 2015). Nanocobalt particles were used for the elicitation of artemisinin in the cell suspension culture of *A. annua*. qRT-PCR and HPLC were used for quantification of the expression levels of *SQS* and *DBR2* genes and artemisinin content in cell suspension culture, respectively. For this purpose, different concentrations (0.25, 2.5, and 5 mg L^{-1}) of nanocobalt particles were used and samples were analyzed after 8, 24, 48, and 72 h. The maximum increase (2.25-fold, i.e., 113.35 mg g^{-1} dw as compared to control) in artemisinin content was recorded when cells were exposed to 5 mg L^{-1} nanocobalt for 24 h. At the same time, suppressed expression of *SQS* and *DBR2* genes was observed. This decline in the expression of *SQS* and *DBR2* genes might be the cause of enhanced production of artemisinin content by high concentrations of the nanocobalt particles. The mechanism of the impact of nanocobalt on enhancing artemisinin content will be unstated with the expression analysis of all genes involved in artemisinin production (Ghasemi et al. 2015). However, to increase the production of a metabolite, enhancing the expression of particular one gene is not sufficient.

Yarizade and Hosseini (2015) examined the effect of nanocobalt and nanozinc (0, 0.25, 0.5, and 1 mg L^{-1}) on the expression levels of *ADS*, *DBR2*, *ALDH1*, and *SQS* genes at 8, 24, 48, and 72 h after treatment in the hairy root culture of *A. vulgaris*. It was reported that cobalt NP at 0.25 mg L^{-1} caused the maximum expression for all genes under investigation, whereas nanozinc particles at 1.0 mg/L caused the maximum gene expression. It was the first report for the use of NPs for increasing the expression level of genes related to artemisinin production. Potential application of nanozinc and nanocobalt oxide as elicitor to increase artemisinin production in biologic systems such as hairy roots was suggested. Nanocobalt was recommended as the better elicitor compared to nanozinc, since concurrent to the increase in the *ADS* upregulation; subsequently, it downregulates its antagonist, the *SQS* gene (Yarizade and Hosseini 2015).

Baldi and Dixit (2008) stated a slight increase in the artemisinin content of artemisia cell suspension upon the addition of yeast extract. This increase was credited to the presence of metal ions Co^{2+} and Zn^{2+}. The mechanism of nanoparticles as elicitors for enhancement in secondary metabolite content is still unrevealed, and more research is required (Zhao et al. 2005a, b).

9.3.2.2 Enhancement of Phenols

Aloe vera is an important medicinal plant from Aloaceae family with African origin. Among 300 *Aloe* species, *A. vera* is considered as an important medicinal plant in many countries (Reynolds 2004; Hasanuzzaman et al. 2008). *A.vera* contains different secondary metabolites, and the most important of them is aloin which is an anthraquinone. Aloin is the active component having medicinal property, displays antimicrobial activity against some bacteria and fungi, and possesses healing ability of skin burns, ulcer, and cutaneous injuries. Raei et al. (2014) investigated the effects of different abiotic elicitors including nano-Ag, nano-TiO_2, NH_4NO_3, and sucrose on cell suspension culture of *A. vera*. The induced calli by elicitors was collected at five intervals (6, 24, 48, 72, and 168 h) and have been analyzed by HPLC. Increased aloin production was recorded in 48 h after elicitation with Ag NPs, but this level was declined gradually with time and reached the control level. The decline might be related to the feedback of aloin on the gene expression, and increased production of aloin is the reason for reduced gene expression (Raei et al. 2014). TiO_2, when used as nanoelicitor, could increase the aloin content in 48 h after elicitation but reduced to a lower level, 8.8 %, than the control. The decline might be related to the toxic effect of nano-TiO_2 in the culture medium or impact of that NP on gene expression. However, both (nano-Ag and TiO_2) of the nanoelicitors enhanced the aloin content 48 h after treatment but after that decreased gradually.

In another study, Krishnaraj et al. (2012) investigated the effect of biologically synthesized SNPs on plant growth metabolism in *Bacopa monnieri* (Linn.) (Brahmi). Total phenol content was assayed in various parts of the plants grown in hydroponic solution, and enhanced total phenol content was recorded in plants treated with Ag NPs. Results revealed that treatment with biologically synthesized Ag NPs exerted a slight stress condition on the growth and metabolism of *B. monnieri*, and therefore, rise in phenol content is one of the mechanisms to mimic mild stress condition.

Enhancement of Polyketides

Hypericum perforatum is a well-known medicinal plant (Deltito and Beyer 1998). Extract of *H. perforatum* is widely used to treat mild-to-moderate depression (Dias et al. 1998). Hypericin and hyperforin are naphthodianthrones and prenylated acylphloroglucinols, respectively, placed under polyketides. Numerous elicitors for the production of hypericin and hyperforin in cell cultures of *H. perforatum* have been examined. Iron- and zinc-nano oxides were used as elicitors for the first time by Sharafi et al. (2013). Different concentrations of zinc- and iron-nano oxides (0, 50, 100, and 150 ppb) were used for the treatment, and samples were analyzed after 72 h. Hypericin and hyperforin were detected, identified, and quantified in cell suspension cultures of *H. perforatum* by HPLC. It was reported that zinc- and iron-nano oxides (100 ppb) promoted the hypericin and hyperforin production in cell suspension culture. In the cultures stimulated by zinc-nano oxide, the hypericin and hyperforin production reached to the maximum (7.87 and 217.45 $\mu g\ g^{-1}$ dry weight, respectively), which were 3- and 13-fold higher than the control. The amount of hypericin and hyperforin was increased from 2.07 and 16.27 $\mu g\ g^{-1}$ dry weight to 11.18 and 195.62 $\mu g\ g^{-1}$ dry weight in cultures treated with iron-nano oxide. The cell cultures treated with zinc- and iron-nano oxides showed increased hyperforin production as compared to the hypericin production. It can be suggested that NPs can be appropriate candidates for elicitation studies of in vitro secondary metabolite production.

Jasmonate (JA), an important stress hormone, triggered various plant defense responses, along with the biosynthesis of defensive secondary metabolites (Menke et al. 2009). Nanoparticles may play an important role in regulating the expression of genes for jasmonate production in treated cells. Induced jasmonate production may be responsible for enhanced production of hypericin and hyperforin. Plant cell wall might act as a barrier for entry of any external materials including NPs. But with diameters less than the pore diameter of the cell wall, nanoparticles can pass through and reach the plasma membrane. They may also cross the membrane by using transport carrier proteins or ion channels. The NPs may bind with different organelles or interfere with the metabolic processes. Studies on the uptake mechanism, transportation, and binding sites of NPs in plant cells are required to elucidate the elicitation mechanism of these in vitro applied NPs for the enhancement of secondary metabolite production. However, higher concentrations of zinc- and iron-nano oxides (150 ppb) showed negative effects on hypericin and hyperforin production (Sharafi et al. 2013).

Enhancement of Flavonoids

Flavonoids and isoflavonoids are the most popular groups of secondary metabolites found in plants. Many legume seeds have been reported to be rich sources of these secondary metabolites (Heiras-Palazuelos et al. 2013). AL-Oubaidi and Kasid (2015) demonstrated the increased production of secondary metabolite (phenolic and flavonoid compounds) in gram on exposure to TiO_2 NPs under in vitro condition. Secondary metabolite contents in the callus were estimated qualitatively and quantitatively using HPLC and compared with the mother plant. TiO_2 NPs at varying concentrations (0.5, 1.5, 3, 4.5, 6) mg L^{-1} were used for an effective increase in secondary metabolites. The results revealed that the secondary metabolite concentration from callus embryo of gram increased to highly significant level at the concentrations of 4.5 and 6.0 mg L^{-1}. The HPLC outcomes confirmed the elevation in the secondary metabolite level under the effect of the TiO_2 NPs when compared with the mother plant.

In a very recent report, Khan et al. (2016) examined the effect of nine types of metal nanoparticles including monometallic and bimetallic alloy nanoparticles [Ag, Au, Cu, AgCu (1:3), AgCu (3:1), AuCu (1:3), AuCu (3:1), AgAu (1:3), AgAu (3:1)] on total phenolic and flavonoid contents in milk thistle plant. The sterilized seeds were soaked in NPs suspensions for 2 h and allowed to grow under in vitro condition. The experiment was conducted for 6 weeks, and samples for total phenolic and flavonoid contents were collected on weekly interval. NPs suspensions affected total phenolic and flavonoid contents in the plant in a different way. It was observed that the amount of phenolics and flavonoids did not show any correlation with the total dry mass of the plant. However, duration of the experiment significantly affected the amount of total flavonoids and phenolics in milk thistle. After 21 days presoaking of seeds in bimetallic alloy, enhanced whereas monometallic NPs suspensions, reduced phenolics and flavonoids content in milk thistle plantlets. After 28 days, Au and Cu NPs caused maximum total phenolic and flavonoid accumulation in milk thistle plants. Therefore, maximum effect on secondary metabolites was recorded with monomatellic NPs. Mainly three factors (size, surface area, and composition of NPs) played a significant role either singly or in combination.

Enhancement of Saponins

The effects of SNPs and methyl jasmonate (MeJA) on secondary metabolites of marigold were studied (Ghanati and Bakhtiarian 2014). When plants were exposed to SNPs, chlorophyll and carotenoid content decreased by 30–50 %, while MeJA increased both of these contents, whereas when plants were treated with 0.4 mM SNPs and 100 μM MeJA, saponin content in the plants augmented by 177 %. Significant reduction in the viability of HeLa cells was recorded when exposed to the extracts of marigold, and this decline was more evident in the plants exposed to MeJA and SNPs.

Enhancement of Phenyl Propanoids and Terpenoids

Aromatic constituents are derived from phenylpropane hydrocarbons. The major identified components of fennel oil are phenyl propanoids and terpenoids. Fennel is annual or biennial aromatic plant, which is widely grown in Mediterranean and some tropical regions, and it is used for herbal drug preparations. One of the major compounds of fennel volatile oil is *trans*-anethole, the amount of which is the major governing factor for the quality of fennel volatile oil (Billia et al. 2002; Gurdip et al. 2006; Chaouche et al. 2011).

Bahreini et al. (2015) analyzed the phytoconstituents of in vitro grown fennel plantlets in normal and nanoelicited (TiO_2 and SiO_2) conditions. A significant difference was observed among the metabolites of normal and elicited conditions. The major components of normal plant were anethole, fenchone and limonene and decane. Some identified constituents of TiO_2-elicited plant extract were dodecane, phytol, and phenol 2,4 *bis* (1,1 dimethyl ethyl), and the most frequent compound was octane. In plants elicited with SiO_2, benzoic acid, jasmonic acid, and hexadecanoic acid were detected as elicited plant components and the major compound was pyrrolidinone. Some of other accumulated metabolites, which appeared by elicitor inductions such as phytol and benzoic acid, can be used as pharmaceutical and industrial precursors (Bahreini et al. 2015).

9.4 Conclusions and Future Perspectives

There is a demanding need for agriculture to produce more output with less input. We are on the edge of time where we have to adopt modern agriculture techniques and technologies as conventional agricultural practices will not be sufficiently able to feed an ever-increasing population with changing climate, depleting resources, and shrinking landscape. Among most recent technical improvements in the field of agriculture, nanotechnology holds an eminent position in remodeling agriculture and food production to fulfill the demands in an efficient and cost-effective way.

Nanotechnology being studied since the last few decades is still in its premature phase of development. However, the whole course of action is very broad and being popularized day by day. Nanotechnology in combination with biotechnology has led to the rapid development of marketable formulations involving deployment of artificially designed nanoparticles for crop improvement. To restrict the indiscriminate use of excess pesticides and fertilizers in plants, nanoparticles are proved to be a gifted tool of this age. Many nanoparticles have been proved to have beneficial role in the case of plant biomass and yield improvement, whereas some nanoparticles have been reported to have a deleterious role regarding reactivity and toxicity in plants. Hence, we are supposed to be very careful during screening and selection of ENPs. Otherwise, they could become the source of potential threat to the whole ecosystem. The interactions between plant cell and nanoparticles modify gene expression in the plants that regulate the overall process of plant growth and development. Nanomaterials that could be used for accelerated plant growth give a new research insight for areas such as biofuels for which the total biomass is important for the final production and yield. Effect of nanoparticles depends on the experimental conditions, and it would change (positive effect to negative or vice versa) if condition varies. Such aspects would include the type of plants and nanomaterials, concentrations of the nanostructures, as well as their chemical and biologic surface functionalizations (Khodakovskaya et al. 2009). Racuciu and Creanga (2007) reported the stimulating effect on growth of maize plants when exposed to low concentrations of aqueous ferro fluid, while its higher concentrations induced an inhibitory effect. However, the future perspectives on nanobiotechnological approaches for the improvement of plant productivity will depend on an in-depth understanding of the molecular mechanisms accountable for the activation of germination potential of seed and plant growth when exposed to complex ENMs (Khodakovskaya et al. 2012).

Nowadays, nanobiotechnology industries are growing very rapidly; however, there is an urgent need to perform profound studies in this field in order to develop comparatively safe and ecofriendly nanoparticles in the long run. The widespread assessment of these ENPs in agri-food sector should also be carried out for public acceptance to prevent them from the unlike challenges as were faced by genetically modified organisms worldwide. The impact of nanotechnology in farmers' field is just in the beginning, but expectations for nanotechnology to help meet the challenges related to food productivity, environment sustainability, and even fossil fuel are still high.

References

Aditya N, Patnakar S, Madhusudan B, Murthy R, Souto E (2010) Artemether loaded lipid nanoparticles produced by modified thin film hydration: pharmacokinetics, toxicological and invivo antimalarial activity. Eur J Pharm Sci 40:448–455

Al Sherbini A, Abd El-Gawad HG, Kamal MA, El-feky SA (2015) Potential of He-Ne laser irradiation and iron nanoparticles to increase growth and yield of pea. Amer-Euras J Agric Environ Sci 15(7):1435–1446

Alidoust D, Isoda A (2013) Effect of gamma Fe_2O_3 nanoparticles on photosynthetic characteristic of soybean (*Glycine max* (L.)Merr.): foliar spray versus soil amendment. Acta Physiol Plant 35 (12):3365–3375

Almeelbi T, Bezbaruah AN (2012) Aqueous phosphate removal using nanoscale zero-valent iron. J Nanopart Res 14:1497–1500

AL-Oubaidi HKM, Kasid NM (2015) Increasing (phenolyic and flavonoids compounds of *Cicer arietinum* L. from embryo explant using titanium dioxide nanoparticle in vitro. World J Pharmaceut Res 4(11):1791–1799

Amuamuha L, Pirzad A, Hadi H (2012) Effect of varying concentrations and time of nanoiron foliar application on the yield and essential oil of Pot marigold. Int Res J Appl Basic Sci 3:2085–2090

Arora S, Sharma P, Kumar S, Nayan R, Khanna PK, Zaidi MGH (2012) Gold nanoparticle induced enhancement in growth and seed yield of *Brassica juncea*. Plant Growth Regul 66:303–339

Asghari GH, Mostajeran A, Sadeghi H, Nakhaei A (2012) Effect of salicylic acid and silver nitrate on taxol production in *Taxus baccata*. J Med Plants 11(8):74–82

Ashrafi SJ, Rastegar MF, Jafarpour B, Kumar SA (2010) Possibility use of silvernano particle for controlling *Fusarium* wilting in plant pathology. In: Riberio C, de-Assis OBG, Mattoso LHC, Mascarenas S (eds) Symposium of international conference on food and agricultural applications of nanotechnologies, São Pedro SP, Brazil. ISBN 978-85-63274-02-4

Aswathy Aromal S, Philip D (2012) Green synthesis of gold nanoparticles using *Trigonella foenum-graecum* and its size-dependent catalytic activity. Spectrochim Acta A Mol Biomol Spectrosc 97:1–5

Bahreini M, Omidi M, Bondarian F, Gholibaygian M (2015) Metabolites screening of nano elicited in vitro Iranian fennel (*Foeniculum vulgare*). Am J Biol Life Sci 3(5):194–198

Bakhtiari M, Moaveni P, Sani B (2015) The effect of iron nanoparticles spraying time and concentration on wheat. Biol Forum 7(1):679–683

Baldi A, Dixit VK (2008) Yield enhancement strategies for artemisinin production by suspension culture of *Artemisia annua*. Bioresour Technol 99:4609–4614

Barrena R, Casals E, Colon J, Font X, Sanchez A, Puntes V (2009) Evaluation of the ecotoxicity of model nanoparticles. Chemosphere 75:850–857

Batsmanova LM, Gonchar LM, Taran NY, Okanenko AA (2013) Using a colloidal solution of metal nanoparticles as micronutrient fertiliser for cereals. In: Proceedings of the international conference on nanomaterials: applications and properties, Crimea, Ukraine, Proc NAP2, 04NABM14

Billia AR, Flamini G, Tagioli V, Morelli I, Vincieri FF (2002) GC–MS analysis of essential oil of some commercial Fennel teas. Food Chem 76(3):307–310

Biswal SK, Nayak AK, Parida UK, Nayak PL (2012) Applications of nanotechnology in agriculture and food sciences. Int J Sci Innov Discov 2(1):21–36

Brennan B (2012) Nanobiotechnology in agriculture. Strategic business insights 2012, Menlo Park, CA, USA, Available from: http://www.strategicbusinessinsights.com/about/featured/2012/2012-10-nanobio-agriculture.shtml. Accessed 18 Apr 2014

Burman U, Saini M, Kumar P (2013) Effect of zinc oxide nanoparticles on growth and antioxidant system of chickpea seedlings. Toxicol Environ Chem 95(4):605–612

Cakmak I (2000) Role of zinc in protecting plant cells from reactive oxygen species. New Phytol 146:185–205

Canas JE, long M, Nations S, Vadan R, Dai L, Luo M, Ambikapathi R, Lee EH, Olszyk D (2008) Effects of functionalized and nonfunctionized single walled carbon nanotubes on root elongation of selected crop species. Nanomat Environ 27:1922–1931

Chaouche T, Haddouchi F, Lazouni HA, Bekkara FA (2011) Phytochemical study of the plant *Foeniculum vulgare* Mill. Pharm Lett 3(2):329–333

Chen HC, Roco MC, Son JB, Jiang S, Larson CA, Gao Q (2013) Global nanotechnology development from 1991 to 2012: patents, scientific publications, and effect of NSF funding. J Nanopart Res 15:1951

Choi HK, Yun JH, Kim SL, Son JS, Kim HR, Kim JH, Choi HJ, Hong SS (2001) Enhanced production of paclitaxel by semi-continuous batch process (SCBP) in suspension culture of *Taxus chinensis*. Enzyme Microb Technol 29:583–586

Chutipaijit S (2015) Establishment of condition and nano particle factors influencing plant regeneration from aromatic rice (*Oryza sativa*). Int J Agric Biol 17:1049–1054

Deltito J, Beyer D (1998) The scientific, quasi-scientific and popular literature on the use of St. John's Wort in the treatment of depression. Affect Disord 51:245–251

Dhoke SK, Mahajan P, Kamble R, Khanna A (2013) Effect of nano particles suspension on the growth of mung (*Vigna radiata*) seedlings by foliar spray method. Nanotechnol Dev 3(1):1–5

Dias ACP, Tomas-Barberan FA, Fernandes-Ferreira M, Ferreas F (1998) Unusual flavonoids produced by callus culturesof *Hypericum perforatum*. Phytochemistry 48:1156–1168

Dimkpa CO, McLean JE, Latta DE, Manangón E, Britt DW, Johnson WP, Boyanov MI, Anderson AJ (2012) CuO and ZnO nanoparticles: phytotoxicity, metal speciation, and induction of oxidative stress in sand-grown wheat. J Nanopart Res 14:1–15

Ditta A (2012) How helpful is nanotechnology in agriculture? Adv Nat Sci Nanosci Nanotechnol 3 (3):033002

Dofing SM (1995) Phenological development-yield relationships in spring barley in a subarctic environment. Can J Plant Sci 75:93–97

Fakruddin MD, Hossain Z, Afroz H (2012) Prospects and applications of nanobiotechnology: a medical perspective. J Nanobiotechnol 10:1–8

Feizi H, Kamali M, Jafari L, RezvaniMoghaddam P (2013) Phytotoxicity and stimulatory impacts of nanosized and bulk titanium dioxide on fennel (*Foeniculum vulgare* Mill). Chemosphere 91 (4):506–511

Ferreira JFS, Simon JE, Janick J (1995) Developmental studies of *Artemisia annua*: flowering and artemisinin production under greenhouse and field conditions. Planta Med 61:167–170

Frewer LJ, Norde W, Fischer ARH, Kampers FWH (2011) Nanotechnology in the agri-food sector: implications for the future. Wiley-VCH Weinheim, Germany. ISBN 9783527330607 p 328

Galbraith DW (2007) Nanobiotechnology: silica breaks through in plants. Nat Nanotechnol 2: 272–273

Gao F, Hong F, Liu C, Zheng L, Su M, Wu X, Yang F, Wu C, Yang P (2008) Mechanism of nano-anatase TiO_2 on promoting photosynthetic carbon reaction of spinach. Biol Trace Elem Res J 111:239–253

Gardea-Torresdey JL, Rico CM, White JC (2014) Trophic transfer, transformation, and impact of engineered nanomaterials in terrestrial environments. Environ Sci Technol 48:2526–2540

Ghanati F, Bakhtiarian S (2014) Effect of methyl jasmonate and silver nanoparticles on production of secondary metabolites by *Calendula officinalis* L (Asteraceae). Trop J Pharmaceut Res 13 (11):1783–1789

Ghasemi B, Hosseini R, Nayeri FD (2015) Effects of cobalt nanoparticles on artemisinin production and gene expression in *Artemisia annua*. Turk J Bot 39:769–777

Giraldo JP, Landry MP, Faltermeier SM, Mc Nicholas TP, Iverson NM, Boghossian AA, Reuel NF, Hilmer AJ, Sen F, Jacqueline AB, Strano MS (2014) Plant nanobionics approach to augment photosynthesis and biochemical sensing. Nat Mater 13:400–408

Gonzalez-melendi P, Fernandez-pacheco R, Coronado MJ, Corredor E, Testillano PS, Risuno MC, Marquina C, Ibarra MR, Rubiales D, Perez-de-luque A (2008) Nanoparticles as smart treatment-delivery systems in plant: assessment of different techniques of microscopy for their visualization in plant tissues. Ann Bot 101:187–195

Gopinath K, Gowri S, Karthika V, Arumugam A (2014) Green synthesis of gold nanoparticles from fruit extract of *Terminaliaarjuna*, for the enhanced seed germination activity of *Gloriosa superba*. J Nanostruct Chem 4:1–11

Gruère G, Narrod C, Abbott L (2011) Agriculture, food, and water nanotechnologies for the poor: opportunities and constraints. Policy Brief 19. International Food Policy Research Institute, Washington DC

Gurdip S, Maurya S, de Lampasona MP, Catalan C (2006) Chemical constituents, antifungal and antioxidative potential of *Foeniculum vulgare* volatile oil and its acetone extract. Food Control 17:745–752

Hafeez A, Razzaq A, Mahmood T, Jhanzab HM (2015) Potential of copper nanoparticles to increase growth and yield of wheat. J Nanosci Adv Technol 1(1):6–11

Haghighi M, Afifipour Z, Mozafarian M (2012) The alleviation effect of silicon on seed germination and seedling growth of tomato under salinity stress. Veg Crops Res Bull 76 (119):126

Hartmann T (2007) From waste products to ecochemicals: fifty years research of plant secondary metabolism. Phytochemistry 68:2831–2846

Hasanuzzaman M, Ahamed KU, Khalequzzaman KM, Shamsuzzaman AMM, Nahar K (2008) Plant characteristics, growth and leaf yield of *Aloe vera* L. as affected by organic manure in pot culture. Aust J Crop Sci 2(3):158–163

Heiras-Palazuelos MJ, Ochoa-Lugo MI, Gutierrez-Dorado R, Lopez Valenzuela JA, Mora-Rochin S, Milan Carrillo J et al (2013) Technological properties, antioxidant activity and total phenolic and flavonoid content of pigmented chickpea (*Cicer arietinum* L.) cultivars. Int J Food Sci Nutr 64:69–76

Hong F, Zhou J, Liu C, Yang F, Wu C, Zheng L, Yang P (2005) Effect of nano-TiO_2 on photochemical reaction of chloroplasts of spinach. Biol Trace Elem Res 105:269–279

Husen A, Siddiqi KS (2014) Carbon and fullerene nanomaterials in plants system. J Nanotechnol 12:1–10

Jaberzadeh A, Moaveni P, Tohidi Moghadam HR, Zahedi H (2013) Influence of bulk and nanoparticles titanium foliar application on some agronomic traits, seed gluten and starch contents of wheat subjected to water deficit stress. Not Bot Horti Agrobot 41(1):201–207

Jayarambabu N, kumari BS, Rao KV, Prabhu YT (2015) Beneficial role of zinc oxide nanoparticles on green crop production. Int J Adv Multi Res Trends 2(1):273–282

Kabera JN, Semana E, Mussa AR, He X (2014) Plant secondary metabolites: biosynthesis, classification, function and pharmacological properties. J Pharm Pharmacol 2:377–392

Karuppanapandian T, Wang HW, Prabakaran N, Jeyalakshmi K, Kwon M, Manoharan K, Kim W (2011) 2, 4-dichlorophenoxyacetic acid-induced leaf senescence in mung bean (*Vignaradiata* L. Wilczek) and senescence inhibition by co-treatment with silver nanoparticles. Plant Physiol Biochem 49(2):168–177

Khan MS, Zaka M, Abbasi BH, Rahman LU, Shah A (2016) Seed germination and biochemical profile of *Silybum marianum* exposed to monometallic and bimetallic alloy nanoparticles. IET Nanobiotechnol. doi:10.1049/iet-nbt.2015.0050

Khodakovskaya M, Dervishi E, Mahmood M, Xu Y, Li Z, Watanabe F, Biris AS (2009) Carbon nanotubes are able to penetrate plant seed coat and dramatically affect seed germination and plant growth. ACS Nano 3:3221–3227

Khodakovskaya MV, de Silva K, Nedosekin DA, Dervishi E, Biris AS, Shashkov EV, Galanzha EI, Zharov VP (2011) Complex genetic, photothermal, and photoacoustic analysis of nanoparticle-plant interactions. Proc Natl Acad Sci USA 108:1028–1033

Khodakovskaya MV, de Silva K, Biris AS, Dervishi E, Villagarcia H (2012) Carbon nanotubes induce growth enhancement of tobacco cells. ACS Nano 6:2128–2135

Khodakovskaya MV, Kim B-S, Kim JN et al (2013) Carbon nanotubes as plant growth regulators: effects on tomato growth, reproductive system, and soil microbial community. Small 9: 115–123

Kisan B, Shruthi H, Sharanagouda H, Revanappa SB, Pramod NK (2015) Effect of nano-zinc oxide on the leaf physical and nutritional quality of spinach. Agrotechnology 5:135. doi:10.4172/2168-9881.1000135

Klaine SJ, Alvarez PJJ, Batley GE, Fernandes TF, Handry RD, Lyon DY, Manendra S, McKaughlin MJ, Lead JR (2008) Nanomaterials in the environment: behavior, fate bioavailability, and effects. Environ Toxicol Chem 27:1825–1851

Kole C, Kole P, Randunu KM, Choudhary P, Podila R, Ke PC (2013) Nanobiotechnology can boost crop production and quality: first evidence from increased plant biomass, fruit yield and phytomedicine content in bitter melon (*Momordica charantia*). BMC Biotechnol 13:37

Krishnaraj C, Jagan EG, Ramachandran R, Abirami SM, Mohan N, Kalaichelvan PT (2012) Effect of biologically synthesized silver nanoparticles on *Bacopa monnieri* (Linn.) Wettst. plant growth metabolism. Process Biochem 47:651–658

Kumar V, Guleria P, Kumar V, Yadav SK (2013) Gold nanoparticle exposure induces growth and yield enhancement in *Arabidopsis thaliana*. Sci Total Environ 461:462–468

Kumari M, Mukherhjee A, Chandrasekaran N (2009) Genotoxicity of silver nano particles in *Allium cepa*. Sci Total Environ 407:5243–5246

Laware SL, Raskar S (2014) Influence of zinc oxide nanoparticles on growth, flowering and seed productivity in onion. Int J Curr Microbiol Appl Sci 3(7):874–881

Lee WM, An YJ, Yoon H, Kweon HS (2008) Toxicity and bioavailability of copper nanoparticles to the terrestrial plants mung bean (*Phaseolus radiatus*) and wheat (*Triticum aestivum*): plant agar test for water-insoluble nanoparticles. Environ Toxicol Chem 27:1915–1921

Lee WL, Mahendra S, Zodrow K, Li D, Tsai YC, Braam J, Alvarez PJJ (2010) Developmental phytotoxicity of metal oxide nanoparticles to *Arabidopsis thaliana*. Environ Toxicol Chem 29:669–675

Lee SY, Chung HI, Kim SY, Lee IS (2012) Assessment of phytotoxicity of ZnO NPs on a medicinal plant, *Fagopyrum esculentum*. Environ Sci Pollut Res 20(2):848–854

Lee SY, Chung HI, Kim SY, Lee IS (2013) The genotoxic effect of ZnO and CuO nanoparticles on early growth of buckwheat, *Fagopyrum esculentum*. Water Air Soil Pollut 224:1668

Lei C, Ma D, Pu G, Qiu X, Du Z, Wang H, Li G, Ye H, Liu B (2011) Foliar application of chitosan activates artemisinin biosynthesis in *Artemisia annua* L. Ind Crop Prod 33:176–182

Li P, Chen J, Wu P (2011) Agronomic characteristics and grain yield in 30 spring wheat genotypes under drought stress and nonstress conditions. Agron J 103:1619–1628

Lin S, Reppert J, Hu Q, Hudson JS, Reid ML, Ratnikova TA, Rao AM, Luo H, Ke PC (2009) Uptake, translocation, and transmission of carbon nanomaterials in rice plants. Small 5: 1128–1132

Linglan M, Chao L, Chunxiang Q, Sitao Y, Jie L, Fengqing G, Fashui H (2008) Rubiscoactivase mRNA expression in spinach: modulation by nanoanatase treatment. Biol Trace Elem Res 122 (2):168–178

Liu R, Lal R (2015) Potentials of engineered nanoparticles as fertilizers for increasing agronomic productions. Sci Total Environ 514:131–139

Liu XM, Zhang FD, Zhang SQ, He XS, Fang R, Feng Z, Wang Y (2005) Effects of nano-ferric oxide on the growth and nutrients absorption of peanut. Plant Nutr Fert Sci 11:14–18

Liu X, Feng Z, Zhang S, Zhang J, Xiao Q, Wang Y (2006) Preparation and testing of cementing nano-sub nano composites of slow or controlled release of fertilizers. Sci Agri Sin 39: 1598–1604

Liu S, Yuan L, Yue X, Zheng Z, Tang Z (2008) Recent advances in nanosensors for organophosphate pesticide detection. Adv Powder Technol 19:419–441

Liu Q, Chen B, Wang Q, Shi X, Xiao Z, Lin J, Fang X (2009) Carbon nanotubes as molecular transporters for walled plant cells. Nano Lett 9:1007–1010

Liu R, Zhang H, Lal R (2016) Effects of stabilized nanoparticles of copper, zinc, manganese, and iron oxides in low concentrations on lettuce (*Lactuca sativa*) seed germination: nanotoxicants or nanonutrients? Water Air Soil Pollut 227:42–54

López-Moreno ML, de la Rosa G, Hernández-Viezcas JA, Peralta-Videa JR, Gardea-Torresdey JL (2010) XAS corroboration of the uptake and storage of CeO_2 nanoparticles and assessment of their differential toxicity in four edible plant species. J Agric Food Chem 58(6):3689–3693

Lu CM, Zhang CY, Wen JQ, Wu GR, Tao MX (2002) Research of the effect of nanometer materials on germination and growth enhancement of *Glycine max* and its mechanism. Soybean Sci 21:168–172

Ma X, Geiser-Lee J, Deng Y, Kolmakov A (2010) Interactions between engineered nanoparticles (ENPs) and plants: phytotoxicity, uptake and accumulation. Sci Total Environ 408(16):3053–3061

Mahajan P, Dhoke SK, Khanna AS (2011) Effect of nano-ZnO particle suspension on growth of mung (*Vigna radiata*) and gram(*Cicer arietinum*) seedlings using plant agar method. J Nanotechnol Article ID 696535, 7 pages

Malakouti MJ, Tehrani MM (2005) The role of micronutrients in yield increasing and improving quality of agricultural products, 3rd edn. Tarbiyat Modarres University, Tehran, pp 273–279

Marchiol L, Mattiello A, Pošćić F, Fellet G, Zavalloni C, Carlino E, Musett R (2016) Changes in physiological and agronomical parameters of barley (*Hordeum vulgare*) exposed to cerium and titanium dioxide nanoparticles. Int J Environ Res Public Health 13:332–350

Menke F, Champion A, Kijne J, Memelink J (2009) A novel jasmonate- and elicitor-responsive element in the periwinkle secondary metabolite biosynthetic gene Str interacts with a jasmonate and elicitor-inducible AP2-domain transcription factor, ORCA2. Eur Mol Biol Org 18:4455–4463

Mishra V, Mishra RK, Dikshit A, Pandey AC (2014) Interactions of nanoparticles with plants: an emerging prospective in the agriculture industry. In: Ahmad P, Rasool S (eds) Emerging technologies and management of crop stress tolerance: biological techniques, vol 1., Elsevier Academic PressNew York, USA, pp 159–180

Mondal A, Basu R, Das S, Nandi P (2011) Beneficial role of carbon nanotubes on mustard plant growth: an agricultural prospect. J Nanopart Res 13:4519–4528

Morales MI, Rico CM, Hernandez-Viezcas JA, Nunez JE, Barrios AC, Tafoya A, Flores-Marges JP, Peralta-Videa JR, Gardea-Torresdey JL (2013) Toxicity assessment of cerium oxide nanoparticles in cilantro (*Coriandrum sativum* L.) plants grown in organic soil. J Agric Food Chem 61:6224–6230

Morteza E, Moaveni P, Farahani HA, Kiyani M (2013) Study of photosynthetic pigments changes ofmaize (*Zea mays* L.) under nano TiO_2 spraying at various growth stages. Springer Plus 2 (1):247

Mulabagal V, Tsay HS (2004) Plant cell cultures–an alternative and efficient source for the production of biologically important secondary metabolites. Int J Appl Sci Eng 2:29–48

Naderi MR, Danesh-Shahraki A (2013) Nanofertilizers and their roles in sustainable agriculture. Int J Agri Crop Sci 5(19):2229–2232

Nair R, Varghese SH, Nair BG, Maekawa T, Yoshida Y, Kumar DS (2010) Nanoparticulate material delivery to plants. Plant Sci 179:154–163

Najafi S, Jamei R (2014) Effect of silver nanoparticles and $Pb(NO_3)_2$ on the yield and chemical composition of mung bean (*Vigna radiata*). J Stress Physiol Biochem 10:316–325

Nanotechnology in Agriculture and Food (2006) European nanotechnology gateway. http://www.nanoforum.org/dateien/temp/nanotechnology%20in%20agriculture%20-and%20food.pdf

Nejatzadeh-Barandozi F, Darvishzadeh F, Aminkhani A (2014) Effect of nano silver and silver nitrate on seed yield of (*Ocimum basilicum* L.). Organic Med Chem Lett 4(1):1–6

Ngo QB, Dao TH, Nguyen HC, TranXT Nguyen TV, Khuu TD, Huynh TH (2014) Effects of nanocrystalline powders (Fe, Co and Cu) on the germination, growth, crop yield and product quality of soybean (Vietnamese species DT-51). Adv Nat Sci Nanosci Nanotechnol 5:1–7

Panwar J, Jain N, Bhargaya A, Akthar MS, Yun YS (2012) Positive effect of zinc oxide nanoparticles on tomato plants: a step towards developing "nano-fertilizers". In: 3rd international conference on environmental research and technology, pp 348–352

Patra P, Choudhury SR, Mandal S, Basu A, Goswami A, Gogoi R, Srivastava C, Kumar R, Gopal M (2013) Effect of sulfur and ZnO nanoparticles on stress physiology and plant (*Vigna radiata*) nutrition. In: Giri PK, Goswami DK, Perumal A (eds) Advanced nanomaterials and nanotechnology. Springer, Berlin, pp 301–309

Pérez-de-Luque A, Hermosín MC (2013) Nanotechnology and its use in agriculture. In: Bagchi D, Bagchi M, Moriyama H, Shahidi F (eds) Bio-nanotechnology: a revolution in food, biomedical and health sciences. Wiley-Blackwell, West Sussex, pp 299–405

Pokhrel LR, Dubey B (2013) Evaluation of developmental responses of two crop plants exposed to silver and zinc oxide nanoparticles. Sci Total Environ 452:321–332

Prasad TNVKV, Sudhakar P, Sreenivasulu Y, Latha P, Munaswamy V, Reddy KR, Sreeprasad TSP, Sajanlal R, Pradeep T (2012) Effect of nanoscale zinc oxide particles on the germination, growth and yield of peanut. J Plant Nutr 35(6):905–927

Racuciu M, Creanga D (2007) TMA-OH coated magnetic nanoparticles internalize in vegetal tissue. Rom J Phys 52:395–402

Raei M, Angaji SA, Omidi M, Khodayari M (2014) Effect of abiotic elicitors on tissue culture of *Aloe vera*. Int J Biosci 5(1):74–81

Raliya R, Tarafdar JC (2013) ZnO nanoparticle biosynthesis and its effect on phosphorous-mobilizing enzyme secretion and gum contents in cluster bean (*Cyamopsis tetragonoloba* L.). Agric Res 2(1):48–57

Rao SR, Ravishankar GA (2002) Plant cell cultures: chemical factories of secondary metabolites. Biotechnol Adv 20:101–153

Razzaq A, Ammara R, Jhanzab HM, Mahmood T, Hafeez A, Hussain S (2016) A novel nanomaterial to enhance growth and yield of wheat. J Nanosci Technol 2(1):55–58

Reynolds T (2004) Aloe chemistry. In: Reynolds T (ed) The genus *Aloe*. CRC Press, Boca Raton, pp 39–74

Rico CM, Lee SC, Rubenecia R, Mukherjee A, Hong J, Peralta-Videa JR, Gardea-Torresdey JL (2014) Cerium oxide nanoparticles impact yield and modify nutritional parameters in wheat (*Triticum aestivum* L.). J Agric Food Chem 62:9669–9675

Rico CM, Barrios AC, Tan W, Rubenecia R, Lee SC, Varela-Ramirez A, Peralta-Videa JR, Gardea-Torresdey JL (2015) Physiological and biochemical response of soil-grown barley (*Hordeum vulgare* L.) to cerium oxide nanoparticles. Environ Sci Pollut Res. doi:10.1007/s11356-015-4243

Salama HMH (2012) Effects of silver nanoparticles in some crop plants, Common bean (*Phaseolus vulgaris* L.) and corn (*Zea mays* L.). Int Res J Biotechnol 3(10):190–197

Scott N, Chen H (2002) Nanoscale science and engineering for agriculture and food systems. National planning workshop, 18–19 Nov 2002, Washington DC, USA. Available from: http://www.nseafs.cornell.edu/web.roadmap.pdf

Seif SM, Sorooshzadeh A, Rezazadehs H, Naghdibadi HA (2011) Effect of nanosilver and silver nitrate on seed yield of borage. J Med Plants Res 5(2):171–175

Serag MF, Kaji N, Venturelli E, Okamoto Y, Terasaka K, Tokeshi M, Mizukami H, Ugent KB, Bianco A, Baba Y (2011a) A functional platform for controlled subcellular distribution of carbon nanotubes. ACS Nano 5:9264–9270

Serag MF, Kaji N, Venturelli E, Okamoto Y, Terasaka K, Tokeshi M, Mizukami H, Ugent KB, Bianco Baba Y (2011b) Trafficking and subcellular localization of multiwalled carbon nanotubes in plant cells. ACS Nano 5:493–499

Serag MF, Kaji N, Tokeshi M, Baba Y (2012a) Introducing carbon nano tubes into living walled plant cells through cellulase-induced nanoholes. RSC Adv 2:398–400

Serag MF, Braeckmans K, Habuchi S, Kaji N, Bianco A, Baba Y (2012b) Spatiotemporal visualization of subcellular dynamics of carbon nanotubes. Nano Lett 12:6145–6151

Serag MF, Kaji N, Habuchi S, Bianco A, Baba Y (2013) Nanobiotechnology meets plant cell biology: carbon nanotubes as organelle targeting nanocarriers. RSC Adv 3:4856–4862

Servin A, Elmer W, Mukherjee A, Torre-Roche RD, Hamdi H, White JC, Bindraban P, Dimkpa C (2015) A review of the use of engineered nanomaterials to suppress plant disease and enhance crop yield. J Nanopart Res 17:92–112

Shah V, Belozerova I (2009) Influence of metal nanoparticles on the soil microbial community and germination of lettuce seeds. Water Air Soil Pollut 197:143–148

Shakeran Z, Keyhanfar M, Asghari G, Ghanadian M (2015) Improvement of atropine production by different biotic and abiotic elicitors in hairy root cultures of Datura metel. Turk J Biol 39:111–118

Sharafi E, Nekoei SMK, Fotokian MH, Davoodi D, Mirzaei HH, Hasanloo T (2013) Improvement of hypericin and hyperforin production using zinc and iron nano-oxides as elicitors in cell suspension culture of St John's wort (*Hypericum perforatum* L.). J Med Plants By-prod 2: 177–184

Sharma P, Bhatt D, Zaidi MG, Saradhi PP, Khanna PK, Arora S (2012) Silver nanoparticle-mediated enhancement in growth and antioxidant status of *Brassica juncea*. Appl Biochem Biotechnol 167:2225–2233

Sharon M, Choudhry A, Kumar R (2010) Nanotechnology in agricultural disease and food safety. J Phytol 2:83–92

Sheykhbaglou R, Sedghi M, Shishevan MT, Sharifi RS (2010) Effects of nano-iron oxide particles on agronomic traits of soybean. Not Sci Biol 2:112–113

Siddiqui MH, Al-Whaibi MH (2014) Role of nano-SiO_2 in germination of tomato (*Lycopersicum esculentum* seeds Mill.). Saudi J Biol Sci 21:13–17

Siddiqui MH, Al-Whaibi MH, Faisal M, Al Sahli AA (2014) Nano-silicon dioxide mitigates the adverse effects of salt stress on *Cucurbita pepo* L. Environ Toxicol Chem 33:2429–2437

Snow RW, Guerra CA, Noor AM, Myint HY, Hay SI (2005) The global distribution of clinical episodes of *Plasmodium falciparum* malaria. Nature 434:214–217

Sonkaria S, Ahn SH, Khare V (2012) Nanotechnology and its impact on food and nutrition: a review. Recent Pat Food Nutr Agric 4(1):8–18

Srinivasan C, Saraswathi R (2010) Nano-agriculture-carbon nanotubes enhance tomato seed germination and plant growth. Curr Sci 99:273–275

Stampoulis D, Sinha SK, White JC (2009) Assay-dependent phytotoxicity of nanoparticles to plants. Environ Sci Technol 43:9473–9479

Sugunan A, Dutta J (2008) Pollution treatment, remediation and sensing. In: Harald K (ed) Nanotechnology, vol 3. Wiley-VCH, Weinheim, pp 125–143

Tahmasbi D, Zarghami R, Azghandi AV, Chaichi M (2011) Effects of nanosilver and nitroxin biofertilizer on yield and yield components of potato minitubers. Int J Agric Biol 13:986–990

Taran NY, Gonchar OM, Lopatko KG, Batsmanova LM, Patyka MV, Volkogon MV (2014) The effect of colloidal solution of molybdenum nanoparticles on the microbial composition in rhizosphere of *Cicer arietinum* L. Nanoscale Res Lett 9:289

Tiwari DK, Dasgupta-Schubert N, Villasenor Cendejas LM, Villegas J, Carreto Montoya L, Borjas Garcia SE (2014) Interfacing carbon nanotubes (CNT) with plants: enhancement of growth, water and ionic nutrient uptake in maize (*Zea mays*) and implications for nanoagriculture. Appl Nanosci 2:577–591

Torney F, Trewyn BG, Lin VS, Wang K (2007) Mesoporous silica nanoparticles deliver DNA and chemicals into plants. Nat Nanotechnol 2:295–300

Vakhrouchev AV, Golubchikov VB (2007) Numerical investigation of the dynamics of nanoparticle systems in biological processes of plant nutrition. J Phy Conf Sr 61:31–35

Villagarcia H, Dervishi E, Silva K, Biris AS, Khodakovskaya MV (2012) Surface chemistry of carbon nanotubes impacts the growth and expression of water channel protein in tomato plants. Small 8:2328–2334

Welch RM, Webb MJ, Loneragan JF (1982) Zinc in membrane function and its role in phosphorus toxicity. In: Scaife A (ed) Proceedings of the ninth plant nutrition colloquium. CAB International, Wallingford, UK, pp 710–715

Yang F, Hong F, You W, Liu C, Gao F, Wu C, Yang P (2006) Influence of nano-anatase TiO_2 on the nitrogen metabolism of growing spinach. Biol Trace Elem Res 110(2):179–190

Yang F, Liu C, Gao F, Su M, Wu X, Zheng L et al (2007) The improvement of spinach growth by nano-anatase TiO_2 treatment is related to nitrogen photoreduction. BiolTrace Elem Res 119:77–88

Yarizade K, Hosseini R (2015) Expression analysis of *ADS, DBR2, ALDH1* and *SQS* genes in *Artemisia vulgaris* hairy root culture under nano cobalt and nano zinc elicitation. Ext J App Sci 3(3):69–76

Yin L, Cheng Y, Espinasse B, Colman BP, Auffan M, Wiesner M et al (2011) More than the ions: the effect of silver nanoparticles on *Lolium multiflorum*. Environ Sci Technol 45:2360–2367

Yugandhar P, Savithramma N (2013) Green synthesis of calcium carbonate nanoparticles and their effects on seed germination and seedling growth of *Vigna mungo* (L.) Hepper. Int J Adv Res 1 (8):89–103

Yuvakumar R, Elango V, Rajendran V, Kannan NS, Prabu P (2011) Influence of nanosilica powder on the growth of maize crop (*Zea mays* L.). Int J Green Nanotechnol 3(1):180–190

Zhang C, Yan Q, Cheuk WK, Wu J (2004) Enhancement of tanshinone production in *Salvia miltiorrhiza* hairy root culture by Ag+ elicitation and nutrient feeding. Planta Med 70(2): 147–151

Zhang B, Zheng LP, Yi Li W, Wen Wang J (2013) Stimulation of artemisinin production in *Artemesia annua* hairy roots by Ag-SiO_2 core shell nanoparticles. Curr Nanosci 9:363–370

Zhao DX, Fu CX, Han YS, Lu DP (2005a) Effects of elicitation on jaceosidin and hispidulin production in cell suspension cultures of *Saussurea medusa*. Process Biochem 40(2):739–745

Zhao J, Davis LC, Verpoorte R (2005b) Elicitor signal transduction leading to production of secondary metabolites. Biotechnol Adv 23:283–333

Zhao JL, Zhou LG, Wu JY (2010) Effects of biotic and abiotic elicitors on cell growth and tanshinone accumulation in *Salvia miltiorrhiza* cell cultures. Appl Microbiol Biotechnol 87:137–144

Zheng L, Hong F, Lu S, Liu C (2005) Effect of nano-TiO_2 on strength of naturally aged seeds and growth of spinach. Biol Trace Element Res 104:83–91

Zhu H, Han J, Xiao JQ, Jin Y (2008) Uptake, translocation, and accumulation of manufactured iron oxide nanoparticles by pumpkin plants. J Environ Monit 10:713–717

Chapter 10
Role of Nanoparticles for Delivery of Genetic Material

Mariya V. Khodakovskaya and Mohamed H. Lahiani

Abstract Use of nano-sized materials as systems for delivery of genetic material into living cells is new and promising approach. Recent data showed that carbon-based, metal-based, composite nanoparticles and polymer nanoparticles have a potential to carry nucleic acids into plant cells. The unique ability of nanomaterials to penetrate plant cell wall and move inside the cell in fast manner can open ways for improvement of a number of transformation techniques including particle bombardment. However, experimental attempts to use nanomaterials as carriers of DNA/RNA *in planta* are rare. Here, we summarize the reports on successful delivery and integration of genetic material inside plants by using different classes of nanomaterials as delivery systems.

Keywords Genetic material · Nanodelivery · Mesoporous silica nanoparticle system · Multiwalled carbon nanotubes · Plant transformation

Nano-sized materials are unique particles with great potential uses as delivery systems. The different properties including their small size, low toxicity, and conjugation capabilities are unique features that facilitate the delivery of different biomolecules including nucleic acids and proteins into cells. The significant progress in gene delivery using nano-sized materials as nanocarriers was achieved in animal or human cells. For example, successful applications of different nanomaterials including nanoshells, fullerenes, carbon nanotubes, gold nanoparticles, and Fe_3O_4 used as nonviral siRNA delivery agents for cancer therapy were described and discussed (Singh 2013). The nanotechnological approach is beneficial for the area of gene therapy because it can reduce the high risk of infectivity due to the

M.V. Khodakovskaya (✉) · M.H. Lahiani
The University of Arkansas, At Little Rock 2801 S. University Ave, Little Rock, AR 72204, USA
e-mail: mvkhodakovsk@ualr.edu

M.V. Khodakovskaya
Institute of Biology and Soil Science, Far-Eastern Branch of Russian Academy of Sciences, 690022 Vladivostok, Russian Federation

C. Kole et al. (eds.), *Plant Nanotechnology*,
DOI 10.1007/978-3-319-42154-4_10

application of common viral vectors. The intracellular, paracellular, and transcellular pathways were described as major routes for uptake of nanoparticles by animal cells (Kumari et al. 2010; Murugan et al. 2015). It was shown that cylindrical one-dimensional nanomaterials, such as carbon nanotubes, can enter the animal cell by tip recognition through receptor binding with following endocytosis (Shi et al. 2011).

The existence of cell wall is making delivery of DNA and RNA into plant cells more difficult. Traditionally, the cell wall barrier can be overcome in plants by the removal of the cell wall (Ochatt 2013), infection by *Agrobacterium* (Křenek et al. 2015), or through particle bombardment of cells (Taylor and Fauquet 2002). The advantages of the use of nanomaterials for plant transformation were described by researchers led by Torney et al. (2007). They demonstrated that mesoporous silica nanoparticle system (MSNS) was capable of delivering DNA and chemicals into plant cells and plant leaves (Torney et al. 2007). They filled the MSNs with gene as well as with chemical inducer of the gene and the closed end of nanoparticles with gold nanoparticles (Au^{Capped}-MSNs, 3-nm pore size). Such unique construction allowed triggering the expression of a gene by supplied inducer using the process of uncapping. In the following paper, they claimed that it is possible to enhance biolistic delivery of MSNS by increasing the density of MSNS nanomaterials through the application of gold plating (Martin-Ortigosa et al. 2012a). Such an improvement led to the most effective introduction of MSNS into plant cells. Created MSNS was effective for mediated codelivery of protein and plasmid DNA into plant cells (Martin-Ortigosa et al. 2012b). Particularly, 10-nm pore-sized, gold-functionalized MSNSs were loaded with proteins, coated with plasmid DNA, and successfully introduced into plant cells using particle bombardment (Martin-Ortigosa et al. 2012b). Established method is a good foundation for some genomic manipulations/transformations for biotechnology. One of the most promising applications of MSNs nanosystem is genome editing that can be applied for generation of precisely modified "nontransgenic" plants. Thus, it was reported that Cre recombinase protein was delivered into maize (*Zea mays*) cells using MSNs as carrier (Martin-Ortigosa et al. 2014). Described delivery resulted in the removal of a *loxP*-defined DNA fragment from maize genome. Nanobiotechnological approach for the deletion of particular DNA fragment has some obvious advantages. For example, MSNs can be customized (tailored) for a particular enzyme that can be directed to the tissue through the biolistic method. Additionally, controlled release of protein can be achieved by capping of the pore opening of MSNs (Torney et al. 2007; Martin-Ortigosa et al. 2014).

Other types of nanomaterials (carbon-based and metal-based) also exhibited potential for use as carries of genetic material. The ability of single-walled carbon nanotubes (SWCNTs) to penetrate the plant cell wall and the plant cell membrane was documented by Liu et al. (2009). To investigate the ability of carbon nanotubes to deliver nucleic acids into the plant cells (tobacco, *Nicotiana tabacum*, cell culture), authors prepared SWCNT/DNA conjugates by noncovalent binding. Using fluorescein isothiocyanate (FITC) as fluorescence agent fused with SWCNT/DNA conjugates, they observed intracellular fluorescence for about 80 % exposed tobacco cells. Such observation was a good evidence for the ability of SWCNTs carrying nucleic acids into the plant cell. Authors did not notice the presence of SWCNTs–DNA conjugates in plant nucleus. However, they experimentally proved the possibility of delivery of DNA by carbon nanotubes inside plant cell for the first time (Liu et al. 2009). The amazing ability of carbon-based nanomaterials to penetrate even very thick plant tissues was noticed in some research papers. For example, penetration of seed coats of different crops by carbon nanotubes (MWCNTs) and single-walled carbon nanohorns (SWCNHs) was documented using Raman spectroscopy and transmission electron microscopy (TEM) (Khodakovskaya and Biris 2009; Lahiani et al. 2013, 2015). Khodakovskaya et al. (2012) also demonstrated the ability of multiwalled carbon nanotubes (MWCNTs) included in growth medium to penetrate the cell wall of tobacco cells using tobacco callus system. The uptake of MWCNTs by tobacco cells was confirmed by both Raman spectroscopy and TEM. It is important that MWCNTs did not play any negative role in cell culture growth even in highest used doses (100–500 ug/ml) (Khodakovskaya et al. 2012). On the contrary, MWCNTs induced cell division and cell proliferation and activated expression of several genes involved in cell division (*CycB*), cell wall extension (*NtLRX1*), and water transport (*NtPIP1*). They hypothesized that MWCNTs should be tested for ability to carry nucleic acids inside plant cells. It is interesting to note that uptake of MWCNTs can lead to changes in plant gene expression. Results of microarray analysis (*Affymetrix* platform) revealed that expression of a number of genes involved in cellular responses, stress responses, and water relations was affected by carbon-based nanomaterials in treated seeds and plants (Khodakovskaya et al. 2011; Lahiani et al. 2015). Thus, it is important to take into consideration all possible effects of carbon-based materials used as the DNA-delivery machine on plant genome and proteome.

Metal-based nanomaterials, such as gold nanoparticles, also may provide an attractive tool for delivery of proteins and genes to living organisms. Such nanoparticles are biocompatible and can be functionalized easily with different molecules. Nano-sized gold materials were widely used for nucleic acid delivery in animal systems (Ding et al. 2014). Martin-Ortigosa et al. (2012a) bombarded plant tissues (onion, *Allium cepa*, epidermis cells) with DNA-coated gold nanorods (NRs). Authors demonstrated improvement of delivery of DNA inside plant cell by use of NRs as DNA carries (Martin-Ortigosa et al. 2012a). It is logical to expect the appearance of new reports focused on the delivery of nucleic acids and chemicals inside plant cells using composite metal nanoparticles in near future. Recently, Nima et al. (2014) enhanced the growth of tobacco cell culture by successful delivery of growth regulator 2,4-D into cells using plasmonically active nanorods based on gold cores and silver shells. Used multiplex nanosystem (AuNR/Ag) not only worked as a carrier for growth regulator but also allowed detection of particles inside cells using high-sensitive SERS detection. Nima et al. (2014) have concluded that AuNR/Ag nanoparticles are excellent candidates for delivery of different molecules including nucleic acid into plant cells. To prove that suggestion, the appropriate experimental work has to be performed.

Polymer nanoparticles are another class of nano-sized materials with a potential to use as carriers for delivery of nucleic acids. The organic nature of polymer nanoparticles and the ability to overcome the use of traditional viral vectors for silencing are promising traits. Silva and coauthors used fluorescent conjugated polymer nanoparticles (CPNs) to carry siRNA into tobacco BY-2 protoplasts without observation of toxic effects (Silva et al. 2010). In this work, authors provided experimental evidence that polymeric nanoparticles can work as an alternative solution for gene knockout in plant cells. Thus, they were able to detect visually the uptake of siRNA fused with CNPs and demonstrate the effective knockdown of genes involved in tobacco cell wall biosynthesis (*NtCesA-1a*, *NtCesA-1b* genes).

Based on the available limited literature (Table 10.1), we can conclude that nano-sized materials are a promising tool for delivery genetic material inside plant cells. Wide range of nanomaterials should be tested for carrying nucleic acids to plant genome. Efficiency of existing plant transformation techniques can be increased using nanovehicles by more precise delivery of genetic material, ability to control gene expression through release of incorporated chemical inducer, and better detection of nano-delivered genetic material inside particular cell compartments.

Table 10.1 Examples of the use delivery active biomolecules to plants using nano-sized material systems

Nanomaterial used for delivery	Type of plant	Delivered molecules	References
Mesoporous nanoparticles system (MSNS)	Tobacco	DNA and chemicals	Torney et al. (2007)
Gold-plated mesoporous nanoparticles system	Maize	CRE recombinase protein	Martin-Ortigosa et al. (2014)
Single-walled carbon nanotubes (SWCNTs)	Tobacco protoplasts	DNA	Liu et al. (2009)
Gold nanorods (NRs)	Tobacco protoplasts	DNA	Silva et al. (2010)
Gold-functionalized silica nanoparticles (Au-MSN)	Onion epidermis tissue	DNA	Martin-Ortigosa et al. (2012a)
Gold–silica nanoparticle system (Au-MSN) Gold nanorods (NRs)	Onion epidermis tissue	DNA, proteins	Martin-Ortigosa et al. (2012b)
Nanorods with gold cores and silver shells (AuNR-Ag)	Tobacco callus	Growth regulator 2,4-D	Nima et al. (2014)
Polymer nanoparticles (CPNs)	Tobacco protoplasts	siRNA	Silva et al. (2010)

References

Ding Y, Jiang Z, Saha K, Kim CS, Kim ST, Landis RF, Rotello VM (2014) Gold nanoparticles for nucleic acid delivery. Mol Ther 22:1075–1083

Khodakovskaya MV, Biris AS (2009) Method of using carbon nanotubes to affect seed germination and plant growth. WO 2011059507 A1—patent application

Khodakovskaya MV, de Silva K, Biris AS, Dervishi E, Villagarcia H (2012) Carbon nanotubes induce growth enhancement of tobacco cells. ACS Nano 6:2128–2135

Khodakovskaya MV, de Silva K, Nedosekin DA, Dervishi E, Biris AS, Shashkov EV, Galanzha EI, Zharov VP (2011) Complex genetic, photothermal, and photoacoustic analysis of nanoparticle-plant interactions. Proc Natl Acad Sci USA 108:1028–1033

Křenek P, Šamajová O, Luptovčiak I, Doskočilová A, Komis G, Šamaj J (2015) Transient plant transformation mediated by *Agrobacterium tumefaciens*: principles, methods and applications. Biotechnol Adv 33(6 Pt 2):1024–1042

Kumari S, Swetha M, Mayor S (2010) Endocytosis unplugged: multiple ways to enter the cell. Cell Res 20:256–275

Lahiani MH, Dervishi E, Chen J, Nima Z, Gaume A, Biris AS, Khodakovskaya MV (2013) Impact of carbon nanotube exposure to seeds of valuable crops. ACS Appl Mater Interf 5:7965–7973

Lahiani MH, Chen J, Irin F, Puretzky AA, Green MJ, Khodakovskaya MV (2015) Interaction of carbon nanohorns with plants: uptake and biological effects. Carbon 81:607–619

Liu Q, Chen B, Wang Q, Shi X, Xiao Z, Lin J, Fang X (2009) Carbon nanotubes as molecular transporters for walled plant cells. Nano Lett 9:1007–1010

Martin-Ortigosa S, Valenstein JS, Lin VS, Trewyn BG, Wang K (2012a) Gold functionalized mesoporous silica nanoparticle mediated protein and DNA codelivery to plant cells via the biolistic method. Adv Funct Mater 22:3576–3582

Martin-Ortigosa S, Valenstein JS, Sun W, Moeller L, Fang N, Trewyn BG, Lin VS, Wang K (2012b) Parameters affecting the efficient delivery of mesoporous silica nanoparticle materials and gold nanorods into plant tissues by the biolistic method. small 8:413–422

Martin-Ortigosa S, Peterson DJ, Valenstein JS, Lin VS, Trewyn BG, Lyznik LA, Wang K (2014) Mesoporous silica nanoparticle-mediated intracellular cre protein delivery for maize genome editing via loxP site excision. Plant Physiol 164:537–547

Murugan K, Choonara YE, Kumar P, Bijukumar D, du Toit LC, Pillay V (2015) Parameters and characteristics governing cellular internalization and trans-barrier trafficking of nanostructures. Int J Nanomed 10:2191

Nima ZA, Lahiani MH, Watanabe F, Xu Y, Khodakovskaya MV, Biris AS (2014) Plasmonically active nanorods for delivery of bio-active agents and high-sensitivity SERS detection in planta. RSC Advances 4:64985–64993

Ochatt S (2013) Plant cell electrophysiology: applications in growth enhancement, somatic hybridisation and gene transfer. Biotechnol Adv 31:1237–1246

Shi X, von Dem Bussche A, Hurt RH, Kane AB, Gao H (2011) Cell entry of one-dimensional nanomaterials occurs by tip recognition and rotation. Nat Nanotechnol 6:714–719

Silva AT, Nguyen A, Ye C, Verchot J, Moon JH (2010) Conjugated polymer nanoparticles for effective siRNA delivery to tobacco BY-2 protoplasts. BMC Plant Biol 10:291–2229-10-291

Singh S (2013) Nanomaterials as non-viral siRNA delivery agents for cancer therapy. BioImpacts: BI 3:53

Taylor NJ, Fauquet CM (2002) Microparticle bombardment as a tool in plant science and agricultural biotechnology. DNA Cell Biol 21:963–977

Torney F, Trewyn BG, Lin V, Wang K (2007) Mesoporous silica nanoparticles deliver DNA and chemicals into plants. Nat Nanotechnol 2:295–300

Chapter 11
Agri-nanotechniques for Plant Availability of Nutrients

Pabitra Kumar Mani and Sudeshna Mondal

Abstract Nanotechnology has opened up a number of scopes for novel applications in the field of agricultural industries, because of several unique physicochemical properties of nanoparticles (NPs), i.e., high surface area, high reactivity, tunable pore size, and particle morphology. Nanoparticles may be treated as "magic bullets," containing nanopesticides, nanofertilizers, etc., which will trigger specific cellular organelles in the plant to release their contents. So far, little information is available on the behavior of nanofertilzers in soil system, as well as utilization of nanoparticles for smart delivery of fertilizers. Still NPs have already shown promise for their potential utility in crop production in the form of nanofertilizers, nanopesticide, nanoherbicides around the world. The present chapter highlights the key role of nanoparticles in soil systems, their characterization, behavior, mobility, and effective means for the smart delivery of fertilizers that has a strong bearing on the growth and yield of plants. Nano-based slow-release or controlled-release (CR) fertilizers have the potential to increase the efficiency of nutrient uptake. In this chapter, utilization of nanoparticles for delivery of fertilizers in an agricultural production system for the sustainable environment has been described.

Keywords Nanoparticles · Nanofertilzers · Smart delivery · Nutrients

---

P.K. Mani (✉)
Department of Agricultural Chemistry and Soil Science, Bidhan Chandra Agricultural University, Mohanpur 741252, India
e-mail: pabitramani@gmail.com

S. Mondal
AICRP on Soil Testing and Crop Response, Bidhan Chandra Agricultural University, Kalyani 741235, India
e-mail: mondal.sudeshna2010@gmail.com

C. Kole et al. (eds.), *Plant Nanotechnology*,
DOI 10.1007/978-3-319-42154-4_11

11.1 Introduction

Nanoparticles (NPs) may be defined as particulate matter with at least one dimension less than 100 nm (Christian et al. 2008; Auffan et al. 2009) or it may be treated as a colloidal particulate system (Nakache et al. 1999). Materials that are nanoscale in three dimensions are nanoparticles, for example, precipitates, dendrimers, fullerenes, colloids, and tiny particles of semiconductor materials (quantum dots) (Sekhon 2014).

Soils contain many kinds of inorganic particles with at least one dimension in the nanoscale or colloidal range (<100 nm). Apparently, only a small proportion of NPs in soil occur as discrete entities. Organic colloids in soils, for example, are largely associated with their inorganic counterparts or form coatings over mineral surfaces (Oades 1989; Chorover et al. 2007). For this reason, individual NPs are difficult to separate and collect from the bulk soil, and extraction yields are generally low (Banfield and Zhang 2001). Nanoparticles in soil are very reactive toward external solute molecules due to their large surface area as well as the presence of surface defects and dislocations (Hochella et al. 2008). Surface properties of NPs can deviate markedly from their bulk counterparts (Theng and Yuan 2008). As expected with the sharp increase in solubility of mineral, particle size decreases below $\sim$10 nm (Banfield and Zhang 2001; Hochella 2002). Soils are the repository of nanosize particles (Karthikeyan 2014). This nanosize particles have a tendency to aggregate and associate with organic colloids (including dissolved organic matter, polysaccharides, humic materials, and peptidoglycan) and it is suggested that NPs, with their large surface-to-volume ratio, could be highly effective in carbon sequestration (Khedr et al. 2006). NPs represent the most important adsorbents in soil and they may control transport of nutrients and pollutants, regulate organic matter fixation, or catalyze precipitation of new mineral phases. The nanoscale constituents in situ in intact soil structures are of fundamental interest in the future. Many functional NPs have been synthesized as candidates for environmental applications (Garrido-Ramírez et al. 2010). Synthetic nanomaterials could be expensive and would be difficult to obtain due to their narrow size distribution. So, intensive effort is being needed toward naturally occurring nanoclays in soils/sediments ecosystem and used to utilize nanoclays for increased input use efficiency in crop production (Plank et al. 2009; Bendall et al. 2010; Calabi-Floody et al. 2011; Karthikeyan 2014).

From the practical point of view, most of the fertilizers become partially available to plants caused by several inhibiting factors like leaching, photolytic degradation, hydrolysis, and decomposition. So, the minimization of nutrient losses in fertilization and increase in crop yield could be made possible through the exploitation of new applications with the utilization of nanomaterials present in soil (Siddiqui et al. 2015). Nutrients derived from nanofertilzers or nanoencapsulation might have properties that are effective to crops demand due to release of the nutrients on demand and controlled release of chemical fertilizers that regulate plant growth and enhance target activity (De Rosa et al. 2010; Nair et al. 2010).

Both major nutrients such as nitrogen, phosphorous, and potassium compounds and micronutrients may effectively be delivered by controlled-release (CR) fertilizer system. Apart from the reduction of costs by increasing the efficiency of fertilizer use, CR also reduces the negative impact caused by fertilizer application (Gabriels et al. 2001; Gumbo et al. 2008; Anderson 2009; Davidson et al. 2013).

Utilization of nanoscale carriers may be successfully employed for the efficient delivery of fertilizers, pesticides, herbicides, plant growth regulators, etc. (Prasad et al. 2012). The three important mechanisms responsible for efficient delivery, better storage, and controlled release are (i) encapsulation and entrapment, (ii) polymers and dendrimers, and (iii) surface ionic and weak bond attachments (Sawant et al. 2006; Johnston 2010). These mechanisms enhance stability against degradation in the environment by reducing the amount of application and runoff losses (Ditta et al. 2015). These carriers should be designed in such a way that they can anchor the plant roots to the surrounding rhizosphere environment. An understanding of molecular and conformational mechanisms between the nanoscale delivery and targeted structures and the soil fraction is required to reveal the anchoring mechanism (NAAS 2013). These may elucidate slower uptake of active ingredients, thereby reducing the consumption of inputs and also reducing the waste material production.

In smart delivery system, nanoscale devices are used to predict nutrient deficiencies and detect diseases or any other maladies prior to showing of any visual symptoms. “Smart delivery systems” could possibly be advocated agriculture due to their unique modus operandi like self-regulated, spatially targeted, and controlled delivery, in a preprogrammed manner. These multifunctional characteristics may lead to avoiding biological barriers to successful targeting (Subramanian and Tarafdar 2011; Roco 2003). Smart delivery systems can monitor the effects of delivery of nutrients or bioactive molecules or any pesticide molecules (Boehm et al. 2003). Nano-encapsulated fertilizers should be designed in such a way that they possess all necessary properties (effective concentration, stability, and solubility), time-controlled release in response to certain stimuli, with the safe and easy mode of delivery and thus avoiding repeated applications.

11.2 Nanoparticles in Soils

A large number of NPs are present in the soil environment, and understanding the behavior of NPs is very important to a wide variety of soil processes pertaining to plant nutrition and soil reclamation. As a result of chemical weathering of silicates, oxides, and other minerals, a large number of varied NPs are produced in soil. The products are amorphous silica, hydrous aluminosilicates (allophane), clays (halloysite), and oxides (magnetite and hematite) (Nowack and Bucheli 2007). To make a thorough understanding of these NPs, their precise function and effects must be well defined (Adhikari 2014). Microorganisms can also produce NPs through the generation of metabolic energy by pathways involving inorganic ions that participate in redox reactions. Oxidation of Fe (II) results in the formation of iron oxide NPs. Similarly, a variety of different manganese oxide NPs and ZnS NPs are formed by bacteria-mediated oxidation–reduction processes in the soil.

Many iron oxides (hematite, magnetite), oxyhydroxides (goethite, akaganeite, lepidocrocite, and feroxyhyte), hydrous oxides (ferrihydrite, hydrohematite, maghemite), and aluminosilicate clay minerals, as well as amorphous substances, occur in soils as NPs (Maurice and Hochella 2008). Ferrihydrite may take as an example of the nanomineral, whose small particle size (<10 nm individual particles) and high surface area (50–200 m^2/g, (Waychunas et al. 2005) make it an important component of reactive surface area and the subject of many experiments in the field of biogeochemistry. Other iron oxides and hydroxides such as hematite, goethite, maghemite, lepidocrocite, and magnetite can also occur as mineral NPs. Minerals such as goethite may have nanoscale domains or parallel subunits, which may alter their sorption and dissolution behavior. The domain boundaries are dislocations that tend to undergo preferential dissolution in acidic attack (Cornell and Schwertmann 2003). Swelling clays such as montmorillonite may have angstrom (up to 80 Å) to nanoscale (1 nm) interlayers. The structure and properties of water and of absorbed molecules (Haack et al. 2008) in these interlayers can be different from what is seen in bulk solution, and even from species adsorbed to external surfaces of minerals. The structure of nanomaterial ferrihydrite has been the subject of much study and controversy. Michel et al. (2007a) showed 2- to 6-nm ferrihydrite nanoparticles to have a similar composition [$Fe_{10}O_{14}(OH)_2$] and structure. However, the NPs show increasing disorder and distortion of some sites as particle size decreases from 6 to 2 nm (Fig. 11.1).

Organic NPs in soil are mostly associated with their inorganic counterparts or occur as coatings on mineral surfaces (Oades 1989; Chorover et al. 2007). Humic and fulvic acids have molecular weights ranging from the hundreds to thousands of Daltons or more, and many individual humic and fulvic molecules are within the nanometer size range. Aggregates of natural organic matter on mineral surfaces have been observed in the nanometer to micron size range (Namjesnik-Dejanovic and Maurice 2001). The partitioning of hydrophobic organic pollutants into humic substances may be controlled at least in part by nanoscale hydrophobic domains (Pignatello 1998).

With allowance for the ecological functions of soil and its role in substance turnover, the following migration pathways of NP entry into this object (Fig. 11.2) are marked out: (1) the translocation pathway that characterizes the transition of a substance from land plants and NP waste; (2) the water migration pathway that characterizes the capability of a substance to migrate from groundwater, sewage, and water sources; and (3) the air migration pathway that characterizes the transition of a substance from the atmospheric air (Venitsianov et al. 2003).

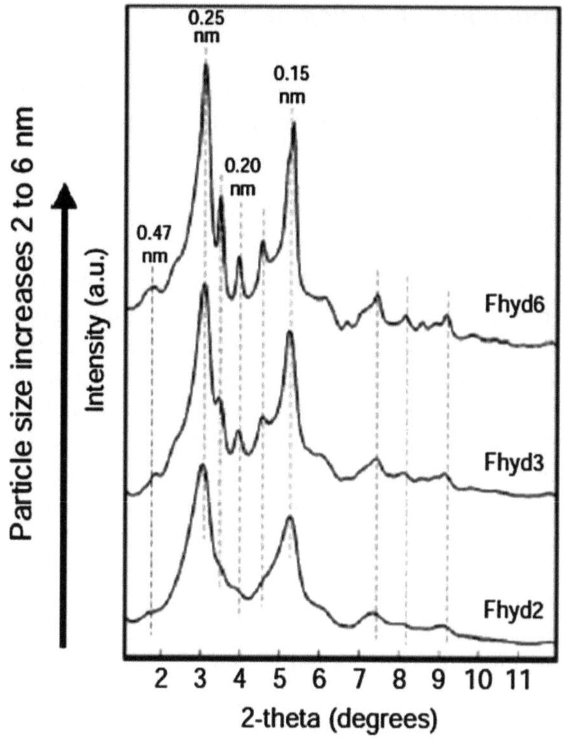

Fig. 11.1 XRD patterns for ferrihydrite particles of size 2, 4, and 6 nm. [Adapted from Michel et al. (2007b)]

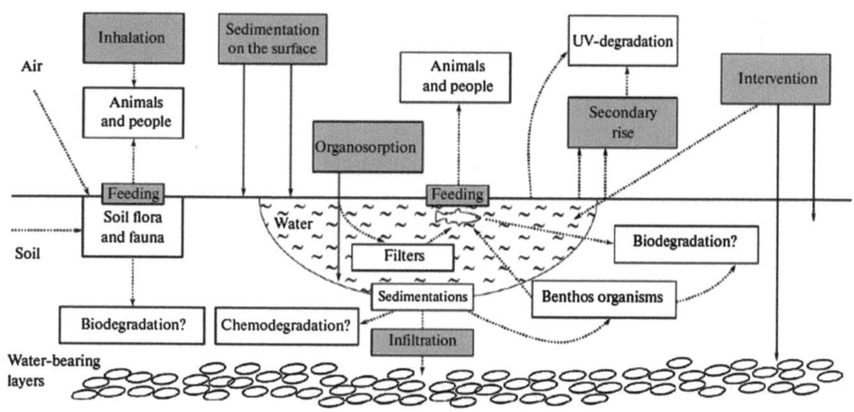

Fig. 11.2 Possible entry and migration pathways of nanoparticles proved experimentally (*solid line*) and supposed pathways (*dotted line*). [Adapted from Krichevskii (2010)]

11.2.1 Classification of Nanoparticles

Nanoparticles may be grouped into two broad groups based on origin: (i) natural nanoparticles and (ii) anthropogenic nanoparticles (ANP).

11.2.1.1 Natural Nanoparticles

Natural NPs have existed since before life began on Earth. All life forms that have developed have been exposed to at least some types of NPs during their evolution, and they have thus developed mechanisms to tolerate their presence (Buffle 2006). The most common natural NPs are soil organic colloids, which include dissolved organic matter, polysaccharides, humic materials, and peptidoglycan, and inorganic particles including clay and ocean salt, which are constituted of silicate clay minerals, iron oxides or aluminum oxides/hydroxides (Nowack and Bucheli 2007). Black carbon is produced during incomplete combustion of fossil fuels, biofuel, and biomass and could exist in soils/sediments in NP size range (Maurice and Hochella 2008).

Biogenically derived NPs are mostly organic colloids, such as polysaccharides, proteins, organisms of nanosize (e.g., viruses), and humic/fulvic acids. These particles are actively involved in biological processes. Researchers even detected carbon-based nanotubes (CNTs) and fullerene in ice core formed 10,000 years ago (Murr et al. 2004). The formation of fullerene and CNTs in the environment was attributed to the metamorphosis of PAHs at 300–500 °C in the presence of sulfur (Heymann et al. 2003) or natural combustion.

11.2.1.2 Anthropogenic Nanoparticles (ANP)

Anthropogenic NPs are of two categories, namely:

(i) Engineered NPs (ENPs) or manufactured NPs (MNPs): particles that are produced by human beings because of possessing specific nanotechnological properties, and
(ii) Accidental NPs: those which are accidentally produced and discharged into the environment during manufacturing and various anthropogenic activities like cooking, electricity generation, industrial boiling, diesel burning, and welding (Murr et al. 2004; Pan and Xing 2010).

11.2.2 Classes of Engineered Nanomaterials

The range of modern nanotechnological products is large and can be represented by different classes according to their physicochemical properties, structure, and form

(Gladkova and Terekhova 2013). Classes of nanomaterials (NMs) are presented as follows: (1) carbon-containing NMs, like fullerenes, single-layer nanotubes, graphenes, and nanodiamonds, (Peralta-Videa et al. 2011); (2) metal-containing NMs, like metal oxides, such as titanium oxide (TiO_2), zinc oxide (ZnO), and cerium dioxide(CeO_2) (Keller et al. 2013);(3) quantum dots (QDs), like semiconductor nanocrystals that have reactive cores, and these cores, can be made of metals or semiconductors (cadmium selenide (CdSe), cadmium telluride (CdTe), indium phosphide (InP), or zinc selenide (ZnSe)). The reactive core is surrounded by a shell (silicon dioxide (SiO_2) or a ZnS monolayer) which protects cores from oxidation (Klaine et al. 2008); (4) inert metals, like nanoscale zerovalent iron, nanosilver, and nanogold; and (5) dendrimers or multifunctional polymers, whose size, structure, and molecular weight can be managed (Klaine et al. 2008).

11.3 Nanoparticle Characterization

Nanoparticle possesses unique properties due to their extremely small size. The most critical characteristics of NPs are their very high surface-to-volume ratio, and owing to large fractions of surface atoms, atomic (and electronic) structure of the surface region is different from the "bulk" material below the surface (Darlington et al. 2009). The high surface area-to-volume ratio of the NPs results in high reactivity, which leads to particle aggregation and settling unless the particles are protected by a capping agent that provides colloidal stability through electrostatic or steric repulsion (Ju-Nam and Lead 2008). Generally, the Gouy–Chapman model of the electrochemical "diffuse double layer" is used in the context of nanoenvironment of charged NPs (Pfeiffer et al. 2014). The thickness of this diffuse double layer is very much dependent on the ionic strength of the solution and characterized by the Debye parameter (κ) or its reciprocal value, i.e., the thickness of the double layer (κ^{-1}) (Pfeiffer et al. 2014). According to the classical Derjaguin, Landau, Verwey, and Overbeek (DLVO) theory (Derjaguin and Sidorenkov 1941; Verwey and Overbeek 1948), colloidal particles are surrounded by a diffuse electrostatic double layer (EDL) and the balance between the van der Waals attraction forces and the electrostatic repulsion forces determines the colloidal stability (Zha et al. 2002; Cosgrove 2005). The magnitude of the electrical charge within and the thickness of the EDL are directly related to solution properties such as pH, ionic strength, and electrolyte ion valence (Jiang et al. 2009).

A change in the concentration of charged species in a solution was observed during the production of charged nanoparticles (Fig. 11.3). For negatively charged NP surfaces (e.g., COO^-stabilized NPs), there is a local depletion of negatively charged analytes (i.e., such as OH^- or Cl^-), as well as a local accumulation of positively charged analytes (i.e., such as Na^+ or H^+, Fig. 11.3a, Zhang et al. 2010). The local ion concentration close to the NP surface will be different from the bulk concentration (Fig. 11.3b), i.e., NPs influence their environment (Pfeiffer et al. 2014).

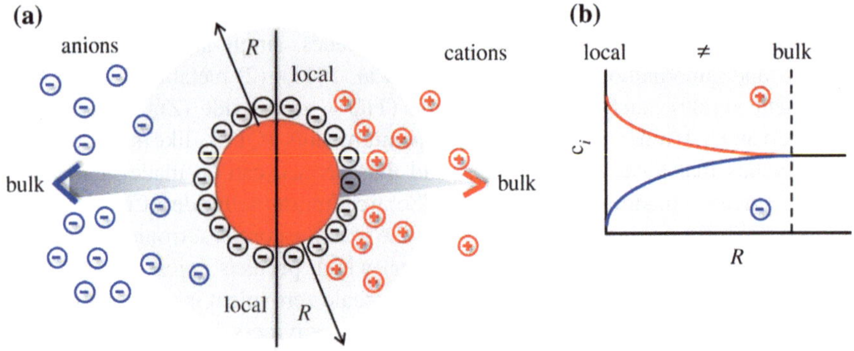

Fig. 11.3 The concentration of ions at the surface of an NP is different from the respective bulk concentration. **a** In the case of negatively charged NPs, there will be a local depletion and accumulation of anions and cations, respectively; **b** the graph demonstrates schematically that the concentration c_i of ions of different species depends on the distance R from the NP surface. [Adapted from Pfeiffer et al. (2014)]

The tendency of particles to sorb to the soil is also influenced by the ionic strength of the soil solution. According to the DLVO theory, higher ionic strength decreases the repulsive forces between particles and the soil surfaces leading to increased aggregation and sorption (Tourinho et al. 2012). For example, with an addition of sodium chlorate, electrostatic repulsion between particles (Fe_2O_3, TiO_2, CuO, and ZnO NPs) is reduced, resulting in aggregation and, thus, mobility in a glass bead column (Ben-Moshe et al. 2010). Fang et al. (2009) also showed that high ionic strength reduced the transport of negatively charged TiO_2 NPs through soil columns (Fang et al. 2009; Tourinho et al. 2012).

11.4 Behavior of Metal-based NPs in Soils

The applicability of colloidal inorganic nanoparticles (INPs) relies on their unique size, morphology, and structure, which determines both their property and reactivity. The confinement of electrons, phonons, and photons at a nanometer scale produces a new generation of materials, which creates different physicochemical properties in comparison with bulk materials (Bastús et al. 2012). The noted examples of such deviation are the size and shape dependence absorption and scattering properties in noble metal inorganic nanoparticles like silver or gold, the enhanced luminescence properties in semiconductor nanocrystals known as quantum dots (CdSe or PbS), and the superparamagnetic moment in magnetic nanoparticles like iron oxide or cobalt (Burda et al. 2005). Moreover, as the size of the material is reduced and percentage of atoms at the surface becomes significant, the entire particles become very reactive (Bastús et al. 2008). Thus, it is expected that in very small crystals, both the thermodynamics and kinetics of reactions could be changed due to the reduction of size leading to the large surface-to-volume ratio which is further accompanied by a lowering of phase transition temperatures (Goldstein et al. 1992). Interestingly, this high fraction of unsaturated atoms at INP's surface may lead to some instabilities which may cause further degradation and corrosion processes. Despite their uncontrollable nature, these secondary processes are extremely useful for catalysis applications (e.g., Pt NPs). These secondary processes may allow reactions at the active sites of their surfaces (Li and Somorjai 2010).

The reactivity of INPs is governed by the size, composition, and structure of its core. More precisely, reactivity of INP depends on the combination of an inorganic core and organic/inorganic shell. This inorganic core plays a vital role in determining physicochemical properties of INP. The shell, which may be either organic or inorganic, dictates the interfacial interactions by the chemical nature of the organic layer (Fig. 11.4). So, the surface coating acts as a major role in controlling and tuning the reactivity of the particle, as well as determining its solubility and selectivity against the desired target (Bastús et al. 2012).

An in-depth study of INP's reactivity is an important prerequisite for boosting the applicability of INPs. In biological systems, this condition is especially important where the interactions and interferences of INPs with cells and tissues determine the potential toxicity of engineered materials as well as its biodistribution, degradation, and biocompatibility (Casals et al. 2008). In biological fluids, the response of nanostructured materials is extremely complex and diverse and depends on various parameters (Bastús et al. 2012). Metal-based NPs possess various physical properties, particularly, size and shape, and chemical properties like acidic/basic properties of the surface and the aqueous solubility of the metal. These special characteristics will determine the extent to which metal-based NPs

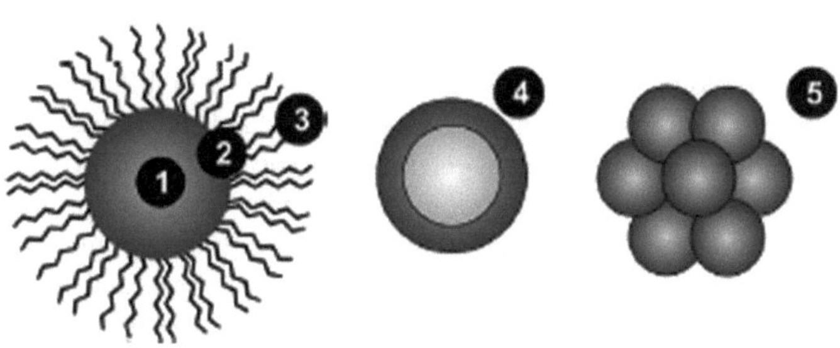

Fig. 11.4 Reactivity of colloidal inorganic nanoparticles are (*1*) effects of inorganic core's size and shape, (*2*) surface chemistry and ligand exchange reaction, (*3*) reactivity of the coating molecule, (*4*) interactions with ions (chemical transformation and degradation) present in the colloidal solution, and (*5*) cooperative effects with other NPs in solution. [Adapted from Bastús et al. (2012)]

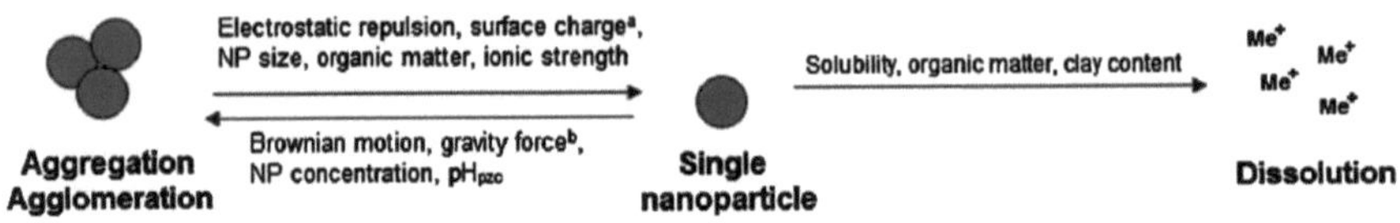

Fig. 11.5 Factors affecting the processes of aggregation/agglomeration and dissolution of single nanoparticles in the environment. [a] Considering similar surface charge. [b] Acting only on larger particles. [Adapted from Tourinho et al. (2012)]

undergo transformations. These transformations will eventually control their fate, performance, and ecotoxicity in the environment involving the processes like aggregation/agglomeration, sorption to surfaces, and dissolution of the ionic metal. Moreover, metal-based NPs are frequently manufactured with surface coatings, which may modify their intrinsic behavior (Tourinho et al. 2012).

Aggregation is the association of primary particles by strong bonding, whereas agglomeration is association by weak bonding caused by van der Waals forces (Jiang et al. 2009). However, in the environment, physical forces (e.g., Brownian motion, gravity, and fluid motion) and unique characteristics of NPs (e.g., surface properties, particle size) will affect agglomeration and aggregation of NP (Farre et al. 2009; Tourinho et al. 2012). The electrostatic surface charge of the particles, affecting agglomeration/aggregation rates and particle stability, is influenced by the chemistry of the medium (Fig. 11.5). In the absence of a surface coating, metal-based NPs have charged surfaces arising due to the presence of hydroxyl (−OH) groups. These charged surfaces then can take up and release protons and consequently take up dissolved chemical species such as metal ions and ligands (Tourinho et al. 2012). The ionic species that detaches from the metal-based NPs and migrates through the electrical double layer into the solution due to dissolution (Borm et al. 2006) may themselves be toxic. Thus, for better understanding the effect of potential NP on organisms over time, one should consider dissolution kinetics and their relative proportion of toxicities produced by both the particulate and dissolved forms (Tourinho et al. 2012).

Study of the biological activity of NPs in connection with different conditions of their entry into the soil is very significant to understand the effect of NPs on the environment (Fig. 11.6). ENPs can get into the soil by three ways: (A) by the dissolved state, (B) by direct absorption of solid NPs, and (C) by direct entry of NPs. NP entry as dissolved state can entail the following: bioaccumulation by plant roots, accumulation and subsequent toxicity by invertebrates, and microbial toxicity. The direct absorption of solid particles can cause the toxicity of the plant roots, invertebrates, and microbes. The direct entry of NPs into the soil can result in their absorption/aggregation or in migration along the profile.

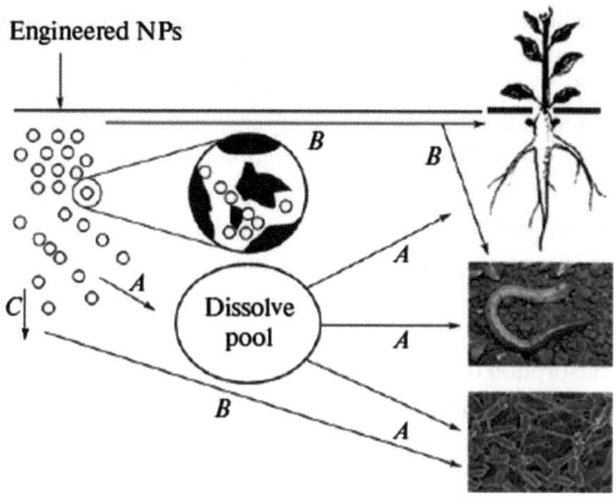

Fig. 11.6 Bioavailability of nanoparticles under different conditions of their entry into the soil environment: *A* is the dissolved pool of NPs; *B* is the direct absorption of solid NPs; *C* is the direct entry of NPs; migration along the profile. [Adapted from Gladkova and Terekhova (2013)]

11.5 Role of Organic Matter on Delivery of Nanoparticles

The soil organic matter provides the energy and nutrients for soil microbes, which ensure high yields of healthy crops due to their enzymatic action. Thus, it is mandatory to conserve it for efficient physical, chemical, and biological soil functioning (Six et al. 2002).

Retention of NPs in soils was studied by Cornelis et al. (2012), wherein the dominant properties that determine the retention of AgNP in natural soil were correlated to negatively charged AgNP, which was found to be adsorbed preferentially at positively charged surface sites of clay-sized minerals. The agricultural soil containing high organic carbon likely contributed to an organic surface coating and resulted in NP mobility through the soil and will come in contact with the soil microbes (Collins et al. 2012).

Dissolved or particulate organic matter presents in soils can sorb to NP surfaces. This sorption may influence particle properties in various ways (Tourinho et al. 2012). Humic substances being negatively charged at environmental pHs, their sorption will make the overall particle–humic conglomerate becomes negatively charged (Ghosh et al. 2008). As a result, it may increase particle stability in solution, reducing aggregation and settling (Fang et al. 2009; Ben-Moshe et al. 2010). Particle affinity for cell membranes may be decreased due to alteration of the surface charge and thus reduces their bioavailability and uptake (Unrine et al. 2008). Steric hindrance effects may likely to contribute to the enhanced stability of humic acid-coated NPs. However, Ghosh et al. (2008) observed a reverse trend and showed that at low pH, humic acid caused aggregation of Al_2O_3 NPs. The plausible reason may be that the charge of the humic acid appeared sufficiently low to allow its aggregation because of hydrophobic interactions. Thus, humic acid-coated particles became susceptible to aggregation at low pH. Based on transmission electron microscopy (TEM) images, Kool et al. (2011) observed that at pH 5.5, ZnO NPs bound to solid-phase organic matter. Their findings suggested that under suitable conditions, organic matter may destabilize particle dispersions. The net effect of sorption of humic substances on particle stability and bioavailability appears to be a function of complex factors, particularly the soil pH and the intrinsic hydrophobicity of the humic substances (Tourinho et al. 2012).

Current studies regarding the effect of NOM (natural organic matter) on MNP (manufactured nanoparticle)–organic chemical interactions mostly depend on the coating of NOM on MNP. As shown in Fig. 11.7 at low NOM concentrations, the adsorption of organic chemicals on MNPs may be increased due to the dispersion of both CNTs and inorganic NPs or NOM coating (for inorganic NPs). However, with further increasing NOM concentration, the adsorption of NOM on MNPs reaches saturation resulting in significant interaction between aqueous NOM and organic chemicals which may finally decrease the adsorption. A number of studies have reported a decreased adsorption of organic chemicals on CNTs (Ji et al. 2009; Wang et al. 2009) or increased adsorption on oxide NPs (Iorio et al. 2008) with the addition of NOM. No further study was conducted to investigate the possible nonmonotonic influences of DOM (dissolved organic matter) on MNP adsorption characteristics (Kumar et al. 2012).

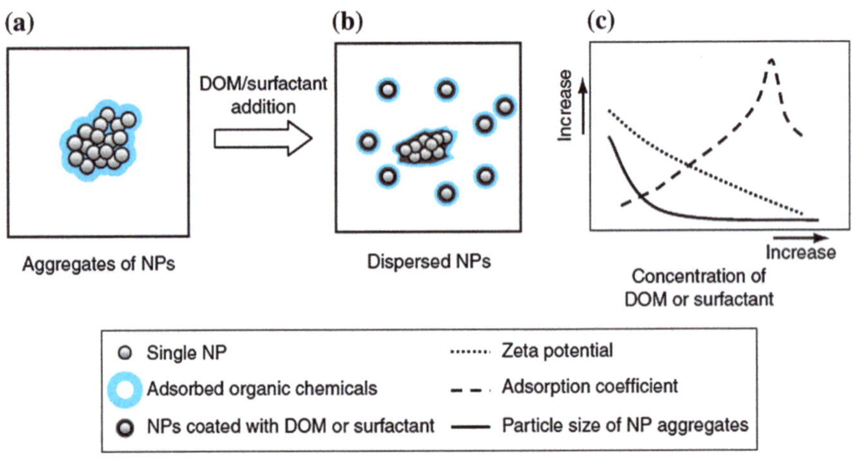

Fig. 11.7 The role of dissolved organic matter (DOM) and surfactants in suspending manufactured nanoparticles (MNPs) and their adsorption for organic chemicals. Surface-coated DOM/surfactant may decrease the zeta potential of MNPs (*C*) facilitating the dispersion of MNP aggregates (*B*). [Adapted from Pan and Xing (2010)]

11.6 Nanoparticle Mobility in Soils

The soil is an important sink for nanoparticles after release into the environment and a possible source of NPs in groundwater. Understanding the transport behavior of NPs in natural soil systems is essential to revealing their potential impact on the food chain and groundwater. Transport of NPs in porous media and soils has attracted increased research attention (Lin et al. 2010).

Mobility or transport of NPs through soils depends on the interaction between particles and solid surfaces (Darlington et al. 2009) which may be influenced by both environmental and physicochemical characteristics of the particles (Tourinho et al. 2012) (Fig. 11.8). The soil is a porous medium consisting both macropores and micropores. The micropores (very small pores that occur within the soil structure) consist of a network of humic materials and soil particles (Kretzschmar and Schafer 2005). Nanoparticles are small enough to fit into these micropore environments, which leads to clogging the micropores which may affect nanoparticle mobility. On the other hand, aggregates of nanoparticles may be too large to fit in the micropores, which may allow them to remain in the macropores (Kumar et al. 2012).

Due to high surface areas, nanoparticles have a strong potential to adsorb to soil and sediment particles (Oberdorster et al. 2005). Nanoparticle sorption to nonmobile particles may inhibit mobility whereas sorption to mobile colloids may enhance mobility. The shape of the nanoparticle, as well as the "collector" surfaces, may also have a considerable effect on adhesion of nanoparticles to surfaces (Nowack and Bucheli 2007; Kumar et al. 2012). Generally, those factors that influence the stability of NPs also tend to influence transport properties. The sedimentation and

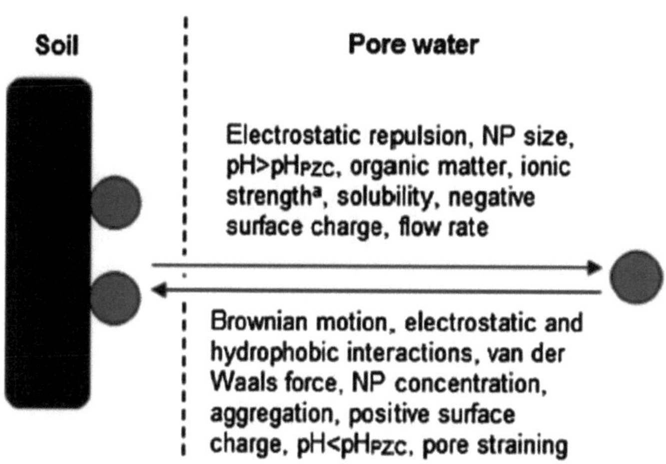

Fig. 11.8 Factors affecting the partitioning of nanoparticles between soil and pore water. [a]Not valid for high values of ionic strength; *dashed line* is the diffusion layer. pH_{PZC} = point of zero charge. [Adapted from Tourinho et al. (2012)]

diffusion of NPs are also influenced by gravity and Brownian motion, respectively (Dunphy Guzman et al. 2006; Tourinho et al. 2012). The transport of NPs in porous media like the soil is controlled by Brownian motion (Lecoanet et al. 2004). Gravitational forces predominate when the particles agglomerate and aggregate, and then the particles are more likely to interact with the surfaces of soil particles (Dunphy Guzman et al. 2006). The interaction between metal-based NPs and soil surfaces is also dependent on the surface charge of both the NPs and the soil. These differences of charges between NPs and soil influence aggregation and electrostatic attraction/repulsion among particles and between particles and soil (Tourinho et al. 2012).

Transport of particles in the porous media depends largely on the rate of their capture or filtration by the stationary grain surfaces. Particle filtration theory can be used to help understand the transport behavior of ENMs in soil (Fig. 11.9). There are three basic mechanisms for the capture of colloidal particles in porous media: gravitational sedimentation, interception, and Brownian diffusion (Schrick et al. 2004). Sedimentation (Fig. 11.9a) and interception (Fig. 11.9b) mechanisms mainly regulate the transport of larger particles and aggregates of pristine particles. For nanoscale particles, the dominant mechanism is diffusion (Fig. 11.9c), as the

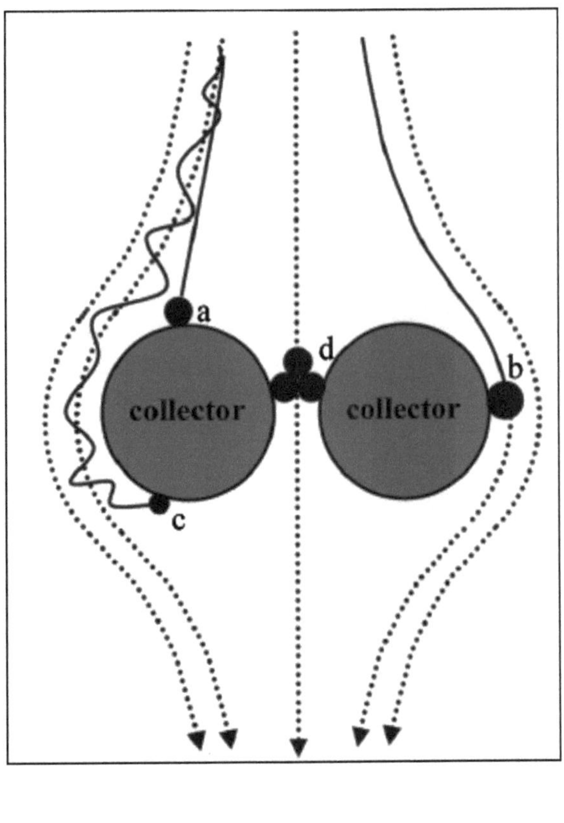

Fig. 11.9 Schematic diagrams for transport of engineered nanoparticles (ENPs) in porous media through filtration mechanisms. The *black spheres, dotted lines,* and *thick lines* stand for NPs, the *fluid streamlines,* and the particle path (trajectory), respectively. The NPs can attach to the collector through (*a*) gravitational sedimentation, (*b*) interception, and (*c*) Brownian diffusion. (*d*) Large aggregates may be physically retained by small pores. [Modified from Lin et al. (2010)]

high diffusivity of nanoparticles leads to a higher incidence of collisions with the surface of soil grains.

In addition, nanoparticles may aggregate to become microscale particles, which can be retained by the media due to physical screening if the size of the aggregate is larger than the pore through which fluid is flowing (Fig. 11.9d). The capture or filtration of particles by the collector surfaces is determined in large part by the chemical–colloidal interaction between particles and surface, which, in turn, is regulated by solution chemistry (e.g., pH, ionic strength, and coexisting organic matters) and chemical characteristics of particles and surfaces. Physical parameters, such as particle size, fluid velocity, grain size, and water temperature, can also play important roles in the filtration of ENMs. The main factors influencing the breakthrough capability of ENMs are illustrated in Fig. 11.9. Physicochemical properties, such as size, morphology, and surface properties, of ENMs, can substantially influence their migration in porous media. The transport of nanoparticles is more complex than microscale particles because of their tendency to aggregate, which strongly depends on particle size in the nanoscale regime. It has been pointed out that 0.1–1 mm is the optimal size range for colloidal particles to migrate through the soil as predicted by the Tufenkji–Elimelech filtration model (Zhan et al. 2008). The transport of nanoscale colloids is theoretically predicted to decrease with increasing particle size (Elimelech et al. 1995). However, nanoparticles are prone to form large aggregates, which can greatly inhibit their transport, and aggregate staining is an important component of particle capture (Hydutsky et al. 2007; Darlington et al. 2009). A number of researchers have revealed that dispersed ENMs can transport through porous media while the pristine ENMs are prone to form aggregates and are filtered by the grains (Schrick et al. 2004; He et al. 2007; Kanel et al. 2007; Yang et al. 2007; Saleh et al. 2008). However, to date, there are different understandings and opinions about the effect of the size of nanoparticles on their transport in porous media.

Surface charge is another important property that can dominate the migration of ENMs in porous media (Darlington et al. 2009). Environmental soil particles are normally negatively charged. Thus, positively charged ENMs will be readily electrostatically attracted to the soil surface. Engineered nanomaterials with higher negative charges are believed more mobile in soil matrix because of the stronger electrostatic repulsion between the nanoparticles and soil particles and between nanoparticles themselves as well. Therefore, various methods have been applied to modify ENM surface properties to control (enhance or restrict) the transport of ENMs in porous media, among which surface functionalization with hydrophilic functional groups (e.g.,−OH and −COOH) and surface physical modification using polymers or surfactants are two commonly adopted methods.

Solution pH and ionic strength can have a dramatic influence on the rate of capture of ENMs by the porous media and, as a result, on the extent of transport of ENMs. Solution pH controls the solubility of metal-based nanoparticles and their surface charges and thus the electrostatic interactions between nanoparticles and between nanoparticles and porous media (Doshi et al. 2008). For electrostatically stabilized ENMs, dissolved counterions in solution will screen the long-range electrostatic interactions and thus decrease the stability and transport of ENMs in the porous media.

There are numerous nano-enhanced products in different countries and nano-based tools and methods with immediate application to addressing the issues pertaining to low use efficiency of inputs like water, fertilizers. These include nano-enhanced products such as nanofertilizers with nano-based smart delivery systems (use of halloysite) to provide nutrients at desired site, time, and rate to optimize productivity. The nanofertilizers can be delivered timely to a rhizospheric target or by foliar spray for higher use efficiency. The emerging literature on nanotechnology has started showing the importance of nanoparticles in increasing bioavailability of nutrient elements and transport of pollutants in soils. Several synthesized NPs like amphiphilic polyurethane, nanoscale zerovalent iron (nZVI), and nano-sized zeolites are widely used for reclamation of heavy metal and poly-aromatic hydrocarbon-contaminated soils.

11.7 Smart Delivery of Fertilizers

Nanofertilizers are nutrient carriers of nanodimensions ranging from 30 to 40 nm (10^{-9}m or one-billionth of a meter) and capable of holding bountiful of nutrient ions due to their high surface area and release it slowly and steadily that commensurate with crop demand (Subramanian et al. 2015). Nanofertilizers have a profound influence on crop production (Priester et al. 2012). There are many issues with the use of traditional chemical fertilizers; however, low use efficiency is the prominent one, which not only increases the cost of production but also causes environmental pollution (Wilson et al. 2008). Nanomaterials with the large surface area could solve this problem due to their nanosize. These could be utilized as nanocoatings, e.g., sulfur nanocoating (≤100-nm layer), ensuring their controlled release, surface protection, and ultimately boosting up their use efficiency (Brady and Weil 1996; Santoso et al. 1995). Nanofertilizers may be looked upon as an alternative and more efficient to the ordinary fertilizers. These nanoformulations of nitrogenous fertilizer minimize nitrogen losses by leaching, emissions, and soil microbial immobilization (Liu et al. 2006). Subramanian et al. (2008) reported that nanofertilizers and nanocomposites can be used to control the release of nutrients from the fertilizer granules so as to improve the nutrient use efficiency while preventing the nutrient ions either get fixed or lost to the environment. Nanofertilizers can be delivered in a timely manner to a rhizospheric target due to their high use efficiency. These are slow-release and super sorbent nitrogenous and phosphatic fertilizers. Some new-generation fertilizers have applications to crop production on long-duration human missions to space exploration (Lal 2008).

Moreover, controlled-release fertilizers may also improve the soil by decreasing toxic effects associated with over-application of traditional chemical fertilizers (Suman et al. 2010). Nanoscale carriers could be utilized for the efficient delivery of fertilizers and pesticides. The common mechanisms to regulate the release of nutrients by the carriers include encapsulation and entrapment and polymers and dendrimers. (Sawant et al. 2006). These mechanisms help to improve their stability against degradation in the environment and ultimately reduce the amount to be applied which reduces chemicals runoff and alleviates environmental problems.

11.7.1 Nanofertilizers

The use efficiency of nutrients of traditional fertilizers is abysmally low. It has been reported that around 40–70 % of nitrogen, 80–90 % of phosphorus, and 50–90 % of potassium content of applied fertilizers are lost in the environment and could not reach the plant which causes significant economic losses (Trenkel 2010; Saigusa 2000; Solanki et al. 2015). The nanofertilizers release the nutrients in a controlled manner in response to the reaction to different signals such as heat, moisture, and other abiotic stress. We know that crops secrete carbonaceous compounds into the rhizosphere under nutrient stress that can consider as environmental signals for incorporation into novel nanofertilizers (Sultan et al. 2009). Novel nanofertilizer application has an edge over traditional methods of fertilizer application by releasing nutrients in a controlled manner, preventing eutrophication and pollution of water resources (Sekhon 2014; Naderi and Abedi 2012). Nano-TiO_2 has a tremendous potential to use as a fertilizer additive due to its photoactivity (Mastronardi et al. 2015). TiO_2 nanoparticles treated on maize had a considerable effect on growth, whereas the effect of TiO_2 bulk treatment was negligible. Titania nanoparticle application caused an elevated level of light absorption and photoenergy transmission (Moaveni and Kheiri 2011). An experiment on soybean revealed that a compound of SiO_2 and TiO_2 nanoparticles increased the activity of nitrate reductase and intensified plant absorption capacity, making its use of water and fertilizer more efficient (Lu et al. 2002). Iranian researchers (INIC 2009) have produced the nano-organic iron-chelated fertilizer that is environmentally sustainable. Nanofertilizers have unique features like ultra high absorption, increase in photosynthesis caused by expansion in surface area of the leaves, etc. (INIC 2009).

The use of nanofertilizer not only causes increased use efficiency of the elements but also reduces the toxicity generated due to over-application in the soil as well as reduces the split application of fertilizers (Naderi and Danesh-Shahraki 2013). The positive effect of the application of zinc oxide nanoparticles on tomato plants opens an avenue for its potential use as a future nanofertilizer. An experiment with foliar application of different concentrations of ZnO NPs (0–100 mg L^{-1}) solution in tomato plants grown in pots revealed that 20 mg mL^{-1} zinc oxide nanoparticle solution recorded maximum growth and biomass production (Panwar et al. 2012; De Rosa et al. 2013).

To improve the nutrient use efficiency, nano-based slow-release or controlled-release fertilizers have the tremendous potential. In arid soil, it was observed that the engineered nanoparticles may be successfully utilized for mitigating the acute problem of moisture retention. Apart from moisture retention, nano-based slow-release fertilizers may augment crop production by mobilizing nutrients in the rhizosphere (Raliya et al. 2013). Nitrogen fertilizer fortified with nanoporous zeolite could be used as an alternative strategy to improve the nitrogen use efficiency in crop production systems (Manikandan and Subramanian 2014). It was observed an improved root development and shoot establishment in rice seedlings grown in carbon nanomaterial-enriched medium compared with the control seedlings by Nair et al. (2012).

In an interesting study, Kottegoda et al. (2011) reported a sustained release of nitrogen into the soil using urea-modified hydroxyapatite nanoparticle which were encapsulated under pressure into cavities of the soft wood of *Gliricidia sepium*. In this study, the release of nitrogen from nanofertilizer followed a sequence of a two-step process: an initial burst and a subsequent slow release up to 60 days. Such release process has an edge over conventional commercial fertilizer, which released heavily at the beginning followed by low and nonuniform quantities until around 30 days.

Subramanian and Rahale (2009) have monitored the nutrient release pattern of nanofertilizer formulations carrying fertilizer nitrogen. The data have shown the nanoclay-based fertilizer formulations (zeolite and montmorillonite with a dimension of 30–40 nm) are capable of releasing the nutrients for a longer period of time (>1000 h) than conventional fertilizers (<500 h). Subramanian and Tarafdar (2009) suggested that clay particles are adsorptive sites carrying a reservoir of nutrient ions. A major portion of nutrient fixation occurs in the broken edges of the clay particles. Zerovalence nanoparticles can adsorb on to the clay lattice, thereby preventing fixation of nutrient ions. Further, nanoparticles prevent the freely mobile nutrient ions from getting precipitated. These two processes assist in promoting the labile pool of nutrients that can be readily utilized by plants. Fertilizer particles can be coated with nanomembranes that facilitate in slow and steady release of nutrients. This process helps to reduce loss of nutrients while improving fertilizer use efficiency of crops. The naturally occurring clay minerals and zeolites have been reduced to the size of nanodimensions using top-down approach and nutrient at desirable proportion have been fortified in the clays after surface modification (Bansiwal et al. 2006). The nanofertilizer formulations before and after loading with nutrients have been characterized using high-resolution microscopes and spectroscopy (Liu et al. 2006). Nanozeolites are capable of retaining nutrients due to its extensive surface area and release slowly and steadily for an extended period of 1000–1200 h while conventional fertilizers could release for about 300–400 h (Subramanian and Rahale 2009). The literature strongly suggests that nanotechnological applications improve the NUE and productivity of crops without associated environmental hazard.

11.7.2 Slow/Controlled-release Nanofertilizers

The most successful use of nanoparticles is as slow-release fertilizers. Because of the high surface tension, they will hold material more strongly from the plant than conventional surfaces. Moreover, nanocoatings can also provide surface protection for larger particles. Direct application of large amounts of fertilizer, in the form of ammonium salts, urea, nitrate, or phosphate compounds, may produce extremely high local concentrations which are harmful. Much of the fertilizer may be dissolved in runoff water and cause adverse effects such as pollution and will not be available to the plants of interest (Wilson et al. 2008).

11.7.3 Mechanism of Controlled Release

Generally, the controlled-release mechanism depends on numerous factors like the nature of the coating material, the type of fertilizer, agronomic conditions, and soil pH. Liu et al. (2008) and Shaviv (2005) postulated a possible release mechanism for coated fertilizers known as the multistage diffusion model. This model described that when applying the coated fertilizer, irrigation water penetrates the solid fertilizer core leading to partial nutrient dissolution (Fig. 11.10). Subsequently, as osmotic pressure builds within the containment, the granule consequently swells and causes two processes. In the first, when osmotic pressure surpasses threshold membrane resistance, the coating bursts and the entire core released spontaneously. This is known as the "failure mechanism" or "catastrophic release" as coined by Goertz (1993). Secondly, if the membrane withstands the developing pressure, core fertilizer is thought to be released slowly via diffusion for which the driving force may be a concentration or pressure gradient, or combination thereof called the "diffusion mechanism". The failure mechanism is generally observed in frail coatings (e.g., sulfur or modified sulfur), while polymer coatings (e.g., polyolefin) are expected to exhibit the diffusion release mechanism.

A pictographic representation of both mechanisms is given in Fig. 11.10. The controlled release of nutrients also depends on ambient temperature and moisture with the release rate increasing at higher temperatures with greater moisture content (Rose 2002). The mechanism of fertilizer release from coating material is driven by water which leads to nutrient movement from the fertilizer–polymer interface to the polymer–soil interface. The sequential process for the release mechanism is diffusion/swelling of the fertilizer followed by degradation of the polymer coating, and subsequent fracture or dissolution. Similar release mechanism was reported by several researchers (Guo et al. 2005; Liang et al. 2007; Liu et al. 2007; Wu and Liu 2008).

Some of the advantages related to transformed formulation of conventional fertilizers using nanotechnology are presented in Table 11.1 (Cui et al. 2010).

11.7.3.1 Zeolites and Nanoporous Zeolites

Use of nano zeolites is an effective alternative strategy for increasing fertilizer use efficiency (Chinnamuthu and Boopathi 2009). Zeolites are known to exist as

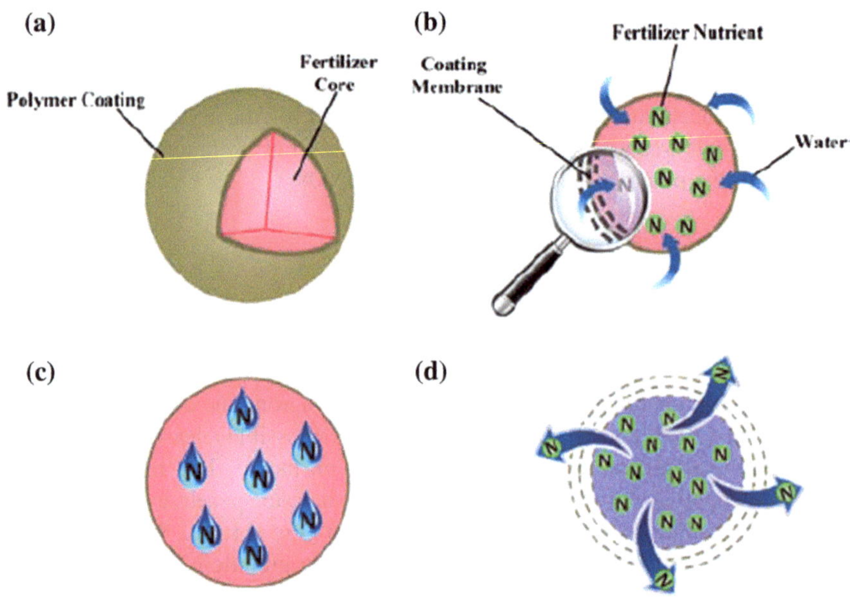

Fig. 11.10 Diffusion mechanism of controlled release; **a** fertilizer core with polymer coating; **b** water penetrates into the coating and core granule; **c** fertilizer dissolution and osmoticpressure development; **d** controlled release of nutrient through swollen coating membrane. [Adapted from Azeem et al. (2014)

Table 11.1 Some of advantages related to transformed formulation of conventional fertilizers using nanotechnology

Desirable properties	Examples of nanofertilizer-enabled technologies
Controlled-release formulation	So-called smart fertilizers might become reality through transformed formulation of conventional products using nanotechnology. The nanostructured formulation might permit fertilizer to intelligently control the release speed of nutrients to match the uptake pattern of crop
Solubility and dispersion for mineral micronutrients	Nano-sized formulation of mineral micronutrients may improve solubility and dispersion of insoluble nutrients in soil, reduce soil absorption and fixation, and increase the bioavailability
Nutrient uptake efficiency	Nanostructured formulation might increase fertilizer efficiency and uptake ratio of the soil nutrients in crop production and save fertilizer resource
Controlled-release modes	Both release rate and release pattern of nutrients for water-soluble fertilizers might be precisely controlled through encapsulation in envelope forms of semipermeable membranes coated by resin polymer: waxes and sulfur
Effective duration of nutrient release	Nanostructured formulation can extend effective duration of nutrient supply of fertilizers into soil
Loss rate of fertilizer nutrients	Nanostructured formulation can reduce loss rate of fertilizer nutrients into soil by leaching and/or leaking

naturally occurring minerals with a honeycomb-like layered crystal structure. Zeolites possess a network of interconnected pores and voids which could be filled up with major nutrients like nitrogen, potassium with slowly dissolving nutrients like phosphorus, calcium, and almost all trace elements. Zeolites may act as a reservoir of nutrients that are slowly released on demand. The main function of zeolites as fertilizer carrier is to capture, to store, and to release nitrogen slowly (Leggo 2000). By using this technique, we can reduce groundwater contamination due to uncontrollable release of soluble nitrogen from conventional fertilizers. Nitrogen release pattern from zeolites is much slower than the ionic form released from conventional fertilizers (Naderi and Danesh-Shahraki 2013).

Zeolites are hydrated aluminosilicates of alkali and alkaline earth cations, with a three-dimensional lattice framework, having an inner network of voids and channels (Preetha et al. 2014). Zeolites have a high cation exchange capacity and have often been used as inexpensive cation exchangers for various applications (Breck 1974). Zeolites possess a series of ordered crystals with complex pores and may be grouped into three categories: microporous, mesoporous, and nanoporous. Ions could be occluded within zeolites by two mechanisms, viz. ion exchange and chemisorption. Creation of weak and strong chemical bond between ions and zeolite plays a major pivotal role in nutrient release and subsequent availability in the soil solution. Natural zeolite (Z) and nanoporous zeolite (NZ) consist approximately 30–40 % of channels of 0.4–1 nm pore diameter (Manikandan and Subramanian 2014). This pore space (35–40 %) could be successfully utilized for loading N and potassium (K) (Bansiwal et al. 2006). Zeolites possess unique preferential ion exchange property by which zeolite may reduce the contamination of natural resources (Wei et al. 2011). Three special properties like mesoporosity, nanoporosity, and high surface area of zeolites could be explored for loading nutrients and it could be used as a slow-release and novel nanofertilizer.

N adsorption on such modified zeolites is mainly influenced by the electrical field generated by the charge created due to the exchange of cations in the pores and also by hydrogen bonding with the surface. In the present context, nanoporous zeolites have drawn much attention due to their unique surface properties, shorter diffusion path lengths, and higher cation exchange capacity (CEC) (Ramesh et al. 2010).

One of the attempts to increase NUE is slow-release or controlled-release fertilizers which release N slowly in available form or to develop materials which control the release of N in available form slowly. The most important slow-release fertilizer is the coating of conventional N fertilizer with sulfur, neem, lac, or clay (Sartain 2010). The idea was due to coating with these materials, urea comes into the soil solutions through diffusion process very slowly, and in this way, they supply nitrogen to the plants at a controlled rate or slow rate but for a longer period. But these attempts to increase the NUE were yielded with little success due to the mismatch between the nutrient release and crop demand. Kundu et al. (2013) developed a protocol to coat urea with oleoresin and the coated urea contained 3.82–4.36 % pine oleoresin and 44.07–44.31 % N. Acidic and antimicrobial properties of pine oleoresin decreased urease activity considerably in the pine oleoresin-treated soils compared to control. Such decreased urease activity was noted irrespective of the soil type. The data fitted to first-order kinetic equation revealed that time required for hydrolysis of 90 % of the applied urea significantly increased from 88.56 to 328.94 h in the presence of pine oleoresin. The behavior of pine oleoresin thus expected that it may be utilized as a potential urease inhibitor by appropriate coating of urea (Kundu et al. 2013).

Nanofertilizer may regulate the release of nutrients and deliver the correct quantity of nutrients required by the crops in suitable proportion and promote productivity while ensuring environmental safety (De Rosa et al. 2010). Millán et al. (2008) stated that NH_4^+ occupying the internal channels of zeolite may be released slowly and freely, thereby allowing the progressive absorption by the crop which is reflected in higher dry matter production of the crop. Dwairi (1998) suggested that zeolite impregnated with urea can be used as slow-release fertilizer carrying the slow and steady release of N from nanozeolite. Perrin et al. (1998) demonstrated that amending sandy soil with ammonium-loaded zeolite can reduce N leaching while sustaining growth of sweet corn and increasing N use efficiency compared to ammonium sulfate. The same result was also demonstrated by Hernandez et al. (1994) that the combination of zeolite and slow-release N fertilizers would increase the N efficiency. Rahale (2010) reported that nanofertilizer increased the NUE up to 45 % over control. She also reported that the release of nitrate from nanozeolite continued even after 1176 h, with concentrations ranging from 110 to 114 mmol L^{-1}. The results clearly demonstrated slow and steady release of N from nanozeolite for more than 45 days while conventional fertilizer does it for only 8 days.

Sheta et al. (2003) suggested that natural zeolites, particularly clintopillolite, have a high potential for Zn and Fe sorption with a high capacity for slow-release fertilizers. Broos et al. (2007) reported that the slow release of Zn is attributed to the sparingly solubility of minerals and sequestration effect of exchanger, thereby releasing trace nutrients to zeolite exchange sites where they are more readily available for uptake by plants. Eeberl (2008) reported that zeolite in soil can aid in the release of some trace nutrients and in their uptake in plants. The release of cationic micronutrients has enhanced by the presence of zeolite in neutral soil. The concentration of Cu and Mn in sudangrass (in mg/kg) was significantly related to the zeolite/P-rock in experimental systems that used two different NH_4 saturated zeolites, two different soils, and two different forms of P-rock. The concentrations of trace elements were also increased by 19 % for iron (Fe^{2+}) and 10 % for manganese (Mn^{2+}).

Synthetic zeolites as an amendment on soil properties and their effect on crop growth were studied by Al-Busaidi et al. (2008). They evaluated in a plot and pot culture experiment with saline water as a source of irrigation water taking barley as a test crop. Results revealed that zeolite could effectively ameliorate salinity stress and improve nutrient balance in a sandy soil. Lin and Xing (2008) reported that zinc oxide nanoparticles were shown to enter the root tissue of ryegrass and improved the germination. Methods of visualization of carbon-coated nanotubes in plant cells using pumpkin plants as the model were reported by Melendi et al. (2008).

Table 11.2 Average particle size distribution (PSD); zeta potential of natural zeolite; nanoporous zeolite and fertilizer formulations at 1:1 ratio (Manikandan and Subramanian 2014)

Source	PSD (nm)	Zeta potential (mV)
Natural zeolite (Z)	794	−45.9
Zeourea (ZU) (1:1)	1120	−49.4
Nanoporous zeolite (NZ)	87	−50.4
Nanozeourea (NZU)(1:1)	366	−64.3

Manikandan and Subramanian (2014) conducted an experiment to study the nitrogen (N) use efficiency of urea using microporous natural zeolite (Z) and nanoporous zeolite (NZ) as substrate. The data revealed that the N release from the urea blended with NZ in a ratio of 1:1 was up to 48 days while the conventional zeolite–urea mix in the same ratio was up to 34 days. The results on particle size distribution (PSD) with zeta potential of adsorbents (Z, NZ) and fertilizer formulations of 1:1 ratio (ZU, NZU) are given (Table 11.2). The mean data on PSD indicated that particle sizes of micro-zeolite (1120) and NZ (87) had increased when N is impregnated into the adsorbent of 1:1 ratio 1120 and 366 nm, respectively. Similar results were obtained by Rahale (2010). This facilitates adsorption processes due to an extensive surface area for adsorption of cationic nutrients and anionic nutrients on surface modification of the zeolite with a cationic surfactant. The value of zeta potentials of the particles in the range of −30 to −65 indicates the stability of the system.

Use of surfactant-modified zeolite with hexadecyl trimethyl ammonium as fertilizer carrier to control nitrate release was demonstrated by Li (2003). Their findings revealed that surfactant-modified zeolite is a suitable sorbent for nitrate, since a slow release of nitrate is achievable. These nature and properties suggest that there is a huge potentiality of surfactant-modified zeolite to be used as a fertilizer carrier to regulate the release of nitrate and other anions (Ramesh et al. 2010).

Regarding improving phosphorus use efficiency, ammonium-charged zeolites have shown their capacity to raise the solubilization of phosphate minerals and thus go to improved phosphorus uptake and yield of crop plants (Ramesh and Reddy 2011). An experiment was conducted by Allen et al. (1993) to evaluate the solubility as well as cation exchange properties in mixtures of rock phosphate and clinoptilolite with NH_4^+ and K^+ saturation. Their research revealed that mixtures of clinoptilolite and rock phosphate media may supply plant nutrition adequately in a slow-release manner. The releases of nutrients were mainly governed by dissolution and ion exchange reaction mechanisms (Allen et al. 1993).

Information regarding slow-release K materials is lacking compared to information available on N materials. According to Sparks and Jardine (1984), a relatively high r^2 values indicated that the Freundlich equation model may successfully describe the kinetics of K adsorption by nanoclays. Pino et al. (1995) conducted an experiment on the slow release of K from K-zeolite and found that it follows first-order kinetics with several stages corresponding to different K fractions.

CEC of the nanoclays becomes high due to substitution of silica (Si^{4+}) by aluminum (Al^{3+}), thereby raising a negative charge of the mineral lattice. This negative charge may be balanced by cations such as ammonium, sodium, calcium, and potassium, which are exchangeable with other cations (Curkovic et al. 1997). Weatherley and Miladinovic (2004) conducted an experiment to study the effect of the presence of individual cations such as Ca^{2+}, K^+, and Mg^{2+} upon NH_4^+ uptake on clinoptilolite and found that the presence of these competing cations could affect K^+ desorption on zeolite.

The unique properties of slower ion exchangeability of the zeolites with selected nutrient cations may serve as an excellent plant growth medium for supplying vital nutrients to plant roots (Subramanian and Rahale 2012). Depending on plant root demand, the nutrients are provided in a slow but regulated-release manner through the process of dissolution and ion exchange reactions. Subsequently, the zeolite will be "recharged" by the addition of more dissolved nutrients, and their selectivity of ion exchange on zeolite was determined in an order of $K^+ > NH_4^+ > Na^+ > Ca^{2+} > Mg^{2+}$ (Guo et al. 2008).

11.7.3.2 Nano Rock Phosphate

The polymer coating of mono-ammonium phosphate (MAP) improved plant recovery of fertilizer phosphorus (P) and provided a modest barley grain yield advantage relative to uncoated MAP (Malhi et al. 2002; Subramanian and Rahale 2012). Coating of P fertilizer could restrict the contact of applied P with the soil, by reducing its precipitation and/or adsorption on soil colloids. Thus, coating leads to increase its availability to plant roots. The development of thin polymer coatings has improved the opportunity to coat fertilizer granules and increased the predictability of nutrients availability from the controlled-release product. To simulate the P release behavior in field conditions by using the constant flow percolation reactor, a comparative study on the release dynamics of P from fertilizer-loaded unmodified zeolite, surfactant-modified zeolite (SMZ), and pure fertilizers was tested by Bansiwal et al. (2006). The results revealed that the P release from fertilizer-loaded SMZ was continued even after 1080 h, whereas P release from KH_2PO_4 was exhausted within 264 h. The results indicated that SMZ is a good sorbent for PO_4^{3-} and has a great potential as the fertilizer carrier for slow release of P (Bansiwal et al. 2006). Eeberl (2008) reported that phosphate (H_2PO_4) could be released to plants from the mixture of phosphate rock (P-rock) mostly as apatite and zeolite having an exchangeable ion such as ammonium. The probable reaction in soil solution is as follows:

$$(\text{P-rock}) + (\text{NH}_4\text{–zeolite}) = (\text{Ca zeolite}) + \left(\text{NH}_4^+\right) + \left(\text{H}_2\text{PO}_4^-\right)$$

The zeolite takes Ca^{2+} from the phosphate rock, thereby releasing both phosphate and ammonium ions. The release of phosphate in a controlled way is governed by a specific chemical reaction in soil. These slowly released phosphate and ammonium ions are either taken up by plants or fixed by soil, thereby shifting chemical equilibria toward forward direction. As a result, more phosphate and ammonium ions are released in the environment to maintain the equilibrium. The rate of phosphate release may be controlled by varying the ratio of P-rock to zeolite. Phosphorus is also released from the rock due to lowering of soil pH as proton is generated during microbial conversion of ammonium ions to nitrate (Subramanian and Rahale 2012).

11.7.3.3 Hydroxyapatite and Synthetic Nano-sized Hydroxyapatite (NHA)

Liu and Lal (2014) studied on synthesis and characterization of the apatite nanoparticles in carboxymethyl cellulose (CMC) solution and assessed the fertilizing effect of the particles on soybean (*Glycine max*) yield through a greenhouse study. The data revealed that the growth rate and seed yield were increased by 32.6 and 20.4 %, respectively, due to application of synthetic apatite nanoparticles. The percentage increase of growth rate and seed yield was compared with respect to treatment having a conventional P fertilizer application [$Ca(H_2PO_4)_2$] (Liu and Lal 2014). Figure 11.11 presents a TEM micrograph of the apatite nanoparticles. TEM images indicated that the nanoparticles were spherical in shape with an average diameter of 15.8 ± 7.4 nm (Liu and Lal 2014). This research showed a future direction that nHA could possibly be used as a P fertilizer in augmenting crop yields and biomass production.

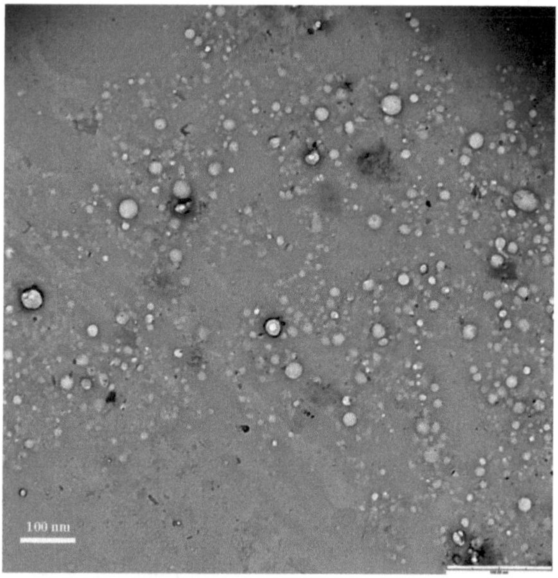

Fig. 11.11 A TEM image of nano-sized hydroxyapatite (nHA). [Adapted from Liu and Lal (2014)]

11.7.3.4 Nanoencapsulation and Slow-release Fertilizer

Encapsulation of fertilizers within a nanoparticle is emerging as a new field in slow-release fertilizer. There are three possible ways for encapsulation of fertilizers within a nanoparticle. The most common technique is an encapsulation of nutrient inside nanomaterials such as nanotubes or nanoporous material. However, the two other methods are coated with a thin protective polymer film, and delivery as emulsions or nanoparticles is also got prominence (Sekhon 2014; Rai et al. 2012; Teodorescu et al. 2009). The encapsulated fertilizers are protected by the nanoparticles for long-term residence in inoculated soils, allowing for their regulated release into the soil (Saigusa 2000; Corradini et al. 2010). The idea of encapsulated fertilizer is relatively novel, so it has a tremendous potential for commercial formulations (De Rosa et al. 2010). The proper coupling of nanodevices with nanofertilizers will synchronize the release of fertilizer N and fertilizer P with their uptake by crops. With the adoption of this technique, it will prevent undesirable nutrient losses to soil, water, and air through direct internalization by crops, and avoid the interaction of nutrients with the biotic and abiotic components of the environment (De Rosa et al. 2010; Naderi and Danesh-Shahraki 2013). It was observed that crops secrete carbonaceous compounds into rhizosphere to enable microbial mineralization of N and/or P from soil organic matter and of P from soil inorganic colloids under stress condition. These root exudates may be treated as environmental signals which could possibly be exploited to prepare nanobiosensors. Nanobiosensors and nano-based smart delivery systems could help the farmers with better fertilization management (Al-Amin Sadek and Jayasuriya 2007; Sultan et al. 2009; Naderi and Danesh-Shahraki 2013). Nanobiosensors may expand the new horizon for basic research and may provide tools for real bio-analytical applications in future.

Sulfur-coated fertilizers are the most attractive of the slow-release fertilizers because the sulfur content may be beneficial, especially for soils low in sulfur (Dana et al. 1994; Lefroy et al. 1994; Singh and Chaudhari 1995; Santoso et al.1995). Figure 11.12 shows a slow-release sulfur-coated fertilizer analyzed by Wilson et al.

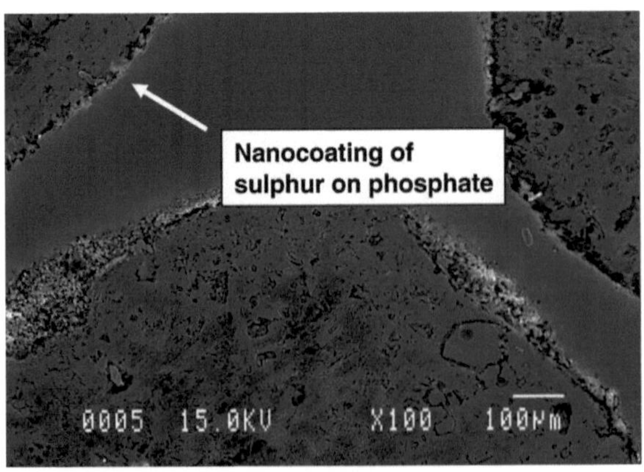

Fig. 11.12 Scanning electron microscopy of sulfur-coated phosphate slow-release fertilizer. [Adapted from Wilson et al. (2008)]

(2008). The coating can be from a few nanometers to 100 nm, so there is room for improvement of uniformity. A successful fertilizer coated assembly can be made which consists of a well-adhered layer of elemental sulfur, and strategically released urea and phosphate to meet the soil and crop demands.

Chitosan (CS) is a polymer of particular interest in the area of nanoencapsulation because it is biodegradable, bioabsorbable, and nontoxic (Coma et al. 2002; No et al. 2007). Chitosan (CS) is a natural polysaccharide produced by deacetylation of chitin, the second most abundant polysaccharide on Earth. Chitosan nanoparticles, as a cationic polymer, may interact with negatively charged molecules and polymers, showing a favorable interaction (Corradini et al. 2010). Wu and Liu (2008) also applied this system in controlled-release fertilizer studies.

A laboratory experiment was conducted to explore the potential of chitosan nanoparticles as controlled release for NPK fertilizers (Corradini et al. 2010). Chitosan nanoparticles were prepared by polymerizing methacrylic acid for the incorporation of NPK fertilizers. The stability and interaction of chitosan nanoparticle suspensions containing NPK were evaluated by FTIR spectroscopy, and the results clearly showed the existence of electrostatic interactions between chitosan nanoparticles and the nutrient elements like N, P, and K (Fig. 11.13) (Corradini et al. 2010). Comparing the FTIR spectrum obtained for CS-PMAA with those obtained for the loaded nanoparticles (urea-loaded chitosan: (CS-PMAA(NH_2CONH_2)), calcium phosphate-loaded chitosan (CS-PMAA($Ca(H_2PO_4)_2 \cdot H_2O$)), potassium chloride-loaded chitosan (CS-PMAA(KCl)), significant differences were observed. The concerned peak of the CS-PMAA nanoparticle appears in the spectra, unveiled that there was no change in the structure of the nanoparticles with the incorporation of N, P, and K fertilizers (Corradini et al. 2010). Other NMs like kaolin and polymeric biocompatible NPs could also be utilized for this purpose (Wilson et al. 2008).

11.7.4 Nanocomposites

Nanocomposites may be defined as dispersed particles having at least one dimension in the nanometer range (Ajayan et al. 2003; Ke and Stroeve 2005; Kumar and Krishnamoorti 2010; Sekhon 2014). The characteristics of nanocomposite materials depend on their morphology and interfacial characteristics. Nanocomposites are nothing but a nanoparticle-reinforced polymer. As a result of reinforcement with small quantities (up to 5 % by weight) of nano-sized particles, the performance of the polymer may be improved. (Tai et al. 2003; Sekhon 2014). Polymer nanocomposites produced by incorporating metal or metal oxide nanoparticles like nanozinc oxide, nanomagnesium oxide may be utilized for their antimicrobial action (Chaudhry et al. 2010).

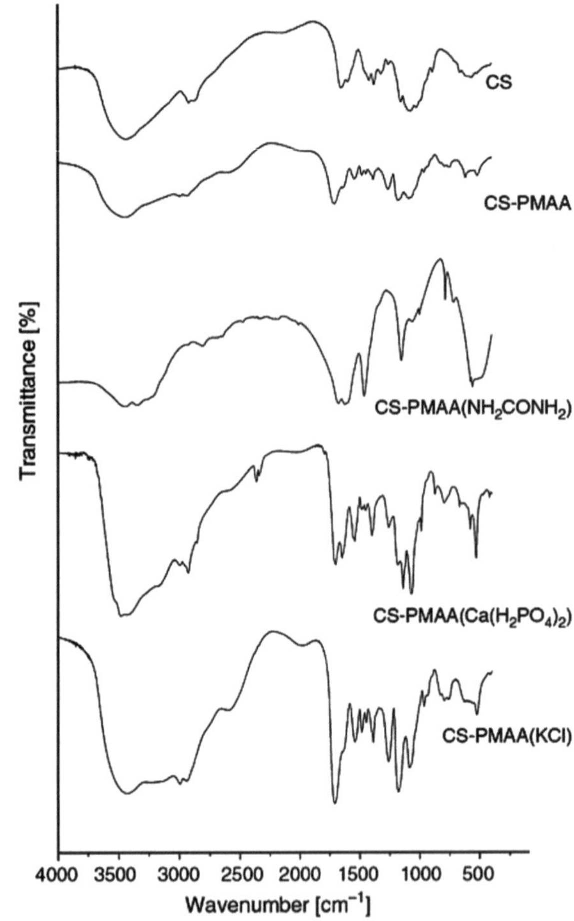

Fig. 11.13 FT-IR transmittance spectra of raw chitosan (CS), chitosan nanopaticles (CS-PMAA), nanoparticles with urea (CS-PMAA(NH_2CONH_2)), nanoparticles with calcium phosphate((CS-PMAA(Ca $(H_2PO_4)_2 \cdot H_2O$)), and nanoparticles with potassium chloride (CS-PMAA(KCl)). [Adapted from Corradini et al. (2010)]

11.7.4.1 Types of Nanocomposites

Three different morphologies of nanocomposites depending on the interphase forces between polymer and clay are thermodynamically accepted (Anadão 2012) (Fig. 11.14). The properties of nanocomposite materials depend both on the properties of their individual parents and also on their morphology and interfacial characteristics.

1. **Intercalated Nanocomposites**: It is the state in which extended polymer chains are intercalated in the interlayer region of the clay. Intercalated nanocomposite consists of good order multiple layer structure with alternating polymeric and inorganic layers at a repetitive distance of a few nanometers between them (Weiss et al. 2006; Mukhopadhyay and De 2014).

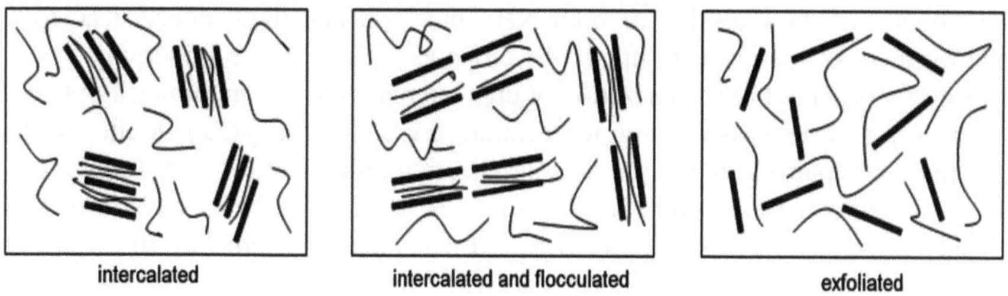

Fig. 11.14 Polymer–clay nanocomposite morphologies. [Adapted fromAnadão (2012)]

2. **Exfoliated Nanocomposites**: The silicate layers are completely separated in a random manner and dispersed in a continuous polymer matrix. The structure and properties of the resulting nanocomposites could be modified by controlling delicate polymer–clay interactions (Oya et al. 2000). It has been reported that exfoliated nanocomposites exhibit the best properties due to their optimal interaction between clay and polymer (Mukhopadhyay and De 2014).
3. **Flocculated Nanocomposites**: Structure is similar to that of the intercalated nanocomposite, except for the formation of floccus due to the interaction between the hydroxyl groups of the silicate (Yeh and Chang 2008; Zaarei et al. 2008; Anadão 2012).

A new class of composites known as clay–polymer nanocomposites draw attention to the users where clay surface treated as a polymerization initiator and monomers could be intercalated between clay mineral platelets. In clay–polymer nanocomposites, the dispersed phase is the silicate mineral composed of particles that have nanometer range (10^{-9} m) dimension. Mineral particles mostly used in these nanocomposites are the smectitic clays, for example, montmorillonite, sepiolite, and hectorite (Alexandre and Dubois 2000; Anadão 2012).

Coating and binding of nanocomposites and sub-nanocomposites are able to regulate the release of nutrients from the fertilizer capsule (Liu et al. 2006). A similar study conducted by Guo (2004) found that application of a nanocomposite composed of major and micronutrients, mannose, and amino acids enhances the nutrient uptake by grain crops. Nanocomposites produced by intercalation of organics with layered double hydroxide (LDH) have drawn considerable attention by researchers. Zinc–aluminum-layered double hydroxide (LDH) nanocomposites showed controlled release of chemical compounds. Such controlled-release behavior was utilized as plant growth regulators and herbicides (Hashim et al. 2007). Studies conducted by De Rosa et al. (2010) revealed that incorporation of fertilizer into cochleate nanotubes (rolled-up lipid bilayer sheets) had improved crop yield (Naderi and Danesh-Shahraki 2013).

A study was conducted by Kim et al. (2011) regarding massive intercalation of urea into montmorillonite (MMT) to find out better urea use efficiency as compared with broadcasted fertilizer. The results indicated that urea intercalates considerable suppression of the emission of both NH_3 and N_2O and these effects lead to an improvement of the nitrogen uptake by crops as well as crop productivity.

An experiment of characterization of urea intercalation into montmorillonite clay by an extrusion process at room temperature reveals that montmorillonite exfoliation into urea matrix can control the solubilization process of urea, leading to delay its release into the environment (Pereira et al. 2012). The results confirmed the effectiveness of this simple process to exfoliate the clay lamellae into the urea matrix by cold extrusion which generated two regions, one comprising the nanocomposite itself (montmorillonite and urea) and the other with urea granules. The release process of urea becomes obstructed by creation of barriers to free diffusion out of the granules (Pereira et al. 2012).

The effects of slow-/controlled-release fertilizers (for regulated, responsive, and timely delivery) cemented and coated by nanomaterials, clay–polyester, humus–polyester, and plastic starch on crops were studied with wheat (Liu et al. 2006; Zhang et al. 2006). It was found that these nanocomposites were safe for wheat seed germination (over 99 % germination), emergence, and growth of seedlings. Leaching experimental results showed that nitrogen release rate of fertilizer coated by plastic starch composites was the lowest and the release rate of coated slow-release fertilizers was lower than that of the cemented slow-release fertilizers.

11.8 Ecotoxicological Impacts of NM in Soil

The ecotoxiclogical impacts of NM and the biokinetics of NPs are dependent on a number of factors including size, shape, surface structure, aggregation, chemical composition, and solubility (Nel et al. 2006). These parameters may influence tissue injury caused due to modification of cellular uptake, protein binding, and translocation from entry point to the target site. (Oberdorster et al. 2005). Nanomaterials may trigger tissue injury either by single or combined mechanisms at the target site. Interactions of nanomaterials with various biological components like cells, body fluids, and proteins play a crucial role in their biological effects. These resultant interaction effects may distribute throughout the body. Proteins may react with nanomaterials as a substrate leading to generation of complex molecules which becomes more mobile and can enter tissue sites. Enzyme malfunction may occur due to structural changes on the nanoparticle surface (Vertegel et al. 2004). This accelerated denaturation or degradation may arise from splitting of intramolecular or intermolecular bonds on the material surface. NM may encounter a number of defenses that can eliminate, or dissolve NPs during their uptake and transport throughout the body (Nel et al. 2006).

Several reports show that inorganic NPs like TiO_2, SiO_2, and ZnO had a toxic effect on bacteria and other organisms. Not much research information is available on interaction of NP with plants. The uptake of many types of NPs in the bacterial cell (prokaryotes) is very much limited as they do not have mechanisms for transport of NPs across the cell wall but in eukaryotes, cellular internalization of

NPs occurs through the process of endocytosis and phagocytosis (Moore 2006). Several workers (Hong et al. 2005; Zheng et al. 2005; Gao et al. 2006; Yang et al. 2006) have shown that nano-sized TiO_2 can have a positive effect on growth of spinach when administrated to the seeds or sprayed onto the leaves whereas nano-TiO_2 was found to show toxic effect in green algae which have a cell wall similar to plants. The most possible interactions of NPs with plant roots are adsorption onto the root surface, incorporation into the cell wall, and uptake into the cell. The NPs could also diffuse into the intercellular space and be absorbed or incorporated into membranes. The interaction of NPs with toxic, organic compound can both amplify and alleviate the toxicity of the compounds. Despite their harmful effects, NPs may also have a positive role in the soil environment.

Arsenic is a well-known groundwater contaminant and exists as both arsenate (As (V)) oxyanions ($H_2AsO_4^{-1}$ and $HAsO_4^{-2}$) at neutral pH under oxidizing conditions (Ferguson and Gavis 1972) and arsenite (As (III)) under mildly reducing conditions. The As (III) species remains protonated as $H_3AsO_3^0$ at pH below 9.2 (Ferguson and Gavis 1972; Manning et al. 2002). Recent investigations have confirmed that INP (iron nanopolymers) along with their corrosion products may be employed for remediation of both As(III) and As(V) (Manning et al. 2002; Kanel et al. 2005, 2006, 2007). There is a tendency for INP to form aggregate during oxidative corrosion to Fe(III) oxide/hydroxide. These corrosion products restrict the effective transport as well as delivery of INP through porous media, which is normally essential for in situ groundwater remediation (O'Hena et al. 2006).

11.9 Conclusion

The long-term deposition of nanomaterials in the form of aggregates and colloids not only threaten the security of soil and water bodies but may prove to be impossible to remediate. The high organic carbon content in the agricultural soil likely contributed to an organic surface coating and resulted in NPs mobility through the porous medium. Soil parameters like salinity, texture, pH, concentration, and nature of organic compounds, and degree of saturation determine nanomaterials bioavailability. In order to get a proper understanding of NPs for delivery of fertilizers to plant, there is a need for a scientific study to evaluate the nanocarriers, nanocomposites, and nano-encapsulated fertilizers on crop plants and their environmental consequences.

Nanocarriers could be designed in such a way that these can anchor the plant roots or the surrounding soil structure and organic matter. This could only be possible through the understanding of molecular and conformational mechanisms between the delivery nanoscale structure and targeted structures and matters in soil.

Several strategies have been focused on technologies to provide fertilizer delivery systems with minimal environmental impacts. The ultimate target is the production of eco-friendly nanofertilizers which will release their nutrient in a regulated manner (either slowly or quickly) in response to different biotic and abiotic signals. The priority of research should include the mechanism of smart delivery of fertilizers in a controlled-release manner as well as to find out the optimum soil condition and other environmental factors which determine the behavior and fate of slow-/controlled-release fertilizers through the novel formulations of nanotechnology.

References

Adhikari T (2014) Implications of nanotechnology in soil science and plant nutrition. In: Proceedings of seventh international conference on smart materials, structures and systems, Bangalore, India, 8–11 July 2014, p 12

Ajayan PM, Schadler LS, Braun PV (eds) (2003) Nanocomposite science and technology. Wiley-VCH Verlag GmbH and Co, KGaA, Weinheim

Al-Amin Sadek MD, Jayasuriya HP (2007) Nanotechnology prospects in agricultural context: an overview. In: Proceedings of the international agricultural engineering conference, Bangkok, 3–6 Dec 2007, p 548

Al-Busaidi A, Yamamoto T, Inoue A, Egrinya Eneji (2008) Effects of zeolite on soil nutrients and growth of barley following irrigation with saline water. In: 3rd international conference on water resources and arid environments and the 1st Arab water forum, Riyadh, Saudi Arabia, 16–19 Nov 2008, p 29

Alexandre M, Dubois P (2000) Polymer-layered silicate nanocomposites: preparation, properties and uses of a new class of materials. Mat Sci Eng Res 28:1

Allen ER, Hossner LR, Ming DW, Henninger DL (1993) Solubility and cation exchange in phosphate rock and saturated clinoptilolite mixtures. Soil Sci Soc Am J 57:1368–1374

Anadão P (2012) Polymer/clay nanocomposites: concepts, researches, applications and trends for the future. In: Ebrahimi F (ed) Nanocomposites—new trends and developments. InTech, Croatia, p 514

Anderson DM (2009) Approaches to monitoring, control and management of harmful algal blooms (HABs). Ocean Coast Manag 52:342–347

Auffan M, Rose J, Bottero JY, Lowry GV, Jolivet JP, Wiesner MR (2009) Towards a definition of inorganic nanoparticles from an environmental, health and safety perspective. Nat Nanotechnol 4:634–641

Azeem B, Kushaari K, Man ZB, Basit A, Thanh TH (2014) Review on materials & methods to produce controlled release coated urea fertilizer. J Control Release 181:11–21

Banfield JF, Zhang H (2001) Nanoparticles in the environment. Chapter 1. In: Banfield JF, Navrotsky A (eds) Nanoparticles and the environment. Mineralogical Society of America, Washington, DC, pp 1–58

Bansiwal AK, Rayalu SS, Labhasetwar NK, Juwarkar AA, Devotta S (2006) Surfactant-modified zeolite as a slow release fertilizer for phosphorus. J Agric Food Chem 54:4773–4779

Bastús NG, Casals E, Vázquez-Campos S, Puntes V (2008) Reactivity of engineered inorganic nanoparticles and carbon nanostructures in biological media. Nanotoxico 2(3):99–112

Bastús NG, Casals E, Ojea I, Varon M, Puntes V (2012) The reactivity of colloidal inorganic nanoparticles. In: Hashim AA (ed) The delivery of nanoparticles. InTech, Croatia, pp 377–400

Bendall JS, Paderi M, Ghigliotti F, Li Pira NL, Lambertini V, Lesnyak V, Gaponik N, Visimberga G. Eychmüller, A, Sotomayor Torres CM, Welland ME, Gieck C, Marchese L (2010) Layer-by-layer all-inorganic quantum dot-based LEDs: a simple procedure with robust performance. Adv Funct Mater 20:3298–3302

Ben-Moshe T, Dror I, Berkowitz B (2010) Transport of metal oxide nanoparticles in saturated porous media. Chemosphere 81:387–393

Boehm AL, Martinon I, Zerrouk R, Rump E, Fessi H (2003) Nanoprecipitation technique for the encapsulation of agrochemical active ingredients. J Microencapsul 20:433–441

Borm P, Klaessig FC, Landry TD, Moudgil B, Pauluhn J, Thomas K, Trottier R, Wood S (2006) Research strategies for safety evaluation of nanomaterials, part V: role of dissolution in biological fate and effects of nanoscale particles. Toxicol Sci 90:23–32

Brady NC, Weil RR (1996) The nature and properties of soils. Prentice-Hall, Upper Saddle River, NJ

Breck DW (1974) Zeolite molecular sieves. Wiley, New York

Broos K, Warne J, Heemsbergen DA, Stevens D, Barnes MB, Correll RL, Mclaughlin MJ (2007) Soil factors controlling the toxicity of Copper and Zinc to microbial processes in Australian Soils. Environ Toxi Chem 26(4):583–590

Buffle J (2006) The key role of environmental colloids/nanoparticles for the sustainability of life. Environ Chem 3:155–158

Burda C, Chen X, Narayanan R, El-Sayed MA (2005) Chemistry and properties of nanocrystals of different shapes. Chem Rev 105(4):1025–1102

Calabi-Floody M, Bendall JS, Jara AA, Welland ME, Theng BKG, Rumpel C, de la Luz Mora M (2011) Nanoclays from an Andisol: extraction, properties and carbon stabilization. Geoderma 161:159–167

Casals E, Vázquez-Campos S, Bastús NG, Puntes V (2008) Distribution and potential toxicity of engineered inorganic nanoparticles and carbon nanostructures in biological systems. TrAC Trends Anal Chem 27(8):672–683

Chaudhry Q, Castle L, Watkins R (eds) (2010) Nanotechnologies in food. Royal Society of Chemistry, Cambridge

Chinnamuthu CR, Boopathi PM (2009) Nanotechnology and agroecosystem. Madras Agric J 96:17–31

Chorover J, Kretzschmar R, Garcia-Pichel F, Sparks DL (2007) Soil biogeochemical processes within the critical zone. Elements 3:321–326

Christian P, Von der Kammer F, Baalousha M, Hofmann T (2008) Nanoparticles: structure, properties, preparation and behaviour in environmental media. Ecotoxicology 17:326–343

Collins D, Luxton T, Kumar N, Shah S, Walker VK, Shah V (2012) Assessing the impact of copper and zinc oxide nanoparticles on soil: a field study. Plos One 1–11

Coma V, Martial-Gros A, Garreau S, Copinet A, Salin F, Deschamps A (2002) Edible antimicrobial films based on chitosan matrix. J Food Sci 67:1162–1169

Cornelis G, Thomas CD, McLaughlin MJ, Kirby JK, Beak DG (2012) Retention and dissolution of engineered silver nanoparticles in natural soils. Soil Sci Soc Am J 76:891–902

Cornell RM, Schwertmann U (2003) The iron oxides: structure, properties, reactions, occurrences, and uses, 2nd edn. Wiley-VCH, Weinheim

Corradini E, De Moura M, Mattoso L (2010) A preliminary study of the incorporation of NPK fertilizer into chitosan nanoparticles. Exp Polym Lett 4(8):509–515

Cosgrove T (2005) Colloid science: principles, methods and applications. Blackwell Publishing, Oxford

Cui HX, Sun CJ, Liu Q, Jiang J, Gu W (2010) Applications of nanotechnology in agrochemical formulation, perspectives, challenges and strategies. In: International conference on Nanoagri, Sao Pedro, Brazil, 20–25 June 2010, pp 28–33

Curkovic L, Cerjan-Stefanovic S, Filipan T (1997) Metal ion exchange by natural and modified zeolites. Water Res 31:1379–1382

Dana M, Lefroy RDB, Blair GJ (1994) A glasshouse evaluation of sulfur fertilizer sources for crops and pastures.1. Flooded and non-flooded rice. Aust J Agril Res 45:1497–1515

Darlington TK, Neigh AM, Spencer MT, Nguyen OT, Oldenburg SJ (2009) Nanoparticle characteristics affecting environmental fate and transport through soil. Environ Toxicol Chem 28:1191–1199

Davidson DW, Verma MS, Gu FX (2013) Controlled root targeted delivery of fertilizer using an ionically crosslinked carboxymethyl cellulose hydrogel matrix. Springer Plus 2:318. doi:10.1186/2193-1801-2-318

De Rosa G, Lopez-Moreno ML, De Haro D, Botez CE, Peralta-Videa JR, Gardea-Torresdey J (2013) Effects of ZnO nanoparticles in alfalfa, tomato, and cucumber at the germination stage: root development and X-ray absorption spectroscopy studies. Pure Appl Chem 85(12):2161–2174

De Rosa MC, Monreal C, Schnitzer M, Walsh R, Sultan Y (2010) Nanotechnology in fertilizers. Nat Nanotechnol 5:91–94

Derjaguin B, Sidorenkov G (1941) Thermoosmosis at ordinary temperatures and its analogy with the thermomechanical effect in helium II. CR Acad Sci 32:622–626

Ditta A, Arshad M, Ibrahim M (2015) Nanoparticles in sustainable agricultural crop production: applications and perspectives. In: Siddiqui MH, Al-Whaibi MH, Mohammad F (eds) Nanotechnology and plant sciences. Springer, Switzerland, pp 55–75

Dunphy Guzman KA, Finnegan MP, Banfield JF (2006) Influence of surface potential on aggregation and transport of Titania nanoparticles. Environ Sci Technol 40:7688–7693

Doshi R, Braida W, Christodoulatos C, Wazne M, O'Connor G (2008) Nano-aluminum: transport through sand columns and environmental effects on plants and soil communities. Environ Res 106:296–303

Dwairi JM (1998) Renewable, controlled and environmentally safe phosphorous released in soil mixtures of NH_4^+-phillipsite tuff and phosphate rock. Environ Geol 34:293–296

Eeberl DD (2008) Controlled release fertilisers using zeolites. USGS science for changing world. Tech Transfer, pp 1–3

Elimelech M, Gregory J, Jia X, Williams RA (1995) Particle deposition and aggregation: measurement, modeling and simulation. Butterworth-Heinemann, Woburn, MA

Fang J, Shan X, Wen B, Lin J, Owens G (2009) Stability of titania nanoparticles in soil suspensions and transport in saturated homogeneous soil columns. Environ Pollut 157:1101–1109

Farré M, Gajda-Schrantz K, Kantiani L, Barceló D (2009) Ecotoxicity and analysis of nanomaterials in the aquatic environment. Anal Bioanal Chem 393:81–95

Ferguson JF, Gavis J (1972) Review of the arsenic cycle in natural waters. Water Res 11:1259–1274

Gabriels W, Goethals P, Hermans P, De Pauw N (2001) Development of short and long-term management options for bergelenput to avoid fish kills caused by algal blooms. Meded Rijksuniv Gent Fak Landbouwkd Toegep. Biol Wet 66:63–70

Gao FQ, Hong FH, Liu C, Zheng L, Su MY, Wu X, Yang F, Wu Yang P (2006) Mechanism of nano-nantase TiO_2 on promoting photosynthetic carbon reaction of spinach–including complex of Rubisco-Rubisco activase. Biol Trace Elem Res 111:239–253

Garrido-Ramírez EG, Theng BKG, Mora ML (2010) Clays and oxide minerals as catalysts and nanocatalysts in Fenton-like reactions- a review. Appl Clay Sci 47:182–192

Ghosh S, Mashayekhi H, Pan B, Bhowmik P, Xing B (2008) Colloidal behavior of aluminum oxide nanoparticles As affected by pH and natural organic matter. Langmuir 24:12385–12391

Gladkovaa MM, Terekhovaa VA (2013) Engineered nanomaterials in soil: sources of entry and migration pathways. Univ Soil Sci Bull 68(3):129–134

Goertz HM (1993) Technology development in coated fertilisers. In: Proceedings Dahlia Greidinger memorial international workshop on controlled/slow release fertilisers, technion-israel institute of technology, Haifa, Israel, pp 102–109, Mar 7–12, 1993

Goldstein AN, Echer CM, Alivisatos AP (1992) Melting in semiconductor nanocrystals. Science 256(5062):1425–1427

Gumbo RJ, Ross G, Cloete ET (2008) Biological control of microcystis dominated harmful algal blooms. Afr J Biotechnol 7:4765–4773

Guo J (2004) Synchrotron radiation, soft X-ray spectroscopy and nanomaterials. Int J Nanotechnol 1(1):193–225

Guo M, Zhu M, Falu WuL (2005) Preparation and properties of a slow-release membrane-encapsulated urea fertilizer with superabsorbent and moisture preservation. Ind Eng Chem Res 44(12):4206–4211

Guo X, Zenga L, Li X, Spark H (2008) Ammonium and potassium removal for anerobically digested wastewater using natural clinoptilolite followed by membrane pretreatment. J Hazard Mater 151:125–133

Haack EA, Johnston C, Maurice PA (2008) Siderophore sorption to montmorillonite. Geochim Cosmochim Acta 72:3381–3397

Hashim N, Hussein MZ, Yahaya AH, Zainal Z (2007) Formation of zinc aluminium layered double hydroxides-4(2,4-dichlorophenoxy)butyrate nanocomposites by direct and indirect methods. Malaysian J Anal Sci 11(1):1–7

He F, Zhao DY, Liu JC, Roberts CB (2007) Stabilization of Fe-Pd nanoparticles with sodium carboxymethy cellulose for enhanced transport and dechlorination of trichloroethylene in soil and groundwater. Ind Eng Chem Res 46:29–34

Hernandez G, Diaz R, Notario del Pino JS, Gonzalez Martin MM (1994) NH_4^+ Na-exchange and NH_4^+—release studies in natural phillipsite. Appl Clay Sci 9:29–137

Heymann D, Jenneskens LW, Jehlicka J, Koper C, Vlietstra E (2003) Terrestrial and extraterrestrial fullerenes. Fullerenes Nanotubes Carbon Nanostruct 11(333):370

Hochella MF Jr (2002) There's plenty of room at the bottom: nanoscience in geochemistry. Geochim Cosmochim Acta 66:735–743

Hochella MF Jr, Lower SK, Maurice PA, Penn RL, Sahai N, Sparks DL, Twining BS (2008) Nanominerals, mineral nanoparticles, and Earth systems. Science 319:1631–1635

Hong FH, Yang F, Liu C, Gao Q, Wan ZG, Gu FG, Wu C, Ma ZN, Zhou J, Yang P (2005) Influence of nano–TiO_2 on the chloroplast aging of spinach under light. Biol Trace Elem Res 104:249–260

Hydutsky BW, Mack EJ, Beckerman BB, Skluzacek JM, Mallouk TE (2007) Optimization of nano- and microiron transport through sand columns using polyelectrolyte mixtures. Environ Sci Technol 41:6418–6424

Iorio M, Pan B, Capasso R, Xing BS (2008) Sorption of phenanthrene by dissolved organic matter and its complex with aluminum oxide nanoparticles. Environ Pollut 156:1021–1029

Iran Nanotechnology Initiative Council (2009) First nano-organic iron chelated fertilizer invented in Iran [webpage on the Internet] Tehran, Iran: Iran Nanotechnology Initiative Council. http://www.iranreview.org/content/Documents/IraniansResearchersProduceNanoOrganicFertilizer.htm

Ji LL, Chen W, Zheng SR, Xu ZY, Zhu DQ (2009) Adsorption of sulfonamide antibiotics to multiwalled carbon nanotubes. Langmuir 25(1608):11613

Jiang J, Oberdorster G, Biswas P (2009) Characterization of size, surface charge, and agglomeration state of nanoparticle dispersions for toxicological studies. J Nanopart Res 11:77–89

Johnston CT (2010) Probing the nanoscale architecture of clay minerals. Clay Miner 45:245–279

Ju-Nam Y, Lead JR (2008) Manufactured nanoparticles: an overview of their chemistry, interactions and potential environmental implications. Sci Total Environ 400:396–414

Kanel SR, Greneche JM, Choi H (2006) Arsenic (V) removal from groundwater using nano scale zero-valent iron as a colloidal reactive barrier material. Environ Sci Technol 40(6):2045–2050

Kanel SR, Manning B, Charlet L, Choi H (2005) Removal of Arsenic(III) from groundwater by nanoscale zerovalent iron. Environ Sci Technol 39(5):1291–1298

Kanel SR, Nepal D, Manning B, Choi H (2007) Transport of surface-modified iron nanoparticle in porous media and application to arsenic(III) remediation. J Nanopart Res 9:725–735

Karthikeyan K (2014) Naturally-occurring nano-clays in Indian soils: their role in plant nutrient management. In: Proceedings of seventh international conference on smart materials, structures and systems, Bangalore, India, pp 33–34, July 8–11, 2014

Ke YC, Stroeve P (2005) Polymer-layered silicate and silica nanocomposites, 1st edn. Elsevier BV, Amsterdam

Keller AA, McFerran S, Lazareva A, Suh S (2013) Global life cycle releases of engineered nanomaterials. J Nanopart Res 15. doi:10.1007/s11051-013-1692-4

Khedr MH, Omar AA, Abdel-Moaty SA (2006) Reduction of carbon dioxide intocarbon by freshly reduced $CoFe_2O_4$ nanoparticles. Mater Sci Eng A 432:26–33

Kim KS, Park M, Choi CL, Lee DH, Seo YJ, Kim CY, Kim JS, Yun S-IN, Ro H-M, Komarneni S (2011) Suppression of NH_3 and N_2O emissions by massive urea intercalation in montmorillonite. J Soils Sediments 11:416–422

Klaine SJ, Alvarez PJJ, Batley GE, Fernandes TF, Handy RD, Lyon DY, Mahendra S, McLaughlin ML, Lead JR (2008) Nanomaterials in the environment: behaviour, fate, bioavailability, and effects. Environ Toxicol Chem 27:1825–1851

Kool PL, Diez Ortiz M, van Gestel CAM (2011) Chronic toxicity of ZnO nanoparticles, non-nano ZnO and ZnCl2 to Folsomia candida (Collembola) in relation to bioavailability in soil. Environ Pollut 159:2713–2719

Kottegoda N, Munaweera I, Madusanka N, Karunaratne V (2011) A green slow-release fertilizer composition based on urea-modified hydroxyapatite nanoparticles encapsulated wood. Curr Sci 101(1):73–78

Kretzschmar R, Schafer T (2005) Metal retention and transport on colloidal particles in the environment. Elements 1:205–210

Krichevskii GE (2010) Nanotechnologies: dangers and risks. Inspecting principles for nano technologies and nanomaterials. Nanotekhnol Okhrana Zdorov'ya 2(3):4

Kumar R, Rawat KS, Mishra AK (2012) Nanoparticles in the soil environment and their behaviour: an overview. J App Natural Sci 4(2):310–324

Kumar SK, Krishnamoorti R (2010) Nanocomposites: structure, phase behavior, and properties. Annu Rev Chem Biomol Eng 1:37–58

Kundu S, Adhikari T, Coumar MV, Rajendiran S, Bhattacharyya R, Saha JK, Biswas AK, Subba Rao A (2013) Pine oleoresin: a potential urease inhibitor and coating material for slow-release urea. Curr Sci 104(8):1068–1071

Lal R (2008) Soils and India's food security. J Indian Soc Soil Sci 56:129–138

Lecoanet HF, Bottero JY, Wiesner MR (2004) Laboratory assessment of the mobility of a nomaterials in porous media. Environ Sci Technol 38:5164–5169

Lefroy RDB, Dana M, Blair GJ (1994) A glasshouse evaluation of sulfur fertilizer sources for crops and pastures.3. Soluble and non-soluble sulfur and phosphorus sources for pastures. Aust J Agric Res 45:1525–1537

Leggo PJ (2000) An investigation of plant growth in an organo–zeolitic substrate and its ecological significance. Plant Soil 219:135–146

Li Z (2003) Use of surfactant-modified zeolite as fertilizer carriers to control nitrate release. Microporous Mesoporous Mat 61:181–188

Li Y, Somorjai GA (2010) Nanoscale advances in catalysis and energy applications. Nano Lett 10 (7):2289–2295

Liang R, Liu M, Wu L (2007) Controlled release NPK compound fertilizer with the function of water retention. React Funct Polym 67(9):769–779

Lin D, Tian X, Wu F, Xing B (2010) Fate and transport of engineered nanomaterials in the environment. J Environ Qual 39:1–13

Lin D, Xing B (2008) Root uptake and phytotoxicity of ZnO nanoparticles. Environ Sci Technol 42:5580–5585

Liu M, Liang R, Zhan F, Liu Z, Niu A (2007) Preparation of superabsorbent slow release nitrogen fertilizer by inverse suspension polymerization. Polym Int 56(6):729–737

Liu LS, Kost J, Fishman M, Hicks KB (2008) A review: controlled release systems for agricultural and food applications. In: Parris N, Liu LS, Song C, Shastri VP (eds) New delivery systems for controlled drug release from naturally occurring materials, ACS Symposium series, vol 992. pp 265–281

Liu RQ, Lal R (2014) Synthetic apatite nanoparticles as a phosphorus fertilizer for soybean (Glycine max). Sci Rep 4:5686. doi:10.1038/srep05686

Liu X, Feng Z, Zhang F, Zhang S, He X (2006) Preparation and testing of cementing and coating nano-subnanocomposites of slow/controlled-release fertilizer. Agric Sci China 5:700–706

Lu CM, Zhang CY, Wen JQ, Wu GR, Tao MX (2002) Research of the effect of nanometer materials on germination and growth enhancement of glycine max and its mechanism. Soybean Sci 21(3):168–171

Malhi SS, Haderlin LK, Pauly DG, Johnson AM (2002) Improving fertiliser use efficiency. Better Crops 86:22–25

Manikandan A, Subramanian KS (2014) Fabrication and characterisation of nanoporous zeolite based N fertilizer. Afr J Agric Res 9(2):276–284

Manning BA, Hunt M, Amrhein C, Yarmoff JA (2002) Arsenic(III) and Arsenic(V) reactions with zero valent iron corrosion products. Environ Sci Technol 36(24):5455–5461

Mastronardi E, Tsae P, Zhang X, Monreal C, DeRosa MC (2015) Strategic role of nanotechnology in fertilizers: potential and limitations. In: Rai M, Ribeiro C, Mattoso L, Duran N (eds) Nanotechnologies in food and agriculture. Springer, Switzerland, pp 25–67

Maurice PA, Hochella MF (2008) Nanoscale particles and processes: a new dimension in soil science. Adv Agron 100:124–153

Melendi G, Fernandez-Pacheo P, Coronado R, Corredor MJ, Testillano E, Risueno PS, Marquina MC (2008) Nanoparticles as smart treatment delivery systems in plants: assessment of different techniques of microscopy for their visualization in plant tissues. Ann Bot 101:187–195

Michel FM, Ehm L, Antao SM, Lee PL, Chupas PJ, Liu G, Strongin DR, Schoonen MAA, Phillips BL, Parise JB (2007a) The structure of ferrihydrite, a nanocrystalline material. Science 316:1726–2729

Michel FM, Ehm L, Liu G, Han WQ, Antao SM, Chupas PJ, Lee PL, Knorr K, Eulert H, Kim J, Grey CP, Celestian AJ, Gillow J, Schoonen MAA, Strongin DR, Parise JB (2007b) Similarities in 2-and 6-line ferrihydrite based on pair distribution function analysis of X-ray total scattering. Chem Mater 19:1489–1496

Millán G, Agosto F, Vázquez M, Botto L, Lombardi L, Juan L (2008) Use of clinoptilolite as a carrier for nitrogen fertilizers in soils of the Pampean regions of Argentina. Ciene Inv Agr 35:245–254

Moaveni P, Kheiri T (2011) TiO_2 nano particles affected on maize (*Zea mays* L.). In: 2nd international conference on agricultural and animal science. IACSIT Press, Maldives, 25–27 Nov 2011, pp 160–163

Moore MN (2006) Do nanoparticles present ecotoxicological risks for health of the aquatic environment? Environ Int 32:967–976

Murr LE, Soto KF, Esquivel EV, Bang JJ, Guerrero PA, Lopez DA, Ramirez DA (2004) Carbon nanotubes and other fullerene related nanocrystals in the environment: a TEM study. J Mater Sci 56:2831

Mukhopadhyay R, De N (2014) Nano clay polymer composite: synthesis, characterization, properties and application in rainfed agriculture. Global J Bio Biotechnol 3(2):133–138

NAAS (2013) Nanotechnology in agriculture: scope and current relevance. Policy Paper No 63, National Academy of Agricultural Sciences, New Delhi, India, pp 1–20

Naderi MR, Abedi A (2012) Application of nanotechnology in agriculture and refinement of environmental pollutants. J Nanotechnol 11(1):18–26

Naderi MR, Danesh-Shahraki A (2013) Nanofertilizers and their roles in sustainable agriculture. Int J Agric Crop Sci 5(19):2229–2232

Nair R, Mohamed MS, Gao W, Maekawa T, Yoshida Y, Ajayan PM, Kumar DS (2012) Effect of carbon nanomaterials on the germination and growth of rice plants. J Nanosci Nanotechnol 12 (3):2212–2220

Nair R, Varghese SH, Nair BG, Maekawa T, Yoshida Y, Kumar DS (2010) Nanoparticulate material delivery to plants. Plant Sci 179:154–163

Nakache E, Poulain N, Candau F, Orecchioni AM, Irache JM (1999) Biopolymer and polymer nanoparticles and their biomedical applications. In: Nalwa HS (ed) Handbook of nanostructured materials and nanotechnology, Academic Press, New York, USA, pp 577–635

Namjesnik-Dejanovic K, Maurice PA (2001) Conformations and aggregate structures of sorbed natural organic matter on muscovite and hematite. Geochim Cosmochim Acta 65:1047–1057

Nel A, Xia T, Madler L, Li N (2006) Toxic potential of materials at the nanolevel. Science 311:622–627

No HK, Meyers SP, Prinyawiwatkul W, Xu Z (2007) Applications of chitosan for improvement of quality and shelf life of foods: a review. J Food Sci 72:87–100

Nowack B, Bucheli TD (2007) Occurrence, behavior and effects of nanoparticles in the environment. Environ Pollut 150:5–22

Oades JM (1989) An introduction to organic matter in mineral soils. In: Dixon JB, Weed SB (eds) Minerals in soil environments, 2nd edn. Soil Science Society of America, Madison, WI, pp 89–159

Oberdorster G, Oberdorster E, Oberdorster J (2005) Nanotoxicology: an emerging discipline evolving from studies of ultrafine particles. Environ Health Perspect 113(7):823–839

O'Hena S, Krug T, Quinn J, Clausen C, Geiger C (2006) Field and laboratory evaluation of the treatment of DNAPL source zones using emulsified zero-valent iron. Remediation 16(2):35–56

Oya A, Kurokawa Y, Yasuda H (2000) Factors controlling mechanical properties of clay mineral/polypropylene nanocomposites. J Mater Sci 35(5):1045–1050

Pan B, Xing B (2010) Manufactured nanoparticles and their sorption of organic chemicals. Adv Agron 108:137–181

Panwar J, Jain N, Bhargaya A, Akhtar MS, Yun YS (2012) Positive effect of zinc oxide nanoparticels on tomato plants: a step towards developing "Nano-fertilizers". In: Proceedings of 3rd international conference on environmental research and technology (ICERT), May 30–June 1, 2012, Penang, Malaysia, pp 348–352

Peralta-Videa JR, Zhao L, Lopez-Moreno ML, de la Rosa G, Hong J, Gardea-Torresdey JL (2011) Nanomaterials and the environment: a review for the biennium 2008–2010. J Hazard Mater 186:1–15

Pereira EI, Minussi FB, Cruz CCT, Bernardi ACC, Ribeiro C (2012) Urea-montmorilloniteextruded nanocomposites: a novel slow-release material. J Agric Food Chem 60:5267–5272

Perrin TS, Drost DT, Boettinger JL, Norton JM (1998) Ammonium-loaded clinoptilolite: a slow-release nitrogen fertilizer for sweet corn. J Plant Nutri 21:515–530

Pfeiffer C, Rehbock C, Hu¨hn D, Carrillo-Carrion C, de Aberasturi DJ, Merk V, Barcikowski S, Parak WJ (2014) Interaction of colloidal nanoparticles with their local environment: the (ionic) nanoenvironment around nanoparticles is different from bulk and determines the physico-chemical properties of the nanoparticles. J Roy Soc Interface 11:20130931

Pignatello JJ (1998) Soil organic matter as a nanoporous sorbent of organic pollutants. Adv Colloid Inter Sci 76–77:445–467

Pino N, Arteaga Padron JS, Gonzdlez Martin IJ, Garcfa Herndndez JE (1995) Phosphorus and potassium release from phillipsite-based slow-release fertilizers. J Control Release 34:25–29

Plank NOV, Howard I, Rao A, Wilson MWB, Ducati C, Mane RS, Bendall JS, Louca RRM, Greenham NC, Miura H, Friend RH, Snaith HJ, Welland ME (2009) Efficient ZnO nanowire solid-state dye-sensitized solar cells using organic dyes and core-shell nanostructures. J Phys Chem C 113:18515–18522

Prasad R, Bagde US, Varma A (2012) Intellectual property rights and agricultural biotechnology: an overview. Afr J Biotechnol 11(73):13746–13752

Preetha SP, Subramanian KS, Sharmila RC (2014) Characterization of slow release of sulphur nutrient—a zeolite based nano-fertilizer. Int J Dev Res 4(2):229–233

Priester JH, Ge Y, Mielke RE, Horst AM, Moritz SC, Espinosa K, Gelb J, Walker SL, Nisbet RM, An YJ, Schimel JP, Palmere RG, Hernandez-Viezcasc JA, Zhaoc L, Gardea-Torresdey JL, Holden PA (2012) Soybean susceptibility to manufactured nanomaterials with evidence for food quality and soil fertility interruption. Proc Nat Acad Sci USA 109:E2451–E2456

Rahale CS (2010) Nutrient release pattern of nano-fertilizer formulations. PhD Thesis, Tamil Nadu Agricultural University, Coimbatore, Tamil Nadu, India

Rai V, Acharya S, Dey AS., N (2012) Implications of nanobiosensors in agriculture. J Biomater Nanobiotchnol 3:315–324

Raliya R, Tarafdar JC, Gulecha K, Choudhary K, Ram R, Mal P, Saran RP (2013) Review article, scope of nanoscience and nanotechnology in agriculture. J Appl Biol Biotechnol 1(03):041–044

Ramesh K, Biswas AK, Somasundaram J, Subbarao A (2010) Nanoporous zeolites in farming: current status and issues ahead. Curr Sci 99:760–764

Ramesh K, Reddy DD (2011) Zeolites and their potential uses in agriculture. Adv Agron 113:219–241

Roco MC (2003) Nanotechnology convergence with modern biology and medicine. Curr Opin Biotechnol 14:337–346

Rose R (2002) Slow release fertilizers 101. In: Dumroese RK, Riley LE, Landis TD (eds) (Technical coordinators) National proceedings: forest and conservation nursery associations-1999, 2000, and 2001. Proceedings RMRS-P-24. USDA Forest Service, Rocky Mountain Research Station, Ogden, UT, pp 304–308

Saleh N, Kim HJ, Phenrat T, Matyjaszewski K, Tilton D, Lowry GV (2008) Ionic strength and composition affect the mobility of surface-modified Fe^0 nanoparticles in water-saturated sand columns. Environ Sci Technol 42:3349–3355

Saigusa M (2000) Broadcast application versus band application of polyolefin-coated fertilizer on green peppers grown on Andisol. J Plant Nutr 23:1485–1493

Santoso D, Lefroy RDB, Blair GJ (1995) Sulfur and phosphorus dynamics in an acid soil/crop system. Aust J Soil Res 33:113–124

Sartain JB (2010) Food for turf: slow-release nitrogen. Grounds Mainten 2:6–14

Sawant R, Hurley J, Salmaso S, Kale A, Tolcheva E, Levchenko T, Torchilin V (2006) Smart drug delivery systems: double-targeted pH-responsive pharmaceutical nanocarriers. Bioconj Chem 17(4):943–949

Schrick B, Hydutsky W, Blough JL, Mallouk TE (2004) Delivery vehicles for zerovalent metal nanoparticles in soil and groundwater. Chem Mater 16:2187–2193

Sekhon BS (2014) Nanotechnology in agri-food production: an overview. Nanotechnol Sci Appl 7:31–53

Shaviv A (2005) Controlled release fertilizers, IFA International Workshop on Enhanced-efficiency Fertilizers, Frankfurt, International Fertilizer Industry Association, Paris, France

Sheta AS, Falatah AM, Al-Sewailem MS, Khaled EM, Sallam ASH (2003) Sorption characteristics of zinc and iron by natural zeolite and bentonite. Microporous Mesoporous Materials 61:127–136

Siddiqui MH, Al-Whaibi MH, Firoz M, Al-Khaishany MY (2015) Role of nanoparticles in plants. In: Siddiqui MH, Al-Whaibi MH, Mohammad F, (eds) Nanotechnology and plant sciences, Springer, Switzerland, pp 19–35

Singh AL, Chaudhari V (1995) Source and mode of sulfur application on ground nut productivity. J Plant Nutr 18:2739–2759

Six J, Feller C, Denef K, Ogle SM, Moraes JC, Albrecht A (2002) Soil organic matter, biota and aggregation in temperate and tropical soils-effects of no-tillage. Agronomie 22:755–775

Solanki P, Bhargava A, Chhipa H, Jain N, Panwar J (2015) Nano-fertilizers and their smart delivery system. In: Rai M, Ribeiro C, Mattoso L, Duran N (eds) Nanotechnologies in food and agriculture. Springer, Switzerland, pp 81–101

Sparks DL, Jardine PM (1984) Comparison of kinetic equations to describe K-Ca exchange in pure and in mixed systems. Soil Sci 138:115–122

Subramanian KS, Paulraj C, Natarajan S (2008) Nanotechnological approaches in nutrient management. In: Chinnamuthu CR, Chandrasekaran B, Ramasamy C (eds) Nanotechnology applications in agriculture. TNAU technical bulletin, Coimbatore, India, pp 37–42

Subramanian KS, Rahale CS (2009) Synthesis of nanofertiliser formulations for balanced nutrition. In: Proceedings of the Indian society of Soil Science-Platinum Jubilee Celebration, December 22–25, IARI, New Delhi, India, pp 85

Subramanian KS, Rahale CS (2012) Nano-fertilizers—synthesis, characterization and applications. In: Proceedings of the application of nanotechnology in soil science & plant nutrition research, September 18–27, IISS, Bhoopal, India, p 107

Subramanian KS, Manikandan A, Thirunavukkarasu M, Rahale CS (2015) Nano-fertilizers for balanced crop nutrition. In: Rai M, Ribeiro C, Mattoso L, Duran N (eds) Nanotechnologies in food and agriculture. Springer, Switzerland, pp 69–80
Subramanian KS, Tarafdar JC (2011) Prospects of nanotechnology in Indian farming. Indian J Agril Sci 81(10):887–893
Subramanian KS, Tarafdar JC (2009) Nanotechnology in soil science. In: Proceedings of the Indian society of Soil Science-Platinum Jubilee celebration, December 22–25, IARI, New Delhi, India, p 199
Sultan Y, Walsh R, Monreal CM, De Rosa MC (2009) Preparation of functional aptamer films using layer-by-layer self-assembly. Biomacromol J 10:1149–1154
Suman PR, Jain VK, Varma A (2010) Role of nanomaterials in symbiotic fungus growth enhancement. Curr Sci 99:1189–1191
Tai WP, Kim YS, Kim JG (2003) Fabrication and magnetic properties of Al_2O_3/Co nanocomposites. Mater Chem Phys 82(2):396–400
Teodorescu M, Lungu A, Stanescu PO, Neamtu C (2009) Preparation and properties of novel slow-release NPK agrochemical formulations based on poly(acrylic acid) hydrogels and liquid fertilizer. Indust Eng Chem Res 48:6527–6534
Theng BKG, Yuan G (2008) Nanoparticles in the soil environment. Elements 4:395–399
Tourinho PS, Van Gestel CAM, Lofts S, Svendsen C, Soares AMVM, Loureiro S (2012) Metal based nanoparticles in soil: fate, behavior, and effects on soil invertebrates. Environ Toxico Chem 31(8):1679–1692
Trenkel ME (2010) Slow-and controlled-release and stabilized fertilizers: an option for enhancing nutrient use efficiency in agriculture. International Fertilizer Industry Association, Paris, France, pp 1–162
Unrine J, Bertsch P, Hunyadi S (2008) Bioavailability, trophic transfer, and toxicity of manufactured metal and metal oxide nanoparticles in terrestrial environments. In: Grassian V (ed) Nanoscience and nanotechnology: environmental and health impacts. John Wiley, New York, pp 345–366
Venitsianov EV, Vinichenko VN, Guseva TV (2003) Ekologicheskii monitoring: shag za shagom (Ecological Monitoring: Step by Step). In: Zaik EA (ed) Moscow University Soil Science Bulletin, Vestnik Moskovskogo Universiteta, Pochvovedenie, Moscow, pp 282–298
Vertegel AA, Siegel RW, Dordick JS (2004) Silica nanoparticle size influences the structure and enzymatic activity of adsorbed lysozyme.Langmuir 20:6800–6807
Verwey EJW, Overbeek JTG (1948) Theory of the stability of lyophobic colloids. Elsevier, Amsterdam
Waychunas GA, Kim CS, Banfield JA (2005) Nanoparticulate iron oxide minerals in soils and sediments: Unique properties and contaminant scavenging mechanisms. J Nanopart Res 7:409–433
Wang XL, Tao S, Xing BS (2009) Sorption and competition of aromatic compounds and humic acid on multiwalled carbon nanotubes. Environ Sci Technol 43:6214–6219
Weatherley LR, Miladinovic ND (2004) Comparison of the ion exchange uptake of ammonium ion onto New Zealand clinoptilolite and mordenite. Water Res 38(20):4305–4312
Wei YX, Ye ZF, Wang YL, Ma MG, Li YF (2011) Enhanced ammonia nitrogen removal using consistent ammonium exchange of modified zeolite and biological regeneration in a-sequencing batch reactor process. Environ Technol 32:1337–1343
Weiss J, Takhistov P, McClements DJ (2006) Functional materials in food nanotechnology. J Food Sci 71:R107–R116
Wilson MA, Tran NH, Milev AS, Kannangara G, Volk H, Lu G (2008) Nanomaterials in soils. Geoderma 146(1):291–302
Wu L, Liu M (2008) Preparation and properties of chitosan-coated NPK compound fertilizer with controlled-release and water-retention. Carbohydr Polym 72(2):240–247
Yang F, Hong FS, You WJ, Liu C, Gao FQ, Wu C, Yang P (2006) Influences of nano-anatase TiO_2 on the nitrogen metabolism of growing spinach. Biol Trace Elem Res 110:179–190

Yang GCC, Tu HC, Hung CH (2007) Stability of nanoiron slurries and their transport in the subsurface environment. Separ Purif Technol 58:166–172

Yeh JM, Chang KC (2008) Polymer/layered silicate nanocomposite anticorrosive coatings. J Indust Engg Chem 14:275–280

Zaarei D, Sarabi AA, Sharif F, Kassiriha SM (2008) Structure, properties and corrosion resistivity of polymeric nanocomposite coatings based on layered silicates. J Coat Technol Res 5:241

Zha L, Hu J, Wang C, Fu S, Luo M (2002) The effect of electrolyte on the colloidal properties of poly (*N*-isopropyl acrylamidecodimethylaminoethylmethacrylate) microgel latexes. Colloid Polym Sci 280:1116–1121

Zhan JJ, Zheng TH, Piringer G, Day C, Mcphérson CL, Lu YF, Papadopoulos K, John VT (2008) Transport characteristics of nanoscale functional zerovalent iron/silica composites for in situ remediation of trichloroethylene. Environ Sci Technol 42:8871–8876

Zhang F, Ali Z, Amin F, Feltz A, Oheim M, Parak WJ (2010) Ion and pH sensing with colloidal nanoparticles: influence of surface charge on sensing and colloidal properties. ChemPhysChem 11:730–735

Zhang F, Wang R, Xiao Q, Wang Y, Zhang J (2006) Effects of slow/controlled-release fertilizer cemented and coated by nano-materials on biology. II. Effects of slow/controlled-release fertilizer cemented and coated by nano-materials on plants. Nanoscience 11:18–26

Zheng L, Hong FS, Lu SP, Liu C (2005) Effect of nano–TiO_2on strength of naturally and growth aged seeds of spinach. Biol Trace Elem Res 104:83–91

Chapter 12
Utilization of Nanoparticles for Plant Protection

Rishu Sharma, Sujaya Dewanjee and C. Kole

Abstract Nanoparticles have wide potential in plant disease diagnosis and as an ecofriendly mode of disease management. Nanosensor, mini-detection instruments could play a vital role in pathogen detection and management of various plant diseases. This chapter is focused on nanoparticles utilized in disease management and possibility of large-scale adaptability of nanoparticles by integrating into present practices, thus avoiding crop loss due to pests and diseases.

Keywords Nanotechnology · Nanoparticles · Plant protection · Plant disease management · Nanosilver

12.1 Introduction

Plant disease caused by parasitic and nonparasitic agents is one of the major factors limiting crop production and productivity. Cultivated plants may be attacked by plant diseases that destroy parts or whole of the plants and reduce much of their produce even before they can be harvested or consumed. Among the total crop losses caused by different sources, 14.1 % are lost due to plant diseases alone and the total annual worldwide crop loss from plant diseases is about $220 billion. Commercial agriculture relies heavily upon inputs of agrochemicals to protect crops

R. Sharma (✉)
Department of Plant Pathology, BC Agricultural University, Mohanpur 741252, India
e-mail: rrishu.sharma90@gmail.com

S. Dewanjee
Department of Genetics and Plant Breeding, BC Agricultural University, Mohanpur 741252, India

C. Kole
Jacob School of Engineering, Sam Higginbottom Institute of Agriculture, Technology and Sciences, Allahabad 211007, India

C. Kole et al. (eds.), *Plant Nanotechnology*,
DOI 10.1007/978-3-319-42154-4_12

against pathogens and pests (Agrios 2005). The continuous and unchecked use of pesticides and fungicides has developed resistance in the pests and plant pathogens, thus leading to serious health hazards (Patel et al. 2014). The pathogen resistance against various fungicides is increasingly becoming a serious threat to crops. So, now is the time to seriously think of the use of these practices because of their critical health and environmental effects. Also, with the increased access to digital technology, consumers are becoming more aware about the use of fungicides and their impact. Thus, scientists and farmers across the globe are trying hard to minimize such hazardous effect of fungicides and other chemical control measures and switching over to other safer technologies. Agri-nanotechnology is a new branch of biology that has originated due to the compatibility of nanosized inorganic and organic particles with biological functions. Based on enhanced effectiveness, the new age drugs are nanoparticles of polymers, metals, or ceramics, which can facilitate several biological applications. Plant diseases are caused by pathogens including fungi, bacteria, mycoplasma, viruses and viroids, nematodes, parasitic plants, and protozoa. The variation in life cycle of these disease-causing organisms produces a variety of responses to micro- and macroclimate, and variation in temperature can lead to differential symptom expression (Bokx and Prion 1977). In the recent times, the increased incidence and severity of several plant diseases is the resultant of wide-scale application of intensive farming and extensive use of high-yielding varieties. The conventional plant disease control measures add toxic chemicals to the environment in addition to the increased cost of food production. And, with the changing environmental conditions, it has been predicted that a large number of new causal organisms of many dreadful diseases will emerge fast due to climate change. Moreover, a number of new physiological races and isolates of the existing pathogens will make them more virulent and destructive. In the changing scenario of climate change, an enormous change in host–pathogen interaction is predicted. Genomics, molecular breeding, and nanotechnology are highly promising emerging techniques to develop disease-resistant crop varieties. Climate change is affecting agriculture due to global warming with an average temperature increase of 0.74 °C since past 100 years, and atmospheric CO_2 concentration has increased from 280 ppm in 1750 to 400 ppm in 2013 (Gautam et al. 2013). These sudden changes will have deep adverse effect on various crops and microorganisms on the earth.

Nanotechnology-derived devices are being explored in the field of plant breeding, genetic transformation, and for reduction in sprayed chemical products using smart delivery of active ingredients leading to low-fertilizer losses and thus leading to yield increase by water and nutrient management (Torney et al. 2007; Gogos et al. 2012). Nanoparticles can act as 'magic bullets,' for targeting particular plant parts to disperse their contents, viz., herbicides, chemicals, or genes into the crops. The effective entry of herbicides can be achieved through the use of nanocapsules in cuticles and deep tissues by slow and moderate release of active substances (Himmelweit 1960; Lugue and Rubiales 2009).

12.1.1 Disease Occurrence and Prior Warning

Plant disease occurrence is usually driven by three factors, viz., a susceptible host, a virulent pathogen, and a conducive environment (Agrios 2005) for a considerable amount of time, which is represented by a 'disease tetrahedron' (Fig. 12.1a). All of these factors must interact, at least to some degree, for an optimum time period, for disease to occur and expression of symptoms (Fig. 12.1b) (Van der Plank 1963). Plant diseases tend to be affected by a number of interactions among host, pathogen, and potential vectors (Fig. 12.1a, b). These interactions can be observed as cycles of biological events including dormancy, reproduction, dispersal, and pathogenesis (Wolf and Isard 2007). Plant diseases have affected human civilization drastically starting very early from the Irish Famine in 1845 due to an epidemic of potato late blight caused by the oomycete, *Phytophthora infestans*, causing socioeconomic impact forcing millions of people to migrate to other parts of the world. This was followed by the Ceylon coffee rust epidemic caused by the fungus *Hemilia vastatrix* in 1875, which forced coffee drinkers to become tea drinkers (Kennelly et al. 2005). Plant disease management is pivotal for feeding the ever-increasing human population. The host–pathogen interactions are in turn largely affected by climate change and thus making pathogen detection even more cumbersome and ambiguous. The amount of disease occurrence in terms of incidence and severity and deterioration of produce quality is directly proportional to the amount of interaction among disease-causing factors. Thus, even slightest changes in host, pathogen, or climate can lead to adverse conditions by aggravating the disease occurrence and spread in both intensity and severity causing severe losses. Fungal pathogens are often strongly dependent on humidity for completing their disease cycles, so changes in these environmental factors are most likely to increase disease risk. Pathogen populations may increase exponentially when weather conditions are favorable for disease development (Agrios 2005). Climate change is also putting wheat stem rust resistance gene (*Sr31*) under the threat of the Uganda 99 (*Ug99*) race of stem rust caused by *Puccinia graminis* f. sp. *tritici* (Hodson 2011). Elevated temperature and

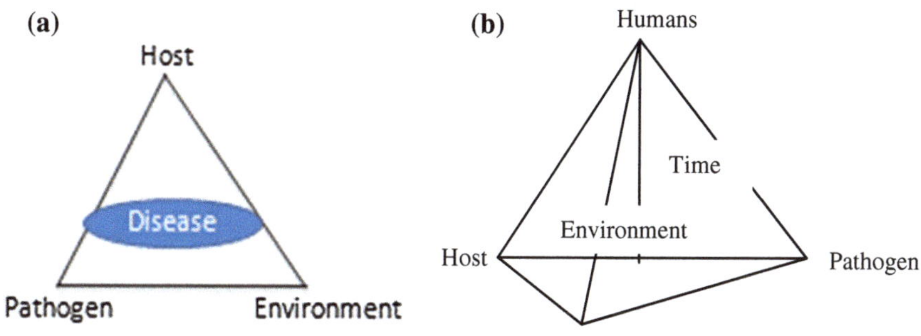

Fig. 12.1 **a** Equilateral disease triangle. The three necessary factors of disease are positioned at the vertices. **b** A disease pyramid or tetrahedron, addition of a fourth factor time depicting the interrelationship among different factors for a specific time period to cause disease epidemics

CO2 concentration are also posing higher threat perception of late blight (*Phytophthora infestans*) disease of potato and many important diseases of rice, such as blast (*Pyricularia oryzae*) and sheath blight (*Rhizoctonia solani*) (Kobayashi et al. 2006). Climate change affects the microclimate of host (local, region, and subcontinent) and in turn brings changes to global climate by incorporating changes in cellular processes and population dynamics (Fig. 12.2). Pathogen genetic variability devises plant disease management more complicated when pathogens overcome host resistance (Strange and Scott 2005). These changes would also affect the reproduction, spread, and severity of many plant pathogens, thus posing a threat to food security. The effective disease management strategies can be developed and implemented keeping in consideration the factors that trigger the development of plant disease epidemics (Campbell and Madden 1990).

Nanotechnology provides efficient tools for early detection of plant diseases by diagnostic tools in managing insects and pathogens by targetted controlled delivery of agri-based chemicals (Sharon et al. 2010; Sharma et al. 2012). Detecting and diagnosing a plant disease at an early stage is vital and has thus tempted scientists to look for a 'nanosolution' for protecting food and agriculture from bacteria, fungus, and viral agents so that tons of food can be protected from possible outbreak. Diagnosis involves pathogen detection and control. An accurate and before-hand detection of pathogens would affect positively in protecting crops from diseases by timely application of pesticides in the field (Bergeson 2010). Nanoparticles (NPs) can be used for delivery of pesticides, fertilizers, and other agrochemicals by the production of nanocapsules being highly stable and biodegradable (Jha et al. 2009). NPs can be used directly or modified in pathogen detection or as a diagnostic tool to detect compounds indicating disease (Ghormade et al. 2011). NPs displaying slow release reduce the frequent application of functional molecules. The discovery of nanosensors has lead to a more precise and quick disease diagnosis and pathogen detection (Khan and Rizvi 2014). Nanosensors can further play role in measuring crop nutrient status, moisture level, soil fertility, etc., which in turn helps in monitoring of crop growth, thus providing inputs for precision farming and

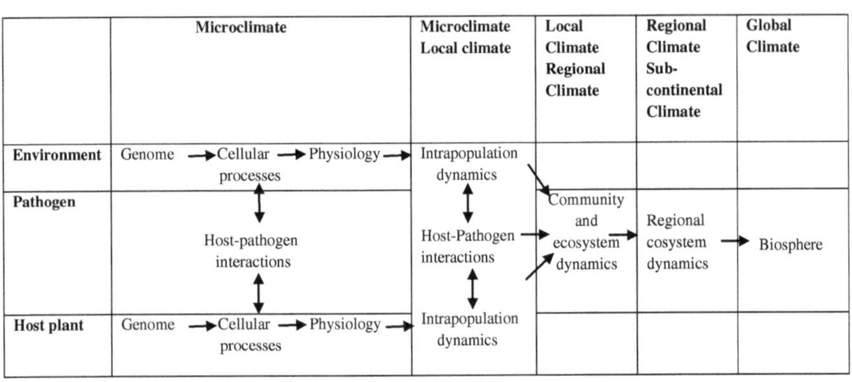

Fig. 12.2 Implications of climate change on plant diseases

maximizing the field outputs (Scott and Chen 2003). These smart systems deliver precise quantities of drugs or nutrients or other agrochemicals required. Thus, application of these systems can monitor and minimize pesticide and antibiotic use. Scientists are working on developing a simple, portable and accurate detection technique for farmers that takes less time and can give results within a few hours. Fluorescent silica nanoprobes have potential for rapid diagnosis of plant diseases. These nanoprobes conjugated with the secondary antibody of goat anti-rabbit IgG (Yao et al. 2009) were used for the detection of a bacterial plant pathogen, *Xanthomonas axonopodis* pv. *vesicatoria* in Solanaceous crops. NPs can act as biomarkers for quick detection of bacteria (Boonham et al. 2008), fungi (Chartuprayoon et al. 2010), and plant viruses (Yao et al. 2009) in plants. NP-based sensors can provide improved detection limits in detecting viral pathogens in plants (Baac et al. 2006). An autonomous nanosensor with GPS system can be very helpful to monitor soil and crop conditions. The equipment with sensors will be of high sensitivity to allow and detect the slightest changes in environment and diseases. Use of single-walled carbon nanotubes (SWCNTs) with metal/metal oxide NPs to deal agricultural by-products such as ammonia and nitrogen oxide are the most recent technologies. These techniques help in the development of nanosensor arrays which are of high density with a potential role in monitoring agricultural pollutants and their impact on biological and ecological health and thus in turn increasing crop productivity. Nanochips containing fluorescent oligo capture probes are microarrays used for disease detection (Lopez et al. 2009). Nanochips are highly specific and have high sensitivity to detect single nucleotide change in bacteria and viruses. Yao et al. (2009) used a NP with fluorescence silica and antibodies to detect pathogen-causing bacterial spot disease caused by *Xanthomonas axonopodis* pv. *vesicatoria*. Singh et al. (2010) used nanogold-based immune sensors using surface plasmon resonance (SPR) for detecting the pathogen *Tilletia indica* causing Karnal bunt disease in wheat (*Triticum aestivum*). SPR's could play a vital and effective role in seed certification and plant quarantine for accuracy in pathogen detection. In depth research on nanosensors is highly important for rapid diagnosis and management of diseases. With the advancement in nanotechnology, a specific biological marker can be detected with the use of quantum dots (QD). QD being very photo-stable and optically sensitive can be used as labeling and can be easily traced with ordinary equipment (Sharon et al. 2010). Thorough understanding of the role of nanosized engineered materials on plant physiology at the molecular level is also lacking (Khodakovskaya et al. 2011). Nugaeva et al. (2005) devised the micromechanical cantilever arrays for detecting fungal spores of *Aspergillus niger* and *Saacharomyces cerevisiae*. Proteins such as concanavalin A, fibronectin, or immunoglobulin G were surface-grafted on microfabricated uncoated as well as gold-coated silicon cantilevers. These proteins were found to have different affinities to bind to the molecular structures present on fungal cell surface. Spore immobilization and germination of the test fungi led to the shift in resonance frequency, which was measured by dynamically operated cantilever arrays. This took only a few hours in contrast to several days in conventional techniques. The finding that shift was proportional to the mass of single fungal spore can be used for quantitative estimation. The biosensors detected the target fungi in the range of 10^3–10^6 cfu ml^{-1}.

12.1.2 Control

Plant diseases can be controlled using nanotechnology by controlled release of encapsulated pesticide, and other agrochemicals in protection against pests and pathogens. The potential application of nanomaterials in crop protection helps in the development of efficient and potential approaches for the management of plant pathogens (Gopal et al. 2011). Nanoparticles remain bound to the cell wall of pathogens and cause deformity due to high-energy transfer leading to its death. Nanotechnological application in plant pathology targets specific agricultural problems in host–pathogen interactions and could provide new avenues for crop protection.

Soil application of encapsulated sulfonylurea herbicides to control *Orobanche* spp. has been conferred (Joel et al. 2007). Nanosized particles have been studied to control various fungal pathogens such as *Pythium ultimum*, *Magnaporthe grisea*, *Colletotrichum gloeosporioides*, *Botrytis cinerae,* and *Rhizoctonia solani*, as well as pathogenic bacteria including *Azotobacter chrococuum*, *Bacillus subtilis*, *Pseudomonas syringae*, *Rhizobium tropici*, and *Xanthomonas compestris* pv. *vesicatoria* (Park et al. 2006). Nanotechnology can also be used for sustainable agriculture in targeted areas, including nanocides (pesticides encapsulated in NPs for controlled release) and nanoemulsions for enhanced efficiency, NPs for soil conservation. The two most important criteria in disease management are met by nanocomposites, i.e., maximum efficacy with minimal ecological impact and less toxicity on humans. Some of the NPs that have entered into the arena of controlling plant diseases are nanoforms of carbon, silver, silica, and alumino-silicates. Scientists are concentrating on carbon nanotubes (CNTs). Also, the use of nanomaterials including copper, zinc, titanium, magnesium, gold, alginate, and silver has been developed, but silver NPs (Nano-Ag) have been proved to be the extraordinarily effective exhibiting antimicrobial efficacy against pathogens, viz., eukaryotic microorganisms, bacteria, and viruses (Guo et al. 2003). An ecofriendly fungicide is underdevelopment that uses nanomaterials to liberate its pathogen-killing properties, only when it is inside the targeted pathogen (Choudhury et al. 2011). Various nanoparticles are used for controlling plant diseases as have been enlisted in Table 12.1.

12.1.2.1 Carbon Nanotubes

CNTs are nanostructures, which are allotropes of carbon with extraordinary mechanical, electrical, thermal, optical, and chemical properties. A nanotube can be visualized as a hexagonal network of hexagonal carbon atoms which upon rolling

Table 12.1 Use of various nanoparticles for controlling plant pathogens

Nanomaterial	Application	Functions	Reference
Thiamine dilauryl sulfate(TDS) nanoparticles	Antifungal activity of TDS NPs (258.6 nm) against *Colletotrichum gloesporoides* associated with pepper anthracnose	NPs at 100 ppm conc. Showed 80 % growth inhibition of *Colletotrichum gloesporoides* compared to control. TDS NPs penetrated inside the hyphal cell membrane and destructed the cells	Seo et al. (2011)
Validamycin loaded with nanosized calcium carbonate	Controlled release of validamycin-loaded nanosized calcium carbonate (50–200 nm) against *Rhizoctonia solani*	The formulation showed better germicidal efficacy against *Rhizoctonia solani* compared to conventional technical validamycin when studied after 7 days. The nanoformulation showed the release of validamycin up to two weeks	Qian et al. (2011)
Chitosan NPs (CS NPs)	Efficacy of CS NPs on fungal growth and chili seed quality	CS NPs at a conc. Of 0.6 % (w/v) significantly delayed mycelia growth of *Rhizopus* spp., *Colletotrichum* spp. (*C. capsici*, *C. gloesporoides*) and *Aspergillus niger* when compared to control	Chookhongkha et al. (2012)
Nanocopper	Antibacterial activity against *Xanthomonas axanopodis* pv. *punicae*, causing bacterial blight of pomegranate	Nanocopper inhibited the growth of *X. axonopodis* at 0.2 ppm, i.e., >10,000 times lower than that usually recommended for copper oxychloride	Mondal and Mani (2012)
Light-activated nanoscale formulations of TiO_2	Nanoscale formulations of TiO_2 with Ag and Zn on *Xanthomonas perforans* toward the control of bacterial spot of tomato	Compared to control, TiO_2/Ag and TiO_2/Zn had high photocatalytic activity against *X. perforans* and the combination significantly reduced bacterial spot severity without causing any adverse effects on tomato yield	Paret et al. (2012)

(continued)

Table 12.1 (continued)

Nanomaterial	Application	Functions	Reference
Copper nanoparticles	Cu-based NPs (11–55 nm) were tested on tomato (*Lycopersicon esculentum*) against *Phytophthora infestans* for their antifungal activity	The synthesized Cu-based NPs were more effective than the commercial agrochemicals at lower concentrations and drastically reduced the active ingredient rate. The particles were not showed phytotoxicity on treated plants	Ginnousi et al. (2013)
DNA-directed silver nanoparticles on graphene oxide	Antibacterial activity of nanoparticles against *Xanthomonas perforans*	Nanoparticles at 100 ppm concentration severely reduced the bacterial spot disease compared to untreated tomato transplants in greenhouse condition without causing phytotoxicity	Ocsoy et al. (2013)
Light-activated nanoparticle formulation of titanium dioxide with Zinc	Nanocomposite for management of bacterial leaf spot on Rosa 'Noare'	Field applications of TiO_2/Zn nanoformulation at H 500–800 ppm Rosa 'Noare' significantly reduced bacterial spot severity compared with the untreated control and other commercial bactericides	Paret et al. (2013)
Chitosan-based nanoparticles (chitosan, chitosan-saponin, and Cu–chitosan nanoparticles)	In vitro evaluation of chitosan-based nanoparticles against *Alternaria alternata*, *Macrophomina phaseolina* and *Rhizoctonia solani*	Cu–chitosan nanoparticles were most effective at 0.1 % concentration and showed 89.5, 63.0 and 60.1 % growth inhibition of *A. alternata, M. phaseolina and R. solani*, respectively, and were effective in controlling spore germination	Saharan et al. (2013)

give a cylindrical structure (Fig. 12.3). CNTs have many applications, especially in the fields of nanotechnology, electronics, and architecture. CNTs can be used during seed germination to deliver desired molecules to provide them protection against diseases (Khodakovskaya et al. 2009). Among the carbon-based (CB) NPs,

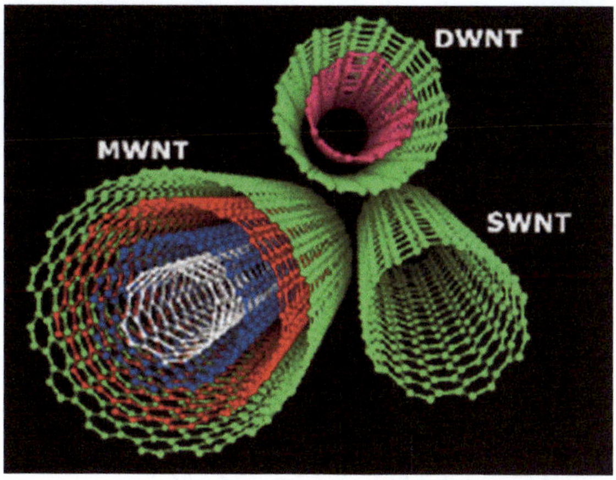

Fig. 12.3 Three-dimensional carbon nanotube structures: **a** Single-walled carbon nanotube (SWCNT), **b** double-walled carbon nanotube (DWCNT), and **c** multiwalled carbon nanotube (MWCNT). *Source* http://voitlab.com/courses/thermodynamics/index.php?title=Carbon_Nanotube_Enhanced_Epoxy (online public access)

the impact of fullerene C_{70} and fullerol in agricultural sciences has been extensively studied (Lin et al. 2009; Kole et al. 2013) and it has been found that these two types of CB NPs get readily accumulated in plants (Rico et al. 2011). Karousis et al. (2008) prepared soluble nanoPd-CNT hybrids by reduction of palladium acetate and in situ stabilization and deposition of Pd NPs onto SDS-solubilized CNTs (Fig. 12.4). The nanoPd-CNT material has shown very good catalytic activity toward hydrogenation of olefinic compounds, which is rationalized in terms of high surface area of nanoPd coated onto the surface of CNTs.

12.1.2.2 Nanosilver

Silver can affect various microorganisms and their biological processes (Donnell and Russell 1999; Pal et al. 2007; Sondi and Salopek-Sondi 2004) and also inhibits protein expression (Yamanaka et al. 2005). The use of nanosized silver particles as antimicrobial agents has become more common as technological advances make their production more economical. Silver has potential applications in management of plant diseases. Since silver exhibits multiple modes of inhibitory action to microorganisms (Park et al. 2006; Roe et al. 2008), it may be used for controlling various plant pathogens in a relatively safer way as compared to synthetic fungicides. Agnihotri et al. (2014) used coreduction method using thermal treatment for controlled synthesis of Ag NPs (Fig. 12.5). The initial reduction was performed using sodium borohydride ($NaBH_4$) at 60 °C (stage I), which induced generation of a large number of Ag NPs. The new silver nuclei are formed by the reduction of silver cations. Silver NPs formed at the initial stage subsequently participated in the growth process, where trisodium citrate (TSC)-mediated reduction of unspent Ag ions was favored at the higher temperature, i.e., 90 °C, prevailing in stage II. In a study, Jo et al. (2009) found it to inhibit colony formation of both *Bipolaris sorokiniana* and *Magnaporthe grisea*. Silver ions and NPs effectively reduced leaf

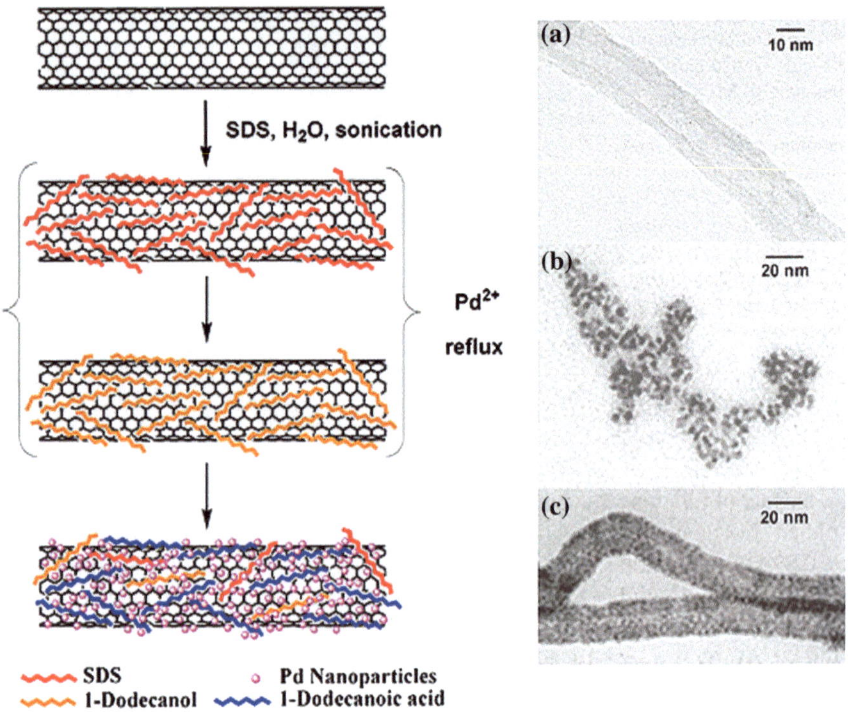

Fig. 12.4 Synthesis of soluble nanopalladium–carbon nanotube hybrids by reduction of palladium acetate and in situ stabilization and deposition of palladium nanoparticles onto sodium dedocyl sulfate-solubilized carbon nanotubes. *Source* Karousis et al. (2008) (Figure adoption Permission obtained)

spot and gray leaf spot on perennial ryegrass (*Lolium perenne*) without noticeable phytotoxicity. The inhibition efficiency of fungus colony exists as long as silver ions do not get neutralized with chloride ions. Silver has strong bactericidal and inhibitory effects. Silver NPs, which have high surface area and high fraction of surface atoms, have high antimicrobial effect as compared to the bulk silver. Kim et al. (2009) studied the *Sphaerotheca pannosa* var. *rosae* causing rose powdery mildew, disease of both greenhouse and outdoor grown roses (*Rosa damascena*). The effectiveness of nanosilver solution against causing rose powdery mildew was observed. Double capsulized nanosilver was prepared by chemical reaction of silver ion using physical method, reducing agent, and stabilizers. A reduction in disease up to 95 % was observed and did not recur for a week. Nanosilver colloid is a well dispersed and stabilized silver NP solution and is more adhesive on bacteria and fungus, hence better fungicide. International Center for Technology Assessment (ICTA) has submitted a petition to Environment Protection Agency (USA) requesting that it regulates nanosilver used in products as a pesticide under the Federal Insecticide, Fungicide, and Rodenticide Act (FIFRA). Silver is now an accepted agrochemical replacement. Silver exhibits excellent qualities, viz., its

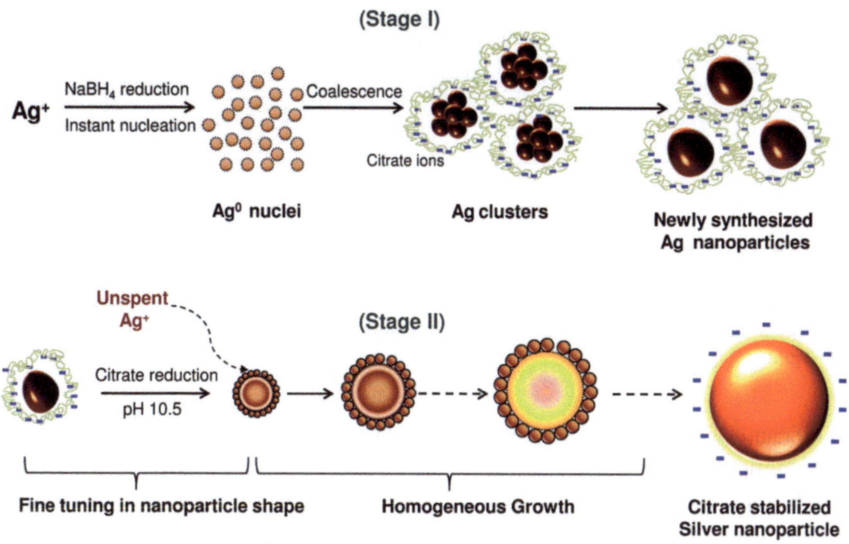

Fig. 12.5 Schematic representation of size-controlled silver nanoparticles synthesized employing the coreduction approach. *Source* Agnihotri et al. (2014) (Accessed online)

tasteless, strong nontoxic disinfectant and good growth stimulator. Silver has been observed to reduce and eliminate unwanted microorganisms from planting systems. The foliar spray of silver is used to stop fungi and various other plant pathogens. Ruffolo et al. (2010) determined the biocidal efficacy of ZnO and $ZnTiO_3$ nanopowders against the fungus *Aspergillus niger*. Ocsoy et al. (2013) observed that DNA-directed silver NPs grown on graphene oxide (GO) composites effectively decrease cell viability in culture and on plants of *Xanthomonas perforans* causing bacterial spot of tomatoes (*Solanum lycopersicum*) in Florida while the pathogen has developed resistance to Cu fungicides. These compounds (Ag@dsDNA@GO) show excellent antibacterial activity in culture at a very low concentration of 16 ppm with higher adsorption rate. Severity of tomato bacterial spot is significantly reduced by application of Ag@dsDNA@GO at 100 ppm in greenhouse when compared to untreated and showed no phytotoxicity.

12.1.2.3 Nanosilica–Silver Composite

Mao et al. (2001) have demonstrated the role silicon (Si) plays in increasing the disease and stress resistance. Aqueous silicate solution has been reported to treat diseased plants to prevent them from pathogens causing powdery and downy mildew (Brecht et al. 2004). Moreover, it promotes the physiological activity and growth of plants and induces disease and stress resistance in plants. But, since silica has no direct disinfection effects on pathogenic microorganisms in plants, it does not exhibit any effect on established diseases. Further, the effects of silica

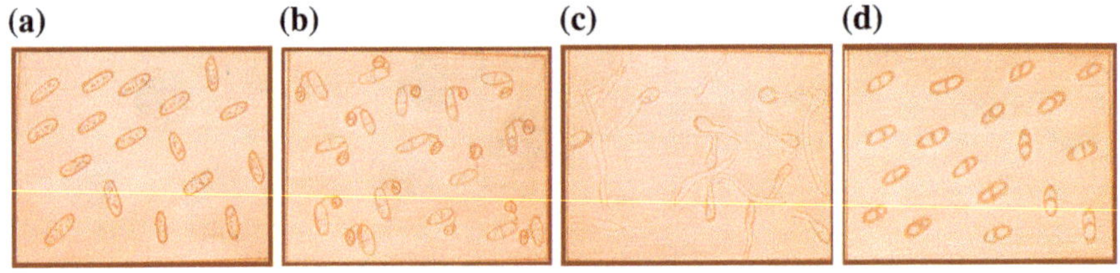

Fig. 12.6 Effect of chitosan–Ag NP composite on conidial germination of *Colletotrichum gloeosporioides*. **a** Normal conidia, **b** water control with 0.1 % (v/v) acetic acid (Appressorium formation), **c** conidial germination in chitosan (100 μg/ml) with 0.1 % (v/v) acetic acid, and **d** complete inhibition of conidia germination by chitosan–Ag NP composite at concentration of 100 μg/ml

significantly vary with the physiological environment, and thus, it is not registered as an agricultural chemical. As mentioned above, silver is known as a powerful disinfecting agent. It kills unicellular microorganisms by inactivating enzymes having metabolic functions in the microorganisms by oligodynamic action (Kim et al. 2009) and is known to exhibit significant inhibitory effects on algal growth also. The ionic state of silver has high antimicrobial potential. Silver in the form of a metal or oxide is stable in the environment, but because of its low antimicrobial activity it is used in relatively increased amount, which is not very desirable. The recent addition to this kitty is nanosized silica–silver which is capable of controlling many plant diseases have been developed by Park et al. (2006), which consist of nanosilver combined with silica molecules and water soluble polymer, prepared by exposing a solution. Park et al. (2006) also studied the 'effective concentration' of nanosized silica–silver in suppressing the fungal growth of many pathogens causing plant diseases and reported that few pathogens showed cent percent(100 %) growth suppression at 10 ppm, viz., *Botrytis cinerea*, *Colletotrichum gloeosporioides*, *Magnaporthe grisea*, *Pythium ultimum,* and *Rhizoctonia solani* while other pathogens like *Azotobacter chrococuum*, *Bacillus subtilis*, *Pseudomonas syringae*, *Rhizobium tropici*, *Xanthomonas compestris* pv. *Vesicatoria* exhibited cent percent (100 %) fungal growth suppression at 100 ppm. They have also reported chemical injuries caused by a higher concentration of nanosized silica–silver on cucumber and pansy plant. Chowdappa and Gowda (2013) observed an inhibition of conidial germination in *Colletotrichum gloeosporioides* (Fig. 12.6) with chitosan–silver NP (chitosan–Ag Np) composite (size distribution from 10 to 15 nm).

12.1.2.4 Nanoalumino-Silicate

Many of the big pesticide industries are now formulating pesticides which are extremely efficient, biologically active, and environmentally safe at nanoscale. The use of alumino-silicate nanotubes with active ingredients is one such type, which has proved advantageous when sprayed on plant surfaces because of its easy pick by insect hairs and lastly these nanotubes get consumed by them leading to their death.

12.1.2.5 Mesoporous Silica Nanoparticles

Mesoporous silica nanoparticles (MSNs) are thermally stable nanoparticles with tractable porosity, having the tendency to deliver DNA and chemicals into plants by targetted release (Wang et al. 2010). The spherical porous silica NP systems with particles having arrays of independent porous channels were developed by Trewyn et al. (2007). A honeycomb-like structure is formed by channels which can be used to fill chemicals. These NPs have a unique 'capping' strategy that seals the chemical inside. They have also demonstrated that the caps can be chemically activated to pop open and release the cargo inside the cells where it is delivered. This potential of NPs for controlled delivery efficiency can be even more increased by controlling time. Plant cells have rigid cell walls and need to be modified with chemical coatings. This technique has been successfully used in Arabidopsis (*Arabidopsis thaliana*), maize (*Zea mays*), and tobacco (*Nicotiana tabacum*) plants to inject DNA and chemicals for desired results.

12.1.2.6 Synthesis of Nanoparticles

The synthesis of NPs from the plant extracts and microbes is a boon for advance research in nanotechnology. Synthesis of extracellular NPs is a novel and economical process for bioprospecting using local biological agents to explore and exploit numerous wide biological species (Sharma et al. 2010). Green synthesis practice helps in reducing generation of hazardous waste by using environmentally safe solvents and nontoxic chemicals. The use of NPs is very important, as several pathogens have developed resistance against various antibiotics, fungicides, and bactericides due to introduction and sexual recombination. For example, *Candida albica* is resistant to fluconazole and *Phytophthora infestans* has developed resistance against metalaxyl (Schaller et al. 2003; Chowdappa and Gowda 2013).

Nanoparticles are synthesized by chemical, physical, and biological methods. The property of the NPs and efficacy of synthesis varies with the procedure of synthesis. The chemical methods have been found to synthesize the NPs more efficiently than other methods. The production and synthesis of NPs on commercial scale is dominated by chemical methods. There are different chemical methods to synthesize the NPs; however, choice of the methods may vary with the material. Some of the important methods are chemical reduction method, microemulsion/colloidal method, sonochemical method, microwave method (electrochemical method and solvothermal decomposition). The physical methods include magnetic NP synthesis (physical vapor deposition, mechanical attrition, and chemical routes from solution). Biological methods such as biosynthesis of NPs (as microorganisms and plant-based materials) and microbial synthesis of NP are among the most common production methods.

12.1.3 Applications of Nanoparticles in Plant Protection

Plant pathologists across the globe are working toward developing an effective solution for food and agricultural products protection from various pathogens. Thus, to integrate nanomethods with existing ones is utmost important to protect crops. Another upcoming field is myconanotechnology, which promotes the synthesis of NPs from fungi with contractible shape and size. Myconanotechnology with potential applications provides exciting waves of transformation in agriculture and captivates microbiologists to develop techniques for assurance in food security using green chemistry (Kashyap et al. 2013).

Nanotechnological approaches have potential to curb crop loses in a specific frame of time (Luque and Rubiales 2009). Nanotechnology applications present significant opportunities to more proficiently and unhazardous treat fungicides, herbicides, and fertilizers, by controlling their release (Li et al. 2007). Now, efforts are being made to develop disease management strategies which are relatively less hazardous to humans and animals with minimum use of synthetic fungicides. The antifungal effect of silver NPs has received only minor attention (Roe et al. 2008). Silver NPs have strong inhibitory potential against fungi *Botrytis cinerea* causing gray mold (Oh et al. 2006). Kirby–Bauer disk diffusion method was used to evaluate the effect of fluconazole and silver NPs for their antifungal activity against three fungal pathogens including *Phoma glomerata*, *Phoma herbarium*, and *Fusarium semitectum* (Gajbhiye et al. 2009). Ag_2S nanocrystals on amorphous silica particles showed antifungal activity against *Aspergilus niger*. Silver ions and NPs were evaluated to determine the antifungal action on *Bipolaris sorokiniana* and *Magnaporthe grisea* (Jo et al. 2009). The laboratory and field evaluations of NPs and silver ions evidenced for decreased disease development of phytopathogenic fungi (Jo et al. 2009). Min et al. (2009) evaluated the antifungal effects of silver NPs, especially on sclerotia-forming phytopathogenic fungi. The antifungal activity of silver NPs was assessed against filamentous ambrosia fungi in South Korea (Kim et al. 2009). The effect of silver NPs on plant pathogenic spores of *Fusarium culmorum* was studied by Kasprowicz et al. (2010). The silver NPs were also found to exhibit antifungal activity against *Fusarium oxysporum* (Musarrat et al. 2010). Silver NPs strikingly decreased the number of germinating fragments and sprout length, relative to the control. *Botrytis cinerea* growth was restricted using zinc oxide NPs causing deformed mycelial pattern and similar inhibition pattern was observed in *Penicillium expansum* leading to death of fungal mats. The mycosynthesized silver NPs may be nontoxic to human and animals than synthetic fungicides. Moreover, in addition to the toxicity that NPs may cause on algae, plants, and fungi, they may also have some positive effects.

The antifungal activity of the silver NPs was evaluated on *Colletotrichum gloeosporioides*, which is responsible for causing fruit anthracnose. Silver NPs significantly reduced the mycelia growth of *Colletotrichum gloeosporioides* in a dose-dependent manner (Aguilar-Méndez et al. 2011). Antifungal properties of silver NPs, silver ions, acrylate paint, and cotton fabric impregnated with silver NPs were assessed against *Aspergillus niger*, *Aureobasidium pullulans,* and *Penicillium phoeniceum* (Khaydarov et al. (2011). Bioassay of elemental and nanosulfur against *Aspergillus niger* showed that nanosulfur was more efficient than its elemental structure (Choudhury et al. 2011). 'Smart Delivery Systems' for agriculture can possess timely controlled, spatially targeted, self-regulated, remotely regulated, preprogramed, or multifunctional characteristics to avoid biological barriers to successful targeting. In order to develop smart treatment–delivery system in plant, González-Melendi et al. (2008) worked with zucchini (*Cucurbita pepo*) plants, which were treated with carbon-coated Fe NPs in vitro. The magnetic cores consisting of Fe NPs allow themselves to be guided to a place of interest in the body (affected part) of an organism using small magnets that create a magnetic field. The carbon coating provides biocompatibility and acts as a surface for adsorption where various types of molecules of interest (drug/DNA/chemical/enzyme) can be adsorbed. González-Melendi et al. (2008) were the first to report the penetration and transport of NPs inside whole plant.

Plants are responsive to various stress conditions and they combat it through physiological changes (Khan and Anwer 2011), viz., induction of systemic defense (Khan and Haque 2013). Wang et al. (2010) developed a sensitive electrochemical sensor by harnessing this indirect stimulus, and the use of gold electrode with copper NPs has been demonstrated to check the levels of salicylic acid and pathogen detection. An extensive research on similar sensors and sensing techniques, however, needs to be expanded for detecting pathogens, their by-products, or monitoring physiological changes in plants.

Nanodispersed formulations can be prepared in a simple cost-effective manner and are suited for developing new forms of fungicidal materials. Among the most widely used nanoformulation for the treatment of diseases in turf and ornamental plants, soil, and seeds is Syngenta's Banner MAXX™. It is commercialized as a microemulsion concentrate formulation, providing excellent tank mix compatibility and stability (Gogoi et al. 2009). Another nanoformulation 'Nano-5' is a marketed product and acts as a natural mucilage organic solution to control many plant pathogens (Table 12.2).

Resistance in plants would help in management of the above-mentioned agents to overcome the problem of economic loss. Nanoparticle-mediated plant transformation has the potential for genetic modification of plants for further improvement. In particular, employment of nanoparticles in agriculture could target specific problems in plant protection, pathogen detection, and deciphering plant–pathogen interactions and offers new methods for plant disease management (Mahendra et al. 2012). For example, introduction of resistance genes in plant cells using nanotechnological approaches for developing resistant varieties will curtail the expenses incurred on agrochemicals. Nanophytopathology can be applied as a tool to understand plant–pathogen interactions, which will provide new methods for crop protection.

Table 12.2 Prevention of plant diseases with nanoformulation Nano-5

Plant disease	Mode of application	Killing time
Gray mold, blast, fusarium wilt, early blight	Spray Nano-5 on leaf surface every 3 days	1–2 h
Late blight, phytophthora diseases, southern blight, white root rot, blister blight of tea, rust	Apply to roots twice	1–2 h
Sclerotinia rot, ergot, powdery mildew, fusarium root rot, downy mildew	Spray on leaf surfaces once in 5–7 days	Stops infection within 1–2 h
Bakanae disease, white rust, leaf blight, soft rot	Apply to roots twice	Stops infection within 1–2 h
Bacterial wilt, leaf spot, rot, brown leaf spot, black rot	Apply to roots twice	Stops infection within 1–2 h
Mosaic, ring spot, transitory yellowing, tristeza virus, exocortis viroid	Spray on leaf surfaces and to roots once in 3 days	Complete control in 7 days

Source http://www.unfortune.com.tw/index.htm (Open Public Access)

12.1.4 Can Nanoparticles Replace Pesticides?

Nanotechnology has entered into the world of pesticides and is capable of pest control by overcoming the use of pesticides such as insecticides, fungicides, and herbicides as in the present times. With new nanoscale techniques of mixing and harnessing genes, genetically modified plants would be replaced by 'atomically modified' plants (Biswal et al. 2012). Pesticides can be more precisely packaged to knock out unwanted pests. The crux of nanotechnology remains in compressing the materials to nanometer range by keeping the size small (Sharon et al. 2010). The application of silicon is integrated into field management of pests and disease due to its non-residue property. A reduction in pest infestation and increased resistance has been observed by Basagli et al. (2003) in wheat cultivars after silicon application. Application of nanosilica to the tomato plants may minimize the problems caused by *Spodoptera littoralis*, leading to development of average resistance and in turn affecting the feeding preference of this pest (El-bendary and El-Helaly 2013). Lastly, it deeply impacts the longevity and fecundity of the insect, thus decreasing the insect population density and ultimately decreasing the yield losses. Though the use of NPs can control the pest populations, further study needs to validate their viability and environmental impacts in comparison with the chemical pesticides.

Potential adaptations for dealing with impacts of climate change on plant diseases are achieved by applying genetic changes, laws, and economic changes and mitigation practices. The amount of GHG can be decreased by adopting new cropping systems and crops that reduce the GHG production by emitting less nitrous oxide.

NPs could be used for site-targeted delivery of nanoparticles to avoid damage to plant parts that might happen by various conventional agrochemical application methods and also to support environmental protection by avoiding soil and water pollution (Jha et al. 2009). Nanoparticle-mediated plant genetic engineering and use of nanosensors in agriculture are frontier areas to increase crop production by avoiding pest and disease incidences, which will be a boon to farmers. Some of the NPs that have entered into the arena of controlling plant diseases are nanoforms of carbon, silver, silica, and alumino-silicates. Nanoparticles with similar mode of action to chemical pesticides can be used as carrier of active ingredients of pesticides, host defense-inducing chemicals to the target pathogens. Nanoparticles may hit/target virus particles due to their ultra-small size and may open a new field of virus control in plants. Plant ion concentrations are affected by environmental conditions (Taiz and Zeiger 1998; Kinnersley and Scott 2001; Isla and Aragues 2010). Plants absorb essential and nonessential elements under specific growing environments. An uptake of these ions above certain concentrations can cause toxicity (Ke et al. 2007; Rico et al. 2011). Plants produce natural nanomaterials under certain necessary for their growth (Wang et al. 2006).

In order to mitigate impacts of climate change, understanding the genetic constraints for pathogen and plant adaptation will allow better mechanistic models by studying ecology and genetic interaction. The possible changes in precipitation, temperature, concentration of CO_2, CH_4, nitrous oxide, and O_3 are expected to have significant effect on crops and pathogen development, survival, and spread. The changes in climate have affected various pathogens differently; for example, with the increased CO_2, the fecundity of *Colletotrichum gloesporoides* increases (Chakraborty and Datta 2003). Increased CO_2 has led to increased pathogen load in prairie (Mitchell et al. 2003). Similarly, under drought conditions (Pritchard et al. 1999) observed plant structural changes, leaf inhibition and stomatal closure (Chaves et al. 2003). With the changed temperature and precipitation pattern, there is an expansion of *Phythophthora cinnamoni* in Europe (Bergot et al. 2004).

12.2 Conclusion and Future Road map

Despite the numerous potential advantages of nanotechnology in plant protection, it has not yet made its way commercially into our diseased fields. First, nanotech products require high initial investments and secondly large-scale field use is a prerequisite for its application. And there are numerous reports of nanomaterials' biosynthesis from plant pathogens. Nanotechnology can play as a catalyst for enhancing agricultural growth rate. Many countries across the globe are pursuing Research & Development for nanotechnological application in agriculture to nullify the toxic effects of chemicals used in field.

In the future, nanoscale devices with novel properties could be used to make agricultural systems 'smart.' A recent study by toxicologists reported that nanomaterials may be toxic (Kahru and Dubourguier 2010) but still there are no regulations to their exposure till date (Powell et al. 2008). Nanomaterials have an immense potential in providing crop protection methodologies cost-effective and environmental friendly with very few ignorable limitations. Also as observed that the use of CNTs has enhanced the plant growth in tomato, and another study using CNTs depicted that it had inhibitory effect on root elongation in tomato, whereas in onion (*Allium cepa*) and cucumber (*Cucumis sativus*) it showed enhancement in root elongation (Canas et al. 2008). Thus, large-scale application of CNTs still needs review and further experimentation. Some other studies have also depicted the toxic effect of multiwalled carbon nanotubes (MWCNTs) in plant cells and application of MWCNTs was found to be responsible for accumulation of reactive oxygen species (ROS) and subsequently decreased cell proliferation and cell death (Tan and Fugetsu 2007; Tan et al. 2009). Based on the positive as well as negative effects of CB NPs, it can be stated that the response of plants or plant cells to NPs varies with the plant species, stages of growth, and the nature of the NPs. Further research on nanosciences is needed to reveal the most efficient and useful combinations of NPs for their safe and efficient use for crop protection in the coming years.

References

Agnihotri S, Mukherji S, Mukherji S (2014) Size-controlled silver nanoparticles synthesized over the range 5–100 nm using the same protocol and their antibacterial efficacy. RSC Adv 4:3974–3983

Agrios GN (2005) Plant pathology, 5th edn. Elsevier Academic Press, San Diego, USA

Aguilar-Méndez MA, Martín-Martínez ES, Ortega-Arroyo L, Cobián-Portillo G, Sánchez-Espíndola E (2011) Synthesis and characterization of silver nanoparticles: effect on phytopathogen *Colletotrichum gloesporioides*. J Nanopart Res 13:2525–2532

Baac H, Hajos JP, Lee J, Kim D, Kim SJ, Shuler ML (2006) Antibody-based surface plasmon resonance detection of intact viral pathogen. Biotechnol Bioeng 94:815–819

Basagli MA, Moraes JC, Carvalho GA, Ecole CC, Gonçalves-Gervásio RD (2003) Effect of sodium silicate application on the resistance of wheat plants to the green-aphids, *Schizaphis graminum* (Rond.) (Hemiptera: Aphididae). Neotrop Entomol 32:659–663

Bergeson LL (2010) Nanosilver pesticide products: what does the future hold? Environ Qual Manag 19:73–82

Bergot M, Cloppet E, Pérarnaud V, Déqué M, Marçais B, Desprez-Loustau ML (2004) Simulation of potential range expansion of oak disease caused by *Phytophthora cinnamomi* under climate change. Glob Change Biol 10:1539–1552

Biswal SK, Nayak AK, Parida UK, Nayak PL (2012) Applications of nanotechnology in agriculture and food sciences. Int J Sci Innov Discov 2:21–36. www.ijsidonline.info

Bokx JA, Prion PGN (1977) Effect of temperature on symptom expression and relative virus concentration in potato plants infected with potato virus Y^N and Y^O. Potato Res 20:207–213

Boonham N, Glover R, Tomlinson J, Mumford R (2008) Exploiting generic platform technologies for the detection and identification of plant pathogens. Eur J Plant Pathol 121:355–363

Brecht MO, Datnoff LE, Kucharek TA, Nagata RT (2004) Influence of silicon and chlorothalonil on the suppression of gray leaf spot and increase plant growth in St. Augustine grass. Plant Dis 88:338–344

Campbell CL, Madden LV (1990) Introduction to plant disease epidemiology. Wiley, New York, pp xvii + 532

Canas JE, Long M, Nations S, Vadan R, Dai L, Luo M, Ambikapathi R, Lee EH, Olszyk D (2008) Effects of functionalized and nonfunctionalized single-walled carbon nanotubes on root elongation of select crop species. Environ Toxicol Chem 27:1922–1931

Chakraborty S, Datta S (2003) How will plant pathogens adapt to host plant resistance at elevated CO_2 under a changing climate? New Phytol 159:733–742

Chartuprayoon N, Rheem Y, Chen W, Myung N (2010) Detection of plant pathogen using LPNE grown single conducting polymer nanoribbon. In: Proceedings of the 218th electrochemical society meeting, Las Vegas, Nevada, USA, pp 2278–2278, 10–15 Oct 2010

Chaves MM, Maroco JP, Pereira JS (2003) Understanding plant response to drought: from genes to the whole plant. Funct Plant Biol 30:239–264

Chookhongkha N, Sopondilok T, Photchanachai S (2012) Effect of chitosan and chitosan nanoparticles on fungal growth and chilliseed quality. International Conference on Postharvest Pest and Diseas e Management in Exporting Horticultural Crops-PPDM2012 973:231–237

Choudhury SR, Nair KK, Kumar R, Gogoi R, Srivastava C, Gopal M, Subhramanyam BS, Devakumar C, Goswami A (2011) Surface-modified sulfur nanoparticles: an effective antifungal agent against *Aspergillus niger* and *Fusarium oxysporum*. App Microbiol Biotechol 90:733–743

Chowdappa P, Gowda SK (2013) Nanotechnology in crop protection: status and scope. Pest Manag Hortic Ecosyst 19:131–151

Donnell MG, Russell AD (1999) Antiseptics and disinfectants: activity, action, and resistance. Clin Microbiol Rev 12:147–179

El-bendary HM, El-Helaly AA (2013) First record nanotechnology in agricultural: silica nano-particles a potential new insecticide for pest control. Appl Sci Rep 2310–2940

Gajbhiye M, Kesharwani J, Ingle A, Gade A, Rai M (2009) Fungus-mediated synthesis of silver nanoparticles and their activity against pathogenic fungi in combination with fluconazole. Nanomedicine 5:382–386

Gautam HR, Bhardwaj ML, Kumar R (2013) Climate change and its impact on plant diseases. Curr Sci 25(105):125

Ghormade V, Deshpande MV, Paknikar KM (2011) Perspectives for nano-biotechnology enabled protection and nutrition of plants. Biotechnol Adv 29:792–803

Giannousi K, Avramidis I, Dendrinou-Samara C (2013) Synthesis, characterization and evaluation of copper based nanoparticles as agrochemicals against Phytophthora infestans. RSC Adv 3:21743–21752

Gogoi R, Dureja P, Singh PK (2009) Nanoformulations a safer and effective option for agrochemicals. Indian Farming 59:7–12

Gogos A, Knauer K, Bucheli TD (2012) Nanomaterials in plant protection and fertilization: current state, foreseen applications, and research priorities. J Agric Food Chem 60:9781–9792

González-Melendi P, Fernández-Pacheco R, Coronado MJ, Corredor E, Testillano PS, Risueño MC, Pérez-de-Luque A (2008) Nanoparticles as smart treatment-delivery systems in plants: assessment of different techniques of microscopy for their visualization in plant tissues. Ann Bot 101:187–195

Gopal MA, Gogoi RO, Srivastava CH, Kumar RA, Singh PK, Nair KK, Yadav SA and Goswami AR (2011) Nanotechnology and its application in plant protection. In: Plant pathology in India: vision 2030.Indian Phytopathological Society, Indian Agricultural Research Institute, New Delhi, pp 224–232

Guo W, Li JJ, Wang YA, Peng X (2003) Conjugation chemistry and bioapplications of semiconductor box nanocrystals prepared via dendrimer bridging. Chem Mat 15:3125–3133

Himmelweit F (ed) (1960) The collected papers of Paul Ehrlich, vol 3. Pergamon Press, London, UK

Hodson DP (2011) Shifting boundaries: challenges for rust monitoring. Euphytica 179:93–104

Isla R, Aragues R (2010) Yield and plant ion concentrations in maize (*Zea mays* L.) subject to diurnal and nocturnal saline sprinkler irrigations. Field Crops Res 116:175–183

Jha AK, Prasad K, Prasad K (2009) A green low-cost biosynthesis of Sb_2O_3 nanoparticles. Biochem Eng J 43:303–306

Jo YK, Kim BH, Jung G (2009) Antifungal activity of silver ions and nanoparticles on phytopathogenic fungi. Plant Dis 93:1037–1043. doi:10.1094/PDIS-93-10

Joel DM, Hershenhorn J, Eizenberg H, Aly R, Ejeta G, Rich PJ, Ransom JK, Sauerborn J, Rubiales D (2007) Biology and management of weedy root parasites. Hort Res 2:33–267

Kahru A, Dubourguier HC (2010) From ecotoxicology to nanoecotoxicology. Toxicology 10:105–119. doi:10.1016/j.tox.2009.08.016

Karousis N, Tsotsou GE, Evangelista F, Rudolf P, Ragoussis N, Tagmatarchis N (2008) Carbon nanotubes decorated with palladium nanoparticles: synthesis, characterization, and catalytic activity. J Phys Chem C112:13463–13469

Kashyap PL, Kumar S, Srivastava AK, Sharma AK (2013) Myconanotechnology in agriculture: a perspective. World J Microbiol Biotehnol 29:191–207. doi:10.1007/s11274-012-1171-6 Epub 2012 Sep 22

Kasprowicz MJ, Kozioł M, Gorczyca A (2010) The effect of silver nanoparticles on phytopathogenic spores of *Fusarium culmorum*. Can J Microbiol 56:247–253

Ke W, Xiong ZT, Chen S, Chen J (2007) Effects of copper and mineral nutrition on growth, copper accumulation and mineral element uptake in two *Rumex japonicas* populations from a copper mine and an uncontaminated field sites. Environ Exp Bot 59:59–67

Kennelly MM, Gadoury DM, Wilcox WF, Magarey PA, Seem RC (2005) Seasonal development of ontogenic resistance to downy mildew in grape berries and rachises. Phytopathology 95:1445–1452

Khan MR, Anwer A (2011) Fungal bioinoculants for plant disease management. In: Paul M, Clinton M, Ahmad I (eds) Microbes and microbial technology. Springer, New York, USA, pp 447–448

Khan MR, Haque Z (2013) Morphological and biochemical responses of five tobacco cultivars to simultaneous infection with *Pythium aphanidermatum* and *Meloidogyne incognita*. Phytopathol Mediter 52:98–109

Khan MR, Rizvi TF (2014) Nanotechnology: Scope and application in plant disease management. Plant Pathol J 13:214–231. doi:10.3923/ppj.2014

Khaydarov RR, Khaydarov RA, Evgrafova S, Estrin Y (2011) Using silver nanoparticles as an antimicrobial agent. In: Nato science for peace and security series a: chemistry and biology, pp 169–177

Khodakovskaya M, Dervishi E, Mahmood M, Xu Y, Li Z, Watanabe F, Biris AS (2009) Carbon nanotubes are able to penetrate plant seed coat and dramatically affect seed germination and plant growth. ACS Nano 3:3221–3227

Khodakovskaya MV, de Silva K, Nedosekin DA, Dervishi E, Biris AS, Shashkov EV, Galanzha EI, Zharov VP (2011) Complex genetic, photothermal, and photoacoustic analysis of nanoparticle-plant interactions. Proc Natl Acad Sci USA 108:1028–1033

Kim SW, Kim KS, Lamsal K, Kim YJ, Kim SB, Jung M, Sim SJ, Kim HS, Chang SJ, Kim JK, Lee YS (2009) An in vitro study of the antifungal effect of silver nanoparticles on oak wilt pathogen *Raffaelea* sp. J Microbiol Biotechnol 19:760–764

Kinnersley RP, Scott LK (2001) Aerial contamination of fruit through wet deposition and particulate dry deposition. J Environ Radioact 52:191–213

Kobayashi T, Ishiguro K, Nakajima T, Kim HY, Okada M, Kobayashi K (2006) Effects of elevated atmospheric CO2 concentration on the infection of rice blast and sheath blight. Phytopathology 96:425–431

Kole C, Kole P, Randunu KM, Choudhary P, Podila R, Ke PC, Rao AM, Marcus RK (2013) Nanobiotechnology can boost crop production and quality: first evidence from increased plant biomass, fruit yield and phytomedicine content in bitter melon (*Momordica charantia*). BMC Biotechnol 13:37

Li ZZ, Chen JF, Liu F, Liu AQ, Wang Q et al (2007) Study of UV-shielding properties of novel porous hollow silica nanoparticle carriers for avermectin. Pest Manag Sci 63:241–246

Lin S, Reppert J, Hu Q, Hudson JS, Reid ML, Ratnikova TA, Rao AM, Luo H, Ke PC (2009) Uptake, translocation, and transmission of carbon nanomaterials in rice plants. Small 5: 1128–1132

Lopez MM, Llop P, Olmos A, Marco-Noales E, Cambra M, Bertolini E (2009) Are molecular tools solving the challenges posed by detection of plant pathogenic bacteria and viruses? Curr Iss, Mol Biol 11:13–46

Luque AP, Rubiales D (2009) Nanotechnology for parasitic plant control. Pest Manag Sci J 65:540–545

Mahendra R, Shivaji D, Aniket G, Elsalam KA (2012) Strategic nanoparticle-mediated gene transfer in plants and animals—a novel approach. Curr Nanosci 8:170–179

Mao HQ, Roy K, Troung-Le VL, Janes KA, Lin KY, Wang Y, August JT, Leong KW (2001) Chitosan-DNA nanoparticles as gene carriers: synthesis, characterization and transfection efficiency. J Control Release 70:399–421

Min JS, Kim KS, Kim SW, Jung JH, Lamsal K, Kim SB, Jung M, Lee YS (2009) Effects of colloidal silver nanoparticles on sclerotium-forming phytopathogenic fungi. Plant Pathol J 25:376–380

Mitchell CE, Reich PB, Tilman D, Groth J (2003) Effects of elevated CO2, nitrogen deposition, and decreased species diversity on foliar fungal plant disease. Glob Change Biol 9:438–451

Mondal K and Mani C (2012) Investigation of the antibacterial properties of nanocopper against Xanthomonas axonopodis pv. punicae, the incitant of pomegranate bacterial blight. Ann Microbiol 62:889–893

Musarrat J, Dwivedi S, Singh BR, Al-Khedhairy AA, Azam A, Naqvi A (2010) Production of antimicrobial silver nanoparticles in water extracts of the fungus *Amylomyces rouxii* strain KSU-09. Bioresour Technol 101:8772–8776

Nugaeva N, Gfeller KY, Backmann N, Lang HP, Duggelin M, Hegner M (2005) Micromechanical cantilever array sensors for selective fungal immobilization and fast growth detection. Biosens Bioelectron 21:849–856

Ocsoy I, Paret ML, Ocsoy MA, Kunwar S, Chen T, You M, Tan W (2013) Nanotechnology in plant disease management: DNA-directed silver nanoparticles on graphene oxide as an antibacterial against *Xanthomonas perforans*. ACS Nano 7:8972–8980

Oh SD, Lee S, Choi SH, Lee IS, Lee YM, Chun JH, Park HJ (2006) Synthesis of Ag and Ag-SiO2 nanoparticles by γ-irradiation and their antibacterial and antifungal efficiency against *Salmonella enteric* serovar *Typhimurium* and *Botrytis cinerea*. Colloids and Surfaces A: Physicochem Eng Aspects 275:228–233

Pal S, Tak YK, Song JM (2007) Does the antibacterial activity of silver nanoparticles depend on the shape of the nanoparticle? A study of the gram-negative bacterium *Escherichia coli*. Appl Environ Microbiol 73:1712–1720

Park HJ, Kim SH, Kim HJ, Choi SH (2006) A New Composition of nanosized silica-silver for control of various plant diseases. J Plant Pathol 22:295–302

Paret ML, Vallad GE, Averett DR, Jones JB, Olson SM (2012) Photocatalysis: Effect of light-activated nanoscale formulations of TiO_2 on Xanthomonas perforans and control of bacterial spot of tomato. Phytopathology 103:228–236

Paret ML, Palmateer AJ, Knox GW (2013) Evaluation of a light-activated nanoparticle formulation of titanium dioxide with zinc for management of bacterial leaf spot on Rosa ëNoareí. HortScience 48:189–192

Patel N, Desai P, Patel N, Jha A, Gautam HK (2014) Agronanotechnology for plant fungal disease management: a review. Int J Curr Microbiol App Sci 3:71–84

Powell MC, Griffin MP, Tai S (2008) Bottom-up risk regulation? How nanotechnology risk knowledge gaps challenge federal and state environmental agencies. Environ Manag 42: 426–443

Pritchard S, Rogers H, Prior S, Peterson C (1999) Elevated CO_2 and plant structure: a review. Glob Change Biol 5:807–837

Qian K, Shi T, Tang T, Zhang S, Liu X, Cao Y (2011) Preparation and characterization of nano-sized calcium carbonate as controlled release pesticide carrier for validamycin against Rhizoctonia solani. Microchimica Acta. 173:51–57

Rico CM, Majumdar S, Duarte-Gardea M, Peralta-Videa JR, Gardea-Torresdey JL (2011) Interaction of nanoparticles with edible plants and their possible implications in the food chain. J Agric Food Chem 59:3485–3498

Roe D, Karandikar B, Bonn-Savage N, Gibbins B, Roullet JB (2008) Antimicrobial surface functionalization of plastic catheters by silver nanoparticles. J Antimicrob Chemother 61:869–876

Ruffolo SA, La Russa MF, Malagodi M, Rossi CO, Palermo AM, Crisci GM (2010) ZnO and $ZnTiO_3$ nanopowders for antimicrobial stone coating. Appl Phys A 100:829–834

Saharan V, Mehrotra A, Khatik R, Rawal P, Sharma SS, Pal A (2013) Synthesis of chitos an bas ed nanoparticles and their in vitroevaluation against phytopathogenic fungi. Int J Biol Macromol 62:677–683

Schaller M, Laude J, Bodewaldt H, Hamm G, Korting H (2003) Toxicity and antimicrobial activity of a hydrocolloid dressing containing silver particles in an ex vivo model of cutaneous infection. Skin Pharamacol Physiol 17:31–36

Scott N, Chen H (2003) Nano scale science and engineering for agriculture and food systems. In: National Planning Workshop, Washington, DC, USA, 18–19 Nov 2003, pp 1–61

Seo Y, Cho J, Jeong H, Yim T, Cho K and Lee T (2011). Enhancement of antifungal activity of anthracnose in pepper by nanopaticles of thiamine di-lauryl sulfate. Korean J Med Crop Sci 19:464–471

Sharma S, Ahmad N, Prakash A, Singh VN, Ghosh AK, Mehta BR (2010) Synthesis of crystalline Ag nanoparticles (AgNPs) from microorganisms. Mater Sci Appl 1:1–7

Sharma K, Sharma R, Shit S, Gupta S (2012) Nanotechnological application on diagnosis of a plant disease. In: International conference on advances in biological and medical sciences, Singapore, 15–16 July, pp 149–150

Sharon M, Choudhary AK, Kumar R (2010) Nanotechnology in agricultural diseases and food safety. J Phytol 2:83–92

Singh S, Singh M, Agrawal VV, Kumar A (2010) An attempt to develop surface plasmon resonance based immuno sensor for Karnal bunt (*Tilletia indica*) diagnosis based on the experience of nano-gold based lateral flow immune-dipstick test. Thin Solid Films 519: 1156–1159

Sondi I, Salopek-Sondi B (2004) Silver nanoparticles as antimicrobial agent: a case study on *E. coli* as a model for Gram-negative bacteria. J Colloid Interf Sci 275:177–182

Strange RN, Scott PR (2005) Plant disease: a threat to global food security. Annu Rev Phytopathol 43:83–116

Taiz L, Zeiger E (1998) Plant physiology, 2nd edn. Sinauer, Sunderland

Tan XM, Fugetsu B (2007) Multi-walled carbon nanotubes interact with cultured rice cells: evidence of a self-defense response. J Biomed Nanotechnol 3:285–288

Tan XM, Lin C, Fugetsu B (2009) Studies on toxicity of multi-walled carbon nanotubes on suspension rice cells. Carbon 47:3479–3487

Torney F, Trewyn BG, Lin VSY, Wang K (2007) Mesoporous silica nanoparticles deliver DNA and chemicals into plants. Nat Nanotechnol 2:295–300

Trewyn BG, Slowing II, Giri S, Chen HT, Lin VSY (2007) Synthesis and functionalization of a mesoporous silican based on the Sol-Gel process and applications in controlled release. Acc Chem Res 40:846–853. doi:10.1021/ar600032u

Van der Plank JE (1963) Plant diseases: epidemics and control. Academic Press, New York, UK, p xvi

Wang YH, Chen PL, Liu MH (2006) Synthesis of well-defined copper nanocubes by a one-pot solution process. Nanotechnology 17:6000

Wang Z, Wei F, Liu SY, Xu Q, Huang JY (2010) Electrocatalytic oxidation of phytohormone salicylic acid at copper nanoparticles-modified gold electrode and its detection in oilseed rape infected with fungal pathogen *Sclerotinia sclerotiorum*. Talanta 80:1277–1281

Wolf ED, Isard SA (2007) Disease cycle approach to plant disease prediction. Annu Rev Phytopathol 45:203–220

Yamanaka M, Hara K, Kudo J (2005) Bactericidal actions of a silver ion solution on *Escherichia coli*, studied by energy-filtering transmission electron microscopy and proteomic analysis. Appl Environ Microbiol 71:7589–7593

Yao KS, Li SJ, Tzeng KC, Cheng TC, Chang CY (2009) Fluorescence silica nanoprobe as a biomarker for rapid detection of plant pathogens. Adv Mater Res 79–82:513–516

Chapter 13
Nanotechnology in Soil-Plant System

Siddhartha Sankar Mukhopadhyay and Nirmaljit Kaur

Abstract Opportunities for applications of nanotechnology in soil-plant system are fast emerging as an alternative to Green Revolution technologies, which need to be phased out due to their limitations in breaking yield barriers and environmental compliances, and ever escalating shortage of farming inputs, especially P- and K-containing fertilizers and irrigation water. Literature and patent applications on nanotechnology applications in soil-plant system encompass novel materials containing nutrients and stimulators of plants, and pesticides. Compatibility of nanomaterials to farming, food and environment is essential because agricultural production functions in open system, where both energy and matter are freely exchanged in the realm of geosphere–biosphere–atmosphere. Apart from nanomaterials intended for farming, thousands other engineered nanoproducts are entering in soil-plant system, which have been altering the pristine state of soil-plant continuum, and therefore calls for framing of regulations on their use. One of the treasures in soil-plant system could be nanofabricated materials containing plant physiologically suitable nutrient ion(s) in clay minerals receptacles. The areas that need further attention in the success of nanotechnology applications in soil-plant system are founding of impeccable paradigms for concepts that govern farm production system, nanofabricating novel materials so as to improve input use efficiency and environmental compliance, interventions in soil fertility and damaged ecosystems, nutrient and water transport mechanisms in soil-plant system, and biosafety of engineered nanomaterials.

Keywords Nanomaterials · Biosafety · Clay mineral · Environmental compliance, farming system · Nano-inputs · Physiological response · Nutrient use efficiency · Ecosystems · Soil fertility

---

S.S. Mukhopadhyay (✉)
Department of Soil Science, Punjab Agricultural University, Ludhiana 141004, India
e-mail: siddharthasm1@gmail.com; siddhartha_soil@yahoo.co.in

N. Kaur
Department of Botany, Punjab Agricultural University, Ludhiana 141004, India

C. Kole et al. (eds.), *Plant Nanotechnology*,
DOI 10.1007/978-3-319-42154-4_13

13.1 Introduction

Application of nanotechnology in the discipline of agriculture is about a decade old initiative. As of now, research works are largely confined in literature and patent applications. But nevertheless, it promises revolutionary changes in farming with time. Encouraging results have already been noticed for a few novel materials based on clays (Table 13.1). The promise could be judged from the fact that there are a good number of claims of utility of the novel nanomaterials for their use as fertilizers, and surely, it is a matter of little more time for at least a few of them to be approved by the fertilizer regulatory authorities for their use in farmers' field.

For agricultural production system, compatibility of nanomaterials to farming and environment is essential because of unique nature of agriculture, which functions as open system, where both energy and matter are freely exchanged in the realm of geosphere–biosphere–atmosphere. Due to the rigors of agricultural production system, neither the manufacturing processes nor their applications meant for industrial purposes can be copied for agriculture. Also economic returns from agriculture are low. It is therefore understandable that nanotechnology research aimed agricultural applications could be minuscule in volume and its translation to reality could be slow compared to other disciplines such as electronics, drug development and security devices. It is worthwhile to remember that the editors of

Table 13.1 Some clay-based nanotechnology ventures for agriculture (adapted after modification from Kalpana-Sastry 2007)

Product	Application	Institution
Nanocides	Pesticides encapsulated in nanoparticles for controlled release	BASF
	Nanoemulsions for greater efficiency	Syngenta
Buckyball fertilizer	Ammonia from buckyballs	Kyoto U., Japan
Nanoparticles	Adehesion-specific nanoparticles for removal of *Campylobacter jejuni* from poultry	Clemson U.
Food packaging	Airtight plastic packaging with silicate nanoparticles	Bayer
Use of agricultural waste	Nanofibres from cotton waste for improved strength of cotton	Cornell U.
Nanosensors	Contamination of packaged food	Nestle, Kraft
	Pathogen detection	Cornell U.
Precision agriculture	Nanosensors linked to GPS for real-time monitoring of soil	USDA
Livestock and fisheries	Nanoveterinary medicine (nanoparticles, buckyballs, dendrimers, nanocapsules for drug delivery, nanovaccine, smart herbs, nanocheck for cleaning fish ponds, iron nanoparticle-feed for fish	Cornell U., Nanovic, Australia

Nature (2011) estimated that any technology takes some 20 years to emerge from the laboratory and be commercialized.

Historically, scientific discoveries happened in those areas, where sound theoretical frameworks, called paradigms had already laid the foundations (Kuhn 1970). This incidentally is a serious hardship for revolutionizing farming. For example, many concepts (e.g., available plant nutrient, soil health, pest control, nutrient transport) are still rudimentary in nature and do not conform to the scientific diligence. Some other important issues pertaining to agriculture are that use of nanomaterials does not guarantee a control framework that we have in electrical machines, or in satellites, or in chemical or synthetic reactors. However, it could possibly be an erudition-predicated passive framework. Similarly, the requirement of inputs in agriculture is gigantic compared to industrial use. For example, requirement of carbon nanowire for 50 million cell phones might be restricted to 50 mg, but for every hectare of land, requirement of nitrogen fertilizer could be 100 kg at optimum level (Mukhopadhyay 2014a). This is true for all inputs (seed, fertilizer, water, pesticide etc.). Whether we use nanomaterials or bulk materials, the optimum requirements of plants to achieve yield cannot be curtailed (First law of thermodynamics). At the same time, it must also be kept in mind that 100 % use efficiency of inputs is not achievable (Second law of thermodynamics).

A few things that are important for a productive venture into the nanotechnology are as follows: (i) knowledge and understanding of nanomaterials and ability to nanofabricate novel materials, (ii) understanding behavior of nanomaterials in soils, and their interaction with plants, (iii) improving use efficiency and availability of plant nutrients, especially phosphorus and micronutrients, and (iv) biosafety and environmental compliance.

13.2 Defining Nanotechnology

Nanotechnology is defined by the US Environmental Protection Agency (2007) as a science of understanding and control of matter at least in one dimension of roughly 1–100 nm where unique physical properties make novel applications possible. The inherent problem-solving capacity of the material is the driving force in imploring nanotechnology in applied disciplines such as agriculture. Therefore, Nakache et al. (1999) and USDA (2002) suggested that the size dimensions of nanoparticles could be between 10 and 1000 nm that are simultaneously colloidal particulates. More appropriate definition could be the one offered by Hall (2006), who described nanotechnology as the science of designing and building machines in which every atom and chemical bond is specified precisely. To him, it was not a set of particular techniques, devices, or products, but the set of capabilities that we will have when our technology gets near the limits set by atomic physics. He emphasized that nanotechnology aims at achieving for control of matter what computers did for our control of information.

13.3 Limitations of Green Revolution Technologies

Green Revolution Technologies have manifold farm productivity to the extent that it successfully addressed global food demand. But, it also brought along with it, degradation of environmental qualities such as causing excessive load of toxins in soils and water bodies, damage to C, N, and P cycles in soil-plant system, loss of biodiversity, declining groundwater levels, rapid weathering of soil minerals, soil acidification, and salt buildups. Together, they have threatened life and life-supporting systems (Mukhopadhyay 2005; Bhalla and Mukhopadhyay 2010; Mukhopadhyay and Sharma 2013; Mukhopadhyay 2014a). One of the root causes of this alarming situation is unrestraint and inappropriate use of farm inputs.

On the other hand, alternate farming ventures such as "Conservation Agriculture," "organic farming," "rainfed/dryland farming," and similar attempts have fallen short of our productivity expectations. The situation becomes worrisome with rising population and accelerated pace of climate change, depleting resources, and shrinking landscape. Under such situation, there exists opportunities for transcending existing farming technologies by novel use of nanotechnology that complies with the farm production system. The novelty of farming-compliance nanotechnology lies in the transcribing of nature (Naik and Stone 2005; Kuzma and VerHage 2006; Mukhopadhyay and Brar 2006; Khot et al. 2012; Mukhopadhyay 2014a).

13.4 Nanomaterials and Their Application Potential in Agriculture

Classes of materials: Four classes of nanomaterials, viz., (i) carbon based, (ii) metal based, (iii) dendrimers, and (iv) composites are recognized by US Environmental Protection Agency (2007). In this chapter, we would try to highlight what could possibly be done with these materials with special emphasis on clay minerals.

Advantages of nanomaterial over its corresponding bulk material: At the nanoscale, matter shows extraordinary properties that are not exhibited by bulk materials. For example, in the case of clay; surface area, cation exchange capacity, ion adsorption, complexation, and many more functions of clays will multiply, if they are brought to nanoscale. One of the principal ways in which a nanoparticle differs from bulk material is that a high proportion of the atoms that are associated with a nanoparticle are present at the surface (Maurice and Hochella 2008). Nanoparticles may thus have different surface composition, different types and densities of sites, and different reactivities with respect to processes such as adsorption and redox reactions (Waychunas et al. 2005).

Nanofabrication with clays as receptacles: Clays have become popular material for nanofabrication. For centuries, they are used in ceramics, computer hardware, oil refinery, filtration, pharmaceuticals, pesticide, cosmetics, construction, cement, and paper industries, to name a few. In nanoform, they have been used since ages in ceramics in China. Since time immemorial, naturally occurring nanoparticles found in soil system have been cleaning surface and ground water bodies, and regenerating life by changing harmful substances dumped on it (Sparks 2004; Maurice and Hochella 2008). Clays are inherently self-regulatory (functions in non-linear dynamical system). They are safe to life and environment, and address many other challenges of synthesis and applications in nanotechnology. Chien et al. (2009) described soils as nature's great electrostatic chemical reactor and suggested that clays could be used to clean up high pollution load in soils and water bodies arising from low use efficiency of fertilizers. The advantages of use of clay in nanofabrication are because of their: (i) ordered structural arrangements, (ii) huge adsorption capacity, (iii) capacity to shield from sunlight (ultraviolet radiation), (iv) ability to concentrate organic chemicals, and (v) ability to serve as polymerization templates. Clay-nanofabricated materials could be used in controlling release of nitrogen by trapping N in zeolites and hydroxylapatites, and in controlling release of plant nutrients, nature, and population of microflora in soil rhizosphere, ion transport in soil-plant system, and emission of dusts and aerosols from agricultural soil. They are potentially useful in precision water farming and zeoponics.

Mukhopadhyay et al. (2014a) have developed a novel process of nanofabrication involving clay minerals as receptacles, developed four advanced Zn-based nanomaterials containing zinc in plant-available form (Zn^{2+}) using clay minerals as receptacles, and then embedded them into a polymer matrix. The four nanoproducts were nanofabricated Zn^{2+} in fuller's clay receptacles, in fuller's clay nanocomposite, in kaolin clay, and in kaolin clay nanocomposite for their use as novel fertilizers (Singh et al. 2013). Mukhopadhyay (2014b) also made novel nanophosphorous products by intercalating phosphate ion (PO_4^{3-}) in kaolin clay mineral. The nanoproduct, when applied to soil as fertilizer, would release either phosphate ions (PO_4^{3-}), or get converted to hydrogen phosphate ions (HPO_4^{2-}) or dihydrogen phosphate ions (H_2PO^{4-}). All three forms are available to plants, and the release of phosphate ion would be through diffusion process. The novelty in this invention lies in the use of clay minerals as receptacles (Mukhopadhyay 2014a) and extracting phosphates from phosphorous minerals free from toxic materials (Mukhopadhyay 2014c). These and many other clay-based nanoproducts are expected to work in the soil system due to their colloidal and charge properties (Singh et al. 2013; Mukhopadhyay 2014a, b, c).

Nanotechnology involves synthesis of nanomaterial and its application to achieve a particular task. For example, if it is nanofertilizer, it has to supply nutrient ions in plant-available forms, and would remain in the soil system for a long time. Similarly, a nanoherbicide has to be specific against targeted weeds, but safe to the crop, and so on. Other requirements of nanosynthesis for agricultural applications would be that the raw materials and processes of synthesis must be cheap because farming requires large amount of inputs, and farmers must be able to afford them! Such scenario calls for innovative routes that are tuned to agriculture, and in all probability, they are likely to be different from the courses followed for manufacture of industrial nanoproducts. The routes of fabrication could rely on charge properties: (a) such as density, origin, and nature of charges, (b) intensity and degree of manifestation of charge in nanoscale, and (c) the nature (geometry) and extent of interface available for reaction (Mukhopadhyay 2013). Historically, nanosynthesis has come a long way from the top-down and bottom-up approaches to what Zubarev (2013) has titled, "Any way you want it". This should be the essence for nanofabrication as well.

Manipulation of bonds in clays: Manipulation of bonds could be an effective means for nanofabrication involving clays. Such manipulations are commonly observed in soil clays, especially in silicates. Covalent and ionic bonds simultaneously coexist in many clay minerals. They can be changed from one form to other through isomorphous substitution or insertion of small ions (e.g., Li^+), or by the use of organic compounds (for transforming of van der Waal's force to a stronger covalent or, metallic bond). Such phenomenon could be implored for improving nutrient supply mechanism in soil-plant continuum through passive control system. It is worthwhile to recognize that the control system that we observe in electrical machines, or in satellites, or in chemical reactors is not implementable in soil-plant system. The only viable system for them could be founding knowledge-based passive control system so as to create millions of rhizospheres in an acre of land to support the growth of millions of plants of a crop; a breakthrough to transcend agriculture into new millennium.

13.5 Fundamental Concepts

The origins of nanoscience can be traced to clay mineralogy and crystallography as clay minerals are crystalline and of nanometer size in all three axes (x, y, and z) (Lower et al. 2001). In nature, transformation and supply of nutrient ions to plants and microbes, fate of accreted materials that get into soil system, and purity of ground and surface water are regulated by clay minerals. The use advantage of clay minerals in nanofabrication over other colloidal material lies in their structural properties such as high anisotropic and often irregular particle shape, broad particle size distribution, different types of charges within the unit cells, heterogeneity of layer charges, pronounced Cation Exchange Capacity (CEC), disarticulation and flexibility of layers, different modes of aggregation, and high surface/mass ratio. Colloidal properties of clay minerals enhanced when bulk material is scaled down to nanoform. Another key feature of use of clay minerals for nanofabrication is their compatibility to life forms, because of their well-documented roles in the genesis of life on Earth and evolutionary diversification of Neoproterozoic life.

13.6 Public Acceptance

Risks and benefits analysis showed that nanotechnology (NT) is neutral, and better placed than genetically modified organisms (GMO), stem cell, biotech, nuclear power etc. (Currall et al. 2006).

13.7 What Are the Opportunities in Soil-Plant System

Improving farm-input use efficiency: Nanotechnology has opened up new opportunities to improve nutrient use efficiency and minimize costs of environmental protection. A disturbing fact is that the fertilizer use efficiency is only 20–50 % for nitrogen, and 10–25 % for phosphorus (<1 % for rock phosphate in alkaline calcareous soils). DeRosa et al. (2010) opined that emergence of nanofertilizers as alternative to conventional fertilizers would eliminate nutrient build ups in soils, which would eventually do away with eutrophication and drinking water contamination. Recent findings have established that plant roots and microorganisms can directly lift nutrient ions from solid phase of minerals (that includes so-called susceptible (i.e., easily weather able), as well as non-susceptible minerals (Mukhopadhyay and Brar 2006). The low solubility and the consequent excess application of P fertilizer, which have been practiced, have led to its build-up in soils and surface water bodies. This problem can be addressed by use of phosphorus-containing nanofertilizers, which have opened new avenues to improve P use efficiency (Mukhopadhyay 2014b, c, d). The central idea is to keep the ratio of applied and plant uptake P around unity by improving efficiency of native and applied phosphorus in soils, regulation of essential and toxic elements associated with phosphorous in pedosphere–hydrosphere continuum as almost all P fertilizers contain heavy metals, and most importantly, supply P to plants in available forms. Singh et al. (2013) developed a high-charge Zn^{2+} nanofertilizer in clay mineral receptacles. Similar to P, Zn^{2+}nanofertilizer would release Zn^{2+} in plant-available form. Micronutrient-based nanoproducts (i) can be applied at the time of sowing by which there will not be any need to wait for deficiency symptoms to appear and thereby no loss of potential yield, (ii) will supply nutrients directly to plant-available forms, and (iii) because of compatible capacity of holding ions on receptacles and their transfer to crops, untargeted loss will minimized (Mukhopadhyay 2014a, b, c). Similarly, nitrogen nanofertilizers will be able to check nitrate pollution and emission of green house gases. Tarafdar et al. (2013) found that nanofertilizers improved quality of agricultural products, removed environmental hazards, and required in lesser amount than conventional fertilizers. Another advantage of nanofertilizers is that the elemental purity in them is very high.

Chinnamuthu and Boopathi (2009) observed that the honeycomb-like layered crystal network of zeolites when filled with nitrogen, potassium, phosphorous, calcium, and a set of minor and trace nutrients slowly released nutrient ions "on demand." They felt that this property of zeolite can be employed to increase fertilizer use efficiency and eliminating nitrate contamination of ground and surface water bodies that arise from the application of soluble N fertilizers and their fast mineralization. Leggo (2000) opined that zeolites could be used for nitrogen capture and storage because the rate of release of absorbed nitrogen (or, fertilizers compounds) is much slower than the adsorbed ionic forms of N. Millan et al. (2008) observed that zeolite chips containing urea in their cavities can be used as slow-release nitrogen fertilizer material. They also found that there was synergic relationship of ammonium ion holding zeolites in solubilization of phosphate minerals that led to improved uptake of phosphorus and crop yield. Li (2003) used hexadecyltrimethylammonium surfactant to modify zeolites. He observed that zeolites were effective fertilizer carrier and suitable sorbent for nitrate, whose rate of release was slowed down.

Slow/controlled release-nanofertilizers: Jinghua (2004) showed that application of a nanocomposite containing of N, P, K, some micronutrients, mannose, and amino acids enhanced the uptake of nutrients in grain crops. Liu et al. (2006) implored zinc–aluminum-layered double-hydroxide nanocomposites containing plant growth regulators and found that the products released chemicals in controlled manner. DeRosa et al. (2010) also observed that that fertilizer incorporated into cochleate nanotubes (rolled-up lipid bilayer sheets) improved crop yield. These reports demonstrated that nanotechnology could be successfully applied to develop advance supply tools, and materials can be synchronized to comply the mechanisms of nitrogen release from nanofertilizer materials in accordance with the demand of crop provided that the nanomaterials must supply chemicals that can be directly internalized by the plants.

Al-Amin Sadek and Jayasuriya (2007) and Sultan et al. (2009) aimed to manufacture nanofertilizers and to develop their delivery mechanisms in such a manner that they would release nutrient ions in controlled manner (slowly or quickly) in response to environmental signals such as heat and moisture. In nutrient-deficient environment in soils, crops secrete carbonaceous compounds into rhizosphere that enable biotic mineralization of N and/or P from soil organic matter, and P from soil inorganic colloids. Authors suggested that since these root exudates can be considered as environmental signals, they may be used for making nanobiosensors, which may be incorporated into novel nanofertilizers.

Opportunities of nanotechnology intervention in salt-affected soils: Some of the possible areas, where research may be initiated are as follows:

(i) *Reducing salt concentration in soil solution*: Nanocomposites and nanopolymers may be explored.
(ii) *Improving drainage*: Sodium salts are most soluble. If drainage can be facilitated, then leaching of these salts will help crop growth. Drainage may be improved by improving structure in subsurface and sub soils. Nano-Ca^{2+}, nanoferrites, and biofriendly nanopolymers might be exploited for this purpose.
(iii) *Replacing Na^+ by Ca^{2+}*: All clay minerals show high to very high selectivity and spontaneous reactions (negative ΔG^0) for Ca^{2+} over Na^+. Can then nano-Ca^{2+}, nano-Mg^{2+}, and nano-K^+ be useful to remove Na^+ from exchange complex?
(iv) *Changing carbonate chemistry*: Sodium carbonates are more soluble than other carbonates. Capping/encapsulating sodium carbonates with nanocomposites and nanopolymer selectively may yield products which are insoluble and may be leached through preferential flows. It is a common knowledge that organic materials are strongly selective on clays (inorganic

substances). Formation of nanoorganic carbonates could be other possibility.

(v) *Prevention of Na_2CO_3 formation*: Nanomaterials such as nanocalcium carbonates, nano-Ca^{2+}, nano-Mg^{2+}, nano-K^+, and nano-iron oxides may be explored to prevent Na_2CO_3 formation in soils.

(vi) *Addition of K^+*: It is proven that potassium has beneficial effect in countering adverse effect of sodium. Illite is most common clay mineral in most of salt-affected soils of India, US, Australia, and many other places in the world. Illite has a very strong preference for K^+ over Ca^{2+}, Mg^{2+}, Na^+, and some other ions. Therefore, nano-K^+ in illite receptacle could be beneficial for accelerating ion exchange reactions to reduce exchangeable sodium saturation.

(vii) *Solubilizing of $CaCO_3$*: Calcium carbonate could possibly be solubilized by the application of nano-Ca^{2+}, nano-Mg^{2+}, and nano-iron oxides.

(viii) *Precipitation*: Nanopolymers and nanoorganic substances may be used so that they form complexes or insoluble salts with harmful ions and precipitated.

(ix) *Common ion effect*: This is a well-known phenomenon. Nano-Ca^{2+}, nano-Mg^{2+}, and nano-iron oxides may be applied to counter adverse effect of Na^+.

Fabricating novel fertilizer materials: Existing fertilizers are known to cause soil acidity, damage soil carbon profile, harm beneficial microflora, weather clay minerals, and accumulate heavy metals resulting in irreparable damage to soils and food quality, and as a consequence, it is detrimental to human health. Literatures is flooded with the reports of adverse consequences of fertilizer use (Khan et al. 2013), but without suggesting viable long-term alternate pathways. One of the key problems of existing fertilizer materials is that most of them are salts, one component consisting of plant-nutrient ion(s), while counter component is not very useful or toxic. It is possible to manufacture novel nanofertilizers using plant-nutrient ions intercalating or adsorbing on clay minerals, which function as receptacles.

13.8 Observing Behavior of Nanomaterials in Soils

Nanotechnology could be a useful venture to obviate some unique problems of soils. In soil system, nanoparticles may be introduced intentionally (e.g., nanofertilizer) or unintentionally (e.g., TiO_2). In soils, the nanomaterials are perceived to move to rhizosphere, because of small size and direction of force from soil to plant, but their participation in ion exchange, adsorption–desorption, and other reactions, and complexation with organic matter cannot be ruled out. Ming and Boettinger (2001) opined that nanoscale ion capture–release mechanism would not let nutrient ions to be lost due to leaching. Also, the ions would be permanently fixed or adsorbed or precipitated and thereby would be retained on soils (including clays) surface by sequestration inside their porous structures. There is no control on the fate, transformations, reactions, and mobility of the nanoparticles once they enter in soil system. However, this need not worry us, because right from the inception of agriculture, knowledge-based passive control system has been pursued to the height of today's level of sophistication to achieve productivity in conformation with environmental standards. The effect of nanomaterials in the physiology of plants is not yet studied. Some nanoform metal oxides increased yield of crops (Tarafdar et al. 2013). The study of effect of nanomaterials in soils and plants remains unresolved because of technological limitations. For example, at one hand, radioisotopes cannot be studied under conventional nanotechnology tools and instruments such as electron microscopes and spectroscope, and on the other hand, radioisotope measuring counters cannot measure nanoparticles.

13.9 Biosafety and Environmental Compliance

Our expanding ability to synthesize nanoparticles for use in electronic, biomedical, ceramic, pharmaceutical, cosmetic, energy, environmental, catalytic, and similar materials has caused concern over the role of these particles in environmental safety. The gravity of situation may be assessed from the data provided by Nowack and Bucheli (2007), who expected use of engineered nanomaterials to the tune of 58000 tons during 2011–2020 from mere 2000 tons in 2004. Apart from native soil materials, many new nanoproducts are entering into soil system, some of which are used for agricultural production and some others for many other purposes. All these materials eventually land on soil. Bernhardt et al. (2010) advocated that nanotechnology interventions must adhere to environmental ethos to be useful to the society.

13.10 Opportunities for Application of Nanotechnology and Nanoscience in Environmental Cleanup Operation

Nanoscience (also nanotechnology) has found applications in controlling release of nitrogen, understanding weathering of soil minerals, soil development, and nutrient ion transport in soil-plant system, nature of dusts and aerosols from agricultural soil, zeoponics, and precision water farming. As it strides forward, nanotechnology has converged soil mineralogy with imaging techniques and artificial intelligence. A fascinating aspect of remediation of pollution is how nanoparticles may affect the fate, transport, and bioavailability of pollutants in soils. There is a long-standing and rich literature on the importance of Fe-hydroxide nanoparticles and nanomaterials interactions with nutrients and pollutants in the subsurface, including sorption and redox phenomena (Brown and Parks 2001; Brown et al. 1998). Cheng et al. (2009) found that depending on the conditions, nanosized carbon such as C_{60} or nanotubes could either enhance or inhibit the mobility of organic pollutants. Hydrophobic organic compounds could potentially sorb into C-based nanoparticles such as fullerenes, thus affecting fate, transport, and other processes such as biodegradation pathways. Amendments to nanoparticles to alter their surface properties may affect interactions with pollutants. In addition, fullerol ($C_{60}(OH)_{24}$) has been demonstrated to produce reactive oxygen species (ROS), which may affect redox processes and stabilities of organic pollutants (Pickering and Wiesner 2005). This may also play a role in disinfection capabilities of engineered nanoparticles. Nanophase minerals also influence the movement of heavy metals in surface and shallow subsurface environments through complex biogeochemical interactions.

13.11 Behavior of Nanomaterials in Plants

During the last decade, advances have been made in the study of fundamental characteristics of nanomaterials and their utilization for many applications. There is, however, scanty information available on the effect of nanomaterials on plant cells, and the way they influence the physiology and development of plants. Plants are the producers of food and oxygen that sustain life. The plants are also the most affected by the unprecedented human activities leading to environmental degradation. The area of nanoscience that has its implications in plant growth and development and ultimate productivity is of current interest (Srinivasan and Saraswati 2010).

The effect of nanoparticles on plant growth is relatively less explored and is an emerging area of research which needs to be meticulously explored. Recent research has focused on engineered nanoparticles as potential candidates for improvement of crop yield (Barik et al. 2008), and their use has been made particularly for efficient nutrient utilization, disease resistance, and enhancement of growth (Nair et al. 2010). However, limited information is available on the mode of action of these nanoparticles on crop plants.

The metal nanoparticles provide more surface area for valence electron exchange with the biomolecules, due to more surface area-to-volume ratio. These metal nanoparticles therefore pose changes in the antioxidant status of plant treated with it as they can participate in cellular redox reactions (Arora et al. 2012). Water-soluble nanotubes can become aligned due to endo-osmotic root pressure in the xylem vessel of plants that enhances water and nutrient uptake capacity. In the presence of CNTs, lignin biosynthesis suggests the formation of more biomass of xylem vessels than is shown to be directly related to the growth of the plant. The essential nutrients required for the plant interact with the hydrophilic groups attached to the surface of carbon nanomaterials by H-bonds and by electrostatic interaction in the temporal periphery of carbon nanoparticles and remain attached there on a temporal basis; thus, the carbon nanomaterial works as storage house for micronutrients. Such retention allows a sustained and slow release of these micronutrients for the facile transport inside the xylem vessel.

Nanoscale carriers: The nanoscale carriers can be used for efficient delivery of fertilizers as well as plant growth regulators. The mechanisms involve mainly encapsulation and entrapment. It may also be in the form of polymers or dendrimers. Such mechanisms help to reduce the input amount and also in alleviating environmental load.

Why nanoparticles?: Usually a very low amount of growth-promoting chemical reaches the target site of plants. This concentration is much lower than the concentration required for plant growth promotion. This happens due to leaching of chemicals, its degradation by photolysis or hydrolysis, or its degradation due to microbial activity. Hence, the repeated applications are required which may cause soil or water quality degradation. Therefore, there is need of nanoencapsulated agrochemicals which may have high stability and effectiveness along with being highly soluble. Such nanoencapsulated agrochemicals must be released in response to specific stimulus and must be safe ecologically (Boehn et al. 2003).

Nanoparticle entry into plants: The plant cells provide a barrier for entry of any external agent into it. This is because of pore diameter of cell wall that ranges between 5 and 20 nm (Fleischer et al. 1999). Hence, nanoparticle aggregates with diameter less than the size of pore diameter of cell wall can pass through cell wall and reach plasma membrane (Moore 2006). The engineered nanoparticles may interact with the pores of cell wall and increase the size of the pores of cell wall or may also give rise to new pores. These engineered nanoparticles may cross the plasma membrane using embedded transport carrier proteins or though ion channels. In the cytoplasm, nanoparticles may bind with different cytoplasmic organelles and interfere with the metabolic processes at the site (Jia 2005). When nanoparticles are applied though leaves, they enter through stomatal opening or through the base of the trichome and then get translocated to the other tissues (Eichert et al. 2008). The accumulation of nanoparticles on the photosynthetic surface may cause foliar heating, which may lead to changes in gas exchange.

13.12 Impact of Nanoparticles on Plants

Carbon-based NPs: The effects of nanoparticles on plants can be beneficial (e.g., seedling growth and development) or non-beneficial as they have been reported to prevent root growth (Zhu et al. 2008). Carbon nanotubes (CNT) have single or multiple layer of carbons established in a cylinder (Wz et al. 1996). Carbon nanotubes behave as fibers, and the properties of CNT are different from the properties of bulk carbon and graphite. They are the strongest small fiber and have been reported to transport to systemic sites, viz., fruits, leaves, and roots, and thus cause a significant change in gene expression. It is important to pre-establish the optimum dose of CNTs because these may have phytotoxic effects on plant cells and may cause death by causing excessive electrolyte leakage. The multiwalled CNTs (MWCNTs) are taken up by the roots and the seeds through the creation of new pores and water uptake for the development of tomato seedlings (Checkin et al.

2012). They improved the seed water uptake, whereas no seed water uptake with single-walled CNTs (SWCNTs) was observed in cucumber seedlings after 84 h of treatment. A number of reports are published on the effects of MWCNTs on seed germination and plant growth. The stimulation of growth of tomato seeds has been reported by Villagarcia et al. (2012), whereas water-soluble MWCNTs have been reported to improve growth in gram plants (Tripathi et al. 2011). Tripathi et al. (2011) and Villagarcia et al. (2012) observed that CNTs stimulated water uptake and thus improved growth of plants, while Saxena et al. (2014) explained that SWCNTs, MWCNTs, and carbon nano-anions readily penetrated plants. The SWCNTs and MWCNTs enhanced rice seed germination when the seeds were germinated in the presence of these nanoparticles (Nair et al. 2012). In zucchini plants, there was no negative effect of MWCNT on seed germination, whereas decrease in biomass of plant was observed during further growth in the presence of SWCNTs (Stampoulis et al. 2009). The response of plants to nanomaterials depends on species of plant, the growth stage, and nature of nanomaterial. Some studies reported potential toxicity of MWCNTs in plant cells. The MWCNTs resulted in accumulation of reactive oxygen species causing increase in oxidative stress, decrease in cell proliferation, and thus cell death (Tan et al. 2009). The carbon-based nanomaterials are highly hydrophobic and thereby they interact with organic substances. Some plants take up specific carbon-based nanoparticles with specific uptake mechanism and accumulation. The toxic effects of CNTs have been reported by some workers. Begum et al. (2012) reported reduction in root fresh weight in rice and cucumber seedlings with the application of MWCNTs. They also induced reduction of germination rate in maize (Lin and Xing 2007) and increased fresh weight and root length in wheat seedlings (Wang et al. 2012). The effects of MWCNTs vary from one report to another because of involvement large number of factors such as concentration of MWCNTs and the process of obtaining it. The effect also depends upon the type of medium of growth used and type of plant material under study. It was concluded by Tiwari et al. (2014) that pristine MWCNTs at low concentration promote the growth of maize seedlings by enhancing water and nutrient transport, but their potency could be reduced by higher concentration of ions or polar species in the medium. They suggested that CNTs can be used for water transport in arid-zone agriculture and for the improvement of crop yields. The SWCNTs remain adhered to external surface of main and secondary roots as reported by Lou et al. (2011). Graphene, another carbon nanoparticle, is a two-dimensional allotrope of carbon. It can also be described as one atom layer of graphite. Graphene can cause phytotoxic effects in plant cells due to its accumulation and may lead to cell death.

The presence of black aggregate of fullerene is reported in seeds and roots of rice as compared to stem and leaves of rice (Torre-Roche et al. 2013). In mature plants, there is translocation of fullerene from roots to aerial parts. Fullerene aggregate is located near vascular system of stem and leaves. The roots do not show presence of fullerene aggregates. This suggested that the fullerene adopts the route of nutrients for translocation and make way through xylem (Torre-Roche et al. 2012). Fullerenols accumulate at interface between the cell wall and plasma membrane.

The accumulation was also between adjacent epidermal cells, which showed that fullerol had the apoplectic mode of transport (Gao et al. 2011).

Metal- and Metal Oxide-based nanomaterials: Metal and metal oxides show size-dependent properties such as fluorescence and photocatalytic degradation. They are used as agrochemicals (Franke et al. 2006). Tarafdar et al. (2012a) reported significant increase in yields with the foliar application of nanoparticles. They insisted on the scope of balanced nutrient uptake by the plants through the nanoproducts obtained through nanotechnology. Tarafdar et al. (2012b) observed increased uptake of nutrients, when fertilizers were encapsulated in nanoparticles.

The most studied metal-based nanomaterials are TiO_2, CeO_2, Fe_3O_4, and ZnO. Higher aggregation of Fe_3O_4 nanoparticle was observed by increasing pH of humic acid. Similar effect was also observed in CeO_2. The response of plant to metal nanoparticle application depends on the nature of plant, type of plant species, and stage of growth.

Titanium oxide (TiO_2): Titanium oxide nanoparticles are used in daily life, but information on their uptake and translocation in the plants is scanty. Titanium oxide nanoparticles produce reactive oxygen species when they interact with UV radiation (Feizi et al. 2013). Titanium oxide nanoparticles showed increase in enzyme nitrate reductase in soybean. They also enhanced the ability to absorb water and stimulated antioxidant system. The seeds which were treated with nano-TiO_2 produced plants with higher dry weight (73 %), increased photosynthetic rate, and 45 percent rise in chlorophyll 'a' formation than control (Mingfeng et al. 2013). Titanium oxide increases plant growth by improving nitrogen metabolism and promotion of adsorption of nitrate. On the contrary, negative effects of TiO_2 nanoparticles on seed germination and number of roots were observed in rice (Folete et al. 2011). The presence of TiO_2 does promote growth of plants through an involvement in nitrogen metabolism and photosynthetic rate. The nanoparticles improved the light absorbance and promoted the activity of Rubisco activity and thus accelerated growth in spinach (Lei 2007). Those nanoparticles enhanced nitrogen metabolism and promoted absorption of nitrate and thus increased fresh and dry weights. There was a decrease in the accumulation of superoxide radicals and promotion of antioxidant stress.

Zinc oxide: Among the metal- and metal oxide-engineered nanomaterials, zinc (Zn) and zinc oxide (ZnO) are commonly applied on plants. One of the widely spread micronutrient deficiency in soil is zinc deficiency, and Stella et al. (2010) reported that it is the fourth most important yield-limiting nutrient after nitrogen, phosphorus, and potassium. Due to its extensive utilization in consumer products, it is likely that either through accidental release or deliberate applications, the Zn or ZnO might enter into atmospheric environments. This may further lead to considerable effect on many organisms, particularly plants which are the essential base component of all ecosystems (Dwivedi and Randhawa 1974). Zinc-containing nanomaterials are needed for chlorophyll production, fertilization, pollen function, and synthesis of auxins. Among the micronutrients, it is Zn that protects the plants from drought stress (Sharma et al. 2009). Zinc and ZnO may also affect the germination rate of the seeds. The effect of ZnO on root germination was observed for the species of Buck wheat (*Fagopyrum esculentum*) (Sooyeon et al. 2013). The ZnO nanoparticles had pronounced effect on onion (*Allium cepa*) root elongation, genetic composition, and metabolism. The seed soaking and incubation in the suspension of Zn/ZnO nanoparticles halted the growth of roots in corn. The toxicity of ZnO nanoparticle and Zn^{2+} could be driven by different theories, either it could be due to the chemical toxicity based on chemical composition or it could be due to the stress or stimuli imposed by size, shape, and surface of the ZnO nanoparticles. Both the theories affected the cell culture response of the plants. Depending on the plant species and the experimental conditions, the most important mechanism of action may be internal efficiency, i.e., Zn/ZnO utilization in tissues, or Zn/Zn uptake which is regarded as external efficiency (Dwivedi and Randhawa 1974). This deliberated the ZnO nanoparticles to enter the root cells and inhibit seedling growth.

The seed germination and root growth study of zucchini seed in hydroponic solution containing ZnO nanoparticles showed no negative response (Stampoulis et al. 2009), whereas seed germination in rye grass and corn was inhibited by nanoscale zinc and ZnO, respectively. It was confirmed by electron microscopy that the uptake of nanoparticles ZnO damaged epidermal and cortical cells and could also injure the endodermal and vascular cells causing growth inhibition in rye grass (Lin and Xing 2007).

Use of nanoparticles for seed quality enhancement has been achieved by Shyla and Natarajan (2014). The beneficial effects of ZnO NPs in improving seed germination could be due to higher precursor activity of nanoscale zinc in auxin production. Moreover, zinc is required for plant growth and is essential for various enzymes catalyzing various steps.

Iron oxide: The supraoptimal amount of iron oxide (Fe_3O_4) as a magnetic nanomaterial results in adverse effects on plant growth. The "Chlorophyll a" levels were amplified at low nano-Fe_3O_4 fluid concentrations, whereas at higher concentration of Fe_3O_4 nanoparticles fluid, the "Chlorophyll a" levels were inhibited. An inhibitory effect was discerned on the growth of the plantlets that led to brown spots on leaves at higher volume fractions of Fe_3O_4 nanoparticle fluids (Stephan 2004). The excess Fe_3O_4 nanoparticles treatment produced some oxidative stress, which in turn affected photosynthesis and resulted in decreased rates of metabolic process. The oxidative stress was induced by nano-Fe_3O_4 fluid concentration in the tissues of living plants (John 1988). Therefore, to overcome such limitations, the coating provides Fe_3O_4 nanoparticles a large adsorption surface as well as biocompatible properties. In pumpkin (*Cucurbita pepo*), the presence of carbon-coated Fe_3O_4 at specific concentrations within some cells or in extracellular space could reduce oxidative stress as well as the amount of chemicals released into the environment (Ionnis and Anastasios 2002). Further, a study on the effect of tetramethyl ammonium hydroxide-coated Fe_3O_4 nanoparticles on the growth of corn depicted that the chlorophyll level increased at low-Fe_3O_4 nanoparticle fluid, whereas, at higher concentration, the chlorophyll level was inhibited. Nanoparticles of Fe_3O_4 fluid induced oxidative stress in the living plant tissue, affected photosynthesis, and resulted in reduced metabolic rates (Ma et al. 2010).

Cerium oxide: The cerium oxide (CeO_2) nanoparticles have received attention due to their excellent catalytic activities. These nanoparticles have various industrial applications and serve as potential antioxidants toward intercellular reactive oxygen species. Nanoparticles of CeO_2 could possibly have dual role as an oxidation catalyst as well as reduction catalyst, depending upon the conditions of the reaction (Chekin et al. 2012). The natural environment may get an exposure to CeO_2 nanoparticle from exhaust catalysts after deposition on plant, when they get collected with road run off or, by industrial waste waters that contain CeO_2 nanoparticles. The CeO_2 nanoparticles are the only tetravalent metal oxides that exhibited different effects on various plant species. However, the possible causes of its toxicity, transport, and fate needs to be further investigated (Chekin et al. 2012).

13.13 Conclusions

Generation of data in most of the disciplines in agriculture is time-consuming and expensive, which is especially true for soil-plant system. In the farm production system, complex intrinsic relationship of nanomaterials with nature and involvement of large number of variables make success of nanotechnology intervention uncertain. Therefore, foresight and patience would be essential for applying nanotechnology in agriculture and assessing of effect of engineered nanomaterials that are steadily entering into soil-plant system. For sure, craving for environmentally clean, highly productive agriculture to mitigate crisis-ridden farm production system looks to nanotechnology. It is pertinent to remember that a large number of nanomaterials existed since time immemorial in the soils, plants, and atmosphere (Theng et al. 2008; Wilson et al. 2008; Li et al. 2012), and played their role in soil-plant system. Therefore, nanotechnology is not new to nature, which calls for inventing and expanding it for agriculture and food systems.

References

Al-Amin Sadek MD, Jayasuriya HP (2007) Nanotechnology prospects in agricultural context: an overview. In: Proceedings of the international agricultural engineering conference, Bangkok, 3–6 December 2007, p. 548

Arora S, Sharma P, Kumar S, Nayar R, Khanna PK, Zaidi MGH (2012) Gold nanoparticle induced enhancement in growth and seed yield of *Brassica juncea*. Plant Growth Regul 66:303–310

Barik TK, Sahu B, Swain V (2008) Nanosilica—from medicine to pest control. Parasitol Res 103:253–258

Begum P, Ikhtiari R, Fugetsu B, Matsuoka M, Akasaka T, Watari F (2012) Phytotoxicity of multiwalled carbon nanotubes assessed by selected plant species in seedling stage. Appl Surf Sci 262:120–124

Bernhardt ES, Colman BP, Hochella MF, Cardinale BJ, Nisbet RM, Richardson CJ, Yin L (2010) an ecological perspective on nanomaterial impacts in the environment. J Environ Qual 39:1–12

Bhalla D, Mukhopadhyay SS (2010) Eutrophication: can nanophosphorous control this menace?—a preview. J Crop Weed 6:13–16
Boehn AL, Martinon I, Zerrouk R, Rung E, Fessi H (2003) Nano precipitation technique for encapsulation of agrochemical active ingredients. J Microencapsul 20:433–441
Brown GE Jr, Parks GA (2001) Sorption of trace elements on mineral surfaces: modern perspectives from spectroscopic studies, and comments on sorption in the marine environment. Int Geolog Rev 43:963–1073
Brown GE Jr, Heinrich VE, Casey WH, Clark DL, Eggleston C, Felmy A, Goodman DW, Grätzel M, Maciel G, McCarthy MI et al (1998) Metal oxide surfaces and their interactions with aqueous solutions and microbial organisms. Chem Rev 99:77–174
Checkin F, Bagheri S, Aroj AK, Hamid SBA (2012) Preparation and characterization of Ni (II)/polyacrylonitrile and carbon nanotube composite modified electrode and application for carbohydrates electrocatalytic oxidation. J Solid State Electrochem 16(10):3245–3251
Cheng C, Muller K, Koziol K, Porter AE (2009) Toxicity and imaging of multi-walled carbon nanotubes in human macrophage cells. Biomaterials 30:4152–4160
Chien SH, Prochnow LI, Cantarella H (2009) Recent developments of fertilizer production and use to improve nutrient efficiency and minimize environmental impacts. Adv Agron 102:267–322
Chinnamuthu CR, Boopathi PM (2009) Nanotechnology and agroecosystem. Madras Agric J 96:17–31
Currall SC, King EB, Lane N, Madera J, Turner S (2006) What drives public acceptance of nanotechnology? Nat Nanotechnol 1:153–155
De-Rosa MC, Monreal C, Schnitzer M, Walsh R, Sultan Y (2010) Nanotechnology in fertilizers. Nat Nanotechnol 5:91
Dwivedi RS, Randhawa NS (1974) Evaluation of a rapid test for hidden hunger of Zn in plants. Plant Soil 40:445–451
Editors (2011) Nature. The long game. Nature 473:419
Eichert T, Kurtz A, Steiner U, Goldback HE (2008) Size exclusion limits and lateral heterogeneity of the stomatal foliar uptake pathway for aqueous solutes and water suspended nanoparticles. Physiol Plant 134:151–160
Feizi H, Kamali M, Jafari L, Moghaddam PR (2013) Phytotoxic and stimulatory impacts of nanosized titanium dioxide on fennel (*Foeniculum vulgare*). Chemosphere 91:506–511
Fleischer A, O'Neill MA, Ehwald R (1999) The pore size of non-graminaceous plant cell wall is rapidly decreased by borate ester cross linking of pectic polysaccharide rhamnogalactouron II. Plant Physiol 121:829–838
Folete A, Masfarand J, Bigorgne E, Nahmani J, Chaurand P, Botta C, Labille J, Rose J, Férard JF, Cotelle S (2011) Environmental impact of sunscreen nanomaterials ecotoxicity and genotoxicity of altered TiO_2 nanocomposites on *Viciafaba*. Environ Pollut 159(10):2515–2522
Franke ME, Koplin TJ, Simon U (2006) Metal and metal oxide nanoparticles in chemiresistors: Does the nanoscale matter? Small 2:36–50
Gao J, Wang Y, Folta KM (2011) Polyhydroxy fullerenes: Beneficial effects on growth and lifespan in diverse biological models. PLOS ONE 6(5): Article Is e19976
Hall JS (2006) Nanofuture: what's next for nanotechnology?. Manas, New Delhi, India 333p
Ioannis AK, Anastasios IZ (2002) Removal of arsenic from contaminated water sources by sorption onto iron oxide coated polymeric materials. Water Res 36:5141–5155
Jia G (2005) Cyto-toxicity of carbon nanomaterials: single walled nanotube, multiwalled nanotube and fullerene. Environ Sci Technol 44:1036–1042
Jinghua G (2004) Synchrotron radiation, soft X-ray spectroscopy and nano-materials. J Nanotechnol 1:193–225
John AR (1988) The iron and molybdenum use efficiencies of plant growth with different energy, carbon and nitrogen sources. New Phytol 109(3):279–287
Kalpana-Sastry R, Rao NH, Cahoon R, Tucker T (2007) Can nanotechnology provide the innovations for a second green revolution in Indian agriculture? In: Proceedings of NSF science & engineering grantees conference, Paris, 3-6 December 2007

Khan SA, Mulvaney RL, Ellsworth TR (2013) The potassium paradox: implications for soil fertility, crop production and human health. Renew Agri Food Syst 29:3–27. doi:10.1017/S1742170513000318

Khot LR, Sankaran S, Maja JM, Ehsani R, Schuster EW (2012) Applications of nanomaterials in agricultural production and crop protection: a review. Crop Protec 35:64–70

Kuhn TS (1970) The Structures of Scientific Revolutions. University of Chicago Press, Illinois, USA

Kuzma J, VerHage P (2006) Nanotechnology in Agriculture and Food Production: Anticipated Applications. Project on Nanotechnology, Woodrow International Centre for Scholars

Leggo PJ (2000) An investigation of plant growth in an organo–zeolitic substrate and its ecological significance. Plant Soil 219:135–146

Lei Z (2007) Effect of nanoanatase TiO_2 on the photosynthesis of spinach chloroplasts under different light illuminations. Biol Trace Elem Res 119:68–76

Li W, He Y, Wu J, Xu J (2012) Extraction and characterization of natural soil nanoparticles from Chinese soils. Eur J Soil Sci 63:754–761. doi:10.1111/j.1365-2389.2012.01480.x

Wz Li, Xie SS, Xian LX (1996) Large scale synthesis of aligned carbon nanotubes. Science 274:1703–1707

Li Z (2003) Use of surfactant-modified zeolite as fertilizer carriers to control nitrate release. Micropor Mesopor Mater 61:181–188

Lin D, Xing B (2007) Phytotixicity of nanoparticles: Inhibition of seed germination and root growth. Environ Pollut 150:243–250

Liu X, Feng Z, Zhang S, Zhang J, Xiao Q, Wang Y (2006) Preparation and testing of cementing nano-subnano composites of slower controlled release of fertilizers. Sci Agri Sin J 39:1598–1604

Lou JC, Jung MJ, Yang HW, Han JY, Huang WH (2011) Removal of dissolved organic matter from raw water by single walled carbon nanotubes. J Environ Sci Health—Part A Toxic/Hazardous Substances and Environ Engg 46(12):1357–1365

Lower SK, Hochella MF Jr, Beveridge TL (2001) Bacterial recognition of mineral surfaces: nanoscale interactions between Shewanella and α-Fe00H. Science 292:1360–1363

Ma Y, Kuang L, He X (2010) Effects of rare earth oxide nanoparticles on root elongation in plants. Chemospehere 78:273–279

Maurice PA, Hochella MF (2008) Nanoscale particles and processes: a new dimension in Soil Science. Adv Agron 100:123–153

Millan G, Agosto F, Vazquez M (2008) Use of clinoptilolite as a carrier for nitrogen fertilizers in soils of the Pampean regions of Argentina. Cien Inv Agr 5(3):293–302

Ming DW, Boettinger JL (2001) Zeolites in soil environments. In: Bish DL, Ming DW (eds) Natural zeolites: occurrence, properties, applications. reviews in mineralogy and geochemistry, Vol 45. Mineralogical Society of America and Geochemical Society Washington DC, USA, pp 323-345

Mingfeng Q, Yufeng L, Tianlai K (2013) Nano-TiO_2 improve the photosynthesis of tomato leaves under mild heat stress, biological trace element research. Biol Trace Elem Res 156:323–328

Moore MN (2006) Do nanoparticles present eco-toxicological risks for the health of aquatic environment. Environ Int 32:967–976

Mukhopadhyay SS (2005) Weathering of soil minerals and distribution of elements: Pedochemical aspects. Clay Res 24:183–199

Mukhopadhyay SS (2013) Clay Minerals in Nanotechnology Venture in Agriculture. In: Adhikari T, Kundu S, Subba-Rao A (eds) Nanotechnology in Soil Science and Plant Nutrition. New India, New Delhi, India, pp 175–190

Mukhopadhyay SS (2014a) Nanotechnology in agriculture: prospects and constraints. Nanotechno Sci Appl 7: 63–71

Mukhopadhyay SS (2014b) Inventor; Indian Council of Agricultural Research (New Delhi) assignee. Nanofabrication process involving clay minerals as receptacles for manufacturing advanced nanomaterials including novel fertilizers. Indian Patent Application# 959/DEL/2014 dt. April 02,

2014. PCT International Patent Pub No WO/2015/150903; International Application No: PCT/IB2015/000433; Publ Date 08.10.2015; International Filing Date: 01.04. 2015

Mukhopadhyay SS (2014c) Inventor; Indian Council of Agricultural Research (New Delhi) assignee. Nanofabrication of phosphorus on kaolin mineral receptacles. Indian Patent Application# 989/DEL/2014 dt. April 07, 2014. PCT International Patent Pub No WO/2015/155585; International Application No: PCT/IB2015/000439; Publ Date: 15.10.2015; International Filing Date: 02.04. 2015

Mukhopadhyay SS (2014d) Inventor; Indian Council of Agricultural Research (New Delhi) assignee. Beneficiation of phosphate rock for the segregation of phosphorus containing heavy metal free minerals. Indian Patent Application# 1042/DEL/2014 dt April 16, 2014. PCT International Patent Pub No WO/2015/159140; International Application No.: PCT/IB2015/000437; Publ. Date: 22.10.2015.; International Filing Date: 02.04. 2015

Mukhopadhyay SS, Brar MS (2006) Mineralogy and management of soils rich in potassium containing minerals. Proceedings of the International Symposium on Balanced Fertilization, Ludhiana, India, during 22–25 November 2005, vol 1. International Potash Institute, Berne, Switzerland, pp 95–114

Mukhopadhyay SS, Sharma S (2013) Nanoscience and nanotechnology: Cracking prodigal farming. J Bionanosci 7(1):497–502

Naik RR, Stone MO (2005) Integrating biomimetics. Mater Today 8:18–26

Nair R, Varghese SH, Nair BG, Maekawa T, Yoshida Y, Kumar DS (2010) Nanoparticulate material delivery to plants. Plant Sci 179:154–163

Nakache E, Poulain N, Candau F, Orecchioni AM, Irache JM (1999) Biopolymer and polymer nanoparticles and their biomedical applications. In: Nalwa HS (ed) Handbook of Nanostructured Materials and Nanotechnology. Academic, New York, USA, pp 577–635

Nowack B, Bucheli TD (2007) Occurrence, behavior and effect of nanoparticles in environment. Env Pollut 150:5–22

Pickering KD, Wiesner MR (2005) Fullerol-sensitized production reactive oxygen species in aqueous solution. Environ Sci Technol? 39: 1359–1365 doi: 10.1021/es048940x

Nair R, Sheikh Wei Gao, Maekawa T, Yoshida Y, Ajayan PM, Sakthi-Kumar D (2012) Effect of carbon nanomaterials on the germination and growth of rice plants. J Nanosci Nanotechnol. doi:10.1166/jnn.2012.5775

Saxena M, Maity S, Sarkar S (2014) Carbon nanoparticles in 'biochan' boost wheat (*Triticum aestivum*) plant growth. RSC Advances 4:39948–39954

Sharma V, Shukla RK, Saxena M, Parmar D, Das M, Dhawan A (2009) DNA damaging potential of ZnO nanoparticles in human epidermal cells. Toxicol Lett 185(3):211–218

Shyla KK, Natarajan N (2014) Customising zinc oxide, silver and titanium dioxide nanoparticles for enhancing groundnut seed quality. Indian J Sci Tech 7(9):1376–1381

Singh M, Mukhopadhyay SS, Kiran-Jeet Kaur, R, Sharma S (2013) Inventors; Indian Council of Agricultural Research (New Delhi) assignee. Zinc in clay-mineral receptacles in nanoforms for their use as advanced materials including novel fertilizer. Indian Patent Application# 2093/DEL/2013 dt 11 July 2013. Patent Pub No 20925; The Indian Patent Office Journal Publ Date: 16.01.2015

Sooyeon L, Sunghyun K, Saeyeon K, Insook L (2013) Assessment of phytotoxicity of ZnO NPs on medicinal plant *Fagupyrumes culentum*. Environ Sci Pollut Res 20:848–854

Sparks DL (2004) Grand challenges and opportunities in basic soil science research and the discipline of soil science. ISBN 1 920842 26 8 Super Soil 2004 Published by the Author

Srinivasan C, Saraswati R (2010) Nanoagriculture-carbon nanotubes into living walled plant cells through cellulase-induced nanoholes. RSC Advances 2:398–400

Stampoulis D, Sinha SK, White JC (2009) Assay dependent phytotoxicity of nanoparticles to plants. Environ Sci Technol 43:9473–9479

Stella WYW, Priscilla TYL, Djurisic AB, Kenneth MYL (2010) Toxicities of nano zinc oxide to five marine organisms: influences of aggregate zinc size and iron solubility. Analyt Bioanalyt Chem 396(2):609–618

Stephan MK (2004) Iron oxide dissolution and solubility in the presence of siderophores. Aquat Sci 66:3–18

Sultan Y, Walsh R, Monreal CM, DeRosa MC (2009) Preparation of functional aptamer films using layer-by-layer self-assembly. Biomacromol J 10:1149–1154

Tan XM, Lin C, Fugetser B (2009) Studies on toxicity of multi walled carbon nanotubes on suspension of rice cells. Carbon 47:3479–3487

Tarafdar JC, Raliya R, Rathore I (2012a) Microbial synthesis of phosphorous nanoparticles from Tri-calcium phosphate using *Aspergillus tubingensis* TFR-5. J Bionanosci 6:84–89

Tarafdar JC, Xiang Y, Wang WN, Dong Q, Biswas P (2012b) Standardization of size, shape and concentration of nanoparticles for plant application. Appl Biol Res 14:138–144

Tarafdar JC, Devakumar C, Chhonkar PK (2013) National Academy of Agricultural Sciences Policy Paper#63 Nanotechnology in Agriculture: scope and current relevance. National Academy of Agricultural Sciences New Delhi India, p 20.

Theng BKG, Yuan G (2008) Nanoparticles in the soil environment. Elements 4:395–400

Tiwari DK, Dasgupta-Schubert N, Villasenor Cendejas LM, Villegas J, Carreto Montoya L, Borjas Garcia SE (2014) Interfacing carbon nanotubes (CNT) with plants: enhancement of growth, water and ionic nutrient uptake in maize (*Zea mays*) and implications for nano agriculture. Appl Nanosci 4:577–591

De La Torre-Roche R, Hawthorne J, Deng Y (2012) Fullerene enhanced accumulation of p, p'—DDE in agricultural crop species. Environ Sci Technol 46:9315–9323

De La Torre-Roche R, Hawthorne J, Deng Y (2013) MU carbon nanotubes and C_{60} fullerenes differentially impact the accumulation of weathered pesticides in farm agricultural plants. Environ Sci Technol 47:12539–12547

Tripathi S, Sonkar SK, Sarkar S (2011) Growth stimulation of gram (*Cicer arietinum*) plant by water soluble carbon nanotubes. Nanoscale 3:1176

United States Department of Agriculture (2002) Nanoscale science and engineering for agriculture and food systems. Report submitted to Cooperative State Research, Education and Extension Service, United States Department of Agriculture, National Planning Workshop. Washington DC, USA, 18–19 November 2002

United States Environmental Protection Agency (2007) Nanotechnology white paper. US Environmental Protection Agency Report EPA 100/B-07/001, Washington DC 20460, USA, p 135

Villagarcia H, Dervishi E, Silva K, Biris AS, Khodakovskaya MV (2012) Surface chemistry of carbon nanotubes impacts the growth and expression of water channel protein in tomato plants. Small 8:2328–2334

Wang X, Han H, Liu X, Gu X, Chen K, Lu D (2012) Multiwalled nano tubes can enhance root elongation of wheat (*Triticum aestivum*) plants. J Nanopart Res 14:841–851

Waychunas GA, Kim CS, Banfield JA (2005) Nanoparticulate iron oxide minerals in soils and sediments: unique properties and contaminant scavenging mechanisms. J Nanopart Res 7: 409–433

Wilson MA, Tran NH, Milev AS, Kannangara GSK, Volk H, Max Lu GQ (2008) Nanomaterials in soils. Geoderma 146:291–302

Zhu H, Han JQ, Jin Y (2008) Uptake, translocation, accumulation of manufactured iron oxide nanoparticles by pumpkin plants. J Environ Monit 10:713–717

Zubarev ER (2013) Nanoparticle synthesis: any way you want it. Nat Nanotechnol 8:396–397. doi:10.1038/nnano.2013.109

Chapter 14
Concerns About Nanoparticle Hazard to Human Health and Environment

Mohamed H. Lahiani and Mariya V. Khodakovskaya

Abstract The number of nanosized products has increased substantially during the last decade. A significant part of these products was developed for human health and fitness. Other nanoproducts belong to areas of automotive, food and beverage, cross-cutting, home and garden, electronics, computers, and appliances. Each year, concern over the exhaustive fate and behavior of nanoparticles (NPs) is increasing. To date, little is known about the safety of using and introducing NPs into the environment. Researchers have tackled this problem by focusing on the interactions of NPs with plants, animals, and human, by studying their behavior in aquatic, soil, and air systems. With the rapid advance of nanotechnology in different fields, regulation measures of the NPs face many challenges in front of contradictory reports and the complexity of properties of NPs.

Keywords Nanoparticles · Nanotoxicity · Human health · Environmental impact of nanoparticles · Genotoxicity

14.1 Introduction

The introduction of nanoparticles (NPs) in biological system applications has opened a new field of research named nanoscience. Specific properties of synthesized NPs (usually within the range of protein's size) played an important role in the development of biotechnological applications such as drug delivery, antigen

M.H. Lahiani (✉) · M.V. Khodakovskaya (✉)
The University of Arkansas at Little Rock, 2801 S. University Ave, Little Rock, AR 72204, USA
e-mail: mhlahiani@ualr.edu

M.V. Khodakovskaya
e-mail: mvkhodakovsk@ualr.edu

M.V. Khodakovskaya
Institute of Biology and Soil Science, Far-Eastern Branch of Russian Academy of Sciences, Vladivostok 690022, Russian Federation

C. Kole et al. (eds.), *Plant Nanotechnology*,
DOI 10.1007/978-3-319-42154-4_14

detection, shape recognition, bacterial detection, nucleic acid purification, DNA hybridization detection, and soil decontamination (Chan and Nie 1998; Wang et al. 2002). The fluorescent properties of certain metallic NPs or quantum dots have been used widely for the creation of biomedical sensors, and diagnostic and imaging tools (Chan and Nie 1998; Daniel et al. 2010; Schirhagl et al. 2012). For example, iron oxide NPs were used specifically for the detection of adenovirus-5 and herpes simplex in different cell lysates without the need of extensive sample preparation (Perez et al. 2003). Recently, a successful attempt to synthesize artificial antibodies by the double-imprinting process was done. It was shown that synthesized NPs had a higher selectivity and sensitivity than natural antibodies (Schirhagl et al. 2012). The research group stated that such biomimetic sensors could be useful in the biotechnology of insulin monitoring as well. Besides the use of nanosized particles as biological labels in different biomedical applications, NPs are being tested for other environmental, agricultural, and industrial purposes.

Agrochemicals and industrial wastes are composed of a variety of chemicals with toxic, carcinogenic, mutagenic, and teratogenic potentials, which can affect the ecosystem. The detoxification of soil and water resources remains one of the biggest challenges to ensuring safe water for all organisms. Nanotechnology has shown promising applications in this area ranging from accelerated decontamination processes (Guix et al. 2012; Soler et al. 2013) to water quality screening (Orozco et al. 2012). For example, carbon nanotubes demonstrate a better capability to remove heavy metal cations than activated carbon when added to media (Mubarak et al. 2014). Moreover, the carbon nanotubes can work as nanosorbent and remove organic pollutants or biological impurities such as bacterial spores (Upadhyayula et al. 2009). In another study, amphiphilic polyurethane (APU) NPs have shown a high affinity to phenanthrene and were able to cure soils contaminated with PAHs (Tungittiplakorn et al. 2004).

To reduce the adverse impacts of herbicides and insecticides on plant growth, new nanoformulations were prepared with the goal to carry the herbicide and release it over a long period. Particularly, alginate/chitosan NPs used in conjugation with paraquat, a quick-acting and nonselective herbicide, have shown a more target selectivity to plant with less contamination to the soil (dos Santos Silva et al. 2011).

While there is evidence in the potential uses of the NPs to solve problems in diverse fields, concerns related to their toxicity remain unresolved. Indeed, NPs, unlike conventional chemicals, have unique properties that can interfere with the toxicity assays. Any characteristics of NPs such as size, shape, surface charge, agglomeration, persistence, mobility, and bioavailability can create difficulties in the evaluation of their chemical behavior, solubility, oxidation, and reduction (López-Serrano et al. 2014). Moreover, different external factors could change the initial properties of NPs and, therefore, interfere with the response of these NPs to different organisms (Tejamaya et al. 2012).

The diverse applications of NPs will be possible solely if the safety of the nanosized materials is proved. There are many reports that contribute to our understanding of the impact of NPs on the environment and the human health. Here, we made an attempt to overview the recent reports on the hazard of NPs to the environment

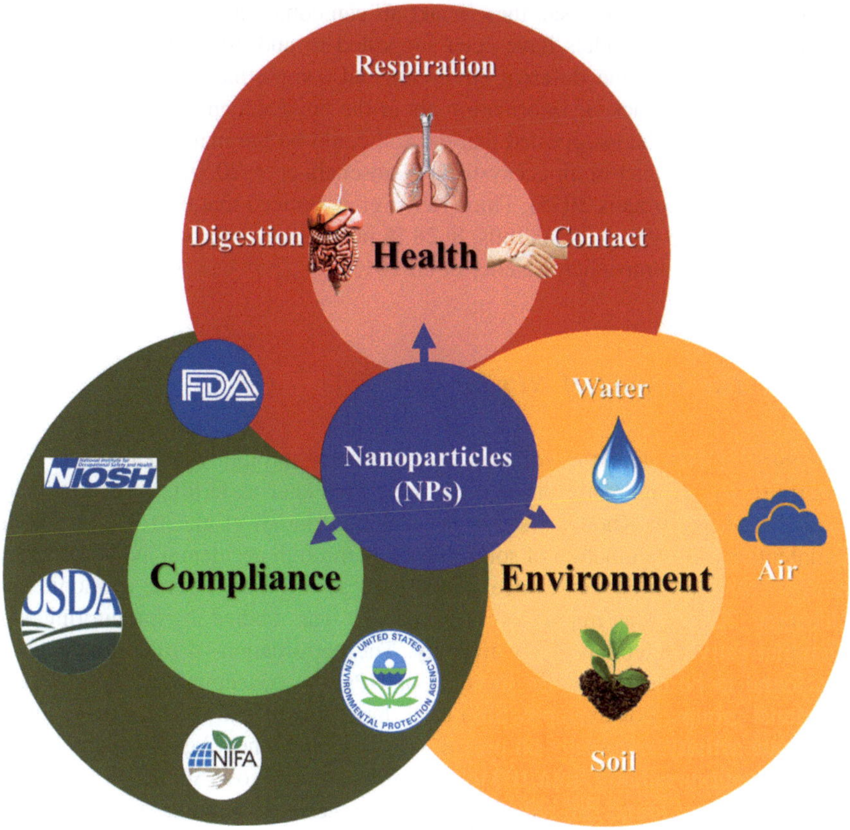

Fig. 14.1 Interconnection between all different areas affected by NPs' exposure. NPs can affect the environment through water, air, and soil contamination. Human health could be significantly impacted by NPs introduction. Compliance and regulation agencies are in proximity to the latest reports on NPs–human and NPs–environment interactions to create a new model that regulates NPs usage

through potential exposure to the air, soil, and water. Furthermore, the impact of NPs on human health through intentional or nonintentional exposure will be discussed. This chapter will also discuss the present compliances used by the regulatory agency to try to regulate the applications of NPs, and the challenges faced. Figure 14.1 shows a recapitulative structure of the different areas covered in this chapter.

14.2 Impact of NPs on the Environment

With the increasing applications of NPs in many commercially available products, new concerns about NPs' environmental implications are raised. The synthesized NPs can be released to the environment by different routes. From the various phases of production to their disposal, these materials can contaminate the soil, water, and air. Many exposure models have been suggested to understand the flow of these materials throughout the different environmental compartments (Gottschalk et al. 2009). Despite the lack of information about the environmental concentrations of NPs, many reports had suggested that the soil could be the major route of entry of NPs compared to water or air ecosystems (Gottschalk et al. 2009; Keller et al. 2013; Sun et al. 2014; Cornelis 2015). Thus, Keller et al. noticed that the release of NP is estimated to be 63–91 % in landfills, 8–28 % in soil, 0.4–7 % in water bodies, and 0.1–1.5 % in the atmosphere (Keller et al. 2013).

14.2.1 The Impact of NPs on Soil

The entry of NPs into the soil can be intentional through applications of NPs directly to soil or indirectly through the use of the later as fertilizers or pesticides (Gajjar et al. 2009; Khodakovskaya et al. 2013; Hamdi et al. 2014). Moreover, nonintentional contamination of soil can happen through accidental industrial spills, atmospheric deposition, or sewage sludge used as soil amendments (Simonin and Richaume 2015). The impact of NPs on soil was studied by monitoring soil microbial activity, biomass, and diversity. The perturbation of microbial activity was found to be dependent on the type of NPs, the concentration, the size, and functional groups. For example, Shin et al. have shown that a larger reduction in enzymatic activity, especially on urease, was observed when Ag NPs were applied at concentrations ranging from 100 to 1000 mg per kg of soil (Shin et al. 2012). Ag NPs have been found to reduce substrate-induced respiration and enzymatic activities at 0.14 mg per kg of soil when applied to mesocosms via sewage sludge (Colman et al. 2014). Carbon-based NPs including fullerenes (Tong et al. 2007), single-walled carbon nanotubes (Jin et al. 2013), and multi-walled carbon nanotubes (Khodakovskaya et al. 2013; Shrestha et al. 2013) were shown to have low-to-no toxicity toward soil microorganisms compared to other metal-based NPs. In addition to soil microbial activity, researchers rely on measuring microbial biomass to study the effect of NPs on soil microorganisms (Brookes 1995). When CeO_2 NPs (50–105 nm nominal size) have been added to soil, a significant impact on microbial biomass was detected (Antisari et al. 2011). Other metal-based NPs including Fe_2O_3, Fe_3O_4, TiO_2, and ZnO had no effect on microbial abundance at the corresponding concentrations tested (Antisari et al. 2013; Simonin et al. 2015). Similarly, carbon-based NPs were reported to have no effect on microbial biomass (Johansen et al. 2008), except for concentrations exceeding 250 mg NPs per kg of soil (Rodrigues et al. 2012; Jin et al. 2013). In contrast, the presence of nZVI in soil had a significant effect on denitrifying bacteria abundance and chloroaromatic mineralizing microorganisms (Fajardo et al. 2012; Tilston et al. 2013).

Even though the microbial biomass can remain constant, the ecosystem composition could have changed in response to soil contamination with NPs. That is why, the microbial genetic variability is a subtle methodology used to study the hazard of NPs to soil and the environment. Chronic contamination of soil with metal-based NPs including Ag, Fe_3O_4 had significantly altered the soil composition by the stimulation of specific groups (Ben-Moshe et al. 2013; Colman et al. 2013, 2014). A longer incubation (60 days) with the same NPs was not enough to recover the complete soil microbial community. Indeed, Ge et al. have shown that TiO_2 and ZnO NPs have decreased soil bacterial diversity by promoting organic pollutant decomposers (e.g., *Sphingomonadaceae, Streptomycetaceae,* and *Streptomyces*) and reducing nitrogen fixer species (*Rhizobiales, Bradyrhizobiaceae,* and *Bradyrhizobium*) (Ge et al. 2011). Carbon-based NPs can only significantly alter bacterial diversity and richness at high concentrations (>250 mg per kg of soil).

14.2.2 *The Impact of NPs on Aquatic Systems*

The emergence of engineered NPs into aquatic systems raised a number of concerns. Specific properties of nanoparticles such as shape, functional groups, agglomeration, chemistry, and capping agents may play an important role in the stability, solubility, and the availability of NPs in any aquatic system. Moreover, the synthesis of NPs was shown to differ from batch to batch preparations, which create a handicap in data reproducibility (Xia 2014). The characterization of most NPs is usually performed before their applications. However, once inside an aquatic system or inside a testing media, NPs can go through a transformation that could change their physiochemical properties. For instance, many NPs tend to agglomerate in solutions due to biotic or abiotic factors (Fatisson et al. 2012). Certain forces (e.g., Van der Waals) attract colloidal particles to each other, which further can change the solution's properties such as pH (Derjaguin and Landau 1993; Buffle et al. 1998; Domingos et al. 2009; Jiang et al. 2009). NPs are also known to interact with other molecules and proteins to form a "corona" (Cedervall et al. 2007; Casals et al. 2010), which can change their agglomeration kinetics among other properties. The size of the NPs is another major factor in the distribution and availability of the NPs to organisms. Rawson et al. showed that NPs at a specific size cannot cross protective membranes, such as the embryonic fish chorion, and translocate within the organism (Rawson et al. 2000). All such characteristics of NPs have to be considered in time of planning experiment in aquatic systems.

The impact of NPs on aquatic systems is complex. In fact, NPs can enter aquatic systems directly through aerial deposition and indirectly via river systems and affect a variety of organisms. Marine organisms are classified into 32 phyla of the animal kingdom and other representatives of the Archae, Bacteria, Chromista, Fungi, Plantae and Protozoans kingdoms. Many toxicology studies were performed on model representative water organisms. It was found that most of the toxicological tests were particles-specific (Bhatt and Tripathi 2011). Few studies on bacteria have suggested that Gram-positive bacteria are more susceptible to antimicrobial activity of metal oxide NPs than that of Gram-negative bacteria (Azam et al. 2012). Algae toxic levels were mainly dependent on particle type, the dissolution, and the propensity to compete with other ionic nutrients (Peng et al. 2011; Manzo et al. 2013). Moreover, toxic effects were noted by reduced swimming in *Daphnia magna* in the presence of silver NPs (Asghari et al. 2012), reduced growth and reproduction (Zhao and Wang 2011), bioaccumulation (Rosenkranz et al. 2009), digestive stress, and reduced feeding (Croteau et al. 2011). It is important to note that mortality, in the discussed studies, was mainly found at metal oxides' concentrations exceeding the environmental relevance (Baker et al. 2014). However, given that release of NPs can dramatically increase in the coming years, these studies could become significant. Therefore, the emission of the NPs into the air or our food chain should be controlled.

14.2.3 The Impact of NPs on the Air and the Atmosphere

The release of NPs to the atmosphere can happen nonintentionally (forest fires, volcanic activities, weathering, formation from clay minerals, soil erosion by wind and water, or dust storms from desert) or intentionally (various industrial and mechanical processes) (Smita et al. 2012; Keller et al. 2013). Natural or engineered NPs can be transported over thousands of kilometer and remain suspended in the air (Kellogg and Griffin 2006). While there are few efforts to estimate the amount of NPs released to the environment, it remains difficult to have an accurate estimation of such amount (DEFRA 2007; Keller et al. 2013). Some analytical evidence of the release of TiO_2, ZnO, and Ag NPs from paints, coating, and pigments has been demonstrated (Kägi et al. 2008; Gottschalk et al. 2009, 2013; Keller et al. 2013). The global material flows for ZnO-engineered NPs, considering the maximum production and emission rate estimates, have projected that the emission of ZnO NPs to the atmosphere is in the order of 90–578 tons/year (Keller et al. 2013).

14.3 Hazard of Nanoparticles to Human Health

Although NPs existed in nature for years, it is only recently that nanoscience started to identify them and study their impact on living organisms in terms of exposure and possible toxicity. Newly synthesized NPs have shown great potential for improving environmental quality or human health; however, their huge number and complex characteristics make it difficult to estimate their potential human hazard. There have been different in vitro and in vivo experiments that studied the mode of transmission of NPs on the cellular and organismal levels. Many research papers questioned the impact of NPs at three levels of exposures: air exposure, skin contact, and ingestion. Other exposures can be linked to engineered medical devices inside the body, which will not be specifically discussed here.

14.3.1 Impact of NPs on the Respiration System

The penetration of NPs to the lungs is relatively easy due to their small size. NPs can either be deposited in the lungs (Donaldson et al. 2006; Oberdörster 2010) or be translocated into other organs (Kreyling et al. 2002; Oberdorster 2004; Borm et al. 2006; Donaldson et al. 2006). Working in vitro, researchers were able to identify some of the potential effect of NPs on the lungs including cytotoxicity, genotoxicity, apoptosis, necrosis, inflammation, and cancer (Oberdorster 2004; Donaldson et al. 2006). The in vivo models further proved the potential translocation of NPs from the lungs to the vascular system and then to other organs (Kreyling et al. 2002; Sumner et al. 2010; Yamashita et al. 2011; Reijnders 2012). Kreyling et al. showed that the translocation rate of Ir NPs was dependent on the size of the particles. Indeed, the highest rate of NPs' translocation was in the order of 1–2 % (Kreyling et al. 2002). Studies on the uptake and translocation of gold NPs inside primary human dermal microvascular endothelial cells showed that the positively charged NPs were internalized to a higher extent compared to neutral or negatively charged NPs (Freese et al. 2012). Once inside the vascular system, NPs can reach the brain, liver, spleen, testis, stomach, and kidney (Aillon et al. 2009; Oberdörster 2010; Hubbs et al. 2011; Ngwa et al. 2011). Recent reports were able to show how NPs can reach the placenta and complicate pregnancy (Braydich-Stolle et al. 2010).

Once inhaled, NPs usually interact with the first line of immune defense, i.e., macrophages (Sibille and Reynolds 1990). The surface characteristics of the NPs are the determinants of how macrophages will respond to NPs. Indeed, macrophages clean up NPs from the lungs by engulfing them and transferring them to the vascular system to clear them from the body. However, certain properties of NPs such as agglomeration, shape, and rigidity can lead to frustrated phagocytosis and eventually become hazardous. For example, carbon nanotubes are few nanometers in width, but can reach few micrometers in length. The invagination of the nanotubes inside the macrophages can happen through different cell entry mechanisms. They can be taken up by cells via diffusion through pores or by endocytosis, or via ion transport systems (Shi Kam et al. 2004; Kam et al. 2005; Raffa et al. 2008; Murugan et al. 2015). However, nanotubes can agglomerate, and therefore, macrophages can fail to engulf the entire aggregates, leaving the potentially hazardous material behind. In fact, Shvedova et al. showed that single-walled carbon nanotubes were detected in mice lungs after one-year post-exposure and led to bronchopneumonia, lymphadenitis, and pulmonary fibrosis. However, the same group noted that the nanotubes did not cause any chronic inflammation as measured by BAL levels of PMN, AM, and cytokines (Shvedova et al. 2014).

Many surface characteristics of NPs were identified as relevant in inhalational hazard studies. These include surface area, surface charge, hydrophobicity, and surface chemistry (Reijnders 2012). Choi et al. showed that the charge of NPs is an important factor in their translocation from the lung to the vascular system. Indeed, noncationic NPs < 34 nm were able to pass blood barrier while cationic NPs failed to (Choi and Frangioni 2010). Other NPs such as Ag and CdSE quantum dots can release ions and, therefore, may become hazardous (Ahamed et al. 2010). Surface reactivity has been suggested to be used as a metric to NPs' hazard, especially that it is highly correlated with the generation of reactive oxygen species (Fubini et al. 2010; Oberdörster 2010; Maynard 2011). The correlation between NPs' properties and the observed inhalational hazard was observed in many model organisms. However, NPs remain prone to interact with other molecules and particles through the formation of a "corona" and, therefore, could be a good carrier of hazardous molecules deep into the lung (Choi and Frangioni 2010; Lesniak et al. 2012; Monopoli et al. 2012).

14.3.2 Impact of NPs During Dermal Exposure

Human skin is a good barrier for many foreign substances that tend to diffuse through the skin. The possibility of NPs to breach the stratum corneum (skin's outermost layer) and enter the epidermis and then to the dermis has been suggested. The hazard of NPs is dependent on their penetration to the living part of the skin. Initial reports on the hazard to TiO_2 and coated ZnO during dermal exposure showed no sign of penetration of the stratum corneum. The authors mentioned, however, that during dermabrasion and sunburn, NPs can breach the skin barrier and lead to the production of reactive oxygen species and the interaction with other cellular components such as proteins. Other NPs such as quantum dots were found in the dermis and epidermis after dermal exposure to intact porcine skin. The NPs can penetrate blood vessels, and this will lead to systemic exposure. Quantum dots with a cationic or neutral coating were reported to penetrate much faster than anionic ones.

Some report using carbon-based NPs showed that pristine fullerene (c60) in organic solvent and unpurified SWCNTs may penetrate the dermal barrier and be translocated within the skin tissues. In fact, mice SKH-1 exposed to unpurified SWCNTs for five days caused oxidative stress, depletion of glutathione, oxidation of protein thiols and carbonyls, elevated myeloperoxidase activity, an increase of dermal cell numbers, and skin thickening. Even though the author reported that 30 % of the SWCNTs mixture included iron contamination, it was hard to conclude whether the toxicity is related to the carbonaceous material or the metal contamination (Murray et al. 2009).

Reports on the toxicity of NPs during dermal exposure are scarce. However, they identified important scenarios where extreme precaution should be considered when contacting the NPs. Indeed, NPs should be considered as hazardous as any biological hazard materials (toxins, viruses, etc.) in the case of dermatitis, psoriasis, abrasion, or sunburn.

14.3.3 Impact of NPs on the Digestive System

The ingestion of NPs can be voluntary through the addition of NPs to food/cosmetic products or involuntary when NPs are emitted into the environment. NPs were shown to attract many food industries due to their capabilities to enhance color and taste. For instance, nano-SiO_2 is a good inorganic glidant that was used in many foodstuffs and pharmaceuticals. TiO_2 is another NP that was used as a whitening agent of the powdered donuts. Some personal products such as lip gloss were found to contain TiO_2-based nanoparticulate sunscreen (Lohani et al. 2014). With the lack of studies concerning the fate of NPs in the human gut, concerns related to food-containing NPs started to flourish.

The initial work of investigating the effect of ingestion of large colloidal silver was linked to dysfunctioning of the central nervous system, liver, kidneys, and the immune system (Sharma and Sharma 2007; Panyala et al. 2008). Rats gavaged with silver NPs for 13 weeks were shown to induce changes in the ileal mucosal microbial population, the intestinal gene expression, and the gut microbiota (Williams et al. 2014). The researchers have shown that the highest shifts in the gut microbial population were recorded with the ingestion of the smallest silver NPs used (10 nm). Moreover, the analysis of the host gene expression has shown a decrease in the expression of important immunomodulatory genes including MUC3, TLR2, TLR4, GPR43, and FOXP3 (Williams et al. 2014).

Few studies using carbon-based NPs showed that these materials after purification can have low-to-no toxicity after ingestion. For instance, rats gavaged for 28 days with SWCNTs or MWCNTs have survived and shown no adverse effects (Matsumoto et al. 2012). In another study, pristine (nonpurified) SWCNTs led to the oxidative damage of DNA as the premutagenic 8-oxo-7,8-dihydro-2'-deoxyguanosine (8-oxodG) in the liver and lungs of rat model (Folkmann et al. 2009). Reports on the effects of MWCNTs on gut bacteria are limited and sometimes contradictory. In a recent study, it has been shown that the CNTs could modify the fatty acid composition of the bacterial membranes (Zhu et al. 2014). Bacteria adapted this mechanism as a mode to mitigate the toxic effects of CNTs. Another study reported the antibacterial effect of MWCNTs against human commensal bacterial population that included *Lactobacillus acidophilus, Escherichia coli, Bifidobacterium adolescentis, Staphylococcus aureus,* and *Enterococcus faecalis* (Chen et al. 2013). Other studies showed that MWCNTs had no significant antibacterial effect against *Salmonella enterica* typhimurium and *Bacillus subtilis* (Arias and Yang 2009). Moreover, only high concentrations of MWCNTs in the order of 1–3 mg/ml were able to show a significant decrease in cell count when tested on *Methylobacterium* spp. and *Sphingomonas* spp. (Kang et al. 2008; Murugan and Vimala 2011; Seo et al. 2014).

Depending on the size and the properties of the NPs, the translocation of the NPs from the intestine to the cardiovascular system remains a possibility (Florence 2005). The presence of the NPs in the blood system may lead to their deposition into other organs and subsequently possible inflammation. In the case of ingestion of ion-releasing NPs such as Ag, ZnO, Cu, and CdSe quantum dots, the assessment of risks becomes more complicated (Van der Zande et al. 2012; Wang et al. 2008). After 28-day oral exposure of Ag NPs, Van der Zande et al. showed that unlike Ag NPs, Ag ions were able to pass the intestinal barriers and reach other organs (Van der Zande et al. 2012).

The risk assessment studies usually choose dose–response relationship according to the NPs concentration, which can be irrelevant sometimes according to the surface area or the reactivity of the NPs. Moreover, many of current researches do not pay attention to the ways in which the NPs are exposed to the organisms. While there is no doubt about the release of NPs to the environment, risk assessment measures should be consistent with the magnitude and the frequency of this flow. Nanoscience has brought new dimensions to how hazard to inorganic particles is viewed. Previously, the chemical composition of the compound determined its risks or hazard; however, it becomes obvious that other factors such as size, composition, and shape should be considered by government agencies in setting the correct compliances that regulate NPs' uses.

14.4 Current Compliances

The current rise of nanotechnological industry and variety of nanobiotechnological applications are raising legitimate questions about the evaluation of risks associated with production of different NPs as well as products containing nanosized compounds. In fact, the interaction of nanosized materials with organisms is dramatically different compared with identical regular-sized materials (Vishwakarma et al. 2010). Additionally, NPs can easily overcome cellular barriers, enter organism, and stay in cells for a long time (Nelson et al. 1993; Zhao and Wang 2011; Shvedova et al. 2014). In the recent past, worldwide governmental organizations and major nanotechnological companies made significant efforts for regulation of flow of nanoproducts and creation of the first nanopolices. Such efforts were based on new scientific discoveries of possible toxicity of NPs for humans, animals, plants, microbes, and other organisms. In the United States, several government organizations including US Department of Agriculture (USDA-NIFA), US Environmental Protection Agency (EPA), US Food and Drug Administration (FDA), Consumer Product Safety Commission, and National Institute for Occupational Safety and Health (NIOSH) are providing guidance on the safety of nanotechnological products and produce recommendations for regulatory aspects. On June 24, 2014, FDA announced three key guidance documents focused onto the use of nanotechnology in regulated products, especially cosmetics and food. On August 5, 2015, FDA published one final guidance paper associated with the use of nanotechnological substances in food for animals (http://www.fda.gov/ScienceResearch/SpecialTopics/Nanotechnology/default.htm). Scientists of EPA are working on the development of scientific base for prediction of how NPs will behave during manufacturing, product use, and end-of-life disposal. EPA is making available information about environmental fate, transport, transformation, biodistribution, exposure, and toxicity of NPs and nanoproducts to humans and other species (http://www2.epa.gov/chemical-research/research-evaluating-NPs-chemical-safety). Both organizations (FDA and EPA) indicated that current law is sufficient for regulation of nano-based-products. However, all government organizations agree that some regulatory mechanisms including particular testing protocols may require constant updates (Watson et al. 2011). NIOSH is at the forefront of US research to understand the occupational health implications of NPs for researchers, product innovators, industry employers, and workers exposed to nanorisks. Particularly, NIOSH offers guidelines for working with specific NPs and provides a global online library on NPs as a working resource for researchers and public. It is important that NIOSH publishes new findings and recommendations consistent with the best scientific knowledge (http://www.cdc.gov/niosh/docs/2009-125/; http://goodnanoguide.org/). The industry can also play an important role in the development of frameworks and practices to protect workers, the public, and the environment. Many US nanotechnological companies are trying to work at the forefront of developing best and transparent practices of protection of workers.

The efforts to understand and regulate production as well as the fate of nanoproducts were made in different countries including India (Sahoo 2013) and European Union (Watson et al. 2011). For example, general principles and obligations for nanotechnology in EU were described in Food Law Regulations (EC) 178/2002 (Watson et al. 2011). It is obvious that an international collaboration and cooperation of worldwide organizations dedicated to the establishment of nanopolices are the promising approaches for significant reduction of environmental nanorisks.

14.5 Conclusion

The global socioeconomic value of nanotechnology is increasing with the increase of applications of NPs in different fields. NPs can be released to different environmental compartments such as air, water, soil, and landfills, and affect human health directly or indirectly. With the increasing growth of human population and the necessity to have clean water, air, and soil, many efforts have to be implemented to keep the environment safe. Nanotechnology has shown a significant impact in the environment field by providing new technologies and innovative solutions for monitoring and solving related environmental issues (Karn et al. 2009; Qu et al. 2012). On the other hand, NPs can impact the environment negatively upon their release intentionally or nonintentionally. Taken into consideration the fact that NPs have unique properties, it becomes necessary to evaluate the influence of these materials on the environment by assessing the risks associated with each type of NPs independently. Not all NPs can have the same characteristics of degradation, biotransformation, coronas-forming ability, tendency, toxicity, etc., once released to the environment. Therefore, the generalization of NPs' effects should be avoided. An explicit evaluation of the NPs' characteristics remains the basis of each rigorous research work to assure reproducibility of the results.

References

Ahamed M, AlSalhi MS, Siddiqui M (2010) Silver nanoparticle applications and human health. Clin Chim Acta 411:1841–1848

Aillon KL, Xie Y, El-Gendy N, Berkland CJ, Forrest ML (2009) Effects of nanomaterial physicochemical properties on in vivo toxicity. Adv Drug Deliv Rev 61:457–466

Antisari LV, Carbone S, Fabrizi A, Gatti A, Vianello G (2011) Response of soil microbial biomass to CeO2 nanoparticles. EQA-Int J Environ Qual 7:1–16

Antisari LV, Carbone S, Gatti A, Vianello G, Nannipieri P (2013) Toxicity of metal oxide (CeO2, Fe3O4, SnO2) engineered nanoparticles on soil microbial biomass and their distribution in soil. Soil Biol Biochem 60:87–94

Arias LR, Yang L (2009) Inactivation of bacterial pathogens by carbon nanotubes in suspensions. Langmuir 25:3003–3012

Asghari S, Johari SA, Lee JH, Kim YS, Jeon YB, Choi HJ, Moon MC, Yu IJ (2012) Toxicity of various silver nanoparticles compared to silver ions in Daphnia magna. J Nanobiotechnol 10:1–14

Azam A, Ahmed AS, Oves M, Khan MS, Habib SS, Memic A (2012) Antimicrobial activity of metal oxide nanoparticles against Gram-positive and Gram-negative bacteria: a comparative study. Int J Nanomed 7:6003

Baker TJ, Tyler CR, Galloway TS (2014) Impacts of metal and metal oxide nanoparticles on marine organisms. Environ Pollut 186:257–271

Ben-Moshe T, Frenk S, Dror I, Minz D, Berkowitz B (2013) Effects of metal oxide nanoparticles on soil properties. Chemosphere 90:640–646

Bhatt I, Tripathi BN (2011) Interaction of engineered nanoparticles with various components of the environment and possible strategies for their risk assessment. Chemosphere 82:308–317

Borm P, Robbins D, Haubold S, Kuhlbusch T, Fissan H, Donaldson K, Schins R, Stone V, Kreyling W, Lademann J, Krutmann J, Warheit D, Oberdorster E (2006) The potential risks of nanomaterials: a review carried out for ECETOC. Part Fibre Toxicol 3:11

Braydich-Stolle LK, Lucas B, Schrand A, Murdock RC, Lee T, Schlager JJ, Hussain SM, Hofmann MC (2010) Silver nanoparticles disrupt GDNF/Fyn kinase signalling in spermatogonial stem cells. Toxicol Sci 116:577–589

Brookes P (1995) The use of microbial parameters in monitoring soil pollution by heavy metals. Biol Fertil Soils 19:269–279

Buffle J, Wilkinson KJ, Stoll S, Filella M, Zhang J (1998) A generalized description of aquatic colloidal interactions: the three-colloidal component approach. Environ Sci Technol 32:2887–2899

Casals E, Pfaller T, Duschl A, Oostingh GJ, Puntes V (2010) Time evolution of the nanoparticle protein corona. ACS Nano 4:3623–3632

Cedervall T, Lynch I, Lindman S, Berggard T, Thulin E, Nilsson H, Dawson KA, Linse S (2007) Understanding the nanoparticle-protein corona using methods to quantify exchange rates and affinities of proteins for nanoparticles. Proc Natl Acad Sci USA 104:2050–2055

Chan WC, Nie S (1998) Quantum dot bioconjugates for ultrasensitive nonisotopic detection. Science 281:2016–2018

Chen H, Wang B, Gao D, Guan M, Zheng L, Ouyang H, Chai Z, Zhao Y, Feng W (2013) Broad-spectrum antibacterial activity of carbon nanotubes to human gut bacteria. Small 9:2735–2746

Choi HS, Frangioni JV (2010) Nanoparticles for biomedical imaging: fundamentals of clinical translation. Mol Imaging 9:291–310

Colman BP, Arnaout CL, Anciaux S, Gunsch CK, Hochella MF Jr, Kim B, Lowry GV, McGill BM, Reinsch BC, Richardson CJ (2013) Low concentrations of silver nanoparticles in biosolids cause adverse ecosystem responses under realistic field scenario. PLoS ONE 8: e57189

Colman BP, Espinasse B, Richardson CJ, Matson CW, Lowry GV, Hunt DE, Wiesner MR, Bernhardt ES (2014) Emerging contaminant or an old toxin in disguise? silver nanoparticle impacts on ecosystems. Environ Sci Technol 48:5229–5236

Cornelis G (2015) Fate descriptors for engineered nanoparticles: the good, the bad, and the ugly. Environ Sci Nano 2:19–26

Croteau M, Dybowska AD, Luoma SN, Valsami-Jones E (2011) A novel approach reveals that zinc oxide nanoparticles are bioavailable and toxic after dietary exposures. Nanotoxicology 5:79–90

Daniel SK, Tharmaraj V, Sironmani TA, Pitchumani K (2010) Toxicity and immunological activity of silver nanoparticles. Appl Clay Sci 48:547–551

DEFRA (2007) Characterising the potential risks posed by engineered nanoparticles. Department for Environment, Food and Rural Affairs

Derjaguin B, Landau L (1993) Theory of the stability of strongly charged lyophobic sols and of the adhesion of strongly charged particles in solutions of electrolytes. Prog Surf Sci 43:30–59

Domingos RF, Tufenkji N, Wilkinson KJ (2009) Aggregation of titanium dioxide nanoparticles: role of a fulvic acid. Environ Sci Technol 43:1282–1286

Donaldson K, Aitken R, Tran L, Stone V, Duffin R, Forrest G, Alexander A (2006) Carbon nanotubes: a review of their properties in relation to pulmonary toxicology and workplace safety. Toxicol Sci 92:5–22

dos Santos Silva M, Cocenza DS, Grillo R, de Melo Nathalie, Silva Ferreira, Tonello PS, de Oliveira LC, Cassimiro DL, Rosa AH, Fraceto LF (2011) Paraquat-loaded alginate/chitosan nanoparticles: preparation, characterization and soil sorption studies. J Hazard Mater 190:366–374

Fajardo C, Ortíz L, Rodríguez-Membibre M, Nande M, Lobo M, Martin M (2012) Assessing the impact of zero-valent iron (ZVI) nanotechnology on soil microbial structure and functionality: a molecular approach. Chemosphere 86:802–808

Fatisson J, Quevedo IR, Wilkinson KJ, Tufenkji N (2012) Physicochemical characterization of engineered nanoparticles under physiological conditions: effect of culture media components and particle surface coating. Colloids Surf B 91:198–204

Florence AT (2005) Nanoparticle uptake by the oral route: fulfilling its potential? Drug Discov Today Technol 2:75–81

Folkmann JK, Risom L, Jacobsen NR, Wallin H, Loft S, Moller P (2009) Oxidatively damaged DNA in rats exposed by oral gavage to C60 fullerenes and single-walled carbon nanotubes. Environ Health Perspect 5:703–708

Freese C, Uboldi C, Gibson MI, Unger RE, Weksler BB, Romero IA, Couraud P, Kirkpatrick CJ (2012) Uptake and cytotoxicity of citrate-coated gold nanospheres: comparative studies on human endothelial and epithelial cells. Part Fibre Toxicol 9:23

Fubini B, Ghiazza M, Fenoglio I (2010) Physico-chemical features of engineered nanoparticles relevant to their toxicity. Nanotoxicology 4:347–363

Gajjar P, Pettee B, Britt D, Huang W, Johnson W, Anderson A (2009) Antimicrobial activities of commercial nanoparticles against an environmental soil microbe, PSEUDOMONAS putida KT2440. J Biol Eng 3:9

Ge Y, Schimel JP, Holden PA (2011) Evidence for negative effects of TiO2 and ZnO nanoparticles on soil bacterial communities. Environ Sci Technol 45:1659–1664

Gottschalk F, Sonderer T, Scholz RW, Nowack B (2009) Modeled environmental concentrations of engineered nanomaterials (TiO2, ZnO, Ag, CNT, fullerenes) for different regions. Environ Sci Technol 43:9216–9222

Gottschalk F, Sun T, Nowack B (2013) Environmental concentrations of engineered nanomaterials: review of modeling and analytical studies. Environ Pollut 181:287–300

Guix M, Orozco J, García M, Gao W, Sattayasamitsathit S, Merkoçi A, Escarpa A, Wang J (2012) Superhydrophobic alkanethiol-coated microsubmarines for effective removal of oil. ACS Nano 6:4445–4451

Hamdi H, De La Torre-Roche R, Hawthorne J, White JC (2014) Impact of non-functionalized and amino-functionalized multiwall carbon nanotubes on pesticide uptake by lettuce (*Lactuca sativa* L.). Nanotoxicology 1–9

Hubbs AF, Mercer RR, Benkovic SA, Harkema J, Sriram K, Schwegler-Berry D, Goravanahally MP, Nurkiewicz TR, Castranova V, Sargent LM (2011) Nanotoxicology–a pathologist's perspective. Toxicol Pathol 39:301–324

Jiang J, Oberdörster G, Biswas P (2009) Characterization of size, surface charge, and agglomeration state of nanoparticle dispersions for toxicological studies. J Nanopart Res 11:77–89

Jin L, Son Y, Yoon TK, Kang YJ, Kim W, Chung H (2013) High concentrations of single-walled carbon nanotubes lower soil enzyme activity and microbial biomass. Ecotoxicol Environ Saf 88:9–15

Johansen A, Pedersen AL, Jensen KA, Karlson U, Hansen BM, Scott-Fordsmand JJ, Winding A (2008) Effects of C60 fullerene nanoparticles on soil bacteria and protozoans. Environ Toxicol Chem 27:1895–1903

Kägi R, Ulrich A, Sinnet B, Vonbank R, Wichser A, Zuleeg S, Simmler H, Brunner S, Vonmont H, Burkhardt M (2008) Synthetic TiO 2 nanoparticle emission from exterior facades into the aquatic environment. Environ Pollut 156:233–239

Kam NW, O'Connell M, Wisdom JA, Dai H (2005) Carbon nanotubes as multifunctional biological transporters and near-infrared agents for selective cancer cell destruction. Proc Natl Acad Sci USA 102:11600–11605

Kang S, Herzberg M, Rodrigues DF, Elimelech M (2008) Antibacterial effects of carbon nanotubes: size does matter! Langmuir 24:6409–6413

Karn B, Kuiken T, Otto M (2009) Nanotechnology and in situ remediation: a review of the benefits and potential risks. Environ Health Perspect 12:1813–1831

Keller AA, McFerran S, Lazareva A, Suh S (2013) Global life cycle releases of engineered nanomaterials. J Nanopart Res 15:1–17

Kellogg CA, Griffin DW (2006) Aerobiology and the global transport of desert dust. Trends Ecol Evol 21:638–644

Khodakovskaya MV, Kim B, Kim JN, Alimohammadi M, Dervishi E, Mustafa T, Cernigla CE (2013) Carbon nanotubes as plant growth regulators: effects on tomato growth, reproductive system, and soil microbial community. Small 9:115–123

Kreyling W, Semmler M, Erbe F, Mayer P, Takenaka S, Schulz H, Oberdörster G, Ziesenis A (2002) Translocation of ultrafine insoluble iridium particles from lung epithelium to extrapulmonary organs is size dependent but very low. J Toxicol Environ Health Part A 65:1513–1530

Lesniak A, Fenaroli F, Monopoli MP, Åberg C, Dawson KA, Salvati A (2012) Effects of the presence or absence of a protein corona on silica nanoparticle uptake and impact on cells. ACS Nano 6:5845–5857

Lohani A, Verma A, Joshi H, Yadav N, Karki N (2014) Nanotechnology-based cosmeceuticals. ISRN Dermatol, ID 843687. http://dx.doi.org/10.1155/2014/843687

López-Serrano A, Olivas RM, Landaluze JS, Cámara C (2014) Nanoparticles: a global vision. Characterization, separation, and quantification methods. Potential environmental and health impact. Anal Methods 6:38–56

Manzo S, Miglietta ML, Rametta G, Buono S, Di Francia G (2013) Toxic effects of ZnO nanoparticles towards marine algae *Dunaliella tertiolecta*. Sci Total Environ 445:371–376

Matsumoto M, Serizawa H, Sunaga M, Kato H, Takahashi M, Hirata-Koizumi M, Ono A, Kamata E, Hirose A (2012) No toxicological effects on acute and repeated oral gavage doses of single-wall or multi-wall carbon nanotube in rats. J Toxicol Sci 37:463–474

Maynard AD (2011) Don't define nanomaterials. Nature 475:31

Monopoli MP, Åberg C, Salvati A, Dawson KA (2012) Biomolecular coronas provide the biological identity of nanosized materials. Nat Nanotechnol 7:779–786

Mubarak N, Sahu J, Abdullah E, Jayakumar N (2014) Removal of heavy metals from wastewater using carbon nanotubes. Separ Purif Rev 43:311–338

Murray A, Kisin E, Leonard S, Young S, Kommineni C, Kagan V, Castranova V, Shvedova A (2009) Oxidative stress and inflammatory response in dermal toxicity of single-walled carbon nanotubes. Toxicology 257:161–171

Murugan E, Vimala G (2011) Effective functionalization of multiwalled carbon nanotube with amphiphilic poly (propyleneimine) dendrimer carrying silver nanoparticles for better dispersability and antimicrobial activity. J Colloid Interface Sci 357:354–365

Murugan K, Choonara YE, Kumar P, Bijukumar D, du Toit LC, Pillay V (2015) Parameters and characteristics governing cellular internalization and trans-barrier trafficking of nanostructures. Int J Nanomed 10:2191

Nelson MA, Domann FE, Bowden GT, Hooser SB, Fernando Q, Carter DE (1993) Effects of acute and subchronic exposure of topically applied fullerene extracts on the mouse skin. Toxicol Ind Health 9:623–630

Ngwa HA, Kanthasamy A, Gu Y, Fang N, Anantharam V, Kanthasamy AG (2011) Manganese nanoparticle activates mitochondrial dependent apoptotic signalling and autophagy in dopaminergic neuronal cells. Toxicol Appl Pharmacol 256:227–240

Oberdorster E (2004) Manufactured nanomaterials (fullerenes, C60) induce oxidative stress in the brain of juvenile large mouth bass. Environ Health Perspect 112:1058–1062

Oberdörster G (2010) Safety assessment for nanotechnology and nanomedicine: concepts of nanotoxicology. J Intern Med 267:89–105

Orozco J, García-Gradilla V, D'Agostino M, Gao W, Cortés A, Wang J (2012) Artificial enzyme-powered microfish for water-quality testing. ACS Nano 7:818–824

Panyala NR, Peña-Méndez EM, Havel J (2008) Silver or silver nanoparticles: a hazardous threat to the environment and human health. J Appl Biomed 6:117–129

Peng X, Palma S, Fisher NS, Wong SS (2011) Effect of morphology of ZnO nanostructures on their toxicity to marine algae. Aquat Toxicol 102:186–196

Perez JM, Simeone FJ, Saeki Y, Josephson L, Weissleder R (2003) Viral-induced self-assembly of magnetic nanoparticles allows the detection of viral particles in biological media. J Am Chem Soc 125:10192–10193

Qu X, Brame J, Li Q, Alvarez PJ (2012) Nanotechnology for a safe and sustainable water supply: enabling integrated water treatment and reuse. Acc Chem Res 46:834–843

Raffa V, Ciofani G, Nitodas S, Karachalios T, D'Alessandro D, Masini M, Cuschieri A (2008) Can the properties of carbon nanotubes influence their internalization by living cells? Carbon 46:1600–1610

Rawson DM, Zhang T, Kalicharan D, Jongebloed WL (2000) Field emission scanning electron microscopy and transmission electron microscopy studies of the chorion, plasma membrane and syncytial layers of the gastrula-stage embryo of the zebrafish Brachydanio rerio: a consideration of the structural and functional relationships with respect to cryoprotectant penetration. Aquac Res 31:325–336

Reijnders L (2012) Human health hazards of persistent inorganic and carbon nanoparticles. J Mater Sci 47:5061–5073

Rodrigues DF, Jaisi DP, Elimelech M (2012) Toxicity of functionalized single-walled carbon nanotubes on soil microbial communities: implications for nutrient cycling in soil. Environ Sci Technol 47:625–633

Rosenkranz P, Chaudhry Q, Stone V, Fernandes TF (2009) A comparison of nanoparticle and fine particle uptake by *Daphnia magna*. Environ Toxicol Chem 28:2142–2149

Sahoo S (2013) Would you mind, if we record this? Perceptions on regulation and responsibility among Indian nanoscientists. NanoEthics 7:231–249

Schirhagl R, Latif U, Podlipna D, Blumenstock H, Dickert FL (2012) Natural and biomimetic materials for the detection of insulin. Anal Chem 84:3908–3913

Seo Y, Hwang J, Kim J, Jeong Y, Hwang M, Choi J, Seo Y, Hwang J, Kim J, Jeong Y (2014) Antibacterial activity and cytotoxicity of multi-walled carbon nanotubes decorated with silver nanoparticles. Int J Nanomed 9:4621–4629

Sharma HS, Sharma A (2007) Nanoparticles aggravate heat stress induced cognitive deficits, blood–brain barrier disruption, edema formation and brain pathology. Prog Brain Res 162:245–273

Shi Kam NW, Jessop TC, Wender PA, Dai H (2004) Nanotube molecular transporters: internalization of carbon nanotube-protein conjugates into mammalian cells. J Am Chem Soc 126:6850–6851

Shin Y, Kwak JI, An Y (2012) Evidence for the inhibitory effects of silver nanoparticles on the activities of soil exoenzymes. Chemosphere 88:524–529

Shrestha B, Acosta-Martinez V, Cox SB, Green MJ, Li S, Cañas-Carrell JE (2013) An evaluation of the impact of multiwalled carbon nanotubes on soil microbial community structure and functioning. J Hazard Mater 261:188–197

Shvedova AA, Yanamala N, Kisin ER, Tkach AV, Murray AR, Hubbs A, Chirila MM, Keohavong P, Sycheva LP, Kagan VE, Castranova V (2014) Long-term effects of carbon containing engineered nanomaterials and asbestos in the lung: one year postexposure comparisons. Am J Physiol Lung Cell Mol Physiol 306:L170–L182

Sibille Y, Reynolds HY (1990) Macrophages and polymorphonuclear neutrophils in lung defense and injury. Am Rev Respir Dis 141:471–501

Simonin M, Richaume A (2015) Impact of engineered nanoparticles on the activity, abundance, and diversity of soil microbial communities: a review. Environ Sci Pollut Res 22(18):13710–13723

Simonin M, Guyonnet JP, Martins JM, Ginot M, Richaume A (2015) Influence of soil properties on the toxicity of TiO2 nanoparticles on carbon mineralization and bacterial abundance. J Hazard Mater 283:529–535

Smita S, Gupta SK, Bartonova A, Dusinska M, Gutleb AC, Rahman Q (2012) Nanoparticles in the environment: assessment using the causal diagram approach. Environ Health 11:S13

Soler L, Magdanz V, Fomin VM, Sanchez S, Schmidt OG (2013) Self-propelled micromotors for cleaning polluted water. ACS Nano 7:9611–9620

Sumner SC, Fennell TR, Snyder RW, Taylor GF, Lewin AH (2010) Distribution of carbon-14 labeled C60 ([14C] C60) in the pregnant and in the lactating dam and the effect of C60 exposure on the biochemical profile of urine. J Appl Toxicol 30:354–360

Sun TY, Gottschalk F, Hungerbühler K, Nowack B (2014) Comprehensive probabilistic modelling of environmental emissions of engineered nanomaterials. Environ Pollut 185:69–76

Tejamaya M, Römer I, Merrifield RC, Lead JR (2012) Stability of citrate, PVP, and PEG coated silver nanoparticles in ecotoxicology media. Environ Sci Technol 46:7011–7017

Tilston EL, Collins CD, Mitchell GR, Princivalle J, Shaw LJ (2013) Nanoscale zerovalent iron alters soil bacterial community structure and inhibits chloroaromatic biodegradation potential in Aroclor 1242-contaminated soil. Environ Pollut 173:38–46

Tong Z, Bischoff M, Nies L, Applegate B, Turco RF (2007) Impact of fullerene (C60) on a soil microbial community. Environ Sci Technol 41:2985–2991

Tungittiplakorn W, Lion LW, Cohen C, Kim J (2004) Engineered polymeric nanoparticles for soil remediation. Environ Sci Technol 38:1605–1610

Upadhyayula VK, Deng S, Smith GB, Mitchell MC (2009) Adsorption of Bacillus subtilis on single-walled carbon nanotube aggregates, activated carbon and NanoCeram™. Water Res 43:148–156

Van der Zande M, Vandebriel RJ, Van Doren E, Kramer E, Herrera Rivera Z, Serrano-Rojero CS, Gremmer ER, Mast J, Peters RJ, Hollman PC (2012) Distribution, elimination, and toxicity of silver nanoparticles and silver ions in rats after 28-day oral exposure. ACS Nano 6:7427–7442

Vishwakarma V, Samal SS, Manoharan N (2010) Safety and risk associated with nanoparticles-a review. J Miner Mater Charact Eng 9:455

Wang J, Liu G, Polsky R, Merkoçi A (2002) Electrochemical stripping detection of DNA hybridization based on cadmium sulfide nanoparticle tags. Electrochem Commun 4:722–726

Wang L, Nagesha DK, Selvarasah S, Dokmeci MR, Carrier RL (2008) Toxicity of CdSe nanoparticles in Caco-2 cell cultures. J Nanobiotechnol 6:1–15

Watson SB, Gergely A, Janus ER (2011) Where is agronanotechnolgoy heading in the United States and European Union. Nat Resour Environ 26:8

Williams K, Milner J, Boudreau MD, Gokulan K, Cerniglia CE, Khare S (2014) Effects of subchronic exposure of silver nanoparticles on intestinal microbiota and gut-associated immune responses in the ileum of Sprague-Dawley rats. Nanotoxicology 9(3):279–289

Xia Y (2014) Editorial: are we entering the nano era? Angew Chemie Int Ed 53:12268–12271

Yamashita K, Yoshioka Y, Higashisaka K, Mimura K, Morishita Y, Nozaki M, Yoshida T, Ogura T, Nabeshi H, Nagano K (2011) Silica and titanium dioxide nanoparticles cause pregnancy complications in mice. Nat Nanotechnol 6:321–328

Zhao C, Wang W (2011) Comparison of acute and chronic toxicity of silver nanoparticles and silver nitrate to Daphnia magna. Environ Toxicol Chem 30:885–892

Zhu B, Xia X, Xia N, Zhang S, Guo X (2014) Modification of fatty acids in membranes of bacteria: Implication for an adaptive mechanism to the toxicity of carbon nanotubes. Environ Sci Technol 48(7):4086–4095

Chapter 15
Future Roadmap for Plant Nanotechnology

Mariya V. Khodakovskaya

Abstract Nanotechnology has started to play a promising role in agriculture and plant biology in the last few years. The experimental base for "nanoagriculture" is still limited. Several research groups demonstrated that nano-sized materials can be useful for the delivery of nucleic acid, pesticides and fertilizers to plants, activation of seed germination and plant growth, suppression of plant diseases caused by pathogens, and sensing of critical plant molecules with a high level of sensitivity. Success in the development of efficient "nano-agro-technologies" will require the creation of reliable and accurate methods of detection of nanomaterials inside plant cell or tissue, the understanding of the biological mechanisms of effects of nanoparticles in plant systems, and the clarification of properties of nanomaterials that can be associated with observed biological effects. Involvement of nanotechnology in agriculture will eventually enhance the flow of nanomaterials into the food chain. Thus, the risk assessment of agricultural plant products contaminated with different nanoparticles intentionally or nonintentionally is the most important task for future plant nanotechnology.

Keywords Nanodelivery · Nucleic acids · Pesticides · Fertilizers · Growth regulators · Suppression · Nanosensors · Risk assessment

The study of nanoparticle–plant interaction is a new, emerging area of modern nanobiotechnology. However, the number of publications associated with the effects of nanomaterials on plant organisms is dramatically lower compared with articles focused on effects of nanoparticles on animals/humans or animal cells (Fig. 15.1).

M.V. Khodakovskaya (✉)
The University of Arkansas at Little Rock, 2801 S. University Ave, Little Rock, AR 72204, USA
e-mail: mvkhodakovsk@ualr.edu

M.V. Khodakovskaya
Institute of Biology and Soil Science, Far-Eastern Branch of Russian Academy of Sciences, Vladivostok, Russian Federation

C. Kole et al. (eds.), *Plant Nanotechnology*,
DOI 10.1007/978-3-319-42154-4_15

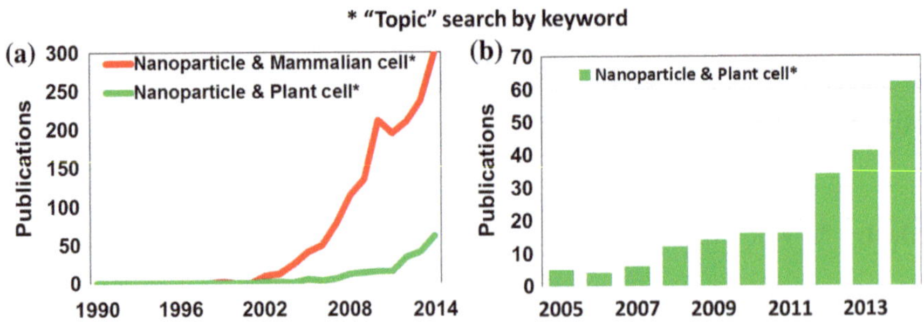

Fig. 15.1 Comparative analysis of some manuscripts published in the area of nanoparticle–plant and nanoparticle–animal interactions during 1990–2014 (**a**). The increase of publications associated with effects of nanoparticles on plants through 2005–2013 years (**b**). The analysis was performed using available data of the Web of Science (WoS) database. Keywords indicated on graphs were used for this search

The interest of research groups to understand how different nanomaterials can affect plant physiology and development significantly increased after 2011 and continues to be elevated (Fig. 15.1b). Such phenomena can be explained by the recent discoveries of the benefits of nanomaterials for fundamental plant biology and applied plant science. During the early years of nanotechnology, investigators mostly focused their efforts on understanding the toxicity of carbon-based (CBNs) and metal-based (MBNs) nanomaterials to different plant species and plant cells. To achieve visual symptoms of phytotoxicity, researchers mostly worked at a range of very high doses (1000–2000 μg/ml) of tested nano-sized materials (Lin and Xing 2007; Stampoulis et al. 2009). As a result, authors noticed no toxicity or visible toxicity of full range nano-sized materials applied at high doses to different plants (Lin and Xing 2007; Stampoulis et al. 2009). However, it was later demonstrated that a significant decrease of working nanomaterial concentrations can change the response of treated plants dramatically. Particularly, it was shown that carbon-based tubular nanomaterials (carbon nanotubes, nanohorns) in concentrations between 10 and 100 μg/ml were sufficient to activate seed germination and plant growth (Khodakovskaya et al. 2011, 2012; Villagarcia et al. 2012; Lahiani et al. 2013; Khodakovskaya et al. 2013; Lahiani et al. 2015).

A range of successful experiments identified the most promising directions of nanomaterial applications for plant improvement and agriculture. Thus, the ability of CBNs to improve cell, seed, and plant performance demonstrates a high potential of CBNs as ***regulators of germination and plant growth*** (Khodakovskaya et al. 2011, 2012; Villagarcia et al. 2012; Lahiani et al. 2013; Khodakovskaya et al. 2013; Lahiani et al. 2015). Studies focused on the use of nanoparticles for targeted delivery of pesticides and fertilizers demonstrated good potential in ***disease suppression and crop yield enhancement*** (Perez-de-Luque et al. 2006; Servin et al. 2015). Particularly, this approach has the potential to provide better penetration through plant tissues and allow the slow and constant release of herbicides (Perez-de-Luque et al. 2006). The ability of gold nanorods to stimulate ***delivery of***

phytohormone 2,4-D to plant cells (tobacco cell culture) and activate cell growth was documented recently (Nima et al. 2014).

Silver nanoparticles (Ag NPs) were described as an active nano-sized material for ***prevention of plant diseases*** caused by wide range of pathogens (Lamsal et al. 2011a, b; Kim et al. 2012). It has been demonstrated that they are very useful for the reduction of plant diseases caused by spores producing fungal pathogens (Jo et al. 2009) or reduction of microbial growth for plant cuttings (Liu et al. 2009; Solgi et al. 2009). Recently, nanosilica was recognized as a powerful nanobiopesticide. Practically, nanosilica can be absorbed into the cuticular lipids of insects and cause the death of insects by desiccation (Barik et al. 2008; Rahman et al. 2009).

Plant genetic engineering can benefit from nanotechnology in the area of improvement of plant transformation. Thus, the ***new technology of nucleic acid delivery to plant cells*** using mesoporous silica system (MSNs) has recently become apparent (Galbraith 2007; Torney et al. 2007; Martin-Gullon et al. 2006; Martin-Ortigosa et al. 2012, 2014). Another promising type of nanomaterials for nucleic acid delivery is polymer nanoparticles. Thus, fluorescent conjugated polymer nanoparticles (CPNs) were used to deliver siRNAs and knockdown specific gene target in plant BY-2 protoplasts (Silva et al. 2010). The big advantages of CPNs are very low toxicity of such material for plants.

Creation of new sensors for plants is new and a promising direction of plant nanotechnology. The number of successful studies is still very limited, but nanosensors can be developed in the very near future. A great example is the recent building of single-walled carbon nanotubes (SWCNTs) radiometric sensors (for H_2O_2 and NO) performed by Giraldo et al. (2015) which proved the efficiency of radiometric nanosensing platform for detecting key compounds in plant tissues.

Without any doubts, the range of possible applications of nanomaterials in plant biology is tremendous. However, there are some factors that can limit wide application of nanomaterials *in planta*. First, it is a significant challenge to compare results of independent research groups because investigators are working in different experimental settings. Properties of applied nanomaterials such as size, purity, presence/type of functional groups, doses, the level of agglomeration, and way of delivery are not precisely identical between presented experiments. Thus, reproducibility of successful experiments *in planta* is not at an acceptable level yet. Secondly, the detection of nanoparticles inside plant tissues or cells is a significant challenge. Transmission electron microscopy and methods of spectroscopy including Raman spectroscopy are efficient confirmation of the presence of particular nanoparticles inside plant sample (Khodakovskaya et al. 2011; Lahiani et al. 2013). However, the quantification of the exact amount of absorbed nanomaterials by plant organs is very challenging. For example, the reliable technique for quantitative analysis of carbon nanotubes located inside exposed plants was developed only very recently (Irin et al. 2012; Lahiani et al. 2015). Future progress in the creation and application of new plant-related nanotechnologies will be dependent on accurate quantitative assays of different nano-sized materials inside plant tissues. Thus, new methods of detection and measurement of absorbed nanomaterial have to be suggested and developed. As shown in this book, some positive effects of nanoparticles on plants were documented up-to-date. In the same time, biological mechanisms of observed effects are not clear. The clarification of biological mechanisms of nanoparticle impact in plant systems will require comprehensive transcriptome/proteome investigations of exposed plants in a combination with high sensitive detection of nanomaterials inside plants. Interdisciplinary collaborations between material scientists, plant biologists, chemical engineers, and physicists can help create new platforms for such studies.

To consider the possible use of nanoparticles in plant systems or plant agriculture, the risk assessment of nanomaterial entering into the food chain should be performed in detail. It is critical to understand the effects of short-term and long-term exposure of CBNs and MBNs delivered to humans or animals through exposed plants. However, the methodology of such risk assessment is not yet fully established. Creation of effective, safe, and simple in vitro and in vivo toxicological experimental protocols for each group of nanomaterials is a major step of risk assessment of plants contaminated with nanomaterials.

References

Barik T, Sahu B, Swain V (2008) Nanosilica—from medicine to pest control. Parasitol Res 103:253–258

Galbraith DW (2007) Nanobiotechnology: silica breaks through in plants. Nat Nanotechnol 2:272–273

Giraldo JP, Landry MP, Kwak S, Jain RM, Wong MH, Iverson NM, Ben-Naim M, Strano MS (2015) A ratiometric sensor using single chirality near infrared fluorescent carbon nanotubes: application to in vivo monitoring. Small 32:3973–3984

Irin F, Shrestha B, Cañas JE, Saed MA, Green MJ (2012) Detection of carbon nanotubes in biological samples through microwave-induced heating. Carbon 50:4441–4449

Jo Y, Kim BH, Jung G (2009) Antifungal activity of silver ions and nanoparticles on phytopathogenic fungi. Plant Dis 93:1037–1043

Khodakovskaya MV, de Silva K, Biris AS, Dervishi E, Villagarcia H (2012) Carbon nanotubes induce growth enhancement of tobacco cells. ACS Nano 6:2128–2135

Khodakovskaya MV, de Silva K, Nedosekin DA, Dervishi E, Biris AS, Shashkov EV, Galanzha EI, Zharov VP (2011) Complex genetic, photothermal, and photoacoustic analysis of nanoparticle-plant interactions. Proc Natl Acad Sci USA 108:1028–1033

Khodakovskaya MV, Kim B, Kim JN, Alimohammadi M, Dervishi E, Mustafa T, Cernigla CE (2013) Carbon nanotubes as plant growth regulators: effects on tomato growth, reproductive system, and soil microbial community. Small 9:115–123

Kim SW, Jung JH, Lamsal K, Kim YS, Min JS, Lee YS (2012) Antifungal effects of silver nanoparticles (AgNPs) against various plant pathogenic fungi. Mycobiology 40:53–58

Lahiani MH, Dervishi E, Chen J, Nima Z, Gaume A, Biris AS, Khodakovskaya MV (2013) Impact of carbon nanotube exposure to seeds of valuable crops. ACS Appl Mater Interf 5:7965–7973

Lahiani MH, Chen J, Irin F, Puretzky AA, Green MJ, Khodakovskaya MV (2015) Interaction of carbon nanohorns with plants: uptake and biological effects. Carbon 81:607–619

Lamsal K, Kim SW, Jung JH, Kim YS, Kim KS, Lee YS (2011a) Application of silver nanoparticles for the control of Colletotrichum species in vitro and pepper anthracnose disease in field. Mycobiology 39:194–199

Lamsal K, Kim S, Jung JH, Kim YS, Kim KS, Lee YS (2011b) Inhibition effects of silver nanoparticles against powdery mildews on cucumber and pumpkin. Mycobiology 39:26–32

Lin D, Xing B (2007) Phytotoxicity of nanoparticles: inhibition of seed germination and root growth. Environ Pollut 150:243–250

Liu Z, Fan AC, Rakhra K, Sherlock S, Goodwin A, Chen X, Yang Q, Felsher DW, Dai H (2009) Supramolecular stacking of doxorubicin on carbon nanotubes for in vivo cancer therapy. Angew Chemie Int Edn 48:7668–7672

Martin-Ortigosa S, Valenstein JS, Lin VS, Trewyn BG, Wang K (2012) Gold functionalized mesoporous silica nanoparticle mediated protein and DNA codelivery to plant cells via the biolistic method. Adv Funct Mater 22:3576–3582

Martin-Gullon I, Vera J, Conesa JA, González JL, Merino C (2006) Differences between carbon nanofibers produced using Fe and Ni catalysts in a floating catalyst reactor. Carbon 44: 1572–1580

Martin-Ortigosa S, Peterson DJ, Valenstein JS, Lin VS, Trewyn BG, Lyznik LA, Wang K (2014) Mesoporous silica nanoparticle-mediated intracellular cre protein delivery for maize genome editing via loxP site excision. Plant Physiol 164:537–547

Nima ZA, Lahiani MH, Watanabe F, Xu Y, Khodakovskaya MV, Biris AS (2014) Plasmonically active nanorods for delivery of bio-active agents and high-sensitivity SERS detection in planta. RSC Advances 4:64985–64993

Perez-de-Luque A, Lozano MD, Cubero JI, Gonzalez-Melendi P, Risueno MC, Rubiales D (2006) Mucilage production during the incompatible interaction between *Orobanche crenata* and *Vicia sativa*. J Exp Bot 57:931–942

Rahman A, Seth D, Mukhopadhyaya SK, Brahmachary RL, Ulrichs C, Goswami A (2009) Surface functionalized amorphous nanosilica and microsilica with nanopores as promising tools in biomedicine. Naturwissenschaften 96:31–38

Servin A, Elmer W, Mukherjee A, De la Torre-Roche R, Hamdi H, White JC, Bindraban P, Dimkpa C (2015) A review of the use of engineered nanomaterials to suppress plant disease and enhance crop yield. J Nanopart Res 17:1–21

Silva AT, Nguyen A, Ye C, Verchot J, Moon JH (2010) Conjugated polymer nanoparticles for effective siRNA delivery to tobacco BY-2 protoplasts. BMC Plant Biol 10(1):2221–2229

Solgi M, Kafi M, Taghavi TS, Naderi R (2009) Essential oils and silver nanoparticles (SNP) as novel agents to extend vase-life of gerbera (Gerbera jamesonii cv. 'Dune') flowers. Postharvest Biol Technol 53:155–158

Stampoulis D, Sinha SK, White JC (2009) Assay-dependent phytotoxicity of nanoparticles to plants. Environ Sci Technol 43:9473–9479

Torney F, Trewyn BG, Lin V, Wang K (2007) Mesoporous silica nanoparticles deliver DNA and chemicals into plants. Nat Nanotechnol 2:295–300

Villagarcia H, Dervishi E, de Silva K, Biris AS, Khodakovskaya MV (2012) Surface Chemistry of Carbon nanotubes impacts the growth and expression of water channel protein in tomato plants. Small 8:2328–2334

Index

C. Kole et al. (eds.), *Plant Nanotechnology*,
DOI 10.1007/978-3-319-42154-4

D

Z

MIX
Papier aus verantwortungsvollen Quellen
Paper from responsible sources
FSC® C105338

If you have any concerns about our products,
you can contact us on
ProductSafety@springernature.com

In case Publisher is established outside the EU,
the EU authorized representative is:
Springer Nature Customer Service Center GmbH
Europaplatz 3, 69115 Heidelberg, Germany

Printed by Libri Plureos GmbH
in Hamburg, Germany